DATE DUE

~~MY 1 ~ '07~~		
~~FE 6 '~~		
~~JE 2 '00~~		
~~MY 10 06~~		
DE - 1 08		

DEMCO 38-296

An Introduction to Neural Networks

James A. Anderson

An Introduction to Neural Networks

A Bradford Book
The MIT Press
Cambridge, Massachusetts
London, England

This book was set in Times Roman by Asco Trade Typesetting Ltd., Hong Kong and was printed and bound in the United States of America.

Library of Congress Cataloging-in-Publication Data

Anderson, James A.
  An introduction to neural networks / James A. Anderson.
     p.   cm.
  Includes bibliographical references and index.
  ISBN 0-262-01144-1
  1. Neural networks (Neurobiology) I. Title.
QP363.3.A534   1995
  612.8—dc20   94-30749
                                                  CIP

Contents

Introduction   *vii*

Acknowledgments   *xiii*

1   *Properties of Single Neurons*   *1*
2   *Synaptic Integration and Neuron Models*   *37*
3   *Essential Vector Operations*   *63*
4   *Lateral Inhibition and Sensory Processing*   *85*
5   *Simple Matrix Operations*   *129*
6   *The Linear Associator: Background and Foundations*   *143*
7   *The Linear Associator: Simulations*   *175*
8   *Early Network Models: The Perceptron*   *209*
9   *Gradient Descent Algorithms*   *239*
10   *Representation of Information*   *281*
11   *Applications of Simple Associators: Concept Formation and Object Motion*   *351*
12   *Energy and Neural Networks: Hopfield Networks and Boltzmann Machines*   *401*
13   *Nearest Neighbor Models*   *433*
14   *Adaptive Maps*   *463*
15   *The BSB Model: A Simple Nonlinear Autoassociative Neural Network*   *493*
16   *Associative Computation*   *545*
17   *Teaching Arithmetic to a Neural Network*   *585*

Afterword   *629*

Index   *631*

Introduction

This book presents an idiosyncratic approach to neurocomputing. I hope that it will sit in a kind of interdisciplinary neutral zone equidistant from neuroscience, cognitive science, and engineering. It is not clear how many native inhabitants this area has, if any, but perhaps tourists from more established disciplines will find the scenery intriguing.

Neurocomputing is sometimes called brainlike computation. It is an attempt to build computers that are designed a little bit like the brain, and that perhaps can do a few of the things the brain can do. Brains are especially good at things such as pattern recognition, motor control, perception, flexible inference, intuition, and good guessing. These are things that we would like to be able to do with artificial systems. Unfortunately, brains are also slow, are imprecise, make erroneous generalizations, are prejudiced, and are usually incapable of explaining their actions. These less desirable properties may also be typical of neurocomputing.

This book introduces some important neural network algorithms and has been used as a textbook for a college-level neural network course. The main emphasis is on the biology and psychology behind the assumptions of the models, and on what the models might be used for, rather than on detailed technical analysis. For a more mathematical discussion, two books in particular can be recommended. *Introduction to the Theory of Neural Computation* by John Hertz, Anders Krogh, and Richard G. Palmer (1991) is a fine review of neural network algorithms at an advanced level. *Neural Networks* by Simon Haykin (1993) is a clear, comprehensive, readable engineering text that is highly recommended for textbook use. Several other good books are also now available or will be shortly.

Most of these introductory books are designed for an audience of engineers or physicists. They focus on basic algorithms and their variants. They are mathematical, and are concerned with the analysis of networks and algorithms. They spend little time on the applications and function of neurocomputing. Beyond a few comments about pattern recognition and some simple examples that understate the complexity and subtlety of real applications, they contain little on what to use a neural network for. Finding the problems of and applications for a neural network is far more difficult than formally understanding algorithms.

Formal mathematics is a language, and, like all languages, it has words and a rigid grammar. Some sentences are correct, whereas others make no sense. The mathematics that is usually applied to neural networks is that of classic engineering, the kind that an engineer or physicist would pick up for a bachelor's degree; that is, a minimum of two and a half years of calculus, perhaps some linear algebra, linear systems analysis, perhaps even some abstract algebra, and number theory.

The marvelous thing about this kind of mathematics, either because God is a Platonist or because the natural selection of concepts made it turn out that way, is that many aspects of the physical world can be described by short pithy sentences. The sentences are short, but they are so elegant and powerful that in a few lines they accurately describe the behavior of large chunks of the physical world. Obvious examples are Maxwell's equations, Schrodinger's equation, Newtonian mechanics, and so on. When these compact formulas are expanded into essay-length particularizations, they describe how the planets revolve, how atoms behave, and how a microwave oven works.

Mathematical analysis of the familiar kind is an indispensible tool; unfortunately, however, it is not entirely clear if it is completely suited to neurocomputing. To analyze a system, we have to have a clear idea about what the system is going to do and the problems we want to solve with it. We have not properly formulated the problems yet for neurocomputing. In fact, we often do not even know what the problems are. Therefore, careful mathematical analysis of simple network models—the kind found in the texts mentioned and that many believe is the proper way to analyze networks—is useful, but may be providing definitive answers to questions that no one is really interested in asking.

In linguistics, a venerable hypothesis recurs over and over, and is reinvented periodically by exitable undergraduates and literary critics. It is called the Whorf hypothesis after Benjamin Whorf, a linguist who worked in the first part of this century and who was primarily known for his work on Native American languages (Whorf, 1956). To oversimplify somewhat, the hypothesis says that what you can think is determined to a considerable degree by the language you think in. For example, classic physics, with discrete objects causally interacting, is supposedly consistent with the structure of Indo-European languages. Many Native American languages, the story goes, tend to be less concerned with causality and have a much wider "present," thereby providing a poor home for causal Newtonian mechanics but one that might be, one could speculate, well suited to quantum mechanics.

Unfortunately, the evidence we have for human languages suggests this provocative hypothesis is not true. It seems that any language can express any human concept, although some can do it in fewer words than others. Navajos make first-class engineers, and Europeans can be nature mystics. However, there is a sense in which the Whorf hypothesis might be true for mathematical analysis such as that applied to neural networks. The traditional approach to mathematical analysis places a premium on shortness and elegance, and uses a restricted range of basic functions and operations. Therefore, the mathematical language used to explain neural networks has a severely limited vocabulary.

It is possible that if we impose the dual constraints of terseness and a limited vocabulary, we may indeed be restricting what we can think about. The computer allows a much different notion of simplicity. A wide range of elementary functions can be inserted into a working simulation program with little effort and their effects on the resulting system investigated. It may not be necessary to analyze a system in closed form. To use an example from later in the book, there is no reason to limit a network to minimizing mean square error, for example, when it is just as easy to let the network minimize absolute value of error, error to the fourth power, or anything else that might be desired. Mean square error is easy to analyze with classic mathematics, probably because of its direct connection with Euclidean length, whereas the others are not. The computer program does not care.

When the program runs, it turns itself into a chain of simple logical statements that may be millions or even billions of operations long. Computing mean square error requires roughly the same number of machine operations as computing absolute value of error. However, the results of one computation can often be stated concisely in formal mathematics, whereas the other cannot. We feel very differently about these two results. With proper training humans can understand a few concise statements in the language of mathematics because they connect directly to our intuitions about complex systems, but they cannot understand long strings of straightforward computer operations. We spend chapter 17 of this book discussing a few of the reasons why simple arithmetic is so difficult to learn for both people and neural networks.

Therefore the emphasis in this book is not so much on the formal analysis of network algorithms as on the *use* of algorithms. What are they good for? What can they do? What is being computed? How robust are they? Neural networks can learn, that is, they are adaptive systems. They provide a hedge against uncertainty. If we knew everything, we would not have to learn anything. If we know nothing, everything must be learned. Most neural network textbooks and much neural network research assume the last statement is true. In fact, we usually know a lot about any particular problem. A practical system can incorporate this information into the design of the network and, in particular, into the way the world is represented by the activity of the elementary units that make up the network. This book devotes some effort to describing the biological representation of data, and giving some examples of biological and cognitive computation using neural networks. These sections are sketchy, speculative, and unsatisfying, largely because enough is not currently known to be more precise.

The nervous system does computation in a very general sense. It receives sensory inputs and it does things with them to generate motor outputs. At the same time, the nervous system is subject to a large number of important constraints that look a lot more like practical engineering than theory of computation. Biology is the science of what you can do with what you have, not what you might want to do in a perfect world. Practical engineering contains a lot of economics: it includes problems with the availability of fuel and building materials, with cost and allowable tolerances in construction. The available components are rarely the ideal ones. A biological neural network has had to deal with all these problems at once. When understood, it will undoubtedly turn out to be an outstanding engineering solution, having a judicious balance of tradeoffs providing effective and reliable performance.

This book tries to explore a few of these issues. Specific applications require an analysis of the problem to be solved, the kind of data available to the network, the best way to represent the data for the network, and the network algorithms to use. Sometimes we may be able to take a degree of inspiration from the way neurobiology suggests the brain is organized. However, we will be a long way away from detailed neurobiological modeling. In the beginning of the book we provide programs for some computer modeling experiments, so students can play with algorithms and theories. Throughout this book we will provide fragments of code, useful Pascal procedures, and functions, and will describe results from network modeling programs. The Afterword tells how to obtain the complete programs, datasets, and further details about the operation and design of the programs.

This book can be usefully supplemented by two others published by MIT Press: *Neurocomputing: Foundations of Research* (Anderson and Rosenfeld, 1988) and *Neurocomputing 2: Directions for Research* (Anderson, Pellionisz, and Rosenfeld, 1990). They contain reprints of a number of important papers relating to the areas covered in *An Introduction to Neural Networks*.

Many of the references given here will be found in full in one of these two books.

An Introduction to Neural Networks evolved from notes for a course I have taught for several years at Brown University, the Neural Modeling Laboratory, Cognitive and Linguistic Sciences 102. Many of the students who took the course seemed to like it and said they found it useful. I hope that others do too.

References

J.A. Anderson and E. Rosenfeld (Eds., 1988), *Neurocomputing: Foundations of Research*, Cambridge: MIT Press.

J.A. Anderson, A. Pellionisz, and E. Rosenfeld (Eds., 1990), *Neurocomputing 2: Directions for Research*, Cambridge: MIT Press.

S. Haykin (1993), *Neural Networks*, New York: MacMillan.

J. Hertz, A. Krogh, and R.G. Palmer (1991), *Introduction to the Theory of Neural Computation*, Reading, MA: Addison-Wesley.

B. Whorf (1956), *Language, Thought, and Reality*, New York: Wiley.

Acknowledgments

Thanks are due to several years of students in the course Cognitive and Linguistic Sciences 102 who received and commented on earlier drafts of some of this material. Many colleagues, undergraduate, and graduate students made suggestions and helped with projects. They are not responsible for the many errors, imprecisions, and general stupidities in the following pages, however; they are all due to me. Particular thanks go to Herve Abdi, Paul Allopenna, David Ascher, Terry Au, Julian Benello, David Bennett, Elie Bienenstock, Sheila Blumstein, Heinrich Bulthoff, Dean Collins, Leon Cooper, Peter Eimas, Mike Gately, Stuart Geman, Richard Golden, Gerry Guralnik, Nicholas Hatsopoulos, Geoff Hinton, Nathan Intrator, Alan Kawamoto, Dan Kersten, Andrew Knapp, David Knill, Phil Lieberman, Art Markman, Jay McClelland, Charlotte Manly, Dick Millward, Paul Munro, Greg Murphy, Menasche Nass, Alice O'Toole, Andras Pellionisz, Andy Penz, Emily Pickett, Ed Rosenfeld, Mike Rossen, Philippe Schyns, Margaret Sereno, David Sheinberg, Jack Silverstein, Steven Sloman, Kathryn Spoehr, Gary Strangman, Susan Viscuso, Bill Warren, Ed Wisniewski, Greg Zelinsky and, in particular, Diana Anderson.

Special thanks to several granting agencies for support, including the Office of Naval Research, the McDonnell-Pew program in Cognitive Neuroscience, and to the National Science Foundation and Dr. Joseph Young, who helped support neural network research when it was definitely not fashionable to do so.

An Introduction to Neural Networks

1 *Properties of Single Neurons*

After a brief introduction to the book as a whole, this chapter discusses a few aspects of single-neuron structure and function. Neurons are believed to be the basic units used for computation in the brain. The chapter presents some essential information about neurons, including rudiments of their anatomy. The Nernst and cable equations are discussed as part of the foundations of neuron electrochemistry. This chapter also discusses some important functional questions about neurons, for example, why they use all-or-none action potentials for transmitting information, and why they operate so slowly.

It is not absolutely necessary to believe that *neural network models* have anything to do with the nervous system, but it helps. Because, if they do, we are able to use a large body of ideas, experiments, and facts from cognitive science and neuroscience to design, construct, and test networks. Otherwise, we would have to suggest functions and mechanisms for intelligent behavior without any examples of successful operation.

This book gives an introduction to neural network modeling. We focus on useful ideas, presenting some that you can try yourself with computer simulations. In general, we will keep as close to the organizational principles of the system that we know works best, the brain, as we are able to infer them. It would be hard to find a subject that is approached from more different directions than the operation of the brain.

First, we must know enough neuroscience to understand why our models make certain approximations. Most important, we must know when the approximations are poor, as they often are, and when they are reasonable.

Second, we must know something of the tools of formal analysis required for the systems we propose. We need some simple mathematics, which we will develop to a limited extent here, as well as access to a computer and the programming language Pascal. We will provide a number of short program fragments that can be incorporated into programs to construct the models. It is easy to convert Pascal code into other languages, particularly C.

Third, we must know enough cognitive science to have some idea about what the system is supposed to do. One of the central themes of this book is that the brain is not a general-purpose computer. It is not even a very powerful computer in the sense that a digital computer is theoretically powerful. The brain gets its computing power from its enormous size, its parallelism, its effective biological preprocessing, its use of memory in place of computing power, and its efficiency when performing a small number of useful operations. We can see traces of this specific architecture, with its associated limitations, in data from human cognition.

Our discussion assumes it is possible to make meaningful simplifications of some aspects of nervous system function. Some would deny this, claiming that brain biology is intrinsically too complicated, or that not enough is known about the software, or that we simply are not yet capable of formulating rules that might capture any of the more complex computational operations our brain performs. These arguments may be true.

History: Function

There are many different ways of organizing a computing system. It is a generally accepted fact in the computer industry that the hardware largely determines the software that runs efficiently on it.

The first attempt at making a general computing "engine" was by Charles Babbage, who, in the nineteenth century, designed a device for doing arithmetic to check mathematical tables. The historical significance of this analytical engine lay in the control structures it incorporated and its degree of programmability.

The first digital computers were also constructed for specific applications. COLOSSUS was produced for cryptographic applications by the British in World War II. ENIAC was developed at the Moore School of Engineering of the University of Pennsylvania to compute ballistic firing tables used for artillery for the United States Army. Both contained features that are now nearly universally applied to the definition of computers: programmable electronic devices performing binary arithmetic and logic functions.

One conclusion that can be drawn from the history of computers is that the driving force for hardware development was usually an important, well-defined, practical problem. The techniques developed for specialized machines turned out to have more general application, most notably in the case of ENIAC, which led in a direct line to much of the modern computer industry. However, the real generality of application possible with this architecture did not show up until later generations of machines.

This is also a reason why practical applications for artificial neural networks have been slow in coming. In this case, the general solution arrived before the specific problems that required networks for their solution.

If one looks at the evolution of the nervous systems of organisms, it is possible that the same thing occurred. First, an inflexible system was formed that fulfilled its function. After long periods of time a degree of learning and flexibility crept into the system to make it more adaptable to the immediate environment. Adaptation and learning are dangerous abilities, because they involve behavioral change, changes in internal wiring, and potential instability. They are kept under tight biological control.

Representations, Computational Strategy, and Practical Applications

Many fondly believe that neural networks are models for animal behavior. The corollary to this is that one can look at animals and see what computational strategies they use.

The problems faced by an animal are similar to the problems found in many proposed practical applications of neural nets: association, classification, categorization, generalization, rapid response, quick inference from inadequate data, and so on. These problems have given rise to intense selective pressure on animals, and nature's biological solutions are very good. We will often refer to biology, possibly in an irritatingly impressionistic way, when we make detailed assumptions in the models discussed later.

Perhaps we should look at the brain through the eyes of an engineer. Engineering considerations are paramount in constructing a device that must function in the real world. Engineers must make devices that work satisfactorily, that are inexpensive to run, that can be made easily, that do not require delicate adjustments, and that are reliable. The solution must balance all these constraints and is generally not the best solution for any one of them. A flexible, adaptable nervous system such as ours, with considerable learning ability, capable of speech and perhaps some degree of abstract reasoning, is not a simple, necessary, or obvious solution.

Developmental constraints are powerful, and the design of our nervous system is not a radical break with that of other vertebrates. In the majority of vertebrates, certainly in mammals, one sees the same basic brain structures. Roughly 99% of our DNA is the same as that of a chimpanzee. Most of the other 1% seems to be in what are called structural genes, that is, genes that modify sizes and shapes of structures. What seems to have happened is that, with modifications of preexisting design, our brain became capable of considerable general computational power and flexibility.

Our brains do not contain any radical new hardware, however, just re-arrangements, inflations, contractions, and modifications of the same structures found in other higher mammals, plus some fine tuning. Biology does not allow the luxury of frequent redesign of the hardware. This is not controversial when one discusses kidneys or the liver, but it must be equally true of the construction of our biological computer.

Also, if computation in the brain has the characteristics of a good engineering solution, this often may mean using specific solutions to specific problems with much less generality than one might like. We are living in one particular world, and we are built to work in it and with it. As William James, America's greatest psychologist, put it in the nineteenth century,

Mental facts cannot properly be studied apart from the physical environment of which they take cognizance. *The great fault of the older rational psychology was to set up the soul as an absolute spiritual being with certain faculties of its own by which the several activities of remembering, imagining, reasoning, willing, etc. were explained, almost without reference to the peculiarities of the world with which these activities deal. But the richer insight of modern days perceives that our inner faculties are adapted in advance to the features of the world in which we dwell ... In the main, if a phenomenon is important for our welfare, it interests and excites us the first time we come into its presence. Dangerous things fill us with involuntary fear; poisonous things with distaste, indispensible things with appetite. Mind and world, in short, have been evolved together, and in consequence are something of a mutual fit.* (James, 1892, p. 11)

One is tempted to replace the words "older rational psychology" with artificial intelligence (AI), at least AI during the glorious days of the 1970s, when it seemed that sheer cleverness could substitute for factual knowledge. If all intelligent organisms must solve the same problems and do it in the most effective way, who needs to know the specific details of the way humans do it?

The evolutionary perspective should not be lost sight of when working with neural networks. The emerging field of neuroethology demonstrates over and over how tightly coupled details of nervous system organization and species-specific behavior are. (See Brown, 1975; Camhi, 1984; and Ewert, 1980, for many wonderful examples of how a nervous system adapts to a particular lifestyle.) We should not be surprised, in fact, we should expect, that brains (or neural networks) cannot compute everything, but only arrive at reasonable solutions for a small but useful set of problems. Details are all important and highly problem specific.

It is worth emphasizing these points because it is sometimes believed that a brainlike computing device, if we could build it, will immediately become a wonder network capable of solving all our problems. We hope to show that there is no magic in neural networks. Networks do suggest, however, a number of fascinating and useful ideas that in the long run are of more value than magic.

Hardware

One recurrent theme of this book is that computer hardware places strong constraints on what can be computed with it. Let us look at the elements of neural hardware to obtain some feeling for the devices.

The elementary computing units of the nervous system are *neurons*, or *nerve cells*. The mammalian brain contains lots of them—it is hard not to be impressed with their sheer numbers. The human apparently has something between 10^{10} and 10^{11} neurons, perhaps more. Somehow these cells are able to cooperate in an effective way.

In neural networks, as we will see, connections are important. Each neuron in the human brain has on the order of hundreds or thousands of connections to other neurons, making the total number around 10^{14} or 10^{15}. This is very many fewer than the number we would get if every neuron connected to every neuron. Considered in relation to the size of the brain, a neuron connects directly to only a small fraction of other neurons.

Other classes of cells are present in the nervous system as well, but it is believed that they do not do important information processing. The most common of these, which are intermixed with neurons in the central nervous system, are the *glia* (glue) cells. Glia perform important support functions for neurons, both metabolic and physical, in that they form a matrix in which neurons grow. Glia occupy space that is not filled by neurons.

Neurons are mechanically highly sensitive. They respond to pressure. This property forms the basis of a number of sensory receptors, but it also causes problems. For instance, direct mechanical stimulation of a neuron anywhere on its surface activates the cell (hitting the "funny bone" is the most familiar example). Therefore the mammalian central nervous system is mechanically protected by extraordinary means. The brain is encased in a hard skull. Since soft tissue can easily be damaged by being in contact with hard bone, the brain is floated in cerebrospinal fluid, which provides a hydraulic suspension system.

Neurons are also metabolically very active. The human nervous system consumes close to 25% of the body's energy. Since it makes up only 1% or 2% of body weight, it requires far more energy than most tissue. Apparently the electrochemistry of neurons requires this high metabolic rate. If it were possible to reduce energy consumption, strong selective pressure would have accomplished it long ago.

High metabolism presents an additional problem. Metabolically active tissue is sensitive to poisons and temporary lack of fuel. Elaborate metabolic means are used to regulate brain chemistry and to insulate it from the rest of the body, which is much less sensitive to bad molecules. The *blood-brain barrier* is a term commonly used to describe one of these filtering mechanisms, which allows only a small number of molecules to enter the nervous system from the blood. It is believed that the glia play an important part in this function.

In mammals, although not in many other vertebrates, central nervous system neurons have an important peculiarity: they do not divide after a time roughly coinciding with birth. When a neuron dies, it is not replaced. Moreover, there is considerable evidence for massive cell death before birth. As many as half the cells may die in parts of the developing nervous system, apparently because cells that die do not form the correct connections. It is not clear how cells know they are correctly "wired up"; perhaps coded molecules are transmitted to them from their target neurons when appropriate connections are made, or perhaps they are not sufficiently activated by their connections. Many neural systems seem to show this pattern: there is competition for connections, and if the appropriate functional contacts are not made, the cells die.

From the network theoretician's point of view, the inability of neurons to divide provides an important ground rule: it is cheating to invent new neurons if the system is to be biologically plausible. The neurons present at the beginning are all you can use. It is all right to reuse old ones, but not to generate new ones. For engineering applications this may seem like an arbitrary assumption, and some useful and practical network models have abandoned it.

Neurons have a lot of biological overhead. Those that do not deliver useful value die. Everyone has been exposed to the statment, usually from an elementary school teacher exhorting students to work harder, that we use only 10% of our brain (or 5%, or 25%, the figure varies). We hope the readers of this book are intelligent enough to realize immediately that this statement must be complete nonsense. We can be assured that if one finds a structure such as the nervous system that uses extravagant amounts of energy, the parts of it must be doing something useful. This suggests the important general rule: *neurons must earn their keep.*

There is strong biological pressure to keep as few neurons as possible because they are expensive. This leads directly to a dismaying and challenging conjecture: it might be necessary to have as many as neurons as we do—10 billion or more—to be as intelligent as we are. For those interested in the practical construction of intelligent machines, the burden of proof may be to show that fewer artificial neurons than this are adequate. Essentially all the simulations of higher mental functions performed to this date use hundreds, or at most thousands, of simulated simple neurons. The huge disparity in number between the real and the simulated biological systems should cause uneasiness that something important related to sheer size is being overlooked.

The Classic Neuron

A great deal of biological terminology has crept into the neural network literature. These terms are worth knowing in any case, together with the other essential parts of a good liberal education, such as quantum mechanics and Shakespeare. Figure 1.1 is a diagram of the generic neuron. The

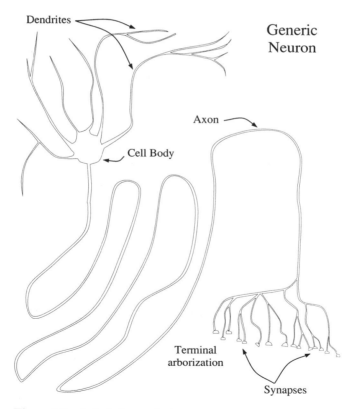

Figure 1.1   Labeled generic neuron.

generic neuron is modeled after spinal motor neurons, one of the best-characterized neurons in mammals. Neurons are cells, and have a nucleus and the related cellular metabolic apparatus.

One end of the cell, the input end, has a number of fine processes called *dendrites* because of their resemblance to a tree (*dendro-* is a Greek root meaning "tree," hence dendrite, dendrochronology, etc.). Figure 1.2 shows the dendritic shapes of a number of real neurons. The variability in shape and size reflects the analog information processing that neurons perform. The cell body is referred to as the *soma*.

Most neurons have a long, thin process, the *axon*, that leaves the cell body and may run for meters. The axon is the transmission line of the neuron. Axons can give rise to collateral branches, along with the main branch, so the actual connectivity of a neuron can be quite complicated. Neurons are among the largest cells in the human body, and are certainly the most extended. For example, single spinal motor neurons in the small of the back can have axons running to the toes and can be well over a meter long. Many even longer ones are known in other animals. When axons reach their final destination they branch again in what is called a *terminal arborization* (*arbor* is Latin for "tree," hence Arbor Day, arboretum, arboreal, etc.). At the ends of the axonal branches are complex, highly specialized structures called *synapses*. In the standard picture of the neuron, dendrites receive inputs from other cells, the soma and dendrites process and integrate the inputs, and information is transmitted along the axon to the synapses, whose outputs provide input to other neurons or to effector organs.

The synapses allow one cell to influence the activity of others. Received dogma in neural network theory believes that synapses vary in strength, and that these strengths, that is, the detailed interactions among many neurons, are the key to the nature of the computation that neural networks perform. Most neurophysiologists agree with this assumption, but, except for a few special cases, such as the *Aplysia* abdominal ganglion and the *Limulus* eye, detailed evidence for it is surprisingly sparse. It is hard to think of plausible alternatives, however.

The nucleus and surrounding machinery have the job of sending nutrients, enzymes, and construction material down the axon to the rest of the cell, which can be some distance away. Paul Weiss, in the 1940s, demonstrated significant axoplasmic flow by simply constricting the axon and showing that it bulged on the nucleus side of the constriction. Retrograde flow of materials back toward the nucleus also occurs. There appear to be a number of intracellular transport mechanisms with different speeds and characteristics. The neuron is a very busy place.

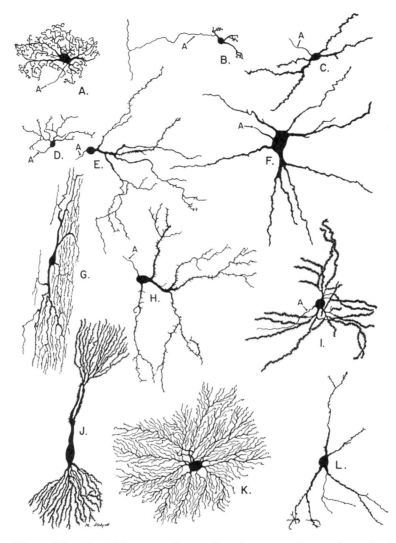

Figure 1.2 Dendritic trees of several real neurons. Savor the majestic ring of the anatomical designations of neural structures. Scaled drawings of some characteristic neurons whose axons (A) and dendrites remain within the central nervous system. A, Neuron of inferior olivary nucleus; B, granule cell of cerebellar cortex; C, small cell of reticular formation; D, small gelatinosa cell of spinal trigeminal nucleus; E, ovoid cell, nucleus of tractus solitarius; F, large cell of reticular formation; G, spindle-shaped cell, substantia gelatinosa of spinal cord; H, large cell of spinal trigeminal nucleus; I, neuron, putamen of lenticular nucleus; J, double pyramidal cell, Ammon's horn of hippocampal cortex; K, cell from thalamic nucleus; L, cell from globus pallidus of lenticular nucleus. Golgi preparations, monkey. (Courtesy of Dr. Clement Fox, Marquette University.) From Truex and Carpenter (1964). Reprinted by permission.

Figure 1.3   Freeze fracture picture of a neuron. Note the protein molecules "float-ing" in the membrane. The membrane broke between the halves of the lipid bilayer. The "bark" in the upper part of the neuron is the other half of the bilayer, which did not detach in this region for some reason. The picture shows a synapse being made on a dendritic spine in the cerebellar cortex of an adult rat. Spines are also common in cerebral cortex and are often held to be involved in the synaptic changes under-lying learning and memory. The figure is figure 3.9 in the third edition (1991) of Peters, Palay and Webster. The original 20 × 27-cm image was at a magnification of 76,000. From Peters, Palay, and Webster (1976). Reprinted by permission.

As in all animal cells, the neuron is covered by a thin membrane with remarkable properties. The function of a membrane is to separate the inside from the outside. In neurons, the inside and the outside are quite different in their chemical and electrical properties. The membrane is only 60 to 70 Å thick, and is composed primarily of lipids and proteins. The lipids are arranged in a bilayer in which the proteins embed themselves; the proteins float in a sort of lipid sea. Figure 1.3 shows a freeze-fracture scanning electron microscope picture of a neuron membrane with proteins embedded in it.

Proteins can be located on the inside or outside face of the membrane, or pass through the membrane. Some of the proteins that pass through the membrane have continuous passages, or pores, through them. Particular ions can pass through the pores and hence through the membrane. The pores can change their conformation under either electrical or chemical control so that ion flow can be regulated; that is, the permeability of the membrane is under the control of the electrical and chemical environment. Figure 1.4 shows a drawing of such a channel that changes its conformation and its permeability to ions when it encounters the chemical acetylcholine. This mechanism for variable ionic conductance forms the basis of the electrical properties of neurons.

It is possible to demonstrate and investigate the properties of single *ionic channels* using an experimental technique called *patch clamping*. In this technique, a fine glass tube, called a *microelectrode*, with an opening a fraction of a micron wide at the tip captures a little piece of membrane at its tip. The membrane actually covers and seals the opening. Sometimes the piece of membrane is so small that it has only one or two ionic channels. Then, the ionic and electrical properties of a single channel can be investigated in detail by studying the minuscule currents passing through it: the channel can be seen to open and close in all-or-none fashion. This technique has confirmed and extended many of our ideas about the operation of neurons.

Each neuron has its own membrane and is separate from other neurons. Information flows from one to another by way of synapses, which let one neuron influence others. Perhaps the simplest way to let one electrically active structure influence another would be by way of a resistor or diode. Such *electrical* synapses are quite common in invertebrates; however, the higher vertebrates make less use of them. The most common synapses that occur in the structures we are trying to model are *chemical* synapses.

Chemical synapses involve several specialized structures and the use of molecules called *neurotransmitters*. It is important to realize the great variety of neurotransmitters and synaptic specializations in the real nervous system. Because of their role in transmitting information, synapses are par-

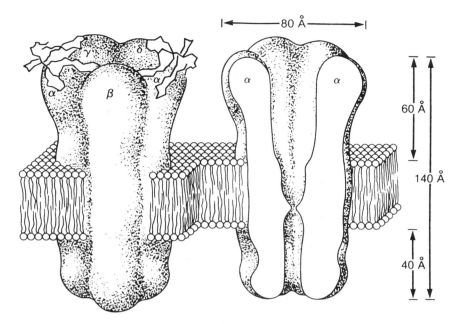

Figure 1.4 Side view of a membrane channel of the nicotinic acetylcholine receptor. The positions of the respective subunits (α, β, γ, and δ) in the pentamer are shown. The two leaflike structures on the extracellular surface of the receptor are molecules of cobra α toxin that are bound primarily to the α subunits. The 40-Å extension into the cytoplasm is accentuated owing to the association of a protein (designated 43K) with the cytoplasmic surface of the receptor. From Taylor (1990). Reprinted by permission.

ticularly likely to be specialized in ways that directly affect their computational functions; these are the functions that directly concern us. Therefore esoteric details of synaptic function may be directly relevant to the neural network modeler in a way that many of the other details of neurophysiology are not.

For example, synaptic activation and operation can show radically different time constants from synapse to synapse, can show long-term changes in coupling between cells, and can respond strongly to modulatory influences, say, specific chemicals in the environment. However, neural network modeling is not generally concerned at present with this level of detail. The vast majority of models assume that a synapse has a strength, like a resistor, that remains stable until it is changed by a well-defined learning rule. It would be nice if it were this simple. We will ignore most of the known complexities of synaptic behavior, but they are without question extremely important in the functioning of real neurons.

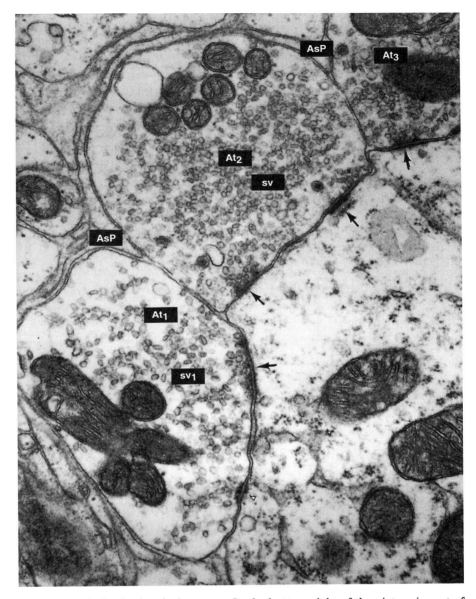

Figure 1.5 A classic chemical synapse. In the bottom right of the picture is part of the perikaryon of a neuron that forms synapses with three axon terminals, At_1 to At_3. One of these terminals (At_2) contains spherical synaptic vesicles (sv), and the lower terminal (At_1) contains predominantly ellipsoidal synaptic vesicles (sv_1). The synaptic complexes (arrows) formed between the plasma membrane of the peri-karyon and the axon terminals are symmetric. From the ventral cochlear nucleus of the adult rat. The original 20×27-cm image was at a magnification of 60,000. From Peters, Palay, and Webster (1976). Reprinted by permission.

Our generic synapse connects the terminal arborization at the end of the axon of the *presynaptic* cell to the dendrite of the *postsynaptic* cell. Essentially all other possible permutations of synaptic connections are seen somewhere in some animal. For example, synapses can make direct contact with the cell body, with axons, and with other synapses.

The incoming presynaptic fiber swells into a typical structure, often called a *bulb* or *bouton*, or, in the case of muscles, which are very much like neurons in their electrical properties, an *end plate*. Postsynaptically, a thickening of the postsynaptic membrane often occurs. A *synaptic cleft*, which is 200 to 300 Å across, separates presynaptic and postsynaptic cells. *Mitochondria* are organelles specialized for oxidative metabolism. Many are present in the presynaptic side of the synapse, suggesting intense metabolic activity in the synapse. Chemical synapses contain large numbers of small, generally round or ovoidal structures, the *synaptic vesicles*. These vesicles contain a few thousand molecules of the neurotransmitter.

At present, many different neurotransmitters are known, both small molecules (acetylcholine, glutamate, gamma-aminobutyric acid, norepinepherine, etc.) and larger polypeptides (endorphins, enkephalins, etc.). Recent evidence suggests that some gas molecules such as carbon monoxide (CO) and nitric oxide (NO) serve as specialized neurotransmitters.

The actual act of information transfer from the presynaptic neuron to the postsynaptic neuron involves release of the neurotransmitter from the vesicles, its diffusion across the synaptic cleft, and its interaction with the postsynaptic membrane. "Used" transmitter can be either rapidly destroyed by enzymes or gathered in and recycled by the synapse. We will discuss the resulting electrical events at the synapse after a brief description of the electrical events in neuron activation. Figure 1.5 shows an electron micrograph of a classic synapse, with the synaptic vesicles, mitochondria, and postsynaptic membrane thickening clearly visible. Neurotransmitters like NO and CO and some of the polypetide transmitters do not fit this picture of release from vesicles.

Neuronal Electrical Behavior

The information-processing abilities of neurons seem to have their most natural interpretation for us in terms of electrical events. We have mentioned that the neuron is surrounded by a thin membrane, and that large electrical and chemical differences exist between the inside and outside of the cell. A recommended source for understanding the *functional* aspects of simple nervous system components and organization, as opposed to biological details, can be found in the first chapters of a clearly written book by Stevens (1966). Neuron biology is well summarized in Nicholls, Martin, and Wallace (1992); Aidley (1992); and Junge (1992).

The Membrane Potential

Suppose we perform the following experiment. We have a sensitive voltmeter and a microelectrode. We can make microelectrodes out of thin glass tubing by heating a small portion of the tube and then quickly pulling it. The opening of the tube is maintained all the way to the end. The microelectrode is filled with a conducting solution such as potassium chloride or potassium citrate, forming a high-resistance (megohms) electrode that responds to voltage differences at its tip (figure 1.6).

Suppose we carefully push the electrode through the membrane. A sudden change in the voltage occurs, seen by the microelectrode tip as the electrode enters the cell. This is the *membrane potential* and it is usually a few tens of millivolts, between 50 and 90 mV would be typical, and depends on the cell type. The inside of the cell is negative with respect to the outside. Although a few tens of millivolts does not seem large, since the membrane is approximately 70 Å thick, a -70-mV membrane potential corresponds to an electrical field across the membrane on the order of 100,000 V per cm. Such a

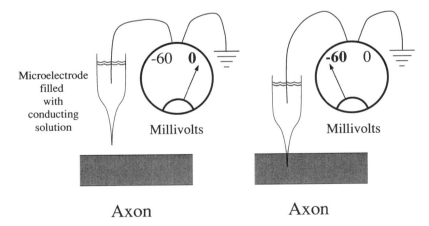

Figure 1.6   A microelectrode is used to measure the electrical properties of a cell. It is made of thin glass tubing, heated at one point, and stretched out so the tip opening becomes very small. It is filled with a conducting solution. When the microelectrode enters an axon, a voltage is recorded, negative with respect to the outside.

high electrical field corresponds to extreme electrical stress on the membrane and the structures in it.

Together with the substantial membrane potential, a number of extreme ionic imbalances exist between the inside and outside of the cell. The concentration of sodium ion (Na^+) is as much as 10 times greater outside the cell than inside it; potassium ion (K^+) is equally out of balance, but its concentration is higher inside than outside the cell. The neuron concentrates K^+ and expels Na^+. One of the mechanisms to accomplish this is the *sodium pump*. This is a complex of large protein molecules that, in exchange for metabolic energy, moves sodium out of the cell and potassium into it. Because the membrane is somewhat leaky, and because activation of neurons causes loss of potassium from inside the cell and allows sodium ion to enter from outside, the pump must operate constantly. This is one important reason for the high metabolic overhead of nerve cells. We will discuss the consequences of these ionic imbalances in the section on the Nernst equation.

With an electrode inside the cell, we can pass current through the membrane if we place a battery in the circuit. The membrane allows some current to pass. Current passing through the membrane causes, by Ohm's law, the voltage to change across the membrane. This allows the experimenter to control the membrane potential of the cell, at least in the immediate vicinity of the electrode that is passing the current.

Some traditional terminology is associated with the membrane potential. When the membrane potential becomes more negative, the cell is referred to as *hyperpolarized*. When the membrane potential becomes less negative, the cell is referred to as *depolarized*. The root *polar* refers to the voltage across the membrane, which polarizes the membrane, and is an old term from electrochemistry that was picked up by neurophysiologists, who added on the prefixes *hyper-*, meaning "greater," and *de-*, meaning "less."

Suppose we put pulses of current into the cell through an electrode. The response of the membrane to a series of small current pulses that hyperpolarize the cell is shown in figure 1.7. The first point of importance is the rounding of edges of the membrane response due to the high *membrane capacity* of the cell membrane. We will discuss the role of membrane capacity in more detail below.

If the current forces the membrane potential to move in the hyperpolarizing direction, and the signals are not very large, all one sees are responses that look like slightly modified versions of the input pulse, with rounded corners due to capacity. If, however, we put in a *depolarizing* current, the picture changes dramatically. As current is increased, large depolarizing changes in voltage start to appear that no longer have a simple relationship to the input signal. Above a certain critical value of input signal, a sudden

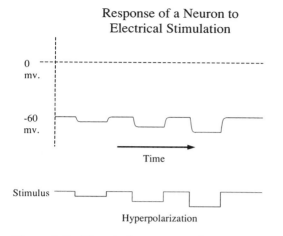

Figure 1.7 Electrical response of a neuron to stimulation. A microelectrode is pushed through a neuron membrane. The inside of the neuron is roughly 60 mV negative with respect to the outside, forming the membrane potential. When a current pulse is passed through a microelectrode so as to make the membrane potential more negative, the cell responds as shown. Note the rounded edges of the response of the membrane potential to the current pulse. This is due to the membrane capacity.

enormous increase in voltage across the membrane occurs, which actually becomes positive by tens of millivolts for about half a millisecond, and then drops back to values closer to the resting membrane potential (figure 1.8).

This is the *action potential*, or *spike*, named for its shape on an oscilloscope screen. It is the most spectacular and obvious aspect of the electrical response of most neurons. Once generated, it does not change its shape with increasing current; it always looks the same. This is the famous *all-or-none* aspect of the action potential—either there or not there—and an aspect that had an important influence on the early history of models of the nervous system. All-or-none is only a tiny conceptual step away from true or false, a step taken by McCulloch and Pitts, among others.

Study of the ionic basis of the action potential led to a large and elegant body of neurophysiology. The classic work was done by Hodgkin and Huxley in the 1950s. They studied the squid (*Loligo vulgaris*) giant axon, a particularly thick axon half a millimeter or more in diameter. The giant fiber controls the escape response of the squid (Brown, 1975). When it is activated, it causes a mantle contraction, and the squid makes a jet-propelled jump backward, perhaps out of danger. Since this cell is controlling an escape response, it has some peculiarities, although its basic physiology is similar to that of other neurons.

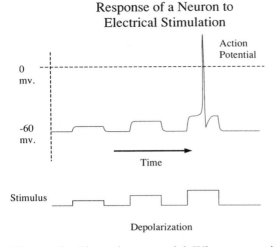

Figure 1.8 The action potential. When current is passed through the membrane so as to make the membrane potential more positive, the neuron will fire an action potential at a critical threshold voltage. The action potential is a spectacular, non-linear electrochemical event, and is the most striking behavior of many neurons.

The mechanism behind the action potential is a regenerative feedback process involving changes in the membrane conductances for sodium and potassium ions. We mentioned that one function of the cell membrane is to separate the inside of the cell from the outside. There are two striking differences between the inside and the outside: the inside is electrically negative with respect to the outside (the membrane potential), and the ionic composition of the inside and outside are very different. These two facts are connected by the Nernst equation, first derived in the nineteenth century.

The derivation of the Nernst equation is short and worth describing, because it shows how electrochemistry gives rise to the electrical events that we observe in the cell.

Figure 1.9 (Katz, 1966) shows the ionic differences between inside and outside of a squid giant axon. Notice the large amount of potassium ion inside the axon and small amount outside, and the large amount of sodium ion outside the axon and small amount inside. Consider a positively charged ion. Positive ions are attracted into a negatively charged region; for example, potassium ions are attracted into the negatively charged axon. Another mechanism affects ion behavior, based on the different concentrations of the ions. If ions are allowed to pass through the membrane, eventually there should be as many ions of a species on one side as on the other. This is because, other things being equal, ions are more likely to move from the high-concentration side to the low-concentration side than the other way around (figure 1.10).

Squid Giant Axon

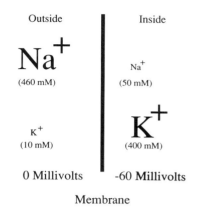

Figure 1.9   Ionic differences between the inside and the outside of a neuron. Concentrations are in millimoles (mM) per liter. Data from Katz (1966).

Concentration Difference

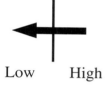

Low   High

Electrical Attraction

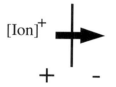

Figure 1.10   Ions prefer to move from a high concentration to a low concentration. Positive ions are attracted to negative potentials. Equilibrium is possible when the two attractions are equal.

Nernst Equation Approximation

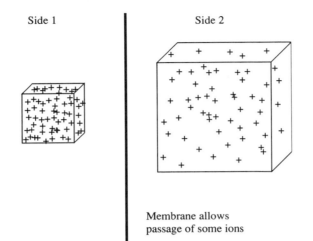

Side 1   Side 2

Membrane allows
passage of some ions

(Concentration) times (Volume) = Number of Ions

Figure 1.11   A very large container divided by a membrane permeable to an ion.

Suppose we could arrange it so that the tendency of a positively charged ion to move from a positively charged region toward a region with a negative charge could be balanced exactly by the tendency of that same ion to move from a high concentration to a low concentration. To keep them in an area of high concentration, we must supply a voltage to attract them, that is, a negative voltage in the case of a positive ion. It is this relationship that forms the basis of the Nernst equation.

Let us consider the situtation diagrammed in figure 1.11. We have a membrane that separates a very large container into two halves. The container has to be so large and so well mixed that small amounts of ionic movement will not affect the overall concentration of the ions. In parts of the real nervous system, where structures are tightly packed, this is sometimes not a good assumption.

We know the amount of electrical work done when a small number of singly charged ions, n, is moved from one side of the membrane to the other, across a voltage difference, E. If n is in moles, and F is that so-called Faraday constant, the total work done is

$$\text{work} = nEF.$$

Note that if we have a doubly charged ion, such as calcium (Ca^{2+}), twice as much work will be done moving the same number of ions.

Piston

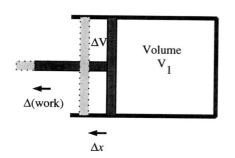

Figure 1.12   When ions in solution move from a high concentration to a low concentration, they can be assumed to act like an expanding gas. The work done in the ionic movement can then be computed.

It is a little more difficult to compute the work done when a small number of ions, n, is moved from the high-concentration side of the membrane to the low-concentration side. Consider a small group of ions. We will make the *approximation* that ions in a solution act like gas molecules. This is the kind of breathtakingly unrealistic assumption that theoreticians are fond of. However, in this case, it turns out to be a very good assumption quantitatively, and it makes analyzing the system easy.

Consider a volume of the solution. This volume contains a certain number of ions. If we multiply the concentration, c, of the ions by the volume, V, it is a constant, the number of ions present. If we take the same number of ions and increase the volume they occupy so it is now V', we reduce the concentration of the ions so it is now c'. In fact,

$cV = $ number of ions $= c'V'$.

There is an inverse relation between concentration and volume.

Consider a small group of ions that pass from the high-concentration side of the membrane to the other side. The volume of the ions will increase. If these ions were really a gas, the expanding gas would do work. Let us see if we can compute the amount of work done when a gas expands from volume V_1 to a larger volume V_2. We invoke the piston arrangement shown in figure 1.12. A frictionless piston of this configuration is familiar to all who suffered through thermodynamics. Assume the pressure is P inside the cylinder and zero outside it. The force on the piston is the pressure, P, times the area, A. If the piston moves a small amount, Δx, the small amount of work Δ (work) done is given by

$\Delta \text{(work)} = PA\,\Delta x.$

The area of the piston, A, times the distance it moved, gives rise to the change in volume, ΔV, so,

$$\Delta \text{ (work)} = P \Delta V.$$

We know from the gas law that

$$PV = nRT, \qquad \text{or} \qquad P = \frac{nRT}{V}.$$

We can assume the temperature is a constant, since the "gas" is actually a solution with lots of other things present, and the numbers of ions moving are very small. Then if we want to know how much work is done when the volume changes from V_1 to V_2, we add up the small changes, Δ (work), and use calculus, so that

$$\text{work} = \int_{V_1}^{V_2} \frac{nRT}{V} dV.$$

The solution to this is the log function,

$$\text{work} = nRT \ln(V_2/V_1).$$

If we convert the volumes into concentrations, we have

$$\text{work} = nRT \ln(c_1/c_2).$$

At equilibrium, the work done by the ions moving down the concentration difference should exactly equal the work done by the ions moving against their electrical gradient. Therefore, from the point of view of the positive ions, the *negatively* charged region (desirable) should contain the *highest* concentration of ions (undesirable), and the two influences should exactly balance at equilibrium.

By equating the work done by the two mechanisms, we end with the final form of the Nernst equation, where E is the voltage across the membrane required for equilibrium, and c_1 and c_2 are the concentrations on either side of the membrane:

$$E = RT \ln \frac{c_1}{c_2}.$$

This is the familiar form of the equation for a singly charged ionic species.

Returning to the concentrations and voltages seen in the axon, the equation indicates that, given the roughly 60-mV negative membrane potential, potassium ions are in approximate equilibrium, but sodium ions are radically out of equilibrium. To obtain the sodium equilibrium suggested by the Nernst equation, the membrane potential would have to become tens of millivolts positive. The resting axon membrane is somewhat permeable to potassium, and perhaps 80 times less permeable to sodium. Electrical behavior is dominated by potassium ion. If the membrane was to become instantaneously completely permeable to sodium ions, sodium ions would flow into the axon, dominate the potassium conductance, and establish the sodium equilibrium potential. Very few ions, only a few thousand per square micron, need be transported across the membrane to shift the membrane potential by this much. Even a few electric charges produce very strong effects.

We can now understand the actual electrochemical mechanism of the action potential: *sodium conductance is a function of the membrane potential.* When the cell is depolarized, the membrane potential becomes more positive and the sodium conductance increases. The increase in sodium conductance further depolarizes the cell since it moves the cell's equilibrium potential toward the sodium equilibrium potential. This further increases the membrane potential, causing a further increase in sodium conductance, and so on. This positive feedback process goes rapidly to a state at which sodium conductance is very large and the cell is near the sodium equilibium potential. This is why the peak of the action potential is several tens of a millivolts positive. After a fraction of millisecond of large sodium conductance, the sodium conductance drops drastically, and potassium conductance experiences a transient increase. The approximate potassium equilibrium potential is restored. If the increased sodium conductance was not shut off, a neuron could change state only once.

A sharp *threshold* value of stimulating current is required to provoke the regenerative process causing the feedback process generating the action potential to go to completion. Below threshold there is no action potential, above threshold there is an action potential. If we look at the threshold as a function of time after the last action potential, we find that for a brief period it is not possible to evoke a second action potential. This is called the *absolute refractory period.* For a somewhat longer period after the action potential, the threshold is elevated. This is called the *relative refractory period.*

The refractory period makes one immediate suggestion about what neurons might be doing. The following argument is imprecise, but the conclusion is correct. Suppose we have a long-lasting stimulus that is above threshold. When the stimulus first appears, an action potential is generated. After the action potential comes a period of declining threshold. At some time the declining threshold meets the input stimulus, followed by a new spike. The

Response of a Neuron to Constant Stimulation

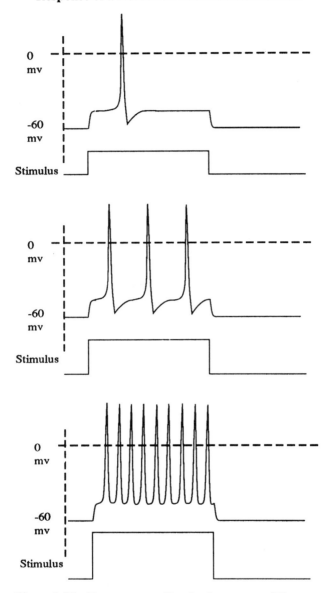

Figure 1.13 Frequency coding in the neuron. When a steady stimulus is applied, the frequency of firing of the neuron is a function of the magnitude of the stimulus. This suggests that what is important in neural activity is not the presence or absence of action potentials, but the frequency of action potentials.

process repeats. This means the cell can fire a regular series of action potentials whose frequency is controlled by the stimulus intensity. This suggests that the neuron is behaving like a *voltage-to-frequency* converter. This kind of coding is often called *frequency coding* in the nervous system (figure 1.13). Real neurons often have complex adaptive processes of various kinds, so firing frequency will drop if a stimulus is maintained. However, for short time periods, there is often quite simple voltage-to-frequency conversion. We will discuss a simple neuron model giving this behavior later. We will also use this property of the neuron when we discuss the *Limulus* eye.

Why are neurons so slow?

The equation that governs the behavior of the bare axon, with its capacitance and resistance, is called the *cable equation* because it was originally developed to explain the difficulties experienced by nineteenth-century transatlantic telegraph cables. The problems faced by a cable thousands of miles in length were exactly those faced by a nerve axon on a much smaller scale. (For a clear discussion of these and other effects in basic electrophysiology, see Aidley, 1992, which the following derivation and notation follows.)

Consider a long insulated conductor. First, the center area is a good, but not a perfect, conductor. Therefore in a sufficiently long run, losses will occur due to the resistance of the center conductor. In the case of the transatlantic cable, even a thick wire of pure copper has significant resistance when miles of it are used. In the case of an axon a few microns in diameter filled with a much poorer conductor than copper, even a few millimeters will have high resistance.

Second, no insulation is perfect. Some current will leak through the membrane in a neuron or the insulator in a cable. Third, and perhaps the most important single reason neurons are slow, is the large membrane capacity. A capacitor is a device that stores energy in the electric field between two conductors. A cell membrane is an insulating sheet tens of Ångstroms thick with conductors on both sides. The membrane material has a high dielectric constant. Therefore we should expect a large membrane capacity, and this is the case: a typical value of membrane capacity is 1 μF per cm^2. In a neuron, associated resistances are also high. Therefore we expect membrane time constants, which, as we will see, are related to the product of capacitance and resistance, to be long. (The time constant in this system is the time required for the voltage to reach $1/e$, about 70%, of its final value.)

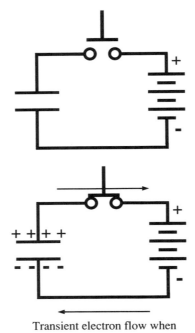

Transient electron flow when
switch is depressed

Figure 1.14 A capacitor in action. A capacitor is schematized as two conducting plates separated by an insulator. When a voltage source is connected to the capacitor, the positive terminal draws electrons from one plate and the negative terminal supplies electrons to the other. A transient current flows. The larger the capacity, the more energy is stored in the electrical field between the plates, and the more current required to change the voltage across the capacitor. Capacitors tend to resist changes in voltage.

Capacitors are open circuits. The capacitive symbol is two parallel plates separated by a gap. A capacitor can store charge because a voltage applied across the capacitor will suck electrons from the plate connected to the positive voltage and put electrons on the plate connected to the negative voltage. Therefore a current flows when the capacitor is first connected to the voltage (figure 1.14). Current flow will stop when the voltage across the capacitor equals the voltage developed between the plates due to the charge imbalance on the two plates.

A standard model of the neuron membrane approximates it by a series of resistors and capacitors (figure 1.15). The network extends to infinity on both ends. The resistive terms r_i and r_o are the resistors associated with the inside and outside resistances of the axoplasm and external medium. The terms r_m and c_m are the resistance and capacitance of the membrane. Note that each repeating element in this membrane model has its own resistance

Electrical Membrane Model

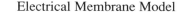

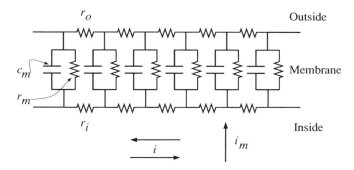

Figure 1.15 Standard passive model of the membrane. Current can leak out through the membrane, and the membrane has high capacity. The axoplasm and the external medium also have resistance. The notation is the same as Aidley (1992).

and capacitance, and each stands for a certain small length of membrane. Therefore, the units of the resistors and capacitors are not really ohms and farads, but ohms per unit of length and farads per unit of length.

Time Domain

First, consider the behavior of this circuit in the time domain. Current through the membrane consists of two parts. The first is a current predictable from Ohm's law, the voltage V is the product of the current, I, and the resistance, R ($V = IR$). With membrane resistance, r_m, and voltage V, then by Ohm's law, the current from this term is (V/r_m).

The membrane capacitor does not act like a simple resistor but responds to the change in voltage. From elementary physics, we know that

$$Q = CV,$$

where Q is the charge stored in the capacitor, C is the capacitance in farads, and V is voltage across the capacitor in volts. When the voltage across the capacitor changes, a current flows. Current is simply the change of charge with time, so we can differentiate the charge and obtain.

$$i = \frac{dQ}{dt} = C\frac{dV}{dt}.$$

Therefore the total current across the membrane will be

$$i_m = \frac{V}{r_m} + c_m\frac{dV}{dT}.$$

The membrane capacitance is c_m. Qualitatively, it is easy to see that the larger the capacitor, the greater the current required to change the voltage in the circuit, as suggested by our earlier analogy.

Space Domain

The space domain analysis is a little more complex. Assume we have the network of resistors and capacitors in figure 1.15. A voltage is established at a particular point on the network. Current flows down the axoplasm, and some will leak out through the membrane. Suppose that flow through the membrane in distance Δx is sufficient to cause a drop in voltage. Suppose the total drop in voltage in this segment of the resistor network is ΔV. Notice that this voltage drop is measured as a function of x. Because we measure resistance in ohms per unit of length, total resistance in the circuit is Δx $(r_i + r_o)$, the product of length times resistance per unit length. Then the current, i, down the axon and returning through the external medium is given by

$$i = \frac{\text{voltage}}{\text{resistance}},$$

$$i = \frac{\Delta V}{\Delta x(r_o + r_i)}.$$

If we make the usual calculus assumption that the distances and voltages involved are very small,

$$i = \left(\frac{1}{r_o + r_i}\right)\frac{dV}{dx}.$$

Current at a point in the network must be equal to the sum of the flow through the membrane and the flow down the axon. Therefore, the change in current going down the axon must be equal to the current through the membrane at that point.

$$\frac{di}{dx} = i_m.$$

If we differentiate the expression for current, we get

$$i_m = \left(\frac{1}{r_o + r_i}\right)\frac{d^2V}{dx^2}.$$

All Together Now

We now have two expressions for membrane current, one in the time domain and one in the space domain. Clearly, membrane current is membrane
current, so we can equate the two. But we derived the space expression
assuming there was no change in the system with time, and vice versa.
Therefore, the final expression for the cable equation uses partial derivatives
for these two terms.

For the purposes of the discussion here, let us define two terms, which are
generally called the space constant, λ, and the time constant, τ

$$\tau = r_m c_m$$

and

$$\lambda^2 = \frac{r_m}{r_o + r_i}$$

Equating the two expressions for membrane current, and doing a little algebra, we have as an expression for the cable equation,

$$V = \lambda^2 \frac{\partial^2 V}{\partial x^2} - \tau \frac{\partial V}{\partial t}.$$

Consider two special cases. First assume that in the dim past we established
a current distribution with voltage V_o at a single point on the membrane,
and the current is no longer changing with time. Then we can easily solve
the resulting differential equation,

$$V = \lambda^2 \frac{\partial^2 V}{\partial x^2}$$

and

$$V(x) = V_o e^{-x/\lambda}.$$

This means that an exponential falloff (figure 1.16) of voltage will occur due
to stimulation at a point. For real neurons, typical values of length constant
would be on the order of a few millimeters.

A similar computation in the time domain yields a similar solution. Let us
assume that there is no change of voltage in the axon with x. In some large
axons this can be done experimentally by inserting a wire into the axon. It
is easy to show that a change in current stimulating the membrane will lead
to a change in voltage with a negative exponential, with time constant τ. If

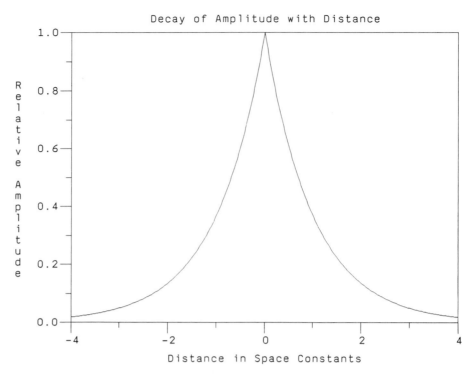

Decay of Amplitude with Distance

Figure 1.16 The cable equation solution for the steady state predicts that a voltage produced at one location will fall off exponentially with distance down an axon. A typical length constant for the axon would be between 2 and 5 mm.

the initial voltage on the membrane is zero and we apply a 1-V step of voltage, the final voltage (after a very long time) will be 1 V. Voltage as a function of time, $V(t)$, will be

$$V(t) = (1 - e^{-t/\tau}),$$

as diagrammed in figure 1.17

A typical value of membrane time constant might be 1 or 2 msec. This suggests that signals changing at a rate of more than 1 kHz will have trouble passing through a neuron, whatever form the actual transmission takes. This is the case.

A realistic chunk of axon with both space and time varying inputs will have a complex spatiotemporal response, but computable from the cable equation. Although we have only discussed a model axon with a constant diameter, dendrites, in which the diameter of the cable may change slowly, give a similar qualitative picture when analyzed. The long time constant acts as a low-pass filter and removes rapidly changing high-frequency compo-

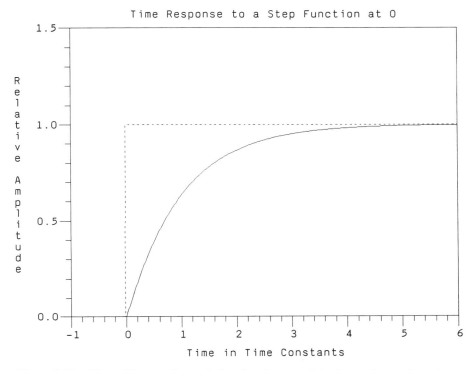

Figure 1.17 The cable equation solution for time predicts that a change in voltage will decay exponentially with time. A typical time constant for a neuron would be 1 msec.

nents from the signal. The farther away the input is from where the voltage is measured, the more capacity is between the observer and the input, and the more low-pass filtering occurs. The space constant will attenuate distant inputs to the cell body.

There are great possibilities for spatiotemporal information processing in the geometry of dendrites. When we look at the shape of a neuron's dendritic tree we are looking at a great deal more than just a convenient way to make connections between neurons. We are looking at a special-purpose analog computer that filters, delays, attenuates, and synchronizes synaptic potentials. If we really understood the cell's function we would understand what this analog computer was doing. We would find that the shape was a reasonably optimum way to arrange the required filtering operations. Unfortunately, this analysis has been carried out for very few cells because of its complexity. The availability of powerful scientific workstations and neuronal modeling programs such as GENESIS and NEURON will change this situation in the near future.

What Good Are Action Potentials?

Now we can return to neuron function. The neuron seems to be signaling a continuous stimulus intensity by converting it into the frequency of occurrence of a remarkable electrical event, the action potential. Therefore, would it not have been more efficient simply to communicate the continuous potential in the first place? That is, why are there action potentials?

The reasons seemed to be based squarely on the physics of axons and the cable equation. An axon is very long and thin. If we consider it as a physical object, it is a thin tube a few microns in diameter, filled with water less salty than the ocean. It has high electrical resistance and a huge amount of capacity, as we have seen. The membrane is leaky. We showed from the cable equation, and it can also be shown experimentally, that the amplitude of voltage caused by the passed current falls off exponentially with distance. The length constant of the decay is a few millimeters. This means that over a distance of a meter, roughly the length of the axon joining the toes with the spinal cord, and for a 2 mm length constant, the attenuation is $e^{(-500)}$. This is a very small number indeed. It means that an enormous voltage could be put in one end of the axon and nothing at all would appear at the other end. The situation is worse than this, because the large capacity and long time constant of a very long axon means that even if the voltage at the far end of the axon was not zero due to attentuation, only very slow changes in voltage could be signaled.

The action potential provides a way to circumvent this problem. When an action potential occurs in a membrane, huge changes occur in membrane permeability, resistance drops, and large ionic currents flow. When a patch of axon fires an action potential, the currents cause other nearby patches of membrane to exceed threshold. The action potential moves to a new patch of membrane and travels down the axon by successively "igniting" parts of the membrane. Since the action potential is all or none, it always looks the same. It travels with constant velocity until it reaches the end of the axon. The traditional image associated with this process is a burning fuse (figure 1.18).

This transmission system works beautifully, and an action potential—a signal—travels from one end of the cell to the other. There is a problem, however. Although the signal travels without attenuation, it moves slowly. It would be desirable to increase the speed of conduction so as to be able to coordinate different parts of large animals or simply to respond faster. The physiology and physics of the axons suggest that the way to increase speed is to make the axon thicker. It can be shown that the speed of conduction

Travelling Action Potential

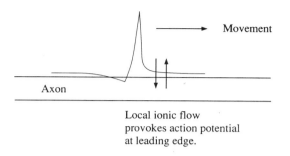

Movement

Axon

Local ionic flow
provokes action potential
at leading edge.

Figure 1.18 A traveling action potential. Once the action potential is initiated, membrane currents excite nearby portions of the membrane to also generate an action potential.

Table 1.1 Conduction Speed

	Diameter	Speed
Crab	30 μ	5 m/sec
Squid	500 μ	20 m/sec (about 16 times thicker, about 4 times faster)
Human	20 μ	120 m/sec

goes up roughly as the square root of the diameter of the bare axon. Table 1.1 lists some speed of conduction data for different neurons (data from Katz, 1966). This is one reason that some invertebrates, for example the squid, have giant axons. However, this is not a fully satisfactory solution since an animal does not have room for very many giant axons.

One trick used by vertebrates to increase conduction speed is to wrap *myelin* around the axon. Myelin is a good insulator, and its layers are far thicker than the membrane (figure 1.19). Every few millimeters on a myelinated axon, a bare patch of axon is exposed at what are called *nodes of Ranvier*. If an action potential occurs at a node, the currents flowing due to the action potential are great enough to trigger an action potential at the next node, since no current can pass through the myelin. The action potential jumps from node to node and goes much faster than if it had to excite all the intervening membrane. This is somewhat like going into hyperspace. The effect is that a 20-μ myelinated mammalian axon can conduct 30 times faster than equivalent-size unmyelinated axon.

In this chapter we have primarily discussed the operation of single neurons. In the next chapter we will discuss how to two neurons interact, and present simple formal models for what neurons are doing.

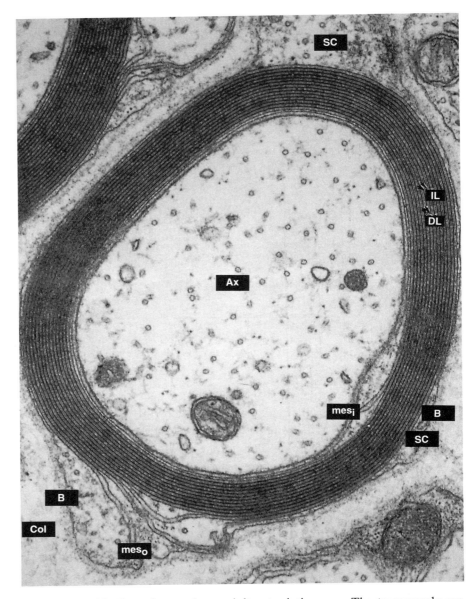

Figure 1.19 Myelinated axon from adult rat sciatic nerve. The transversely sectioned axon (Ax) is surrounded by a myelin sheath composed of lamellae formed from the spiraled plasma membrane of a Schwann cell. In the myelin sheath the alternating major dense lines (DL) and intraperiod lines (IL) are visible. Surrounding the myelin is a thin rim of Schwann cell cytoplasm (SC). The original 20 × 27-cm image was at a magnification of 100,000. From Peters, Palay, and Webster (1976, 1991).

References

D.J. Aidley (1992), *The Physiology of Excitable Cells*, 3rd ed., Cambridge: Cambridge University Press.

J.L. Brown (1975), *The Evolution of Behavior*, New York: Norton.

J.M. Camhi (1984), *Neuroethology*, Sunderland, MA: Sinauer.

J.-P. Ewert (1980), *Neuroethology: An Introduction to the Neurophysiological Fundamentals of Behavior*, Berlin: Springer.

W. James (1892/1984), *Psychology, the Briefer Course*, Cambridge: Harvard University Press.

D. Junge (1992), *Nerve and Muscle Excitation*, 3rd ed., Sunderland, MA: Sinauer.

B. Katz (1966), *Nerve, Muscle and Synapse*, New York: McGraw-Hill.

J.G. Nicholls, A.R. Martin, and B.G. Wallace (1992), *From Neuron to Brain*, 3rd ed., Sunderland, MA: Sinauer.

A. Peters, S.L. Palay, and H. deF. Webster (1976), *The Fine Structure of the Nervous System*, 2nd ed., Philadelphia: W.B. Saunders.

C.F. Stevens (1966), *Neurophysiology: A primer*, New York: Wiley.

P. Taylor (1990), Agents acting at the neuromuscular junctions and autonomic ganglia. In A. Gilman, T.W. Rall, A.S. Nies, and P. Taylor (Eds.), *Goodman and Gilman's the Pharmacological Basis of Theraputics*, New York: Pergamon Press.

R.C. Truex and M.B. Carpenter (1964), *Strong and Elwyn's Human Neuroanatomy*, Baltimore: Williams & Wilkins.

2 *Synaptic Integration and Neuron Models*

In this chapter we will discuss the ways the complex electrochemical structure called the neuron interacts with other neurons. The neuron connects to and receives inputs from other neurons by means of synapses. Different synaptic inputs to a cell interact to produce the activity of the cell. The details of these interactions and how the cell responds to them determine much of the cell's information processing ability. We will discuss three models for neuronal function. The first is the McCulloch-Pitts model in which the neuron is assumed to be computing logic functions. The second is a simple analog integrate-and-fire model. The third is the generic connectionist neuron, which integrates its inputs and generates unit output activity using a nonlinear function.

Synaptic Electrical Events

Consider the chemical synapse. (See Figures 1.3, 1.5, and 2.3 for pictures of real ones.) An extraordinary amount is known about the biochemistry, anatomy, and physiology of synapses. Unfortunately, for modelers there is no generic synapse. Instead, there is a great variety of synapses that differ from each other in ionic mechanism, time constant, tendency to change strength with activity, long- and short-term effects of activation, and many other important properties. We will assume here that we are describing a classic chemical synapse.

A synaptic junction has two sides. The input side, receiving an action potential from the driving cell, is referred to as *presynaptic*. The driven cell is the *postsynaptic* side. We will make the common assumption that the synapse is at the end of a branch of the presynaptic axon, and that the synapse is made to the dendrite of the postsynaptic cell. In fact, a synapse has a great variety of locations: between axons and between dendrites, and even an important class of synapses that are made onto other synapses.

Examples of classic synapses would be those present on a spinal motor neuron or at the neuromuscular junction, both of which have been studied in detail for decades. More recently, it has become clear that a great variety of nontraditional synapses are present even in the mammalian nervous system. The mathematical approximations made by neural network modelers are almost always based on the behavior of the familiar classic synapse. We will follow this tradition and ignore important complexities and qualifications.

When the presynaptic cell becomes active and influences the postsynaptic cell, a characteristic pattern of electrical and ionic activity occurs. If we have an electrode just under the synapse in the dendrite of the postsynaptic cell, when an action potential arrives at the presynaptic side there is initially no response. This characteristic *synaptic delay* is around half a millisecond in mammals. It seems to arise from the need for at least two relatively slow diffusion processes to occur. First and slowest, calcium ions (Ca^{2+}) are required to facilitate release of the neurotransmitter. Calcium ions enters the presynaptic part of the synapse during the increase in conductance associated with the action potential. The calcium ions increase the probability of release of neurotransmitter from the synaptic vesicles. Vesicles fuse with the presynaptic membrane. In the second diffusion process, the fused vesicles release their contents into the synaptic cleft. The neurotransmitter diffuses across the cleft and interacts with *receptors* on the postsynaptic membrane. Figure 2.1 shows a schematic diagram of the synapse and some of these interactions.

Neurotransmitters interact with receptors and cause an increase in the conductance of various ions. Different synapses affect ionic conductance differently. For example, in the neuromuscular junction, an increase in both sodium and potassium conductance is caused by release of the transmitter acetycholine. This increase in conductance depolarizes the cell by electrochemical mechanisms such as those we discussed in chapter 1. Note the difference between the selective increase in the sodium conductance in the action potential, which can cause the inside of the cell to become positive with respect to the outside, and the synaptic increase in conductance to *two* ions, one with an equilibrium potential above the resting membrane potential (Na^+) and one with an equilibrium potential below it (K^+). The potential inside the cell does not become as positive. Inhibitory synapses may increase only potassium conductance, which tends to move the cell closer to the potassium equilibrium potential. Potassium equilibrium is already near the membrane potential and well away from the action potential threshold.

The actual details of ion flows during synaptic events are complex since there can be a number of different ionic channels with different conductances even for the same ion that may become active at various times during activation of a synapse. Nine or more different ionic channels have been described in a single neuron. Most important, some synaptic events seem to

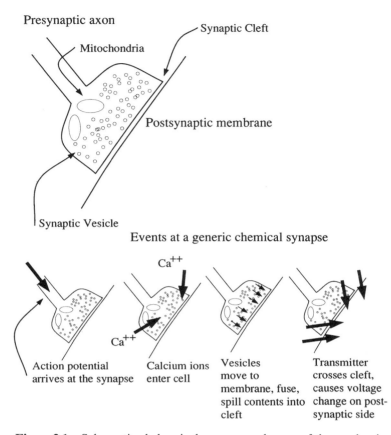

Events at a generic chemical synapse

Figure 2.1   Schematized chemical synapse and some of the mechanisms involved in synaptic transmission.

be able to trigger long-term changes in the efficacy of a synapse. These changes seem to underly the biological changes associated with learning and memory.

After the roughly half-millisecond synaptic delay, the ionic flows caused by the neurotransmitter give rise to an electrical potential in the postsynaptic cell. This is called the *postsynaptic potential* (PSP). There are two general types, *excitatory postsynaptic potentials* (EPSPs), in which the synaptic potential tends to depolarize the cell and move it toward threshold, and *inhibitory postsynaptic potentials* (IPSPs), in which the synaptic potential makes the cell less likely to fire.

Postsynaptic potentials are of much longer duration than the action potentials giving rise to them. A typical fast EPSP would have a rising phase of 1 or 2 msec, and then a more gradual decline, with a decay time of 3 to 5 msec. A wide range of time constants is seen physiologically; some IPSPs

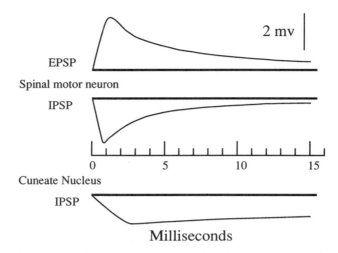

Figure 2.2   Postsynaptic potentials. The top IPSP and EPSP are from spinal motor neurons. Many IPSPs are much longer in duration. The bottom trace shows an IPSP from the cuneate nucleus. The IPSPs in cortex can have durations of hundreds of milliseconds. The synaptic delay is not represented in this figure. Redrawn from Eccles (1964).

may last for tens or even hundreds of milliseconds. The length of a PSP is an obvious mechanism for affecting the degree of interaction between synaptic events. Figure 2.2 shows several examples of postsynaptic potentials.

A single postsynaptic cell can have several thousand synapses on it. The number can range from hundreds, to tens of thousands on cortical pyramidal cells, to as many as 200,000 on Purkinje cells in the cerebellum. Most synaptic contacts are made on dendrites. As we have mentioned, the shapes of dendritic trees are closely coupled to their information-processing function. The action potential is often started at a a structure called the *axon hillock*, where the axon leaves the cell. The passive membrane in the dendrites has important filtering, timing, and shaping effects on electrical signals from the synapses and acts as a complex analog computer, refining and integrating the input signals. The dendrites of a few types of neurons seem to support action potentials, although this is probably not common.

Figure 2.3 shows an electron micrograph of several synapses in the rat brain from a region called the ventral cochlear nucleus. Note that the synapses are made on a small process called a *dendritic spine*, also visible in the freeze fracture picture in figure 1.3. Dendritic spines are strong candidates for involvement with learning. They have been shown to respond to certain kinds of environmental manipulation in ways that suggest they are adaptively modifiable, and seem to contain a type of synapse that has been strongly implicated in the long-term changes in synaptic strength that many

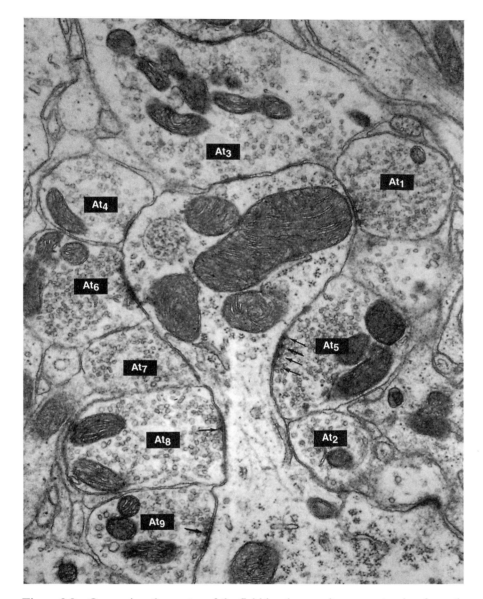

Figure 2.3   Occupying the center of the field is a large spine or protrusion from the perikaryon of a neuron, and surrounding this structure is a variety of axon terminals. Two of the terminals (At$_1$ and At$_2$) contain small spheric synaptic vesicles, and the spheric vesicles in two other terminals (At$_3$ and At$_4$) are somewhat larger. The other axon terminals (At$_5$ to At$_9$) contain some elongate synaptic vesicles. The synaptic junctions in which these various axon terminals participate seem to be symmetrical. The junction formed by At$_5$ is noteworthy because of the regular array of presynaptic densities (arrows). At other synaptic junctions only one or two of these densities (arrows) are apparent. The original 27 × 20-cm size corresponded to a magnification of 44,000. From Peters, Palay, and Webster (1976). Reprinted by permission.

believe are related to memory. Simply by changing the morphology—length or width—of the spine one can control to some degree the electrical influence that the presynaptic cell has on the postsynaptic cell, as well as the ionic environment of the spine. Current thinking, with support from computer simulations, is that the geometry of the spine allows an "isolated biochemical microenvironment" (Koch and Zador, 1993, p. 420) around the synaptic connection, allowing high local concentrations of calcium ion, in particular, to exist. This microenvironment could trigger the events leading to permanent synaptic change. Because of the degree of physical and ionic isolation provided by a spine, the strength of one synapse could be changed independent of others. (Lynch, 1986; Bear, Cooper, and Ebner, 1987).

Interaction of Synaptic Effects

What happens when more than one synapse becomes active at the same time? In some cases, the resulting EPSPs or IPSPs come close to adding their values algebraically. This simple interaction is the approximation that is made in most neural network models, as we will see when we present the generic connectionist neuron at the end of this chapter. The way synaptic interactions actually occur is a function of the cell type and where the synapses are located relative to each other. It might be thought that simple membrane interactions, since they arise from mechanisms that are close to simple circuits of resistors and capacitors, would give rise to algebraic addition. The actual situation is more complex than this because of the changes in membrane conductance associated with synaptic currents, presence of neurotransmitters, and changes in membrane potential. If the membrane resistances do not stay constant, as they do not, complex nonlinear interactions between synaptic potentials can and do occur.

Given these difficulties, the only sure way to verify synaptic interactions is experimentally. Unfortunately, experimental data on synaptic integration are not available for most neural systems. The simple synaptic interactions assumed by many network models must be treated with suspicion, as they are based on the properties of a small number of neural systems that may not be typical.

One of the best-studied systems in terms of synaptic interactions is the spinal motor neuron. When two synapses, for example, an inhibitory synapse and an excitatory synapse, are simultaneously activated, there is a high degree of linearity in the addition of synaptic effects. Even here, however, in a significant number of cases the synaptic interaction is nonlinear to some degree. Figure 2.4 is a well-known figure that gives some idea of the range of

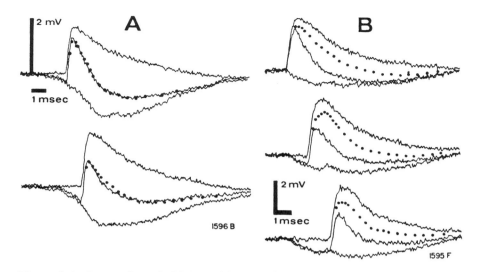

Figure 2.4 Interaction of IPSPs and EPSPs shows linearity (A) and nonlinearity (B) of summation of the PSPs. See text for description. From Rall et al. (1967). Reprinted by permission.

Dendritic Integration

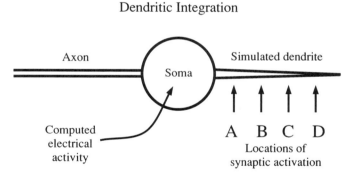

Figure 2.5 Schematic diagram of the simple model neuron simulated by Rall. This model neuron has four input regions at different distances along a single dendrite.

synaptic interactions possible even for one neuron type (Rall et al., 1967). Another thoroughly studied system, the eye of the horseshoe crab, also shows good linear response between synaptic potentials. We will discuss this system in more detail in chapter 4.

Let us give one famous example of the use of synaptic integration for information processing, in which integration is combined with the properties of the cell membrane that we have already discussed. Consider the system studied by Rall (1964), diagrammed in figure 2.5. This model "cell" has a single dendrite with four identical synapses located at various distances from the spike-generating region. This single-dendrite model was

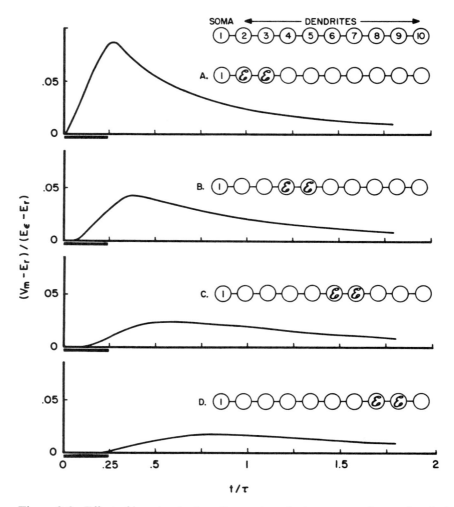

Figure 2.6 Effect of input pulse location on transient soma membrane depolarization. The somatodendritic receptive surface is divided into 10 equal compartments (upper right). The four input pulse locations are shown in the diagrams beside the letters A, B, C, and D. These curves were drawn through computed values for time steps of 0.05τ. To convert to real time, a τ value between 4 and 5 msec is appropriate for mammalian motoneurons. From Rall (1964). Reprinted by permission.

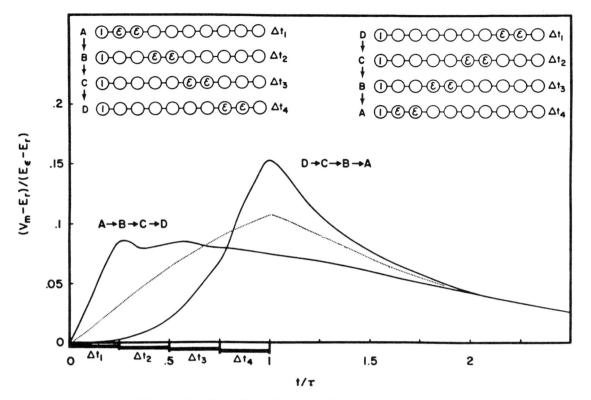

Figure 2.7   The effect of two spatiotemporal sequences on transient soma membrane depolarization. Two pulse sequences, A → B → C → D and D → C → B → A, are indicated by the diagrams at upper left and upper right; the component pulse locations are the same as in figure 2.6. The time sequence is indicated by means of the four successive time intervals, Δt_1, Δt_2, Δt_3, and Δt_4, each equal to 0.25τ. The dotted curve shows the computed effect of pulses in eight compartments (2 through 9) for the period $t = 0$ to $t = \tau$. From Rall (1964). Reprinted by permission.

originally designed to be an approximation to certain classes of dendritic trees, and is more realistic than it first appears. The synapse far away from the cell body has its postsynaptic potential extensively low-pass filtered, so the cell body sees a low-amplitude, slowly rising, and slowly falling potential.

A synapse close to the cell body has a rapidly rising and rapidly falling potential because there is less capacitance and resistance between the two (figure 2.6). If the synapses are fired in temporal sequence toward (D → C → B → A) or away (A → B → C → D) from the cell body, radically different potentials at the cell body are seen (figure 2.7). In one case (toward) the peaks of the PSPs coincide in time, and a large, brief PSP is formed; in the other case the PSPs do not coincide, and a long, low-amplitude PSP is formed from the sum. The first postsynaptic potential might give rise to a high-frequency burst of spikes, whereas the other pattern might correspond to a maintained low, or even absent, discharge, depending on threshold of the postsynaptic cell. This simulation provides one example of the complexities of even the simplest synaptic and dendritic interactions.

Slow Potential Theory of the Neuron

A cell with many inputs may be receiving inputs constantly. Since the PSPs are temporally extended with respect to the action potentials, many opportunities for *temporal integration* exist, as we have just seen, for events occurring at different times to add. The extent of dendritic spread allows significant *spatial integration*, when events at different locations can add. The integrated membrane potential at the cell body varies slowly in a *slow potential*, as the inputs wax and wane. Since the cell acts like a voltage-to-frequency converter, this slow potential is converted into firing frequency and sent down the axon, propagating to the terminal arborization at the far end of the axon. Synaptic temporal integration at the postsynaptic cell causes the presynaptic spikes to be low-pass filtered, and a slow potential, derived from the slow potential in the presynaptic cell, is reconstructed. The idea that the value of slow potential at the axon hillock is what is being signaled by the output of the cell is sometimes called the *slow potential theory of the neuron* (Stevens, 1966). Figure 2.8 suggests the way that presynaptic action potentials are received, low-pass filtered by the synapse, and integrated at the axon hillock, giving rise to a new processed slow potential that is then transmitted to other cells.

The slow potential theory makes one interesting prediction: it suggests the important feature of cell activity is the value of the slow potential, not the presence or absence of an action potential. Action potentials are used in neurons because information must be propagated from one end of a long, poor conductor to the other. If we had neurons very close together, we might not need or use action potentials. This turns out to be the case for the vertebrate retina, in which, of the six cell types, five do not use action potentials, but seem to communicate entirely by slow potentials. Such communication is possible since the distances involved are less than a millimeter, that is, probably less than one membrane space constant. Only the output cells of the retina, the retinal ganglion cells, which must communicate information over several centimeters to the rest of the brain, use action potentials.

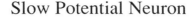

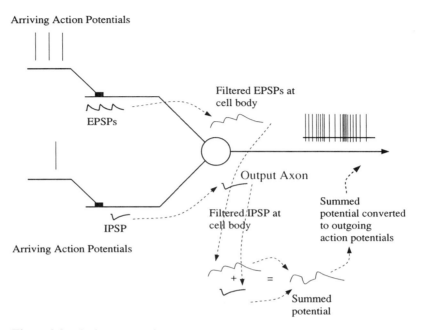

Figure 2.8   A slow-potential neuron. The input to a neuron generates a slow potential that is turned into a train of action potentials, which are transmitted to a distant synaptic terminal. The integrative action of the distant synapse and the membrane of the postsynaptic cell give rise to a reconstructed slow potential. By various combinations of excitation, inhibition, and time and space constants, very complex analog computations are possible with neurons.

Noise

It is often claimed that neurons are noisy. We can see several intrinsic mechanisms causing noise, based on what we have learned up to now.

First, electrochemical events in the cell are due to ion flows through ionic channels that are either open or shut. This discreteness causes a significant amount of quantization noise.

Second, neurotransmitter is released from the synapses in discrete packets because transmitter is stored in synaptic vesicles. The amount of transmitter is not identical in every vesicle. In addition, in some thoroughly characterized synapses, the probability of release of transmitter from a vesicle is given by a Poisson process, with the mean of the process controlled by presynaptic voltage and ionic concentrations (Katz, 1966). A postsynaptic potential is produced by shifting the mean of the Poisson distribution. The variance of a Poisson process is large, equal to the mean.

Third, the neuron communicates over long distances by using pulses—action potentials; that is, it forms a simple pulse code-modulation system with a limited dynamic frequency range, perhaps 100 to 1. Postsynaptic frequency estimation seems to be done for only a few tens or, at most, hundreds of milliseconds. There are numerous other sources of noise as well, but these three are intrinsic to the operation of the the neuron and are not avoidable.

Theoretical Models of the Neuron

Many formal models for the single neuron exist. Since neural networks are built up of interconnected model neurons, it is important to know what these neurons are supposed to do. From our discussion of the cable equation and elementary neurophysiology, it is clear that a real neuron is an immensely complex structure, and it can be modeled at many levels of detail. If we tried to put into a model everything we know, or even a small fraction of what we know about a real neuron, it would be impossible to work with, even with current computer technology. Therefore we must use a model that is adequate for what we want to use it for.

For purely practical reasons, we must discard detail. This, of course, deeply offends many neuroscientists who may have spent their entire careers studying the mechanisms that theorists casually discard. Any simplification can and will be attacked as simplistic. Choosing the level of detail to put in a model is something of an aesthetic judgment. Most people like models that are simple enough to understand, but rich enough to give behavior that is surprising, interesting, and significant. If all goes well, the result will predict things that might be seen in a real organism. Or, if a practical application is in mind, it will be able to perform satisfactorily the function required of it.

Many neuron models are used in the neural network literature. We will present three of them, all of which are used currently. The historically important McCulloch-Pitts neuron, first proposed in 1943, is used by physicists to study certain classes of complex nonlinear networks (see chapter 12). A simple integrator model is a good first approximation for more physiologically based modeling. Probably the most common is the generic connectionist neuron.

Two-State Neurons

In 1943 Warren McCulloch and Walter Pitts wrote a famous and influential paper that was based on the computations that could be performed by two-state neurons. (See the highly recommended collection of McCulloch's work (1965) or the first volume of *Neurocomputing*.) McCulloch and Pitts made perhaps the first attempt to understand what the nervous system might actually be doing, given neural computing elements that were abstractions of the physiological properties of neurons and their connections.

These investigators made a number of assumptions governing the operation of neurons that define what has become known as the *McCulloch-Pitts neuron*, which is familiar to most computer scientists. It is a binary device; that is, it can be in only one of two possible states. Each neuron has a fixed threshold, and receives inputs from excitatory synapses, all of which have identical weights. Excitatory synaptic activations add linearly. A time quantum for integration of synaptic inputs is based loosely on the physiologically observed synaptic delay. The neuron can also receive inputs from inhibitory synapses, whose action is absolute; that is, if the inhibitory synapse is active, the neuron cannot become active.

The mode of operation of the model McCulloch-Pitts neuron is simple. During the time quantum, the neuron responds to the activity of its synapses, which reflect the state of the presynaptic cells. If no inhibitory synapses are active, the neuron adds its synaptic inputs and checks to see if the sum meets or exceeds its threshold. If it does, the neuron becomes active. If it does not, the neuron is inactive.

As an example, suppose we have the simple unit in figure 2.9 with two excitatory inputs, *a* and *b*, and with threshold 1. The notation in figure 2.9

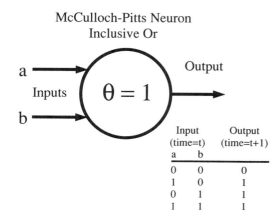

Figure 2.9   A simple McCulloch-Pitts neuron.

follows Minsky (1967) rather than that of the original McCulloch-Pitts paper. Synapses connected to active cells contribute one unit. Suppose at start of the ith time quantum, both a and b are inactive. Then, at the start of the $(i + 1)$st time quantum, the unit is inactive since the sum of two inactive synapses is zero. If a is active but b is inactive, the state of the unit at the start of the $(i + 1)$st time quantum is active since the inactive synapse plus the active synapse is 1, which equals the threshold. Similarly, if b is active and a is inactive, or if both are active, the cell becomes active. This threshold one unit with two excitatory inputs is performing the logical operation INCLUSIVE OR on its inputs, and becomes active only if a or b or BOTH a AND b is active.

If the unit was given a threshold of 2, it would compute the logic function AND, since it would only become active if a AND b are active. Table 2.1 gives the list of possible input and output states for cells with thresholds 1 and 2. A logic function can be characterized by a complete list of its inputs and outputs. This list of responses to inputs makes up what is known as a truth table for a logic function, if we take 0 as false and 1 as true.

The neuron is performing what is called *threshold logic*. McCulloch and Pitts wrote, "The 'all-or-none' law of nervous activity is sufficient to insure that the activity of any neuron may be represented as a proposition. Physiological relations existing among nervous activities correspond, of course, to relations among the propositions; and the utility of the representation depends upon the identity of these relations with those of the logic of propositions" (1943, p. 117).

Therefore, a network of connections between simple propositions can give rise to complex propositions. The central result of the 1943 paper is that any

Table 2.1 Input and output states for cells with thresholds 1 and 2

Input at Time t		Cell Output at Time $t + 1$
a	b	
Unit threshold of 1 (inclusive OR)		
0	0	0
1	0	1
0	1	1
1	1	1
Unit threshold of 2 (AND)		
0	0	0
1	0	0
0	1	0
1	1	1

finite logical expression can be realized by a network of McCulloch-Pitts neurons. This was exciting, since it showed that simple elements connected in a network could have immense computational power. It suggested that the brain was potentially a powerful logic and computational device.

Are McCulloch-Pitts neurons correct approximations to real neurons? Given the state of neurophysiology in 1943, when the ionic and electrical basis of neural activity was unclear, the approximations were much more supportable than they are now. Even now, we find McCulloch-Pitts neurons used in the modeling literature. Occasionally it is suggested that they are adequate brain models and useful approximations to neurophysiology. This is not correct. Our current understanding of neuron function suggests that neurons are not devices realizing the propositions of formal logic. But the work of McCulloch and Pitts gave rise to a tradition, which still exists to some degree, that views the brain as a kind of noisy processor, doing logical and symbolic operations much as a digital computer does.

The immense theoretical impact of the McCulloch-Pitts paper was not among neuroscientists but among computer scientists. The history of this work is encouraging for theoreticians. It is not necessary to be correct in detail, or even in the original domain of application, to create an enduring work of great importance. The McCulloch-Pitts neuron played a role in the history of the computer. John von Neumann was aware of the McCulloch-Pitts paper and cited it in several places. For example, in an influential 1945 technical report, where the single CPU-stored program architecture that has become known as the Von Neumann computer was first described, he wrote

It is worth mentioning that the neurons of the higher animals are (relay-like) elements ... They have all-or-none character, that is two states: Quiescent and excited.... An excited neuron emits the standard stimulus along many lines (axons). Such a line can, however, be connected in two different ways to the next neuron: First, in an excitatory synapse, *so that the stimulus causes excitation of that neuron. Second, in an* inhibitory synapse, *so that the stimulus absolutely prevents the excitation of that neuron by any stimulus on any other (excitatory) synapse. The neuron also has a definite reaction time, between the reception of a stimulus and the emission of the stimuli caused by it, the* synaptic delay.

Following W. Pitts and W.S. McCulloch ... we ignore the more complicated aspects of neuron functioning: Thresholds, temporal summation, relative inhibition, ... etc.... It can easily be seen that these simplified neuron functions can be imitated by telegraph relays or by vacuum tubes. (1982, pp. 387–388)

A More Realistic Class of Models: Integrators

Bruce Knight (1972) analyzed a simple integrator model of a single neuron that was suggested by the neurophysiology of the horseshoe crab *Limulus* (see chapter 4). By making a number of approximations to the Hodgkin-Huxley equation used to describe the behavior of the squid giant axon, it is possible to derive this simple equation. It is sometimes called the *integrate-and-fire* model of the neuron. It and its close relative the *forgetful integrate-and-fire* model are still used as simple, tractable neural elements for some physiologic applications. They catch enough of the flavor of reality to be useful.

Imagine a noise-free neuron containing an internal variable $u(t)$, which might correspond to membrane potential. A stimulus, $s(t)$, might correspond, say, to ionic current. Then,

$$\frac{du}{dt} = s(t). \qquad (s(t) > 0).$$

When u reaches a criterion threshold, θ, a nerve impulse is fired. The system resets u to its starting value after the spike is fired (figure 2.10).

Suppose a spike was fired at time t_1 and the time now is t. Then,

$$u = \int_{t_1}^{t} s(t)\,dt.$$

If the next spike is fired at time t_2, there is a straightforward relation between threshold θ and u at t_2,

$$\theta = \int_{t_1}^{t_2} s(t)\,dt.$$

Integrate-and-fire Neuron

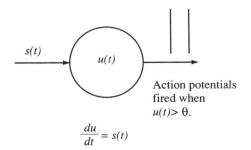

$$\frac{du}{dt} = s(t)$$

Figure 2.10   A simple integrate-and-fire neuron.

Suppose $s(t) = s$, where s is a constant. The solution of the equation is simple. We can derive an expression for the time Δt between the the two spikes at t_2 and t_1,

$$\Delta t = t_2 - t_1$$

as a function of s and θ. We have the integral of a constant, s, so

$$\theta = s(t_2 - t_1) = s\Delta t.$$

Since the instantaneous firing rate is defined as the reciprocal of the interspike interval, we have the spike frequency, f, for the system given by

$$f = \frac{1}{\Delta t},$$

and we end up with the simple relation between stimulus magnitude and threshold as

$$f = \frac{s}{\theta}.$$

This simple linear relation looks something like the relation between input current and resultant firing rate found in a number of sensory receptors (figure 2.11). In particular, it is a good model of some of the behavior of the *Limulus* eye.

A more complex version of the integrator model assumes a decay of membrane potential due perhaps to membrane leakage. Incorporating this assumption lead to a differential equation relating stimulus magnitude, s, and internal variable, u, now, with a decay term,

$$\frac{du}{dt} = -\gamma u + s(t).$$

The solution is now more complex, but the above expression can be integrated, and the final relation between instantaneous firing frequency and stimulus magnitude s for a constant input is given by

$$f = \frac{-\gamma}{\ln(1 - \gamma\theta/s)}.$$

This relation also looks linear except at low values of s, where it shows threshold behavior. Figure 2.11 shows the relations between input signal strength and output firing rate for the two models.

Constant Stimulus Magnitude

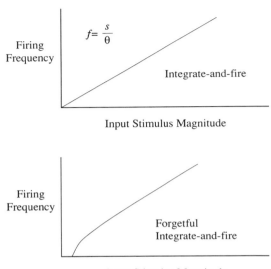

Figure 2.11   Relation between input stimulus magnitude and output firing rate for the integrate-and-fire model and the forgetful integrate-and-fire model.

The Generic Neural Network Neuron

The model-computing element used in most neural networks is an integrator, such as the integrate-and-fire model, and does detailed computation based on its connection strengths—model synaptic weights—like the McCulloch-Pitts neuron. It also contains an important nonlinearity.

The operation of the generic connectionist neuron is a two-stage process. In the first stage, inputs from the synapses are added together, with individual synaptic contributions combining independently and adding algebraically, giving a resultant activity level. This activity level is a metaphorical slow potential. In the second stage this activity level is used to generate the final output activity of the model neuron by using the activity level as the input to a nonlinear function relating activity level (membrane potential) to output value (average output firing rate). *The generic connectionist neuron does not fire action potentials but has an output that is a continuous graded activity level.* Figure 2.12 shows the basic architecture.

Generic Connectionist Neuron

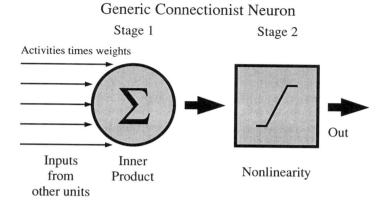

Figure 2.12 The generic connectionist or neural network unit. It has two parts: the first is a simple integrator of synaptic inputs weighted by connection strengths, and the second is a nonlinearity that operates on the output of the integrator.

The First Stage: Linear Addition of Synaptic Inputs

Almost all neural network units assume linear addition of synaptic interactions as the first stage in information processing. Suppose we have a neuron with, say, n inputs. Each input, $1 \ldots n$, has an activity, $f[1] \ldots f[n]$. Each input has a *connection strength*, or *synaptic strength*, $a[1] \ldots a[n]$, that measures degree of coupling between a unit and its inputs. Setting the connection strengths correctly is what network learning is about. The notation used in this book will often denote the connection strengths as either $a[i]$ or $w[i]$, the input activities by $f[i]$, and the output activities by $g[i]$. It is universally assumed that the connection strengths are single numbers; the more complex temporal behavior and nonlinear interactions known to exist in real synapses are ignored in the name of simplicity. Also ignored is the fact that in most real neurons information is signaled by action potentials. Cell activity is approximated by a continuous variable, justified by relating activity to the firing frequency of the cell, integrated over a few tens of milliseconds. A switch in emphasis has occurred with time: from McCulloch and Pitts, where the all-important action potential signaled truth or falsehood of a logical proposition, to the generic connectionist neuron, where discrete action potentials are ignored completely and replaced by a continuous average rate.

Let us assume the first stage of our unit has an internal veriable, called u, which is given by the weighted sum of its inputs, and a bias term, θ, characteristic of the cell, that is,

$$u = a[1]f[1] + a[2]f[2] + a[3]f[3] \ldots + a[n]f[n] + \theta$$

$$= \sum_{i=1}^{i=n} a[i]f[i] + \theta.$$

This quantity, u, is, except for θ, the dot product or inner product (see chapter 3), taken between *vectors* representing the connection strengths, the weights, and the input activities. It is easy to eliminate the threshold term by incorporating it into the dot product as a special weight always connected to a constant input value, and this is often done for mathematical convenience. Suppose, for example, that we invent a new activity, $f[n + 1]$ which always equals 1 and let $\theta = a[i + 1]$. Then,

$$u = a[1]f[1] + a[2]f[2] + a[3]f[3] \ldots + a[n]f[n] + f[n + 1]\theta$$

$$= \sum_{i=1}^{i=n+1} a[i]f[i].$$

This lets us describe the output of the cell entirely as a function of the activity pattern, the $f[i]$, and a set of connection strengths, which are now *augmented* by the value of the bias term, θ. This little bit of redefinition simplifies the mathematics, because the sum describing the unit activity is now given by the inner product between an activity pattern and a set of weights.

Many of the techniques of basic statistics and signal processing can be built from operations that use inner products. If we assume that an inner product is incorporated into the basic parallel hardware of neural networks, it is not surprising that in many cases, proper arrangement of units and setting of weights can do good statistical processing. As we will see, there is nothing mysterious about the operation of neural networks. They often can be shown to perform known, effective statistical computations quickly and in parallel. The purported speed advantages of parallelism arise because this inner product, the sum of products of weights and activities, is assumed to be computed in a single step, independent of the number of connections. Of course, subject to integration delays, real neurons do compute something like a low-precision analog inner product when they integrate their multitude of synaptic inputs. However, a great many more traditional algorithms could also go very fast if they could do single-step inner products. Many computer companies build expensive special-purpose hardware—vector processors—to do exactly these computations quickly.

The Second Stage: The Input-Output Function

The internal variable, u, is useful as it stands, and can be used all by itself to construct interesting neural networks; for example, a model of the *Limulus* eye, and the *linear associator* in this book. For a number of reasons we will discuss in detail later, however, the actual output of the model neuron is usually passed through a nonlinear second stage. The first-stage output is used as the input to a nonlinear function, $f(u)$, relating the internal variable, u, to the actual unit output, which we will call g. That is,

$$g = f(u) = f\left(\sum_{i=1}^{i=n+1} a[i]f[i]\right).$$

We should emphasize again that neuron activity, g, is simply a number, and the presence of discrete action potentials is ignored and replaced by a short-term average rate. One reason for the incorporation of $f(u)$ is that the sum, u, is potentially unbounded. Real neurons are limited in dynamic range from zero-output firing rate to a maximum of a few hundred action potentials per second. Therefore the unit output must be contained within this range. The details of the function, $f(u)$, differ somewhat from model to model, but usually are designed to contain the output within this limited range of possible values.

The PDP group, in their famous collection of papers (McClelland and Rumelhart, 1986; Rumelhart and McClelland, 1986), colorfully referred to a function that compresses the range of values of u from $[-\infty, \infty]$ to a limited range as a *squashing function*. If the function is monotonic, differentiable, and smooth, it is often realized by a *sigmoid* or S-shaped curve, a term whose theoretical significance was pointed out by several in the 1970s (Wilson and Cowan, 1972; Grossberg, 1976). The squashing function was described by the PDP group as *semilinear*, which means it increases continuously and monotonically and does not do anything very strange.

One example of a commonly used sigmoid would be simple *logistic* function,

$$f(u) = \frac{1}{1 + e^{-u}}.$$

The logistic function ranges from 0 to 1 as u ranges from $-\infty$ to $+\infty$.

It is sometimes more convenient to have a function, $f(u)$, that ranges from, say, -1 to $+1$. Allowing negative output values makes analysis much easier, and can often be justified by the observation that many neurons have a nonzero spontaneous firing rate. Thus deviations in both directions from this rate are allowable and, indeed, appear to be used physiologically in some cases, although rarely with such convenient symmetry. A symmetric function of this type is

$$f(u) = \arctan(u),$$

Which ranges from $(-\pi/2, \pi/2)$, a popular function among physicists who have become interested in neural networks in the past few years.

Another simple squashing function is one that we will use sometimes in this book because it is easy and fast to program (see chapter 15). The function, f, is simply the LIMIT function, which clips values above and below some preset value and does nothing in between. Inside the limits, the function is purely linear, outside it, values are simply set to the appropriate limit; that is,

$$f(u) = \begin{cases} \text{IF } u \geq \text{upper limit THEN } f(u) = \text{upper limit.} \\ \text{IF lower limit} < u < \text{upper limit, THEN } f(u) = u. \\ \text{IF } u \leq \text{lower limit THEN } f(u) = \text{lower limit.} \end{cases}$$

A strictly linear portion of the sum is surrounded by hard limits. This function is also not differentiable at two points (the limits), which is sometimes

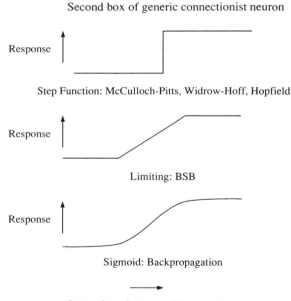

Favorite Nonlinearities
Second box of generic connectionist neuron

Step Function: McCulloch-Pitts, Widrow-Hoff, Hopfield

Limiting: BSB

Sigmoid: Backpropagation

Output from first stage (inner product)

Figure 2.13 Some commonly used nonlinearities for the second box of the generic connectionist computing unit.

Synaptic Integration and Neuron Models

mathematically inconvenient. Figure 2.13 gives a sketch of some commonly used variants of nonlinear functions.

If we ask which of these semilinear functions works best in simulation, the answer seems to be that it makes little difference. If we ask which squashing function looks most like real neurons, the answer is probably a softened version of the hard limiter, where a linear region is surround by limits, but the transition to clipping is not abrupt. True hard limits are never seen in nature; there are graded transition regions. One example of real neuron input-output relations, the crab motor neuron (Chapman, 1966), suggestive of this is shown in figure 2.14. Of course, since observed firing rates are the result of several processes, it is hard to do more than make a kind of vague qualitative observation that a sizeable, smooth, sometimes nearly linear region in the input-output response of a real neuron seems to be a commonly observed pattern. The various smooth nonlinearities used in models are close enough to this pattern to be indistinguishable in their effects from it,

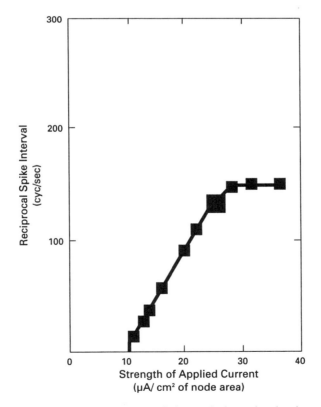

Figure 2.14 Experimental data relating stimulus intensity and firing rate from a crab motor neuron, a biological sigmoid. Modified from Chapman (1966). Reprinted by permission.

given the other uncertainties and gross assumptions involved in network simulations.

Binary Model Neurons

If the slope of the squashing function is very steep, we have a model neuron acting like the two-valued McCulloch-Pitts neuron. Values of u even slightly above zero will be given the maximum value, and even slightly below zero will be given the minimum value of the squashing function. If the synaptic weights, $a[i]$, and input activities, $f[i]$, are restricted to the appropriate discrete values (0 and 1 for $f[i]$ and $-\infty$, 0, and 1 for $a[i]$), we have an analog of the 1943 McCulloch-Pitts neuron. Two-valued neurons are still used in the current neural network literature because of their convenience for many applications and their nice interface with digital electronics and formal logic.

Conclusion

Nowadays, you can go to your local Radio Shack store and buy McCulloch-Pitts neurons, neatly packaged, for a few cents. You can then go home and build a small digital computer on your kitchen table. If you were to go to Radio Shack and buy a bunch of more realistic neurons instead, what would you make when you got home? Our claim is that it would be quite a different device, with quite different properties, ones that we will proceed to explore.

References

M. Bear, L.N. Cooper, and F.F. Ebner (1987), A physiological basis for a theory of synaptic modification. *Science*, *237*, 42–48.

R.A. Chapman (1966), The repetitive responses of isolated axons from the crab, *Carcinus maenas. Journal of Experimental Biology*, *45*, 475–488.

J.C. Eccles (1964), *The Physiology of Synapses*, Berlin: Springer.

S. Grossberg (1976), Adaptive pattern classification and universal recoding. I. Parallel development and coding of neural feature detectors. *Biological Cybernetics*, *23*, 121–134.

B. Katz (1966), *Nerve, Muscle, and Synapse*, New York: McGraw-Hill.

B. Knight (1972), Dynamics of encoding in a population of neurons, *Journal of General Physiology. 59*, 734–66.

C. Koch and A. Zador (1993), The function of dendritic spines: Devices subserving biochemical rather than electrical compartmentalization. *Journal of Neuroscience, 13*, 413–422.

G. Lynch (1986), *Synapses, Circuits, and the Beginnings of Memory*, Cambridge: MIT Press.

J. McClelland, D.E. Rumelhart, and the PDP Research Group (1986), *Parallel Distributed Processing*, Vol. II, Cambridge: MIT Press.

W.S. McCulloch (1965), *Embodiments of Mind*, Cambridge: MIT Press.

W.S. McCulloch and W.H. Pitts (1943), A logical calculus of the ideas immanent in nervous activity. *Bulletin of Mathematical Biophysics, 5*, 115–133.

M. Minsky (1967), *Computation: Finite and Infinite Machines*, New York: Prentice-Hall.

A. Peters, S.L. Palay, and H. deF. Webster (1976), *The Fine Structure of the Nervous System*, 2nd ed., Philadelpha: W.B. Saunders.

W. Rall (1964), Theoretical significance of dendritic trees for neuronal input-output relationships. In R.F. Reiss (Ed.), *Neural Theory and Modelling*, Stanford, CA: Stanford University Press.

W. Rall, R.E. Burke, T.G. Smith, P.G. Nelson, and K. Frank (1967), Dendritic locations of synapses and possible mechanisms for the monosynaptic EPSP in motorneurons. *Journal of Neurophysiology, 30*, 1169–1193.

D.E. Rumelhart, J.L. McClelland, and the PDP Research Group (1986), *Parallel Distributed Processing,* Vol. I, Cambridge: MIT Press.

C. Stevens (1966), *Neurophysiology: A Primer*, New York: Wiley.

J. von Neumann (1982), First draft of a report on the EDVAC. In B. Randall (Ed.), *The Origins of Digital Computers: Selected Papers*, 3rd ed., Berlin: Springer.

H.R. Wilson and J.D. Cowan (1972), Excitatory and inhibitory interactions in localized populations of model neurons. *Biophysical Journal, 12*, 1–23.

3 *Essential Vector Operations*

This chapter starts by briefly discussing data representation, that is, how information is represented as a pattern of unit activities. It also mentions the essential differences between local and distributed representations. In a local representation only one unit of many is active. In a distributed representation, many units are active. The remainder of the chapter discusses one way in which patterns of unit activity can be described mathematically. Some of the basic ideas of linear algebra are presented.

We have suggested that the computational business of the brain involves the activity of many nerve cells. A critical point that we will discuss later in considerable detail involves the *represention* of data in the nervous system (see chapter 10). There are two extreme possibilities for the pattern of cell activations during a computation. In the first, only one unit is active, and this single unit represents the input or output of the system. For example, to represent the 200 or so ASCII characters, we might have that number of units. The activation pattern for a letter or character would correspond to one and only one active unit. Or we could use a data representation, such as that used in computers, where a single character is represented by a pattern of active units. In a computer, an eight-bit byte can represent any one of the over 200 ASCII characters by a specific pattern of 1s and 0s. This means each unit is in the 1 state or the 0 state for roughly half the characters, and a character cannot be determined by a single bit without knowing what state the other bits are in.

The first representation technique is sometimes called a *grandmother cell* representation because, when talking about the brain, it assumes we have a single neuron that becomes active when and only when grandma is seen. The other representation is called a *distributed* representation, because the pattern cannot be unambiguously interpreted by looking at a single unit. Figure 3.1 shows a character-based version of distributed and localized representation. In general, distributed codes are more efficient (for example, eight bits versus a couple of hundred), but grandmother cells are more intuitive and have a nice conceptual clarity to them.

Distributed Coding

```
'T'  =  ASCII  84     + + - + - + - -
's'  =  ASCII  115    + + + + - - + +
'e'  =  ASCII  101    - + + - - + - +
```

Note these two elements are
identical in all three different letters.

Grandmother Cell Coding

```
'a'  =  + - - - - - - - - - - - - - - ...
'b'  =  - + - - - - - - - - - - - - - ...
'c'  =  - - + - - - - - - - - - - - - ...
'd'  =  - - - + - - - - - - - - - - - ...
```

And so on ...
One unit, one character.

Figure 3.1   An example of a distributed versus a grandmother cell (local) representation for characters. Only one unit is active for a grandmother cell representation, whereas all the units are active all the time for a fully distributed representation. Biological representations fall somewhere in between these extremes.

Both representation types are frequently used in models. They may, and often are, mixed in applications such as pattern classifiers. An example would be a neural network in which a distributed input representation, say, values for a number of features, gives rise to a grandmother cell output representation, where each output unit corresponds to a specific classification of the input data.

A major theme of this book is that data representation is the key step in neural network models. Evidence is overwhelming that moderately distributed representations correspond more closely to what is seen biologically than grandmother cells. However, the conceptual clarity of grandmother cells often makes them irresistible to modelers. For either representation, we must have a mathematical formalism that lets us work with activity patterns as primitive entities.

Linear Algebra

When we work with patterns of activity in a neural network, we can make contact with the branch of mathematics called *linear algebra*. When many units such as the generic connectionist computing unit are coupled together with synaptic weights, it is possible to compute many of the resulting relationships using the kinds of operations that are best described in the language of entities called *vectors* and *matrices*. The pattern shown by units in a group of units at a moment in time is called a *state vector*. A great deal is known about how to manipulate and how to understand the behavior of state vectors. What we need for our computations is what would be covered in the first few weeks of a college linear algebra course: only a few of the fundamental definitions, some notation, and a few important results.

Most of the computer time used in network simulations is spent on the simplest vector and matrix operations. Practically, this works to our advantage. Operations of this kind are used extensively in scientific computation, and special-purpose hardware is available for them. Many supercomputers perform with maximum efficiency on long vectors and large matrices, and with a little effort it is possible to get very high computation speed for neural networks. Many of the current generation of high-performance computers contain many CPUs operating simultaneously. Such machines can perform fast vector and matrix operations, although obtaining maximum speed from them can be difficult.

The discussion in this chapter is not very mathematical. We will emphasize useful Pascal PROCEDUREs and FUNCTIONs, and try to give some intuitive idea of what these operations mean. Since so many excellent traditional introductions to linear algebra are available, it is not necessary to duplicate them here. Practically any easily accessible book will be useful and well written, although we strongly recommend texts written for engineers or scientists. Books written for mathematicians are generally useless for anyone else because they are so abstract. We limit our discussion here to the simplest Euclidean vector spaces and their applications to neural networks. A Euclidean space is a modest generalization of the type of mathematical space taught in high school geometry.

State Vectors

A state vector in a neural network is a set of computing unit activities at a particular instant. Units are neurons, or something closely related to them. They are the basic integrating computer element in the neural network. If there were only two units, we would use a set of two numbers to describe the state vector. Activity of a single unit or element, that is, one location in the vector, is called a *component* of the vector. We could identify each pair of numbers with a point on a large piece of graph paper, a point on the plane (figure 3.2).

Vectors

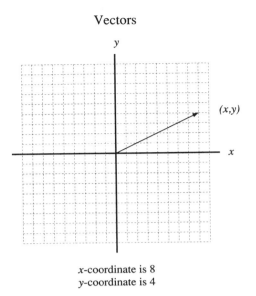

x-coordinate is 8
y-coordinate is 4

Figure 3.2   The vector (8, 4) on a Cartesian coordinate system.

If we had three components, we could identify each vector with a point in a three-dimensional space. Obviously we have a lot more neurons than three, but we do not have to stop our identification with geometry. If we were to have, say, a million element activities, we could consider a particular set of element values a point in a million-dimensional space, or a vector with a million components. Although we have no feeling for the detailed geometry of such a space, nor is it necessary to have one, we can carry over some useful ideas about geometry from more familiar situations. For example, it makes sense to talk about the *length* of a vector or the *angle* between two vectors in a space of any number of dimensions. These quantities are often easy to compute.

Notational and Computer Conventions

If we have a vector of high dimensionality, **f**, the twenty-third component of this vector will be written $f[23]$. This notation is used because we will be using computer programs written in Pascal. In Pascal (and most computer languages) a set of numbers is called an *array*, and a component in an array is denoted by square brackets. We shall use boldface type for vectors and, later, matrices.

One of the great conceptual virtues of Pascal and other modern computer languages is the use of data TYPEs and RECORDs as elegant and simple ways of representing complex entities. We will make extensive use of them in our programs. These programs and bits of code are collected in the appendixes. The program fragments in this chapter are in appendix A.

In this chapter we will start by being concerned with vectors. A vector is an array of real numbers. The *dimensionality* of the vector is the number of elements in the array. So we have our first program fragment: a definition of the TYPE Vector. Suppose our Dimensionality is 200. Suppose we have a PROGRAM Test_vectors, which we will construct during this chapter. Then,

```
                Program Fragment 3.1
PROGRAM Test_vectors (Input, Output);
CONST Dimensionality = 200;
TYPE Vector = ARRAY [1..Dimensionality] OF REAL;
...
```

We must make one point about mathematical notation. We assume vectors are columns of numbers, and we will consistently write them that way. This is of importance in the way some of the mathematical notation is ordered. The mathematical operation *transpose* interchanges rows and columns, and therefore turns a column vector into a row vector. For example,

$$\mathbf{a} = \begin{bmatrix} 1 \\ -3 \\ 2 \\ 1 \\ -2 \end{bmatrix} \qquad \mathbf{a}^T = (1, -3, 2, 1, -2).$$

Vector Addition and Subtraction

For our first procedure, we will describe how to add two vectors. Think back to two dimensions. Suppose we had a two-dimensional vector, a point $\begin{pmatrix} 4 \\ 3 \end{pmatrix}$ and another vector $\begin{pmatrix} 1 \\ 4 \end{pmatrix}$.

We want to add these two vectors together. If we had a one-dimensional system, we would know what to do: simply add the two numbers. A vector is a point in space. A consistent extension of addition to vectors is usually justified something like this. We leave our starting point and walk four miles east and three miles north, to the first point. From this point we walk one

Vector Addition

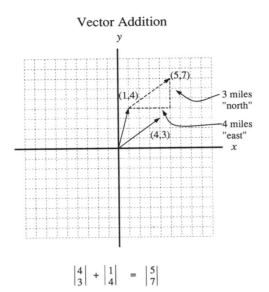

$$\begin{vmatrix} 4 \\ 3 \end{vmatrix} + \begin{vmatrix} 1 \\ 4 \end{vmatrix} = \begin{vmatrix} 5 \\ 7 \end{vmatrix}$$

Figure 3.3   Addition of two vectors (4, 3) and (1, 4).

mile east and four miles north. Where are we? Assuming a flat earth, and one where north is at right angles to east (good approximations to make for many problems), we will be at a position five miles east and seven miles north of our starting point. Figure 3.3 shows this computation on the plane. This is the most useful kind of vector addition, and the one we will use.

Simply adding the components generalizes to any number of dimensions. If we have two vectors, **a** and **b**, in five dimensions, it is easy to form their sum by this rule:

$$\mathbf{a} \qquad \mathbf{b} \qquad \mathbf{a+b}$$
$$\begin{bmatrix} 1 \\ -3 \\ 2 \\ 1 \\ -2 \end{bmatrix} + \begin{bmatrix} 4 \\ 1 \\ 3 \\ -5 \\ 1 \end{bmatrix} = \begin{bmatrix} 5 \\ -2 \\ 5 \\ -4 \\ -1 \end{bmatrix}.$$

We can make this definition formal using standard mathematical notation. If we let n equal the dimensionality, then the components of the sum, vector **c**, are given by

$c[i] = a[i] + b[i].$

A PROCEDURE to add two vectors is straightforward:

```
Program Fragment 3.2
PROCEDURE Add_Vectors
PROCEDURE Add_Vectors (Vector_1, Vector_2: Vector;
                            VAR Vector_sum: Vector);
    VAR I: INTEGER;
    BEGIN
    FOR I:= 1 TO Dimensionality DO
        Vector_sum [I] := Vector_1 [I] + Vector_2 [I];
    END;
```

Subtracting two vectors is equally simple.

```
Program Fragment 3.3
PROCEDURE Subtract_Vectors
PROCEDURE Subtract_Vectors (Vector_1, Vector_2: Vector;
                            VAR Vector_difference: Vector);
    VAR I: INTEGER;
    BEGIN
    FOR I:= 1 TO Dimensionality DO
        Vector_difference [I] := Vector_1 [I] + Vector_2 [I];
    END;
```

One thing to remember in writing a Pascal PROCEDURE or FUNC-
TION is the distinction between the two ways arguments can be passed.
Sometimes one of the arguments is modified and sometimes it is not. For
example, in the PROCEDUREs for vector addition and subtraction, above,
the two inputs, Vector_1 and Vector_2, are not modified. The vectors corresponding to the output, Vector_sum and Vector_difference, are modified,
since those have VAR preceding them. (Remember, VARiables can change.)
The rule to remember is that if the argument is going to be modified by the
PROCEDURE, pass it as a VAR; if it is not going to be modified by the
PROCEDURE, pass it as an argument without the VAR, that is, pass it to
the procedure as what is technically called "by value."

There are important differences in the way Pascal handles these two ways
of passing arguments. When an argument is passed as a VAR, the starting
address of the argument is passed to the procedure and the PROCEDURE
works with the actual argument. When the argument is passed by value, a
copy of it is made so the original is not modified. Copying can take a long
time if the argument is a large data structure.

Ordinarily we would not be concerned with this kind of detail of our
computer language, but this distinction leads to some difficult bugs in

Distance Computation

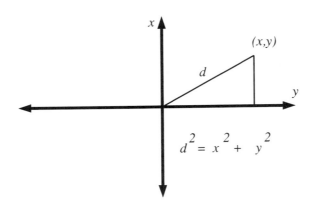

Figure 3.4   Two-dimensional distance computed by the familiar Pythagorean theorem.

Pascal programs. It is easy to leave out a **VAR** before an argument and work for a long time trying to find why the **PROCEDURE** seems to be doing nothing.

Length

We know how to find the length of a two-dimensional vector, thanks to the Pythagorean theorem learned in high school geometry. In figure 3.4, given a point (x, y) we know that its distance from the origin, d, is given by

$$d^2 = x^2 + y^2.$$

If we have a three-dimensional vector (x, y, z) we can compute the new distance between the origin and the point by considering the right triangle formed with one leg of length d and the other leg of length z (figure 3.5). So, the square of the distance, d', becomes

$$d'^2 = d^2 + z^2 = x^2 + y^2 + z^2.$$

We can continue the process for any number of dimensions, and finish with the square of the distance from a point in an n-dimensional space to the origin:

$$d^2 = \sum_{i=1}^{i=n} x^2[i].$$

3-D Distance Computation

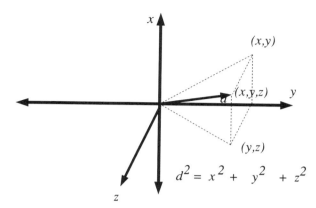

Figure 3.5 Extension of the distance computation to three dimensions.

Once we have the equation, we can write a FUNCTION to calculate it. The length is a single number, so we can use a FUNCTION rather than a PROCEDURE.

```
                Program Fragment 3.4
                FUNCTION Vector_length
FUNCTION Vector_length (Vector_1: Vector): REAL;
  VAR Sum_of_Squares: REAL;
              I: INTEGER;
  BEGIN
  Sum_of_Squares:= 0;
  FOR I:= 1 TO Dimensionality DO
    Sum_of_Squares :=Vector_1 [I]*Vector_1 [I] + Sum_of_squares;
  Vector_length:= SQRT (Sum_of_Squares);
  END;
```

Besides descriptiveness, we have called the FUNCTION Vector_length rather than simply Length, because a number of implementations of Pascal use Length as the name of a useful FUNCTION for string manipulation. Pascal will obligingly replace the built-in FUNCTION with the new FUNCTION for the duration of the program, and the loss of the original FUNCTION is not noticed until it is needed.

Multiplication by a Constant

We would like to be able to scale vectors by multiplying (or dividing) them by a constant. Most reasonably, this takes the form of multiplying each element by the constant, so we can write a PROCEDURE to accomplish this easily. A constant is often referred to as a *scalar* quantity, that is, a simple number.

```
            Program Fragment 3.5
          PROCEDURE Scalar_times_vector

PROCEDURE Scalar_times_vector (Scalar: REAL; Vector_in: Vector;
                                    VAR Vector_out: Vector);
    VAR I: INTEGER;
    BEGIN
    FOR I:= 1 TO Dimensionality DO
        Vector_out [I] := Scalar * Vector_in [I];
    END;
```

It is easy to see that multiplication by a scalar multiplies the length of the vector by that scalar, as we should hope. In a system of dimensionality n, if c is the scalar, and **a** is the original vector, the length of $c\mathbf{a}$, given by d, is

$$d^2 = \sum_{i=1}^{i=n} (ca[i])^2$$

$$= c^2 \sum_{i=1}^{i=n} a[i]^2,$$

and, d is therefore c times the length of **a**.

The Inner Product

Computation of the length of a vector introduces us to a special case of one of the most common operations in vector and matrix computation. Notice that in computing length, we have a term-by-term multiplication of the vector by itself and a summation of the partial multiplication of each term. This is a particular example of the use of a quantity called the *dot product* or *inner product*. We saw this quantity in chapter 2 as the function computed in the first stage of the generic connectionist neuron. More time is probably spent doing inner products on a computer than any other vector operation.

Suppose we have two vectors, **a** and **b**, and a system of dimensionality, *n*. The inner product of **a** and **b** is the single number

$$[\mathbf{a}, \mathbf{b}] = \sum_{i=1}^{i=n} a[i]b[i].$$

Engineers often call this quantity the dot product and denote it as $\mathbf{a} \cdot \mathbf{b}$. Mathematicians usually call this same quanity the *inner product* and denote it $[\mathbf{a}, \mathbf{b}]$, as above. There are other notational variants, but these two are the most common. In this book we will call it the inner product and denote it as $[\mathbf{a}, \mathbf{b}]$. Remember that the inner product of two vectors is a single number.

Program Fragment 3.6

FUNCTION Inner_product

```
FUNCTION Inner_product (Vector_1, Vector_2: Vector): REAL;
   VAR              I: INTEGER;
       Sum_of_products: REAL;
   BEGIN
   Sum_of_products := 0;
   FOR I:= 1 TO Dimensionality DO
     Sum_of_products := Vector_1 [I] * Vector_2 [I] + Sum_of_
       products;
   Inner_product:= Sum_of_products;
   END;
```

The inner product appears in many useful contexts. For example, the vector length of a vector **v** can be written as vector length $(\mathbf{v}) = \sqrt{[\mathbf{v}, \mathbf{v}]}$.

Normalization

It is often convenient to take vectors and make them a standard length, usually 1. This process is called *normalization*. Given the discussion above, it is easy to see how to normalize a nonzero vector: simply compute the length and divide each component by that length. Suppose **b** is to be the normalized vector and **a** is the starting vector. Then,

Vector_length $(\mathbf{a}) = [\mathbf{a}, \mathbf{a}]^{1/2}$.

Remember that we previously mentioned $[c\mathbf{a}, c\mathbf{a}] = c^2[\mathbf{a}, \mathbf{a}]$. So,

$$\text{Vector\_length }(\mathbf{b}) = \left(\left[\frac{\mathbf{a}}{[\mathbf{a},\mathbf{a}]^{1/2}},\frac{\mathbf{a}}{[\mathbf{a},\mathbf{a}]^{1/2}}\right]\right)^{1/2}$$

$$= \left(\frac{[\mathbf{a},\mathbf{a}]}{[\mathbf{a},\mathbf{a}]}\right)^{1/2}$$

$$= 1,$$

which is what we want.

We can write a **PROCEDURE** to normalize a vector quite easily.

```
                 Program Fragment 3.7

                 PROCEDURE Normalize
PROCEDURE Normalize (Vector_in: Vector; VAR Vector_out: Vector);
   VAR Length_of_vector_in: REAL;
       I
       Vector_in_is_zero  : BOOLEAN;
   BEGIN
   Vector_out:= Vector_in;
   Length_of_vector_in := Vector_length (Vector_in);
   IF (Length_of_vector_in = 0) THEN Vector_in_is_zero := TRUE
                                ELSE Vector_in_is_zero := FALSE;
   IF Vector_in_is_zero
      THEN WRITE  ('Normalized zero vector!');
      ELSE FOR I:= 1 TO Dimensionality DO
           Vector_out [I] := Vector_in [I] / Length_of_vector_in;
   END;
```

There is defensive programming in the above **PROCEDURE**, which is often a good idea in the early stages of a programming project. The **BOOLEAN** variable, Vector_in_is_zero, checks to see if we will divide something by zero, that is, if the vector to be normalized is the zero vector. It gives a short error message if it is. If we try to divide by zero when the program is running, results will depend on how the particular implementation of Pascal being used handles run-time errors, but it will almost certainly abort the program, sometimes with mysterious error messages. Internal program error messages are sometimes valuable because even unlikely events can occur when programs are being developed. These messages should be removed when the programs are working properly, since they slow the computation. After the error message, the program returns with the input vector, that is, the output is the zero vector. Many work stations have elaborate debugging facilities that will do a much better job on complex errors than the inserted comment above. For simple bugs, however, and bugs in scientific routines are often simple, straightforward techniques may be easier than learning how to use the debugger.

Angle Between Vectors

We will show next that the angle between two vectors can be computed using the inner product. This is important for our later work, and it is worth going through in a little detail. If the vectors have very large dimensionality, the meaning of the cosine between them might seem obscure, but it actually is not. If we think about the plane containing two vectors, **a** and **b**, and the origin, the problem reduces to a simple picture that we can visualize (figure 3.6). The lengths of the sides of the triangle formed by **a**, **b**, and the origin are easy to compute, given the discussion of length. The angle between **a** and **b**, θ, can be used to give the relations between the lengths using the law of cosines presented in high school geometry as a grown-up version of the Pythagorean theorem. If the lengths of the sides are given by Vector_length (**a**), Vector_length (**b**), and Vector_length (**a**–**b**), the law of cosines gives

$$\text{Vector\_length } (\mathbf{a} - \mathbf{b})^2 = \text{Vector\_length } (\mathbf{a})^2 + \text{Vector\_length } (\mathbf{b})^2$$

$$- 2 \text{ Vector\_length } (\mathbf{a}) \text{ Vector\_length } (\mathbf{b}) \cos \theta.$$

If the angle θ is a right angle, $\cos \theta$ is zero and the expression becomes the Pythagorean theorem.

If we put in the actual expression for the lengths, we can see how to compute the $\cos \theta$, given the inner product [**a**, **b**]:

$$\sum_i (a[i] - b[i]^2) = \sum_i a[i]^2 + \sum_i b[i]^2$$

$$- 2 \text{ Vector\_length } (\mathbf{a}) \text{ Vector\_length } (\mathbf{b}) \cos \theta$$

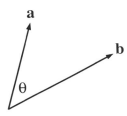

Figure 3.6   The angle between two vectors is defined on the plane containing them.

$$\sum_i a[i]^2 + \sum_i b[i]^2 - 2 \sum_i a[i]b[i])$$

$$= \sum_i a[i]^2 + \sum_i b[i]^2 - 2 \text{ Vector\_length (a) Vector\_length (b) } \cos \theta$$

$$\sum_i a[i]b[i] = \text{Vector\_length (a) Vector\_length (b) } \cos \theta.$$

Therefore, in terms of inner products and lengths,

$$\cos \theta = \frac{[\mathbf{a}, \mathbf{b}]}{\text{Vector\_length (a) Vector\_length (b)}}.$$

Or, if we wanted to put the whole thing in terms of inner products,

$$\cos \theta = \frac{[\mathbf{a}, \mathbf{b}]}{[\mathbf{a}, \mathbf{a}]^{1/2} [\mathbf{b}, \mathbf{b}]^{1/2}}.$$

Two special cases of angle θ are worth mentioning. Suppose $\mathbf{a}$ and $\mathbf{b}$ are the same. Then

$$\cos \theta = \frac{[\mathbf{a}, \mathbf{a}]}{[\mathbf{a}, \mathbf{a}]^{1/2} [\mathbf{a}, \mathbf{a}]^{1/2}} = 1.$$

Suppose $\mathbf{a}$ and $\mathbf{b}$ are in opposite directions. Then it is easy to see from the above expression that $\cos \theta = -1$. This involves realizing that

$$[\mathbf{a}, -\mathbf{b}] = -[\mathbf{a}, \mathbf{b}],$$

and, more generally, that if c is any constant then

$$[c\mathbf{a}, \mathbf{b}] = c[\mathbf{a}, \mathbf{b}],$$

and

$$[c\mathbf{a}, c\mathbf{b}] = c^2 [\mathbf{a}, \mathbf{b}].$$

We can write a program to compute $\cos \theta$ easily. If we have two vectors, $\mathbf{a}$ and $\mathbf{b}$, and we want to compute the angle between them, we can either compute their inner product and divide by the lengths, or normalize $\mathbf{a}$ and $\mathbf{b}$ and then compute the inner product. Program fragment 3.8 gives one way to compute the cosine between two vectors.

```
                Program Fragment 3.8
         Cosine of the Angle Between Two Vectors
FUNCTION Vector_cosine (Vector_1, Vector_2: Vector): REAL;
   VAR Normal_vector_1, Normal_vector_2: Vector;
```

```
BEGIN
Normalize (Vector_1, Normal_vector_1);
Normalize (Vector_2, Normal_vector_2);
Vector_cosine:= Inner_product (Normal_vector_1,Normal_vector_2);
END;
```

Suppose one of the vectors is the zero vector. In that case the PROCE-DURE Normalize given in program fragment 3.7 will give an error message and return with the zero vector in the Normal_vector position. Therefore the Vector_cosine will be zero.

There is a close relation between the cosine and the correlation coefficient of statistics. If we were to subtract the mean of the element values from the elements, or, alternatively, arrange it so the sum of the elements was zero (that is, $\sum a[i] = 0$), the cosine between the two vectors would then be the classic correlation coefficient. This is only one of the many close connections between neural networks and statistics; more will appear shortly.

Orthogonality

An important special case, for reasons that will be made plain later, is where **a** and **b** are at right angles to each other. Then $\cos \theta$ between **a** and **b** is zero, which implies that the inner product $[\mathbf{a}, \mathbf{b}]$ is also zero. When two vectors are at right angles to each other, they are called *orthogonal*.

It is quite easy to come up with vectors that are orthogonal; for example, the vectors that lie along the axes of coordinates, such the x and y axes in a two-dimensional system. In a higher-dimensionality space, for example, eight dimensions, we can immediately find eight orthogonal coordinate vectors, $\mathbf{u}_1$ through $\mathbf{u}_8$:

$$\mathbf{u}_1 = \begin{bmatrix} 1 \\ 0 \\ 0 \\ 0 \\ 0 \\ 0 \\ 0 \\ 0 \end{bmatrix} \quad \mathbf{u}_2 = \begin{bmatrix} 0 \\ 1 \\ 0 \\ 0 \\ 0 \\ 0 \\ 0 \\ 0 \end{bmatrix} \quad \mathbf{u}_3 = \begin{bmatrix} 0 \\ 0 \\ 1 \\ 0 \\ 0 \\ 0 \\ 0 \\ 0 \end{bmatrix} \quad \mathbf{u}_4 = \begin{bmatrix} 0 \\ 0 \\ 0 \\ 1 \\ 0 \\ 0 \\ 0 \\ 0 \end{bmatrix}$$

$$\mathbf{u}_5 = \begin{bmatrix} 0 \\ 0 \\ 0 \\ 0 \\ 1 \\ 0 \\ 0 \\ 0 \end{bmatrix} \qquad \mathbf{u}_6 = \begin{bmatrix} 0 \\ 0 \\ 0 \\ 0 \\ 0 \\ 1 \\ 0 \\ 0 \end{bmatrix} \qquad \mathbf{u}_7 = \begin{bmatrix} 0 \\ 0 \\ 0 \\ 0 \\ 0 \\ 0 \\ 1 \\ 0 \end{bmatrix} \qquad \mathbf{u}_8 = \begin{bmatrix} 0 \\ 0 \\ 0 \\ 0 \\ 0 \\ 0 \\ 0 \\ 1 \end{bmatrix}.$$

In terms of neural network representation, these orthogonal vectors are much like grandmother cells. Much of the conceptual clarity of grandmother cells comes from the fact that they are represented by orthogonal vectors. However, it is not necessary that there be a lot of zeros, and it is sometimes convenient to use an orthogonal set of state vectors that have nonzero values in every component. In neural networks, this might correspond to a more distributed representation. One useful set of such vectors is derived from what are called *Walsh functions*, which are used in a number of contexts and are a simple digital periodic function, something like a squared-off sine wave.

The set of eight orthogonal Walsh functions for an eight-dimensional space is

$$\mathbf{w}_1 = \begin{bmatrix} 1 \\ 1 \\ 1 \\ 1 \\ 1 \\ 1 \\ 1 \\ 1 \end{bmatrix} \qquad \mathbf{w}_2 = \begin{bmatrix} 1 \\ -1 \\ -1 \\ 1 \\ 1 \\ -1 \\ -1 \\ 1 \end{bmatrix} \qquad \mathbf{w}_3 = \begin{bmatrix} 1 \\ 1 \\ -1 \\ -1 \\ -1 \\ -1 \\ 1 \\ 1 \end{bmatrix} \qquad \mathbf{w}_4 = \begin{bmatrix} 1 \\ -1 \\ 1 \\ -1 \\ -1 \\ 1 \\ -1 \\ 1 \end{bmatrix}$$

$$\mathbf{w}_5 = \begin{bmatrix} 1 \\ 1 \\ 1 \\ 1 \\ -1 \\ -1 \\ -1 \\ -1 \end{bmatrix} \qquad \mathbf{w}_6 = \begin{bmatrix} 1 \\ -1 \\ -1 \\ 1 \\ -1 \\ 1 \\ 1 \\ -1 \end{bmatrix} \qquad \mathbf{w}_7 = \begin{bmatrix} 1 \\ 1 \\ -1 \\ -1 \\ 1 \\ 1 \\ -1 \\ -1 \end{bmatrix} \qquad \mathbf{w}_8 = \begin{bmatrix} 1 \\ -1 \\ 1 \\ -1 \\ 1 \\ -1 \\ 1 \\ -1 \end{bmatrix}.$$

Walsh Functions

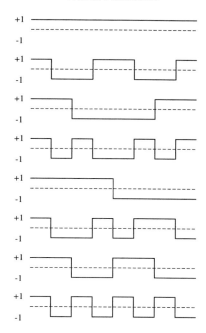

Figure 3.7   The eight, eight-dimensional Walsh functions.

Figure 3.7 shows a graphic presentation of these functions. Walsh functions have many interesting symmetries and patterns. A complete set such as that given is possible only in a space that has a dimensionality that is a power of 2. This set of functions is also interesting because every element is 1 or -1, with no zeros at all. Identification of which Walsh function is present is dependent on the pattern shown by all the values, not by a single element. For example, $w_1[8]$, $w_2[8]$, $w_3[8]$, and $w_4[8]$ all equal 1, and to distinguish them requires checking other elements: knowing that the valule of a single element is 1 only divides the number of possible vectors that it could be in two. It does not identify uniquely the vector. In a neural network context, the coding for Walsh functions is distributed because no single element contains unique identification information, as opposed to the grandmother cell representation found for the coordinate vectors, $u[i]$.

Every one of the eight vectors in the above two sets is orthogonal to every other one. It is reasonable to conclude that there can be at most eight orthogonal vectors in an eight-dimensional system, or, more generally, in a system of n dimensions, there can be at most n mutually orthogonal vectors. This is correct, and can be proved easily. However, consider the set of coordinate axis vectors. We have eight of them. If someone presented us a new nonzero eight-dimensional vector, v, and claimed it was orthogonal to all the vectors in this set, we could easily see that this could not be true just by considering the inner products. The inner product $[\mathbf{u}_i, \mathbf{v}]$ is simply the ith component of $\mathbf{v}$. At least one component of $\mathbf{v}$ must be different than zero, therefore the vector is not orthogonal to one of the $u[i]$.

Linear Independence

If we have a set of vectors, how many different vectors do we have? If none of them are identical, we could, of course, say that they are all different. But suppose we have a situation where one vector in a set can be represented as the sum of several of the others in the set. In this case, the vector that is the sum of the others is somehow superfluous, since it can be replaced by a simple combination of the others. If one or more vectors can be represented by such a sum, the set is referred to as *linearly dependent*. If none of the vectors can be represented as a sum of the others, the set is called *linearly independent*.

A set of orthogonal vectors is linearly independent. One way to see this easily is to assume we have a linearly dependent set of nonzero orthogonal vectors $\{\mathbf{a}\}$. Let us consider the vector $\mathbf{a}_1$, say, which can be represented as a sum of the other members of the set $\{\mathbf{a}_i\}$ with weighting coefficients c. The dimensionality is n.

$$\mathbf{a}_1 = \sum_{i=2}^{i=n} c_i \mathbf{a}_i.$$

We can show this assumption leads to a contradiction by considering the inner product $[\mathbf{a}_1, \mathbf{a}_1]$.

$$[\mathbf{a}_1, \mathbf{a}_1] \overset{?}{=} \sum_{i=1}^{i=N} c_i [\mathbf{a}_i, \mathbf{a}_1].$$

But all the inner products $[\mathbf{a}_i, \mathbf{a}_1]$ are zero, since these vectors are orthogonal. Therefore $[\mathbf{a}_1, \mathbf{a}_1]$ is zero, there is a contradiction, and an orthogonal set $\{\mathbf{a}_i\}$ is linearly independent.

A powerful geometric idea lies behind the notion of linear independence. It is described formally in books on linear algebra, but it is simple. Figure 3.8 shows a version of an example of linear independence and independence. Suppose instead of considering just a set of vectors $\{\mathbf{a}_i\}$, we consider all the vectors that can be made by weighted sums and differences of these vectors.

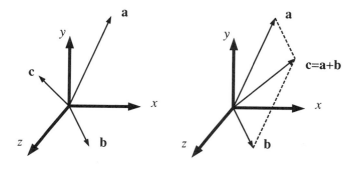

Linear Independence   Linear Dependence

Figure 3.8 Linear independence and dependence. For the lefthand figure, any point in the three-dimensional space can be represented as some sum of **a**, **b**, and **c**. In the righthand figure, vectors **a** and **b** lie in the same plane as their sum, vector **c**. Sums of **a**, **b**, and **c** can only lie in this plane, and points in the space outside this plane cannot be sums of **a**, **b**, and **c**.

All these vectors lie in a particular part of the high-dimensionality space: a subspace. If we consider a set of four of the eight-dimensional orthogonal coordinate vectors mentioned earlier, say, the first four, $\{\mathbf{e}_1, \mathbf{e}_2, \mathbf{e}_3, \mathbf{e}_4\}$, any vector with only the first four elements nonzero can be made from a sum of the four vectors. For example,

$$\mathbf{a} = \begin{bmatrix} c[1] \\ c[2] \\ c[3] \\ c[4] \\ 0 \\ 0 \\ 0 \\ 0 \end{bmatrix} = c[1]\mathbf{e}_1 + c[2]\mathbf{e}_2 + c[3]\mathbf{e}_3 + c[4]\mathbf{e}_4.$$

No vector with nonzero values in the last four components can be made as a sum of those four vectors. So the subspace in this case, generated from all possible sums, can be seen to be a simple four-dimensional subspace. The subspace generated from a linearly independent set is said to be *spanned* by the set.

Intuitively, linearly independent means very different, because no member can be replaced without a loss of size of the subspace generated or spanned by this process. The dimension of the spanned space can be shown to be equal to the maximum number of linearly independent vectors.

Test Program

We have now written a set of useful vector PROCEDUREs. We should give them a test to see if they work. Testing programs is a major practical problem. Indeed, it has been claimed that there is no totally correct (bug-free) software of any size operating anywhere. In our case, we must check to see that the PROCEDUREs and FUNCTIONs actually compute the correct answers, and that they do not do unexpected things when confronted with a wide range of possible inputs.

Our strategy for the test program will be to construct a system of small enough dimensionality so we can compute the correct answers by hand and then check the output. We also must incorporate extra code to write out the results, and a PROCEDURE to initialize the system. This kind of necessary machinery corrupts the easy flow of the program. It would be tempting to suppress it as somehow irrelevant to the operation of the program, but in practical terms, this kind of support often takes more effort and thought than the parts of the program that do all the glamorous computation.

In this book we will give a number of program fragments. Occasionally we will present some inelegant code to test, format, and organize the more theoretically interesting computations.

If we are concerned with machine efficiency, and we should be, because neural net computation places great demands on CPU cycles, we want to write efficient code when it comes time to test our ideas. It is almost always better to write special PROCEDUREs when we know exactly what we want to do. For example, suppose we want to multipy a vector, **a**, by a constant, *c*, and then add it to another vector, **b**, generating a resulting vector, **v**. We could compute this by calling two PROCEDURES, for example,

```
PROCEDURE Scalar_times_vector (c: REAL; a: Vector;
                                        VAR a: Vector);
PROCEDURE Add_Vectors (a, b: Vector;
                       VAR Vector_sum: v).
```

Or, we could write a new PROCEDURE doing both operations at once:

```
        PROCEDURE Special_Purpose
PROCEDURE Special_Purpose (a,b: Vector;
                          VAR v: Vector);
    VAR I: INTEGER;
    BEGIN
    FOR I:= 1 TO Dimensionality DO v[I] := c*a[I] + b[I];
    END;
```

The special-purpose PROCEDURE will usually be faster than the two standard PROCEDUREs in sequence. How much time we will save will depend critically on the compilers and the hardware configuration we use. When one does serious scientific computation it is *essential* to get some feeling for how the hardware and software of a particular machine are organized, and what they do well and poorly. Since current-generation operating systems can be used on many different machines, and the user interface looks the same, it is sometimes not realized how much machine architecture will interact with software to speed up or slow down certain operations. Experimention with code is in order and may pay big dividends.

In PROGRAM Test_vectors we set Dimensionality to eight and simply write in some small vectors in PROCEDURE Initialize. There are two of the Walsh vectors, w_2 and w_3, a coordinate vector, u_1, the zero vector, z, and two arbitrary vectors, a and b. We compute the lengths of all of these, which are easy to check by hand, and the cosines of the angles between pairs of them. (Note that the cosine of the angle between w_2 and w_3 is zero, that is, they are orthogonal. Also note the error message when we try to compute the cosine between the zero vector, z, and a nonzero vector.

We then take sum and difference of a and b, forming c and d, and compute lengths and cosines again, followed by use of the normalization procedure. Note that the lengths of the normalized vectors are all 1, as we would hope. Normalizing two vectors does not change the angle between them.

In the listings for programs, composed of previously given PROCE-DUREs, we will leave out the body of the PROCEDURE to save space. The complete code for the test program, Test_Vectors, can be found in appendix A.

4 *Lateral Inhibition and Sensory Processing*

The first part of this chapter discusses basic sensory processing: how physical stimuli are converted into the discharge of neurons. The rest is devoted to the structure and function of the lateral eye of the horseshoe crab, Limulus polyphemus, *perhaps the best-understood small nervous system. The compound* Limulus *eye is composed of many small independent eyes called* ommotidia. *When active, the ommotidia mutually inhibit each other by a process called* lateral inhibition. *The simplest form of lateral inhibition can be modeled using a linear neural network. This chapter discusses how to model such a network and how the resulting system behaves. One extreme form of computation using lateral inibition is the* winner-take-all *network where the desired outcome is a single active unit from a large group of units.*

Sensory systems contain special structures designed to concentrate, focus, and preprocess physical aspects of the world: the vertebrate visual system has an eye that focuses light on a retina, the mammalian ear has an elaborate set of impedance transformers and other structures designed to convert pressure changes in the air to mechanical vibration of a membrane, and so on. These specialized structures contain *receptors* that are responsive to the physical stimuli. For example, the rods and cones in the eye respond to absorbed photons, the hair cells in the ear and the vestibular organs respond to small mechanical displacements, receptors in the skin respond to pressure, receptors in the olfactory system respond to the shapes of molecules, and so on.

Each sensory system is specialized to have maximum efficiency responding to a particular physical stimulus dimension. Receptors will attain the highest possible level of useful physical sensitivity. For example, rods in the visual system will respond to single absorbed quanta, although two quanta absorbed in a small region of retina seem to be necessary to give rise to the sensation of light, to avoid excessive false signals due to spontaneous decay of the photosensitive pigment molecules. Hair cells are able to detect mechanical displacements on the order of molecular dimensions. Any more sensitivity would be pointless because of the thermal motion in the environment. Specialized chemical sensors in insects can detect the presence of single molecules of behaviorally important chemicals such as pheromones.

Optimality, of course, depends on what is being optimized. In the visual system, extremely simple stimuli—a weak flash of light, for example—are often not detected well, compared with an ideal receiver for the same stimulus. A more complex stimulus pattern, such as a diffuse bar of light, is detected with far better efficiency. One could conjecture that our sensory systems are more interested in analyzing complex stimuli of behavioral relevance than weak, diffuse flashes of light. Fitting sensory transduction to behavior has fascinated biologists for years.

As the specialized receptor structures perform their function, a local change occurs in the ionic permeability of a nerve cell. This change, using

Sensory Transduction

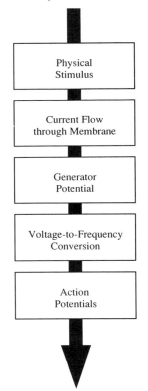

Figure 4.1 Steps in sensory transduction. A physical stimulus gives rise to generator currents passing through the membrane. The generator currents give rise to a generator potential, which can cause action potentials.

mechanisms such as as those we have discussed in earlier chapters, gives rise to *generator currents* passing through the membrane. These currents produce voltage changes called *generator potentials*, local depolarizations that can generate action potentials. Figure 4.1 shows a chain of events converting a physical stimulus into neuron activity.

These steps vary widely because every sensory system is specialized. For example, in the vertebrate eye, the photoreceptors, rods and cones, hyperpolarize when they absorb light. In normal neurons, hyperpolarization would mean that the cell would be less likely than before to produce action potentials. However, probably because of the small distances between units in the retina, action potentials do not appear in the retinal neural circuitry until several stages beyond the receptors.

Tactile Receptors

Vernon Mountcastle and his collaborators recorded from single axons in the median nerve of the arm of the monkey. It is reasonably easy to record single-fiber discharges from this nerve, and a few similar studies were done with human volunteers. The monkey and the human seem to be identical for this sensory system. One class of sensory axons studied came from receptors innervating the pads of the fingers. The experimenters constructed a small probe with a hemispheric tip 2 mm in diameter, driven by a solenoid that could provide small, precisely controlled displacements of the skin. By looking at the response of the nerve fibers, the experimenters could see how the impulses were coding the small poke delivered by the probe. (See Mountcastle, 1967, for a review.)

Like most tactile and, indeed, most sensory receptors, the response of the cell shows strong adaptation. That is, the cell responds most strongly to the onset of stimulation, and then its activity rapidly declines. This means that the firing frequency due to stimulation is not constant, but is continuously declining. The experimenters simply counted the number of spikes in a few hundred milliseconds and used this as a measure of the cell's response. The exact time period used for the spike count made little difference to the final pattern of results. Figure 4.2 shows extracellular records from skin stimulation for different depths of indentation. (Extracellular means that the action potentials are recorded from outside the cell. When an action potential occurs, the large currents involved can be detected at some distance from the cell). If one counts the number of action potentials in response to a 600-msec

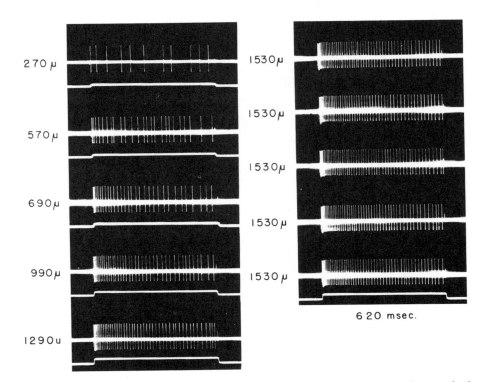

Figure 4.2 Extracellular oscillograph recordings of trains of nerve impulses evoked in a single fiber of the median nerve of a monkey. The axon innervated dermal ridges on the distal pad of the middle finger. Records in the left column are obtained for stimuli of different depths of indentation. Those on the right column are obtained during serial presentation of a stimulus of one intensity to show metronome-like repetition of response. Lower trace of each pair indicates intensity and time course of the stimulus. From Mountcastle (1967). Reprinted by permission of the Rockefeller University Press.

poke, a striking linear relation between depth of indentation and number of action potentials is seen (figure 4.3). This linear relation between physical stimulus magnitude and average firing frequency is not universal. A more common pattern would be a linear relationship between something like the *logarithm* or the *power function* of the physical stimulus magnitude. For example, neurons in the eye seem to code light intensity roughly as the logarithm of the physical intensity. Psychological sound intensity is perceived roughly as the logarithm of the physical sound pressure, which forms the justification for measurement of sound intensity in the log scale, decibels. It is often hard to differentiate experimentally a logarithm from a power function with a fractional exponent. A power function seems to be more accurate than the logarithm for many senses, among them the skin senses.

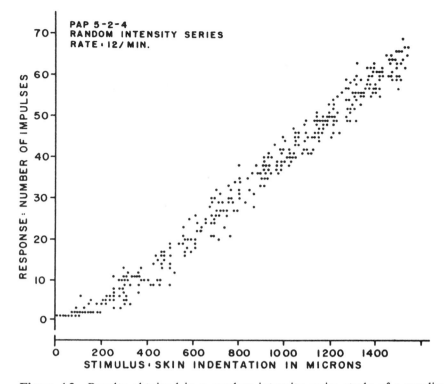

Figure 4.3   Results obtained in a random-intensity series study of a myelinated axon of the palmar branch of the median nerve of a monkey. The axon innervated a field extending across five dermal ridges on the thenar eminence. The stimulator probe tip, 2.0 mm in diameter, approximately covered the receptive field. Intensities were delivered in random order. Skin indentation is plotted against number of spikes in 600 msec. From Mountcastle (1967). Reprinted by permission of the Rockefeller University Press.

In another set of experiments using receptors from the hairy part of the monkey's hand, Mountcastle and collaborators found cells that had a power law relation between displacement and number of impulses (figure 4.4). This system is characterized by an exponent in a power law relationship of about 0.5 between displacement and the number of action potentials. The particular exponent describing the power law function was maintained up to cortex, indicating good retention of the coding for the intensity of the stimulus.

Once nonlinear coding has been performed, often at the level of the specialized receptor cells, there is sometimes faithful transduction of aspects of the nonlinearly coded output signal. Because the neurology of the peripheral skin receptors is so well mapped out, Mountcastle's group was able to fol-

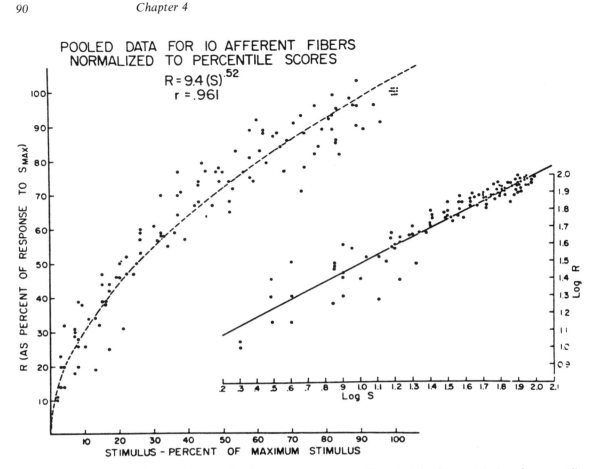

Figure 4.4 Results from intensity series like that in figure 4.3, but from tactile corpuscles of the hairy skin normalized to percentile scores. Graph to the left plots the stimulus response function in linear coordinates, that to the right in log-log coordinates. The relation for each fiber was well fitted by a power function. From Mountcastle (1967). Reprinted by permission of the Rockefeller University Press.

low the intensity signal all the way through the central nervous system, perhaps even into "consciousness." Because monkey and human peripheral nerves seemed to have identical responses, human subjects were asked to rate subjectively the intensity of the same stimuli delivered to the same locations as in the monkeys (figure 4.5). Humans showed the same linear relationship between displacement and intensity estimation shown by the peripheral nerve fibers. Mountcastle was also able to record from monkeys at other way stations for the tactile signal—thalamus and cortex—and show again that the quantitative relation between response and displacement was preserved.

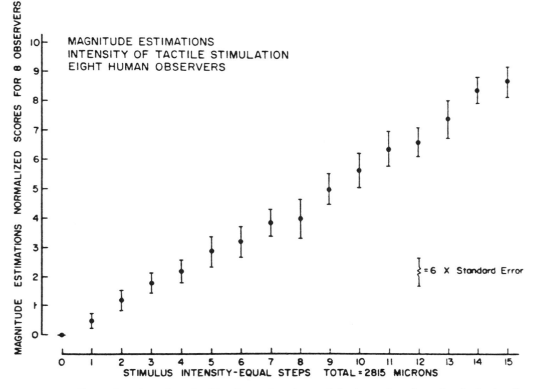

Figure 4.5   Results of subjective estimations of the intensity of mechanical stimuli by human observers. Stimuli were delivered to distal pad of middle finger at rate of 12/minute. The probe tip was 2 mm in diameter. Intensity continuum was divided into zero and 15 equal steps; a maximum 2815-μ indentation was used. Observers usually rated along a scale of 10 to 20 steps; these have been normalized to a scale of 10 for the ordinal values plotted here. Each observer received 5 each of the 16 stimuli, randomly ordered. Step 8 was given three times as an orienting stimulus before the series began, which may account for the slight displacement of the estimated value for 8 from the otherwise nearly perfect linear relationship. From Mountcastle (1967). Reprinted by permission of the Rockefeller University Press.

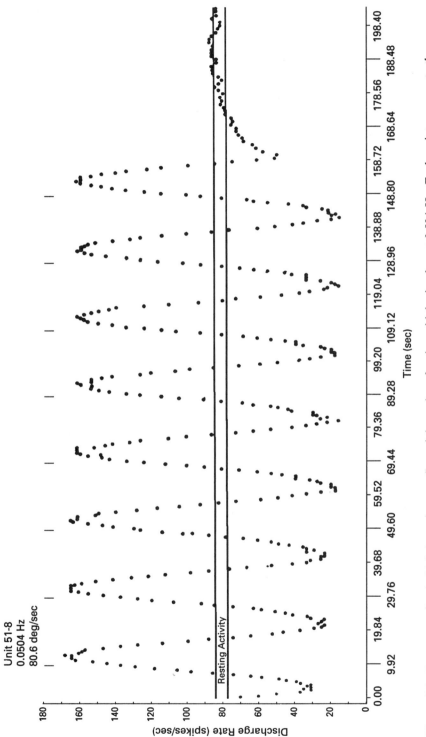
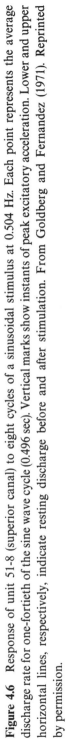

Figure 4.6 Response of unit 51-8 (superior canal) to eight cycles of a sinusoidal stimulus at 0.504 Hz. Each point represents the average discharge rate for one-fortieth of the sine wave cycle (0.496 sec). Vertical marks show instants of peak excitatory acceleration. Lower and upper horizontal lines, respectively, indicate resting discharge before and after stimulation. From Goldberg and Fernandez (1971). Reprinted by permission.

Other Systems

Another example of faithful transduction between a physical stimulus and a neural output signal can be seen in the vestibular system of the squirrel monkey (Goldberg and Fernandez, 1971). These vestibular organs respond to tilt and angular acceleration. If recordings are made from the eighth nerve, and the animal is rocked back and forth with a sinusoidal movement, some units will accurately duplicate the sinusoidal shape in their firing rate (figure 4.6). Of interest, these cells also show a rather high spontaneous activity level, allowing transduction of tilt both in a positive (increased activity) and a negative (decreased activity) direction. When the dynamic range of the neuron is exceeded, the peaks are clipped.

The point of this discussion, which could be extended with other examples, is that many aspects of sensory coding are straightforward and sometimes linear or nearly linear, once past an initial nonlinear transduction. When major nonlinearities are found, they are often simple ones, for example, rectification. Sensory systems are complex, but not impenetrable.

Do It Yourself: Simple Single-Unit Recording

Although the current technology of neuroscience is forbidding, it should not be forgotten that the first single-unit recordings were done with very simple equipment. It is possible to record extracellular action potentials from sensory receptors with available consumer electronics. All that is required are (1) a stereo system with a phono input, (2) a block of wax from the grocery store, (3) two common pins or needles with wires attached to them (alligator clips are adequate and very convenient), (4) a phono plug, and (5) the largest cockroach you can find. The alligator clips, wire, stereo amplifier and speaker, and phono plug can be purchased at the local Radio Shack; you are on your own with respect to the cockroach. The wires from the needles are attached to the phono plug, one to the center conductor and one to the outside of the plug.

Take the cockroach, and cut off one of its large legs at the joint it makes with the body. Lay the leg on the wax block and push the two pins into its large upper portion. The RCA plug should be plugged into the phono input jack for one channel and the stereo amplifier set up for listening to a phonograph record. (You are only using one channel; the other channel can be removed with the balance control if it is distracting.) As you move the lower portion of the leg back and forth, you should hear from the loudspeaker the clicking sound of extracellular single unit potentials from the joint receptors. You can study the properties of the receptor in some detail with this setup, measuring, for instance, how the discharge of the single units responds to joint angle, speed of leg movement, and so on.

Limulus: A Simple Distributed Neural Computation

Perhaps our best-understood small nervous system is lateral inhibition in the lateral eye of *Limulus polyphemus*, the horseshoe crab. This was also the first system that gave some idea of the computational power of a distributed network. The importance of the *Limulus* work for neural networks is considerable, as it showed that a real interacting neural system was understandable. The interactions in the *Limulus* eye turned out to be largely linear, and some of its behavior could be understood almost completely in terms of linear systems analysis. The architecture and analysis of the eye was enormously influential on succeeding modeling work.

We will use the PROCEDUREs and FUNCTIONs we developed in chapter 3 to a demonstrate the properties of this simple network. The program we write will not be very flexible, but we can work with it. More of the fascinating details of the *Limulus* eye are given in the references at the end of this chapter.

Biology

Limulus polyphemus is commonly known as the horseshoe crab or sometimes the king crab, although taxonomically it is not a crab. It is very common on the East Coast of the United States, and is found as far south as Yucatan. The *Limulus* is an arthropod, a member of the same phylum that contains the insects, spiders, and true crabs. It is in the class Cheliceratea, which also contains the spiders, ticks, and mites, and in subclass Merostomata, order Xiphorsura. This subclass contains only three living genera with a total of five species; it has many more species in the fossil record (Bracha and Bracha, 1990). The other living species of merostomate live in Asia, look very similar to *L. polyphemus*, and have similar habits. The term "living fossils" is sometimes given to these animals. Some argue that *Limulus* (and its relatives) is the nearest living relation of the long extinct trilobites because its embryos look very much like trilobites.

Figure 4.7 shows different views of a *Limulus*, and figure 4.8 photographs of a male and female. We will be modeling the behavior of the lateral eyes, marked on the formen and visible in the latter. The animals can be very

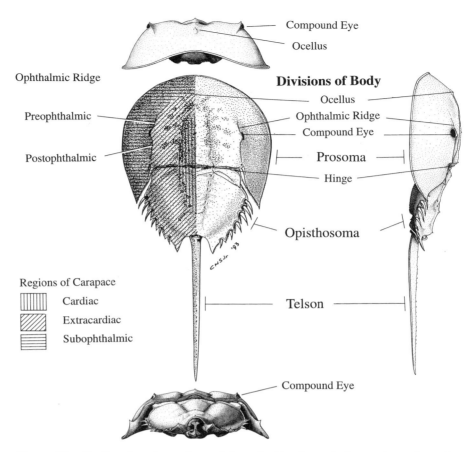

Figure 4.7 Outline drawings of an adult male *Limulus polyphemus* from the Miles River, Chesapeake Bay, Maryland. Prosomal width is 19.0 cm. Labels identify the topographic features associated with the compound lateral eyes. Despite their fixed position, the curvature and location of these eyes give the animal wide, almost hemispherical fields of vision. This is best illustrated by examining the several aspects of the carapace shown here (clockwise, from top) based on figures 1, 2, 6, 7, 8, and 9 from Shuster (1955). (i) The anterior view best shows the curvature of eyes that gives a wide up-down and forward-backward field of vision. (ii) The lateral aspect shows the position of the ophthalmic ridge and the major dorsal spines. These spines, blunt in adults, are relatively much longer and sharper in juvenile horseshoe crabs. (iii) The posterior view shows the extent of the backward-looking curvature of the compound eye, and the articulations of the prosoma and the telson with the opisthosoma. (iv) The dorsal aspect shows the three divisions of the body, *prosoma* (cephalothorax), *opisthosoma* (abdomen), and *telson* (tail). The prosoma and opisthosoma articulate along a straight line (hinge); the telson moves on a universal joint, functioning mainly in locomotion and in aiding an overturned crab to right itself. The dorsal aspect also shows the three regions of the carapace (shell), which overlie, in the main, different organ systems, thus revealing the position of internal structures: the *cardiac* region occupied by a long tubular heart; the *extracardiac* area of muscle attachments, seen on the external surface of the carapace as a series of lobate markings; and the *subophthalmic* area covering the terminal portion of the gonads and the multibranched digestive diverticula (hepatopancreas). Figure and caption courtesy of Dr. Carl N. Shuster, Jr.

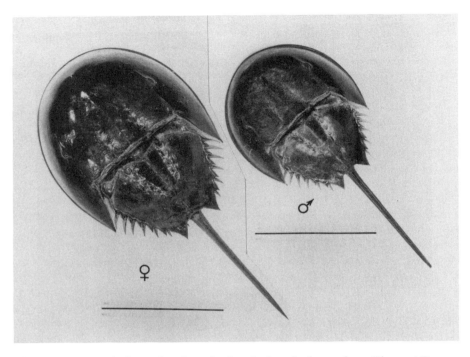

Figure 4.8 Dorsal views of male and a female *L. polyphemus* from Pleasant Bay on Cape Cod, Massachusetts. The female (left) measured 21.0 cm in prosomal width; male (right) had a prosomal width of 17.0 cm (tip of the telson was broken off in handling). Both measurements were to the nearest 0.5 cm. The scale line is 15.2 cm (6 in.) long. Based on figures 29 and 30 in Shuster (1955). Figure and caption courtesy of Dr. Carl N. Shuster, Jr.

large, as much as 60 cm from head to the tip of the tail. They live near offshore in shallow water with clean sandy bottoms. They often bury themselves just beneath the surface. They are scavengers as well as hunting mollusks and worms. They can dig in the sand for prey, and crush shells with small powerful claws on the underside of the animal.

Like many animals with an exoskeleton, *Limulus* must discard its old skeleton whenever it grows out of it. It is common to find what appear at first glance to be dead *Limulus* on East Coast beaches. These are discarded exoskeletons that no longer contain the animal, which may be elsewhere, perfectly healthy, with a new, larger exoskeleton. The animal escapes from its old skeleton by way of a gap in front between its upper and lower parts.

Limulus has a few practical applications. A survey of the invertebrates of Massachusetts from 1841 comments that "The King-Crab or Horse-Shoe (*Limulus polyphemus*) is employed as food for hogs; and many of them are speared by boys for this purpose and sold for a half a cent apiece. It is also regarded as excellent bait for eels" (Anonymous, 1841, p. 361). More re-

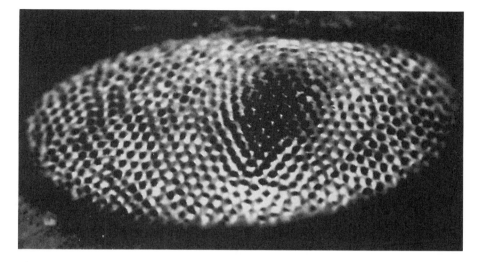

Figure 4.9 The lateral eye of the horseshoe crab *Limulus*, corneal surface. In a medium-size adult each eye forms an ellipsoidal bulge on the side of the carapace, about 12 mm long by 6 mm wide. The facets of the ommatidia are spaced approximately 0.3 mm apart, center to center, on the surface of the eye. The optical axes of the ommatidia diverge, so that visual fields of all those in one eye taken together cover approximately a hemisphere. The optical axes of the ommatidia with the dark circular facets near the center of the eye are oriented in the direction of the camera. From Ratliff (1965). Reprinted by permission.

cently, large numbers of *Limulus* were ground up and used for fertilizer on the East Coast. The animal was exceedingly common at one time. In 1856 over a million were collected within a mile of shoreline, and as late as the 1920s millions were harvested yearly. Although its numbers have dropped somewhat, the animal is still extremely common. They mate in May and June, and thousands of them may congregate on beaches at high tide and full moons (Shuster, 1960, 1982).

The animal also has a more sophisticated practical application. It has a circulatory system. Compounds in its blood are highly sensitive to the presence of the endotoxins of gram-negative bacteria. In the presence of even small amounts of some kinds of bacterial endotoxins its blood will clot. Extracts from the blood of *Limulus* are sold commerically for medical testing. Blood is extracted from the animals, and they are returned to the ocean. More about the biology and ecology of *Limulus* can be found in articles in the collection edited by Bonaventura, Bonaventura, and Tesh (1982).

We are concerned with modeling the behavior of the lateral eye. Figure 4.9 is a picture of the intact lateral eye; figure 4.10 shows a cross section of the eye. The lateral eye is divided up into about 800 *ommatidia*, which

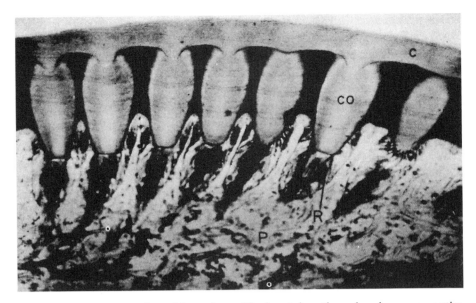

Figure 4.10   Cross section of lateral eye. Horizontal section, showing seven entire ommotidia. Contrast between different corneal layers (C) and crystalline cones (CO) is exaggerated. Sections are moderately bleached to show retinular cells (R). Only a portion of the plexus (P) is shown. Lee Brown modification of Mallory's analine blue stain. From Hartline, Wagner and MacNichol (1952, figure 6). Reprinted by permission.

are physically and optically separated from each other. Each ommatidium, clearly visible in the picture, has its own light-gathering and neural-transduction apparatus and is separated from its neighbor by opaque partitions. Although not strictly lenses, the transparent crystalline cones have a graded index of refraction, directing light to the photoreceptors. The eyes of a number of invertebrates use lenses with a variable index of refraction because these lenses show much less spheric and chromatic aberration in sea water than those with a fixed index of refraction (Land, 1985). Graded-index lenses are not often used in artificial optical systems because they are so hard to construct. Each ommatidium covers a few degrees of visual angle, and the field of view of the eye as a whole covers nearly a hemisphere.

For many years it was not known how, or even if, the *Limulus* used its visual system. This was embarassing because of the immense amount of work on *Limulus* visual physiology. In 1982, however, Barlow, Ireland, and Kass demonstrated that the animal indeed had functional and significant visual behavior. They constructed a painted concrete model of a female and showed that males were attracted to it based strictly on visual cues.

No "simple" animal is really simple. It was discovered recently that the *Limulus* eye has a pronounced circadian rhythm of sensitivity. After sunset,

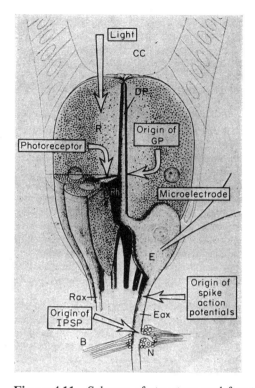

Figure 4.11   Schema of structure and function in the compound eye of *Limulus*. Light enters the ommotidia through the cornea (not shown in this figure; see figure 4.10) and passes through the crystalline cone (CC). The light is partly absorbed by pigment in the photosensitive rhabdomes (Rh) of the retinular cells (R). The eccentric cell (E) produces a generator potential (GP) in response to conductance changes. The IPSP (inhibitory postsynaptic potential) of lateral inhibition is due to activity of neighboring ommotidia. At small distances below the layer of ommotidia, both retinular cell axons (Rax) and eccentric cell axons (Eax) give rise to numerous fine lateral branches (B). The fibers in these bundles form clumps of neuropil (N) around the axons of the eccentric cells. From Ratliff, Hartline, and Lange (1966). Reprinted by permission.

the sensitivity of the visual system increases by a factor of 100,000. This immense increase means that the animal can see reasonably well both night and day. Thus the ability of the males to spot females is roughly the same both day and night (Powers, Barlow, and Kass, 1991). Sensitivity is apparently under the control of *efferent* fibers in the optic nerve, that is, axons going from the brain to the eye (Barlow, 1988; Batra and Barlow, 1990).

Because of the immense range of ambient light level, many animals have complex mechanisms to adjust visual sensitivity over many orders of magnitude. The mammalian eye has two classes of photoreceptors, rods and cones,

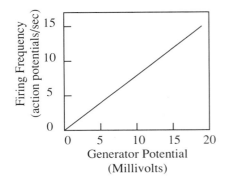

Figure 4.12 Relation between generator potential and frequency of action potential in the *Limulus* eye. Abstracted from data in MacNichol (1955).

each with different sensitivity ranges. The visual system shifts smoothly between them as the ambient illumination changes. The mammalian eye has several other mechanisms as well to adapt sensitivity to light conditions. In this chapter we will only discuss the classic model for the *Limulus* eye, in which complex changes in sensitivity do not occur.

The receptor cells in the ommatidium (*retinula* cells) respond to light with a small *receptor* potential. Then the output cell, the *eccentric* cell (named for its off-center position in the ommatidium, not for any behavioral peculiarities), responds by converting this voltage into firing frequency of action potentials. Figure 4.11 shows the arrangement of cell types in the ommatidium. Figure 4.12 shows the remarkably simple linear relationship between the generator potential and the cell-firing frequency.

It is relatively easy to remove the eye from the carapace (with a hacksaw) and remove as well a small piece of optic nerve. The nerve can be teased apart, and the neural discharge corresponding to the output of a single ommatidium can be recorded, even with quite simple apparatus. As early as the 1930s, H. K. Hartline was able to record trains of action potentials from single ommatidia and studied this small neurooptical system for a number of years. He received the Nobel prize for this work. Figure 4.13 shows a figure taken from a 1932 paper by Hartline and Graham showing the response of a cell to light. A single ommatidium responds to light with an increased number of action potentials (spikes) per second. One of the first significant observations on the ommatidium was that the early and late behaviors of the response to a sudden change in light intensity were different. Although the initial firing rate was logarithmically related to the light intensity, the firing rate after a few seconds was smaller and not so simply related to intensity (figure 4.14).

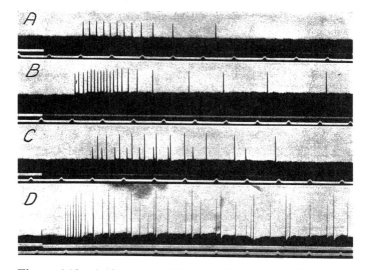

Figure 4.13 Action potentials recorded from optic nerve of *Limulus*. Note the initial rapid firing when the light stimulus is first applied (appearance of dark bar in the trace) and the later slower activity. Trace D contains extracellular records from several active axons; they can be distinguished by their different amplitudes. From Hartline and Graham (1932). Reprinted by permission.

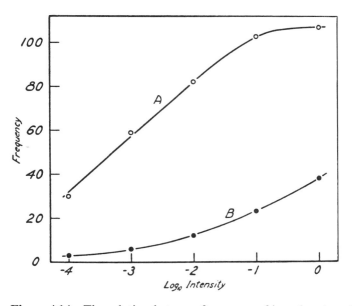

Figure 4.14 The relation between frequency of impulses (number per second) and logarithm of the intensity of stimulating light. Intensity in arbitrary units (1 unit = 630,000 meter candles). Curve A, frequency of the initial maximum discharge. Curve B, frequency of discharge 3.5 seconds after onset of illumination. From Hartline and Graham (1932). Reprinted by copyright permission of the Rockefeller University Press.

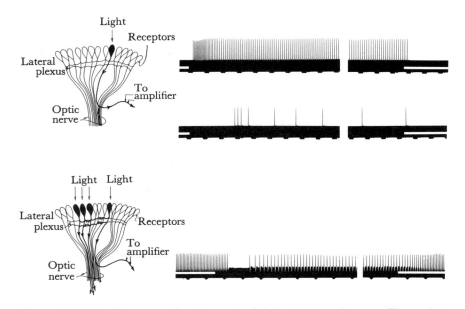

Figure 4.15 Oscillograms of action potentials in single optic nerve fibers of *Limulus*. The experimental arrangements are indicated in the schematic diagrams. (Above) Response to steady illumination of a single ommatidium. For the upper record the intensity of the stimulating light was 10,000 times that used to obtain the lower record. The signal of exposure to light blackens out the white line about the 1/5-second time marks. Each record interrupted for approximately 7 seconds. (Below) Inhibition of the activity of a steadily illuminated ommatidium produced by illumination of a number of other ommatidia near it. The blackening of the white line above the 1/5-second time marks signals the illumination of the neighboring ommatidia. From Ratliff (1965). Reprinted by permission.

When it was technically possible to record from several ommatidia simultaneously, it was discovered that nearby ommatidia were interconnected, so that activity in one *inhibited* activity in its neighbors. This phenomenon was referred to as *lateral inhibition*, lateral because cells at the same level are interacting with each other by means of cross connections. The inhibitory interconnections are short range, being strongest to adjacent cells and less strong as distance between ommatidia increases. In the top trace of figure 4.15 the cell response is an initial burst followed by a steady-state firing frequency. When nearby ommatidia are illuminated, the cell is first so strongly inhibited that it is turned off, and then returns to a low steady-state firing rate.

With care, independently controllable light stimuli can be applied to one or two ommatidia, allowing their interactions to be studied with precision. Figure 4.16 shows that lateral inhibition is reciprocal. Here, two cells are illuminated and recorded separately. Although it would be possible to de-

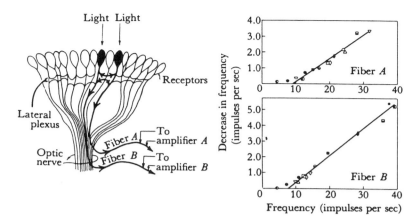

Figure 4.16 Mutual inhibition of two receptor units in the lateral eye of *Limulus.* Action potentials were recorded simultaneously from two optic nerve fibers as indicated in the schematic. In each graph the magnitude of the inhibitory action (decrease in frequency of discharge) of one of the receptor units is plotted on the ordinate as a function of the degree of activity (frequency) of the other on the abscissa. The different points were obtained by using various intensities of illumination on receptor units A and B in various combinations. The slope of the line gives the value of the inhibitory coefficient. From Ratliff (1965). Reprinted by permission.

scribe their mutual interactions in many confusing ways, the picture is simplest if their interaction is described in terms of cell firing rates. In figure 4.16 we think in terms of firing rates. When the discharge rate of fiber A is inhibited by fiber B, a linear relation exists between the inhibition of A in spikes per second and the firing rate of B. There is also a threshold, so that no inhibition occurs until the inhibiting cell is firing more than about 10 spikes per second; the complexities arising from the threshold for inhibition are often ignored in theoretical analyses. (For a discussion of the system when a few of these nonlinearities are included, see Barlow and Quarles, 1975). The effect of fiber A on B and of fiber B on A are about the same on the average. The slope of the two lines in figure 4.16 gives the degree of coupling between the cells; it can be seen that it is similar for these two cells. Although it is usually assumed for modeling purposes that two cells inhibit each other equally, in fact they often do not. Values of inhibitory coupling were never seen experimentally to exceed about 0.2.

If one ommatidium was receiving inhibition from several other cells, the resulting inhibition, in decrease of activity in spikes per second, was given by the simple sum of the inhibitions from the several inhibiting cells; that is, the postsynaptic cell was simply summing inhibition from all its inputs. Best of all, the summation was linear. The definition of a linear system requires that it obeys the principle of superposition. (See chapter 5 for a discussion of the significance of superposition in relation to linear systems.)

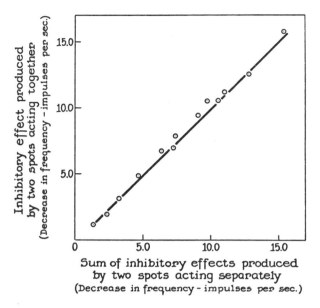

Figure 4.17   The summation of inhibitory effects produced by two widely separated groups of receptors. The sum of the inhibitory effects on a test receptor produced by each group acting separately is plotted as abscissa; the effect produced by the two groups of receptors acting simultaneously is plotted as ordinate. The solid line is drawn through the origin with a slope of 1.0. The two spots of light used to stimulate the two groups of receptors were each 1.0 mm in diameter, each illuminating about a dozen receptors, and were 4.6 mm apart on the eye. The test receptor was located midway between these two spots of light. From Hartline and Ratliff (1958). Reprinted by copyright permission of the Rockefeller University Press.

The recording in figure 4.17 was taken from an ommatidium that received inhibitory input from two groups of cells. In the experiment, the inhibition from each group was measured independently. Then the inhibition arising from the simultaneous illumination of both groups was measured. If inhibition is part of a linear system, the sum of the groups illuminated independently should be equal to the effect of the two groups illuminated together. On the graph, linearity corresponds to a straight line with a slope of 1, almost exactly what was seen. This important observation is used to justify the use of simple linear summation in the *Limulus*. It also serves as important support for the use of a simple additive model of synaptic interactions in the generic connectionist neuron.

Other obvious nonlinearities are present in the system besides the threshold for inhibition. Cells cannot fire at rates less than zero, and there is a maximum firing rate so cells cannot increase rate indefinitely; that is, there is both an upper and lower limit on cell firing rate, as well as on the inhibition.

The cell has a limited dynamic range. As long as these limits are respected, the behavior of the system can be analyzed as a linear system.

Given the experimental data and their interpretation, it is now possible to write a simple approximate equation for the behavior of a single ommatidium in *Limulus*. It is sometimes referred to as the *Limulus* equation. Suppose we have a group of ommatidia, each with activity $f[i]$. Suppose a certain amount of light falls on the eye, causing a discharge of the cell, if there was no inhibition, of $e[i]$ spikes per second. Suppose also that the cells are interconnected by synaptic weights $A[i,j]$, where $A[i,j]$ is the strength of connection of cell j to cell i. A synaptic weight is a measure of the degree of influence of the presynaptic cell on the postsynaptic cell, that is, the strength of coupling between cells or the slope of the lines in figure 4.16. (The rationale behind this notation will become clear later.) Then

$$f[i] = \sum_j A[i,j]f[j] + e[i].$$

For any particular $f[i]$ we can view the first part of this expression as an inner product of the kind we met in chapter 3. Specifically, it is an inner product between a set of synaptic strengths, $A[i,j]$, associated with the ith neuron, and an input activity pattern, f. We will use this equation when we discuss a computer program for lateral inhibition. A slightly more elegant way of writing the same equations, although identical formally, will be introduced in the next chapter when we present matrix notation. However, considering the synapses and activities of a single neuron is often more intuitive, so let us start by using vectors to do our computation.

One other important point arises. Does a cell inhibit itself; that is, is $A[i,i]$ nonzero? In *Limulus*, self-inhibition is present. It is strong, and has a longer time constant than inhibition from other cells. One of the results of self-inhibition is to act as a kind of automatic gain control.

The *Limulus* eye is a visual system, which means ommatidia are arranged in a two-dimensional array and process spatial information. The problems arising in this simple eye are (by the physics of vision) similar to those in the mammalian eye, and some of the biological solutions are the same. Lateral inhibition of a more complex kind is prominent in our own retina, and is found in many other places as well.

A Program Demonstrating Lateral Inhibition

Next we will incorporate the *Limulus* equation and lateral inhibition into a computer program and show some of its spatial-processing properties, as well as constructing our first working neural network program.

One Dimensional Eye
with Wrap-around

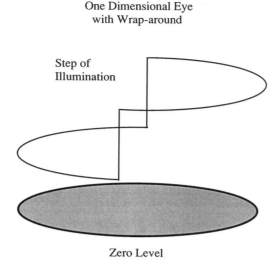

Step of
Illumination

Zero Level

Figure 4.18   Circular retina used for the simulations. This 80-unit retina is given an initial input pattern of activation, and the response of the lateral inhibition network is computed.

Let us use the simplest possible approach to demonstrate the primary results of lateral inhibition. Several practical problems arise immediately when we try to make a program. The eye is two dimensional and has finite extent, so we must figure out what to do with the edges. Let us avoid problems with edges by assuming a one-dimensional eye, with the ommatidia arranged in a circle so the activity wraps around from the last ommatidium to the first. Figure 4.18 shows such an activity pattern, a function with two steps, on a one-dimensional circular eye.

Let us now work out the details. We will be more verbose than usual describing details because this is our first program.

The Main Program

An instructor in a computer science course teaching Pascal once said that a good way of grading programs was to look at the length of the main program. The main program section of A programs was rarely longer than ten lines. Ideally, in a Pascal program it should look like a high-level block diagram, where the names of the PROCEDUREs express the operation of each block.

```
                    Program Fragment 4.1
         PROGRAM Lateral_inhibition: The Main Program
BEGIN {Main Program.}
Initialize_parameters;
Initialize_state_vector;
Initialize_display;
Display_state_vector (Initial_state_vector, 12, '+');
Make_inhibitory_weights;
Compute_inhibited_state_vector;
Display_state_vector (State_vector, 12, '*');
READLN;                          {Keeps display on screen until
                                  hit <RET>.}
END.  {Main Program.}
```

The complete version of the program is found in appendix B. We will briefly describe some of the functions of the procedures necessary to let us experiment with simple lateral inhibition. The first five PROCEDUREs are to initialize the program. The simulation parameters are read in first, then the state vector is initialized with a simulated pattern of light falling on the retina. Next, the display is initialized and the initial state vector is displayed on the screen. Then, the actual computation is performed and the result displayed, superimposed on the initial state vector so the initial and final states can be compared. The READLN statement keeps the program from ending until ⟨RETURN⟩ is entered so the display is not altered.

Types, Constants, and Variables

We have to define a number of global constants and variables for the program. Since we use vectors we will define a TYPE Vector defined as a real array.

```
                    Program Fragment 4.2
                    Global Declarations
PROGRAM Lateral_inhibition (Input, Output);
CONST  Dimensionality     =  80;
       Half_dimensionality =  40;
       Number_of_iterations=  50;
```

```
TYPE    Vector              =ARRAY [1..Dimensionality] OF REAL;
VAR     Initial_state_vector,          {Start pattern.          }
        State_vector        : Vector;  {Current state of the
                                        system.                 }
                                       {Vectors containing
                                          inhibitory weights.   }
        Inhibitory_weights: ARRAY [1..Dimensionality] OF Vector;
        Upper_limit,                   {Maximum firing rate of model
                                          neuron.               }
        Lower_limit,                   {Minimum firing rate of model
                                          neuron.               }
        Length_constant,               {Length constant of
                                          inhibition.           }
        Epsilon,                       {Computational constant.}
        Max_strength        : REAL;    {Maximum value of
                                          inhibition.           }
```

The functions of most of the global variables are obvious from the names. We have an Initial_state_vector, the current State_vector, and an array of Inhibitory_weights, one for each neuron, which are the weights of synapses connecting that neuron to all the others. This variable is a vector also. Neurons have Upper_limits of firing, as well as Lower_limits of firing.

We assume for the simulation that inhibition falls off as a negative exponential with a Length_constant and a Maximum_value. This is a first approximation to what is seen in the real animal. The weight of inhibition from a neuron some Distance away from a target neuron is given by

$$\text{Inhibitory\_weight} = \text{Max\_strength} * e^{-(\text{Distance/Length\_constant})}.$$

This expression says that the self-inhibitory coefficient is equal to Max_strength. The values of Dimensionality and Number_of_iterations are a compromise between computing speed and size.

Initialization

Pascal is unforgiving about initialization. Nothing can be assumed about the initial state of any variables in the system, unlike some computer languages. Unless items are specifically initialized, their values can be anything, depending on what the computer was doing in the immediate past. Strange and unreproducible errors can result. The single most common cause of student errors in Pascal programs that compile but do not run correctly is faulty initialization.

We must initialize several different things. First, we must read in the control parameters for the simulation. It is no fun having a computer program that you can't fiddle with. We have left three places to make significant changes in the simulation when the program is run: the two parameters involved with inhibitory fall-off, strength and spatial decay constant, and a constant associated with the numerical solution of the equations involved. (We will discuss this parameter, Epsilon, below.) This version of a lateral inhibition program uses self-inhibition. It is easy to modify the program so a cell does not inhibit itself. Accepting parameters is the job of PROCEDURE Initialize_Parameters: it is straightforward. Something like this PROCEDURE occurs in every program we will write in this book, so it is worth making a few points about it.

Program Fragment 4.3

PROCEDURE Initialize_Parameters

```
PROCEDURE Initialize_parameters;
    VAR I, J     : INTEGER;
        Input_char: CHAR;
    BEGIN
    ClrScr;
    GoToXY (1,1);
    WRITELN('Lateral Inhibition Demonstration.');
    WRITELN('Using a One Dimensional Circular Eye, with
                            wraparound.');
    WRITELN('Parameters of Lateral Inhibition.');
    WRITELN;
    WRITE  ('Inhibitory Maximum Strength: ');
                      READLN (Max_strength);
    Max_strength:= ABS (Max_strength);
                      {Assures inhibition.}
    WRITE  ('Inhibitory Length Constant : ');
                      READLN (Length_constant);
    WRITE  ('Epsilon                  :')
                      READLN (Epsilon);
    Upper_limit := 60;     {Neuron discharge is limited above
                                and below.}
    Lower_limit :=  0;
    END;
```

When the program calls this PROCEDURE, two screen control PROCEDUREs clear the screen and position the cursor at the upper left corner.

Then it writes on the screen the name of the program and a short description of its function. When it reads in a value that is restricted to a particular range, it checks to make sure that the variable cannot get out of that range. In this simulation, lateral inhibition must always have a negative sign. The program takes the negative absolute value of the input value, ensuring it is always negative, no matter what is typed in. A more complex checking system than used here would not only check range but TYPE acceptability; for example, using a character when a number is required by the program. Making such an error will terminate execution of the program in all Pascal systems.

We must initialize the state vector. This is done with PROCEDURE Initialize_state_vector. This PROCEDURE sets the values of the state vector to display a rectangular step of intensity, going from 10 units at neuron 20 to 40 units at neuron 21, and from 40 units at neuron 60 to 10 units at unit 61. It is easy to change starting patterns by modifying this PROCEDURE, and we will use another starting pattern at the end of this chapter. An interactive PROCEDURE for constructing state vectors could be written, or a menu using interesting demonstration waveforms (i.e., ramps, spikes, etc.).

```
                Program Fragment 4.4
            PROCEDURE Initialize_state_vector
PROCEDUE Initialize_state_vector;
        VAR I: INTEGER;
        {The initial input is a rectangular pattern, changing
        from 10 to 40}
        {at neuron 20 and 40 to 10 at neuron 60.}
        BEGIN
        FOR I:=  1 TO 20 DO Initial_state_vector [I]:= 10;
        FOR I:= 21 TO 60 DO Initial_state_vector [I]:= 40;
        FOR I:= 61 TO 80 DO Initial_state_vector [I]:= 10;
        State_vector:= Initial_state_vector;
        END;
```

We must next initialize the inhibitory weights. This is done in PROCEDURE Make_inhibitory_weights. As mentioned above, we are assuming inhibition falls off exponentially with distance from the target neuron. We are also assuming a circular retina. This means if we are looking at the inhibition received by neuron 1, the synaptic weight coupling neuron 1 and neuron 2 is the same as the weight coupling neuron 1 and neuron 80; that is, they are both at a distance of 1 unit from the neuron 1.

We are also assuming every neuron receives the same pattern of inhibition; that is, the inhibitory weights impinging on neuron 2 are the same values as on neuron 1, just shifted over one unit. An efficient way to generate the weights uses this observation.

PROCEDURE Make_inhibitory_weights contains two local PROCE-DUREs. The first PROCEDURE Make_first_vector, generates the synaptic weights for neuron 1, as we discussed above. We use a constant called Half_dimensionality, which is, logically, half the dimensionality. The range of inhibition is half the retina. That is, neurons from 2 to Half_dimensionality are assumed to be on one side of neuron 1, and neurons from Half_Dimensionality +1 to Dimensionality are assumed to be on the other side of neuron 1. The second local PROCEDURE is called PROCEDURE Shift_elements_right, which does what its name implies: it takes a vector of inhibitory weights and shifts all the elements one position right, wrapping the last element around to the first element.

Given these two procedures, the overall PROCEDURE Make_inhibitory_weights consists of generating the vector of weights for the first neuron and sucessively shifting it to get the values for the rest of the neurons. (See appendix B for the full listings.)

Displays and Screen Control Procedures

Even though most computers nowadays have good graphics capability, there is so much variation from system to system that we will only assume an alphanumeric screen with simple ANSI screen control, for example, a terminal such as a VT-100, or a system such as Turbo Pascal or equivalent. We will assume that the computer used has available two PROCEDUREs for screen control: PROCEDURE ClrScr, which clears the screen, and PROCEDURE GoToXY, which sends the screen cursor to location (X, Y), where the column number is X and the row number is Y (see appendix A).

This will let us construct simple alphanumeric graphics and displays. It has the obvious drawback of restricting us to a screen with only 24 by 80 pixels, but this is adequate for what we want to see here. My experience has been that often more time is spent programming fancy graphics than their importance actually warrants. (This is a special difficulty for computer science majors). We are interested in experimenting with a simulation, not in making the output look breathtaking. In the case of PROGRAM Lateral_inhibition, we want to display the activity of the neurons in a portion of the retina. It is more convenient to let the retina be arranged vertically (in the Y direction) and let the activity be arranged horizontally (in the X direction). Activity in our model retina then can display a range from zero to about 60 spikes per second, and we can display about 20 neurons on one screen, which is enough to show most of the interesting effects we want to simulate.

PROCEDURE Initialize_display clears the screen and writes enough information to describe the display. A legend containing a scale is written just above the display. PROCEDURE Display_state_vector is used to display the initial and final state vectors, so it has to have a slight degree of generality. We cannot put the whole state vector on the screen, so we give the PROCEDURE a starting point, and then display activity of that element and the next 18 element activities. The display character can be changed so we can visually separate different conditions on the display. In the program we use this PROCEDURE to plot both the initial state ('+') and the final state ('*'). We will use this minimal graphics display in all our examples of lateral inhibition. Graphics routines are very machine dependent, and the particular implementation we use is of little general interest.

The Computation

One PROCEDURE, Compute_inhibited_state_vector, does essentially all the computational work in this program. We will first discuss the strategy we will follow for the computation, and then sketch the PROCEDUREs used to realize it.

The amount of inhibition received by a unit is a function of the activities of the other units in the system (including itself, because there is self-inhibition) and the inhibitory synaptic weights. We will be given a pattern of light intensity falling on the one-dimensional eye. We wish to see what the resulting activity pattern looks like.

The initial pattern, due strictly to light intensities, must change into a new pattern that includes the effects of lateral inhibition. We want to find this new pattern, the solution vector, s. It is tempting, especially if we know matrix algebra, to solve the equations describing the inhibition and assume the system is therefore solved. However, such solutions are unlikely to be the way the *Limulus* does it.

It helps sometimes to think like the retina, and see how the computational problem might be posed by a real light stimulus. Suppose initially the retina shows zero activity. A pattern of light appears on the retina. If the receptors are assumed to be fast relative to the inhibition, the cells immediately take activities corresponding to the light intensity, the $e[i]$ in the equation.

As soon as neurons in the eye show nonzero activities, they will start to inhibit their neighbors. If we were to apply the inhibition all at once in full strength the system will often oscillate. The activity pattern would initially be inhibited. Then it would rebound, since the activity of all the cells causing inhibition would be depressed because of their sudden large inhibition, and so on.

Computers must approximate continuous time in discrete intervals. The state of the system at one time gives rise to the state of the system at a later time. These times are separated by an interval called the *step size* or *time quantum*. Consider a discrete form of the *Limulus* equation. We want to find the activity of the *i*th cell, $f[i]$, at a time $t + 1$, as a function of the values of the activity at time t. We will denote this quantity $f[i](t)$, which is awkward to write but is consistent with our past notation.

$$f[i](t + 1) = e[i] + \sum_{j=1}^{n} A[i,j]f[i](t),$$

where $A[i,j]$ is the weight connecting the *j*th cell to the *i*th cell, and $e[i]$ is the discharge that would be caused by the light in the absence of inhibition. We will assume the light pattern, **e**, is always constant and not a function of time during the settling time of the system state.

We want to find the solution vector, **s**, where the state of the system no longer changes with time. Suppose the system is in system state **f**(*t*). Let us define a vector Δ**f**, which is the difference between the state vector at the *t*th and (*t* + 1)st state. Then, for a particular unit *i*,

$$\Delta f[i] = f[i](t + 1) - f[i](t) = e[i] + \sum_{j=1}^{n} A[i,j]f[i](t) - f[i](t).$$

One way to find the solution vector, **s**, is by what is called fixed-point iteration (Kreysig, 1988). The solution, **s**, is a fixed point of Δ**f**, because repeated iterations do not change the solution. That is, the vector **f**(*t* + 1) = **f**(*t*). This implies Δ**f** = 0. A solution state $s[i](t)$ will be reached when for all the units,

$$s[i](t + 1) = e[i] + \sum_{j=1}^{n} A[i,j]s[i](t) = s[i](t).$$

Suppose we start at a point **f**(*t*). We can compute Δ**f** at that point. Suppose the starting point for the next iteration, **f**(*t* + 1), is given by,

$$\mathbf{f}(t + 1) = \mathbf{f}(t) + \varepsilon\Delta\mathbf{x},$$

where ε is a small constant. A theorem (Kreysig, 1988) states that after repeated iterations, the system state will eventually end up at a fixed point as long as certain conditions on the function Δ**x** are satisfied. (Technical comment: For simple functions the condition requires that the derivative must always be less than one in a region. This quantity is not so easy to define in the case of vectors and matrices. However, for the specific case of lateral inhibition, this condition corresponds to having all the eigenvalues be less than one in magnitude. See chapter 5. This condition can be ensured by making ε small enough.) It is easy to perform the iterations, since they only require knowing the starting point, $\mathbf{f}(t)$, the connection strengths, and the initial activity due to the pattern of light.

One general problem of practical concern is whether or not there is more than one solution, that is, more than one point where $\Delta\mathbf{f}$ is zero. This can be a very difficult question to answer in general. However, in this simple case we are looking for the solution to a large number of simultaneous linear equations, that is, the point at which a number of hyperplanes intersect. The systems we are interested in will have only one solution.

In the PROCEDURE Compute_inhibited_state_vector, we have also included a minor nonlinearity that has little effect on the dynamics as analyzed above, but makes our computing units somewhat more realistic. PROCEDURE Limit_state_vector simply restricts the state vector to have values at or above zero or less than or equal to a maximum firing rate. We will make extensive use of nonlinearities like this later in this book. A simple nonlinearity can have major effects on system behavior (see chapter 15). The stable state we are searching for is usually away from the limits.

We have also included a PROCEDURE Convergence_test to warn of difficulties caused by the value of Epsilon in the program being too large. If Epsilon is not sufficiently small, the state vector may oscillate with period 2. For example, one iteration has too much inhibition, causing the next iteration to have too little, since the units are heavily inhibited, and so on. Or, perhaps the iterations have not proceeded to completion. This PROCEDURE simply computes the difference in length between the last two iterations and comments (with a BEEP) if the length is large.

```
                Program Fragment 4.5
PROCEDURE Compute_inhibited_state_vector;
    VAR I,J             : INTEGER;
        New_state_vector,
        Difference      : Vector;
    FUNCTION Inner_product (Vector_1, Vector_2: Vector): REAL;
        . . .
    PROCEDURE Subtract_Vectors (Vector_1, Vector_2: Vector;
                                VAR Vector_difference: Vector);
        . . .
```

```
FUNCTION Vector_length (Vector_1: Vector): REAL;
    ...
PROCEDURE Limit_state_vector (VAR V: Vector);
    ...
PROCEDURE Convergence_test;
    ...
Begin {Compute_inhibited_state_vector.}
FOR I:= 1 TO Number_of_iterations DO
    BEGIN
    FOR J:= 1 TO Dimensionality DO
      BEGIN
      {Add a small amount of Delta_State_Vector to the
        State_vector
      New_state_vector[J]:= State_vector[J] +
          Epsilon * ( Initial_state_vector[J] +
              Inner_product (Inhibitory_weights[J],
              State_vector) - State_vector[J]);
      END;
    Limit_state_vector (New_state_vector);
    IF (I = Number_of_iterations) THEN Convergence_test;
    State_vector:= New_state_vector;
    END;
END;   {Compute_inhibited_state_vector.}
```

Operation and Examples

This program is computation intensive, spending almost all its time doing floating point arithmetic; however, modern personal computers can do the arithmetic without difficulty. Problems with personal computer implementations of neural networks likely have to do more with the limitations some compilers put on the size of arrays than with speed. This problem shows itself in the inability to run networks with more than a small number of units even though raw computer speed is more than adequate. One point we try to make over and over in this book is that sheer size in a network strongly affects the way it is used, its architecture, and its potential power and scalability. Big networks operate in quite different ways than small networks.

Readers are encouraged to play with different sets of parameters to see what happens. Figures 4.19, 4.20, 4.21, 4.22, and 4.23 show several examples taken from the screen of operation of the simulation with two sets of parame-

Simple Lateral Inhibition

```
Initial state: +.  Final state: *
Parameters: Length Constant: 2.00 Maximum Inhibition: 0.10
Neuron  |  Firing Rate: Spikes/Second |          |         |
        0........10........20........30........40........50
```

Neuron		
12	* +	
13	* +	
14	* +	
15	* +	
16	* +	
17	* +	
18	* +	
19	* +	
20	* +	
21		* +
22		* +
23		* +
24		* +
25		* +
26		* +
27		* +
28		* +
29		* +
30		* +

Maximum Inhibition: 0.1

Figure 4.19 Response of the network with inhibitory coefficient of 0.1.

ters, from very weak (0.1) to quite strong (2). (The maximum inhibition experimentally observed in a *Limulus* is less than 0.2.) For the figures, Epsilon was 0.1 and the length constant was 2 units. Figures 4.20 and 4.21 can be compared with experimental data from the *Limulus* when an edge is presented to the eye, shown in figure 4.24.

Note that the activity falls on the dark side of the edge and rises on the light side of the edge. This is the well-known edge enhancement effect of lateral inhibition, and it has the qualitative effect of increasing the contrast of an edge. The larger the amount of inhibition, the larger the edge enhancement. Edge enhancement due to a somewhat more complex version of lateral inhibition is prominent in our own eye, and can be seen dramatically if we reproduce a gray scale (figure 4.25). The steps are uniform in intensity, but one gets a powerful subjective impression that they are not. Lateral inhibition in the human eye is much more complex than in *Limulus*, but

Simple Lateral Inhibition

```
Initial state: +.  Final state: *
Parameters: Length Constant: 2.00 Maximum Inhibition: 0.20
Neuron  |  Firing Rate: Spikes/Second |          |          |
        0........10........20........30........40........50
   12        *    +
   13        *    +
   14        *    +
   15        *    +
   16        *    +
   17        *    +
   18       *     +
   19      *      +
   20    *        +
   21                              *              +
   22                            *                +
   23                          *                  +
   24                          *                  +
   25                        *                    +
   26                        *                    +
   27                        *                    +
   28                        *                    +
   29                        *                    +
   30                        *                    +
```

Maximum Inhibition: 0.2

Figure 4.20   Response of the network with inhibitory coefficient of 0.2.

some of the qualitative effects are similar. Notice also the overall depression of the activity pattern, due to the fact this system is, after all, inhibitory.

Winner-Take-All Networks

Lateral inhibition has long been a favorite deus ex machina for modelers. Many models using what are called *winner-take-all* (WTA) networks have been proposed. In WTA networks the desired final state vector has a single active unit, with all the other activity suppressed by lateral inhibition. This has the supposedly desirable property of selecting a single feature, property, or interpretation from a set of candidate interpretations. This is a desirable result only for those who like tidy systems without ambiguity or alterna-

Simple Lateral Inhibition

```
Initial state: +.  Final state: *
Parameters: Length Constant: 2.00 Maximum Inhibition: 0.50
Neuron  |  Firing Rate: Spikes/Second |         |         |
        0........10........20........30........40........50
    12      *       +
    13      *       +
    14      *       +
    15      *       +
    16      *       +
    17      *       +
    18     *        +
    19   *          +
    20   *          +
    21                    *                    +
    22                 *                       +
    23              *                          +
    24              *                          +
    25            *                            +
    26            *                            +
    27            *                            +
    28            *                            +
    29            *                            +
    30            *                            +
```

Maximum Inhibition: 0.5

Figure 4.21   Response of the network with inhibitory coefficient of 0.5.

tives. Early versions of the Perceptron, one of the first simple learning networks (see chapter 9), used something like lateral inhibition in such a role. The strength of inhibition must be very large in these systems, because *one* active cell must have sufficently strong coupling to other members in its group to turn them off completely. In a WTA network, the length constant must be very large so one unit can strongly inhibit all the other units in the group no matter where they are located. With a smaller length constant, it is possible that two or more well-separated units will remain active since neither can inhibit the other enough to turn it off. For a WTA network to function, it is, of course, necessary to remove self-inhibition. We can do that in our program by putting in zero for all the self-connections, $A[i, i]$.

We can get simple lateral inhibition to act as a WTA network quite easily. Let us use a new initial pattern, a triangular peak, that has a maximum

Simple Lateral Inhibition

```
Initial state: +.  Final state: *
Parameters: Length Constant: 2.00 Maximum Inhibition: 1.00
Neuron  |  Firing Rate: Spikes/Second |           |           |
        0........10........20........30........40........50
   12      *       +
   13      *       +
   14      *       +
   15      *       +
   16      *       +
   17      *       +
   18     *        +
   19    *         +
   20    *         +
   21              *                                +
   22              *                                +
   23        *                                      +
   24        *                                      +
   25        *                                      +
   26        *                                      +
   27        *                                      +
   28        *                                      +
   29        *                                      +
   30        *                                      +
```

Maximum Inhibition: 1.0

Figure 4.22   Response of the network with inhibitory coefficient of 1.0.

value of 50 centered at unit 20, so it will be right in the middle of the display. We will assume other units, except those in the peak, will be zero. With reasonably high coupling coefficients, it is possible for a WTA network to function adequately to suppress the other active cells in the pattern. Figure 4.26 shows the results of lateral inhibition on the peak pattern. Simple peak detection would reliably pick up the maximum value in this favorable case.

This system will also work with a more difficult initial situation that has continuous background activity. Suppose all units had an initial excitation of 10 units. (The modification to PROCEDURE Initialize_state_vector simply involves changing the first line so the initial activity is 10 rather than zero.) The network still functions effectively to suppress the constant activity if the strength of inhibition is sufficiently strong. Figure 4.27 shows a typical example of such a system.

Simple Lateral Inhibition

```
Initial state: +.  Final state: *
Parameters: Length Constant: 2.00 Maximum Inhibition: 2.00
Neuron |  Firing Rate: Spikes/Second |          |          |
         0........10........20........30........40........50
```

Neuron		
12	* +	
13	* +	
14	* +	
15	* +	
16	* +	
17	* +	
18	* +	
19	* +	
20	* +	
21	*	+
22	*	+
23	*	+
24	*	+
25	*	+
26	*	+
27	*	+
28	*	+
29	*	+
30	*	+

Maximum Inhibition: 2.0

Figure 4.23 Response of the network with inhibitory coefficient of 2.0.

The problem is more difficult if we have two separate peaks and wish to suppress the smaller. Two results from such a two-peaked pattern are shown in figures 4.28 and 4.29. (The initial state vector used can be read off the graph.) It is possible to arrange the constants so that the larger peak suppresses the smaller, but it requires first, that the length constant of inhibition be large and second, that the inhibition be very strong. Such a tightly coupled network can be built, but it is sensitive to noise at the input and during the computation. As the networks grow larger, these problems become worse because more and more total inhibition is required. Another problem is that the connection strengths must be specified accurately, because slight differences in, say, the strength of connection between cells i and j and the reverse connection between j and i will magnify and interact with differences in their activities. The slight nonlinearity—PROCEDURE Limit_state_

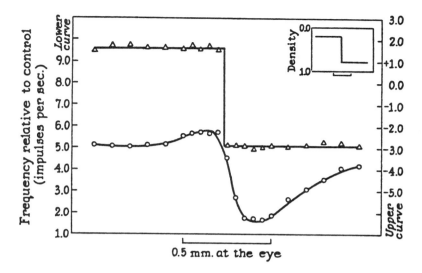

Figure 4.24   The discharge of impulses in response to a simple step pattern of illumination in various positions on the retinal mosaic. The pattern of illumination was rectangular, covering an area 1.65 by 1.65 mm on the eye. The insert shows the relative density of the plate along its length as measured, before the experiment, by means of a photomultiplier tube in the image plane where the eye was to be placed. The upper (rectilinear) graph shows the frequency of discharge of the test receptor, when the illumination was occluded from the rest of the eye by a mask with a small aperture, minus the frequency of discharge elicited by a small control spot of light of constant intensity also confined to the facet of the test receptor. Scale of the ordinate is on the right. The lower (curvilinear) graph is the frequency of discharge from the same test receptor, when the mask was removed and the entire pattern of illumination was projected on the eye in various positions. From Ratliff and Hartline (1959). Reprinted by copyright permission of the Rockefeller University Press.

Gray Scale

Figure 4.25   A gray scale. This figure is composed of strips of different reflectances. The scalloped appearance of each strip is the result of mechanisms in the observer's eye. The nonuniformity is not present in the physical image.

Winner-Take-All Network

```
Initial state: +.  Final state: *
Parameters: Length Constant: 10.00 Maximum Inhibition: 1.00
Neuron  |  Firing Rate: Spikes/Second  |         |         |
        0........10........20........30........40........50
    12  *
    13  *
    14  *
    15  *
    16  *         +
    17  *                   +
    18  *                             +
    19   *                                  +
    20                                         *    +
    21   *                                 +
    22  *                            +
    23  *                    +
    24  *         +
    25  *
    26  *
    27  *
    28  *
    29  *
    30  *
```

Single Peak, No Bias Light Level

Figure 4.26 A winner-take-all network using lateral inhibition. This network uses powerful inhibition to suppress the activity of all but the most active unit. It is sometimes used to produce a grandmother cell output representation, with only one active unit.

vector—that we have built into our program assists in stabilizing the network during the long-range, strong interaction used in WTA networks.

For practical applications of lateral inhibition, or in potential applications to computational neuroscience, time domain behavior can be important. The problem we solved computationally was very simple: a steady light pattern. When we start building time delays into the inhibitory circuitry, stability problems can develop. A system with time delays and powerful inhibition is very likely to oscillate. In a paper on the stability of lateral inhibition, Marcus, Waugh, and Westervelt (1991) discussed criteria for stable networks, and mention in passing that both real *Limulus* eyes and artifi-

Winner-Take-All Network

```
Initial state: +.  Final state: *
Parameters: Length Constant: 10.00 Maximum Inhibition: 1.00
Neuron  |  Firing Rate: Spikes/Second  |            |            |
        0........10........20........30........40........50
    12  *           +
    13  *           +
    14  *           +
    15  *           +
    16  *           +
    17  *                        +
    18  *                              +
    19  *                                      +
    20                                               *      +
    21  *                                      +
    22  *                              +
    23  *                        +
    24  *           +
    25  *           +
    26  *           +
    27  *           +
    28  *           +
    29  *           +
    30  *           +
```

Single Peak, Bias Light Level

Figure 4.27   The winner-take-all network will still work even with a constant light bias.

cial retinas have been observed displaying self-sustained oscillations under some conditions. Coultrip, Granger, and Lynch (1992) discussed a WTA network in the context of cortical information processing.

It is worthwhile experimenting a bit with lateral inhibition to get a feeling for what it can do as a signal-processing technique. It is such a useful sharpening and enhancing tool for networks that some form of it is nearly ubiquitous in both vertebrate and invertebrate nervous systems. Lateral inhibition in WTA networks is valuable, but it can require some care in construction to form networks that are stable and not sensitive to noise. Practical difficulties grow rapidly as the networks increase in size. Conceptual difficulties do as well. It is not clear exactly what a very large WTA network would be good for.

Winner-Take-All Network

```
Initial state: +.  Final state: *
Parameters: Length Constant: 10.00 Maximum Inhibition: 1.00
Neuron | Firing Rate: Spikes/Second |         |        |
         0........10........20........30........40........50
    12   *          +
    13   *          +
    14   *                      *
    15              *                      +
    16   *                      *
    17   *          +
    18   *          +
    19   *                      *
    20   *                              +
    21                                      *          +
    22   *                              +
    23   *                      *
    24   *          +
    25   *          +
    26   *          +
    27   *          +
    28   *          +
    29   *          +
    30   *          +
```

Twin Peaks, Inhibition: 1.0

Figure 4.28 The winner-take-all network has some trouble handling two peaks. The most active unit in the second peak is not fully suppressed.

Bibliographic Note

Several of the figures in this chapter were taken from the early work of H.K. Hartline and F. Ratliff. We have supplied the original journal references, but these papers, as well as others, were collected and reprinted in a book entitled, *Studies in Excitation and Inhibition in the Retina*, edited by Floyd Ratliff (1974). This book collects most of the early papers on *Limulus* from the group at Rockefeller University and is recommended for further reading on this fascinating animal. Ratliff's 1965 classic *Mach Bands* puts *Limulus* research in the more general context of vision research. It is possible to carry the linear analysis of the eye much farther. It was possible to measure both the spatial and temporal transfer functions for the eye. Once the transfer

Winner-Take-All Network

```
Initial state: +.  Final state: *
Parameters: Length Constant: 10.00 Maximum Inhibition: 2.00
Neuron  |  Firing Rate: Spikes/Second |          |        |
        0........10........20........30........40........50
   12   *           +
   13   *           +
   14   *                      +
   15   *                               +
   16   *                      +
   17   *           +
   18   *           +
   19   *                      +
   20   *                               +
   21                                        *      +
   22   *                               +
   23   *                      +
   24   *           +
   25   *           +
   26   *           +
   27   *           +
   28   *           +
   29   *           +
   30   *           +
```

Twin Peaks, Inhibition: 2.0

Figure 4.29 By doubling the amount of inhibition, the second peak (in figure 4.28) can be suppressed.

functions were known, the response of the eye to essentially any pattern of light falling on it could be predicted. This remarkable result means that the response of the simple *Limulus* eye preparation we have discussed is now understood in almost complete detail (see Brodie, Knight, and Ratliff, 1978).

References

Anonymous (1841), *Report on the Invertebrata of Massachusetts, Comprising the Mollusca, Crustacea, Annelida, and Radiata.* Cambridge, MA: Folsom, Wells, and Thurston.

R.B. Barlow, Jr. (1988), Circadian rhythms in the *Limulus* retina nearly compensate for the day-night changes in ambient illumination (abstract). *Investigative Opthalmology and Visual Science, 29,* 350.

R.B. Barlow, Jr., L.C. Ireland, and L. Kaas (1982), Vision has a role in *Limulus* mating behavior. *Nature, 296*, 65–66.

R.B. Barlow, Jr. and D.A. Quarles, Jr. (1975), Mach bands in the lateral eye of *Limulus*: Comparison of theory and experiment. *Journal of General Physiology, 65*, 709–730.

R. Batra and R.B. Barlow, Jr. (1990), Efferent control of temporal response properties of the *Limulus* lateral eye. *Journal of General Physiology, 95*, 229–244.

J. Bonaventura, C. Bonaventura, and S. Tesh, Eds., (1982), *Physiology and Biology of Horseshoe Crabs: Studies on Normal and Environmentally Stressed Animals.* New York: Alan R. Liss.

R.C. Bracha and G.J. Bracha (1990), *Invertebrates.* Sunderland, MA: Sinauer.

S.E. Brodie, B.W. Knight, and F. Ratliff (1978), The response of the *Limulus* retina to moving stimuli: A prediction by Fourier synthesis. *Journal of General Physiology, 72*, 129–166.

R. Coultrip, R. Granger, and G. Lynch (1992), A cortical model of winner-take-all competition via lateral inhibition. *Neural Networks, 5*, 47–54.

J.M. Goldberg and C. Fernandez (1971), Physiology of peripheral nerves innervating semicircular canals of the squirrel monkey. *Journal of Neurophysiology, 34*, 635–684.

H.K. Hartline and C.H. Graham (1932), Nerve impulses from single receptors in the eye. *Journal of Cellular and Comparative Physiology, 1*, 227–295.

H.K. Hartline and F. Ratliff (1958), Spatial summation of inhibitory influences in the eye of *Limulus* and the mutual interaction of receptor units. *Journal of General Physiology, 41*, 1049–1066.

H.K. Hartline, H.G. Wagner, and E.F. MacNichol, Jr. (1952), The peripheral origin of nervous activity in the visual system. *Cold Spring Harbor Symposia on Quantitative Biology, XVII*, 125–141.

E. Kreysig (1988), *Advanced Engineering Mathematics, 6th ed.* New York: Wiley.

M.F. Land (1985), Optics and vision in invertebrates. In: H. Autrum (Ed.), *Handbook of Sensory Physiology*, Volume VII 1B, pp. 471–585. Berlin: Springer Verlag.

E.F. MacNichol, Jr. (1955), Visual receptors as biological transducers. In: R.G. Grenell (Ed.), *Molecular Structure and Functional Activity of Nerve Cells*, pp. 34–53. Washington, DC: American Institute of Biological Sciences.

C.M. Marcus, F.R. Waugh, and R.M. Westervelt (1991), Connection topology and dynamics in lateral inhibition networks. In: R.P. Lippmann, J.E. Moody, and D.S. Touretsky (Eds.), *Neural Information Processing Systems 3.* San Mateo, CA: Morgan-Kauffman.

V.B. Mountcastle (1967), The problem of sensing and the neural coding of sensory events. In: G.C. Quarton, T. Melnechuk, and F.O. Schmitt (Eds.), *The Neurosciences*. New York: Rockefeller University Press.

M.K. Powers, R.B. Barlow, Jr., and L. Kass (1991), Visual performance of horseshoe crabs day and night. *Visual Neuroscience, 7*, 179–189.

F. Ratliff (1965), *Mach Bands: Quantitative Studies on Neural Networks in the Retina*. San Francisco: Holden-Day.

F. Ratliff, Ed., (1974), *Studies in Excitation and Inhibition in the Retina*. New York: Rockefeller University Press.

F. Ratliff and H.K. Hartline (1959), The responses of *Limulus* optic nerve fibers to patterns of illumination on the receptor mosaic. *Journal of General Physiology, 42*, 1241–1253.

F. Ratliff, H.K. Hartline, and D. Lange (1966), The dynamics of lateral inhibition in the compound eye of *Limulus*. In: C.G. Bernhard (Ed.) *The Functional Organization of the Compound Eye*. Oxford: Pergamon Press.

C.N. Shuster, Jr. (1955), On the morphometric and serological relationships within the Limulidae, with particular reference to *Limulus polyphemus*. Ph.D. dissertation, New York University.

C.N. Shuster, Jr, (1960), Horseshoe "crabs." *Estuarine Bulletin of the University of Delaware, 5*(2), 3–8.

C.N. Shuster, Jr, (1982), A pictorial review of the natural history and ecology of the horseshoe crab *Limulus polyphemus*, with reference to the other limulidae. In: J. Bonaventura, C. Bonaventura, and S. Tesh, (Eds.), *Physiology and Biology of Horseshoe Crabs: Studies on Normal and Environmentally Stressed Animals*. New York: Alan R. Liss.

5   *Simple Matrix Operations*

This chapter contains some useful mathematical background. It describes, both mathematically and with fragments of programs, some of the properties of elementary matrix operations.

In the last chapter we described a system in which every neuron was potentially connected to every other neuron. The neural elements were assumed to be simple devices that added up their inputs from other elements, weighted by synaptic strength. Therefore the activity shown by a particular neuron was a function of the activity of every other neuron, weighted by an inhibitory coefficient. If we ignore the limiting nonlinearity we used in the program, this kind of set of interconnected variables forms a set of simultaneous linear equations.

In the program we assumed that a vector of inhibitory weights was associated with every neuron. Any weight was represented by two index numbers, one corresponding to the neuron receiving the connection and one corresponding to the neuron transmitting its activity. A doubly subscripted array of this kind is called a *matrix*. In the neural modeling literature it is sometimes called a *connection matrix* because it describes a set of connections between units.

When we are talking about a matrix representing the coefficients of a set of simultaneous linear equations, we can use a large body of important results and concepts in linear algebra, as we did for vectors in chapter 3. We will follow the accepted conventions of linear algebra in this book, and we will use our linear inhibitory system for occasional examples.

A Matrix

We will most often be concerned with square (i.e., *n* by *n*) matrices. However, matrices can be more general than this. In our simulations, every neuron is potentially connected to every other one. If we have a system of dimen-

```
              Col 1      Col 2      Col 3    ...    Col n
               ↓          ↓          ↓               ↓
  Row 1  →  ⎛  A[1,1]     A[1,2]     A[1,3]   ...   A[1,n]  ⎞
            ⎜                                               ⎟
  Row 2  →  ⎜  A[2,1]     A[2,2]     A[2,3]   ...   A[2,n]  ⎜
            ⎜                                               ⎜
  Row 3  →  ⎜  A[3,1]     A[3,2]     A[3,3]   ...   A[3,n]  ⎜  =  A
            ⎜                                               ⎜
  ...       ⎜  ...        ...        ...      ...   ...     ⎜
            ⎜                                               ⎜
  Row n  →  ⎝  A[n,1]     A[n,2]     A[n,3]   ...   A[n,n]  ⎠
```

Figure 5.1   Matrix-naming conventions.

sionality n, then we have n^2 possible connections. It would also be possible to have a set of m model neurons, each receiving n connections, forming an m by n rectangular array of numbers. Much of the discussion to follow is equally applicable to rectangular matrices. There are exceptions, however, such as our brief mention of eigenvectors and eigenvalues, which are only defined for square matrices.

The general notational convention for matrices that we use is like that we used for vectors in chapter 3, and conforms to Pascal conventions: the ijth matrix element of a matrix **A** is given by $A[i,j]$. These elements, when they are printed, are arranged in an array that can be divided up into rows (across the matrix) and columns (down the matrix). The first index, i, increases as we move down a column and the second index, j, increases as we move across a row from left to right. Even many familiar with matrix algebra have to stop and think for a second to remember what goes where. Figure 5.1 writes out a matrix to show the conventions.

An m by n rectangular matrix would have m rows and n columns. Since we are using Pascal, we can define a new data type called Matrix, as in program fragment 5.1, to let us handle two-dimensional arrays conveniently in our programs. For the program fragments we will assume a square matrix.

Program Fragment 5.1

TYPE Matrix

```
CONST Dimensionality = 100
TYPE Matrix = ARRAY [1..Dimensionality, 1..Dimensionality] OF REAL;
```

Adding, Subtracting, and Multiplying By a Constant

The useful operations of adding and subtracting matrices are almost the same as those we are already familiar with from our discussion of vectors.

One can consider a vector as a special kind of rectangular matrix with a single column (an *m* by 1 matrix). We can add matrices, as in program fragment 5.2.

```
               Program Fragment 5.2
               PROCEDURE Add_matrices
PROCEDURE Add_matrices (Mx_1, Mx_2: Matrix; VAR Matrix_sum: Matrix)
   VAR I,J: INTEGER;
   BEGIN
   FOR I:= 1 TO Dimensionality DO
      FOR J:= 1 TO Dimensionality DO
         Matrix_sum [I,J] := Mx_1 [I,J] + Mx_2 [I,J];
   END;
```

Subtracting matrices is almost identical to adding them, as we see in program fragment 5.3.

```
               Program Fragment 5.3
            PROCEDURE Subtract_matrices
PROCEDURE Subtract_matrices (Mx_1, Mx_2: Matrix;
               VAR Matrix_difference: Matrix);
   VAR I,J: INTEGER;
   BEGIN
   FOR I:= 1 TO Dimensionality DO
      FOR J:= 1 TO Dimensionality DO
         Matrix_difference [I,J] := Mx_1 [I,J] - Mx_2 [I,J];
   END;
```

Multiplying a matrix times a constant is equally straightforward, given our discussion of vectors, as described in program fragment 5.4.

```
               Program Fragment 5.4
            PROCEDURE Constant_times_matrix
PROCEDURE Constant_times_matrix   (C: REAL; Mx_in: Matrix;
                           VAR Mx_out: Matrix);
   VAR I,J: INTEGER;
   BEGIN
   FOR I:= 1 TO Dimensionality DO
      FOR J:= 1 TO Dimensionality DO
         Mx_out [I,J] := Mx_in [I,J];
   END;
```

Multiplying a Vector and a Matrix

The matrix operation that we perform more than any other in this book is multiplying a vector by a matrix. In the *Limulus* eye, the synaptic connection strengths coupling a neuron with the other neurons formed a vector of strengths. The inner product between the vector of strengths and the current activity contained all the term-by-term products, and formed the overall sum that was the total amount of inhibition from the other cells. This is the operation that defines the product of a vector times a matrix. Let us make the computation precise.

We will assume that we are considering the ith unit that is connected to another unit, for example, the jth unit, by weight $A[i,j]$. Then, *the weights associated with a single unit*, i, *lie along the rows of the matrix*. So, the way to do multiplication of a vector times a matrix so as to agree with what we want from our earlier simulation is to ensure that our rules form the inner product of a row of the matrix, the synaptic strengths, and the state vector, a column.

Suppose we have **f** as the input state column vector, and **g** as the resultant output state column vector, with **A** as the matrix. Given the existing mathematical conventions, this is accomplished by forming the product,

g = Af,

and a single term of **g**, say the ith, is given by

$$g[i] = \sum_{j=1}^{j=n} A[i,j]f[j].$$

This forms the basis for our multiplication PROCEDURE, given in program fragment 5.5.

```
                  Program Fragment 5.5
              PROCEDURE Matrix_times_vector
PROCEDURE Matrix_times_vector (A: Matrix; F: Vector;
                               VAR G: Vector);
    {This PROCEDURE forms the product Af = g where }
    {both f and g are column vectors.              }
    VAR I,J: INTEGER;
    BEGIN
    FOR I:= 1 TO Dimensionality DO G[I]:= 0;   {Zero vector G.}
    FOR I:= 1 TO Dimensionality DO
        FOR J:= 1 TO Dimensionality DO G[I]:= G[I] + A[I,J] * F[J]
    END;
```

This definition of multiplication is not the only one possible. Depending on a number of other mathematical conventions, one could use row vectors or put the vector first rather than last in the multiplication.

Transpose

A useful operation and a useful notation, one we will use extensively in the rest of this book, come together in the idea of *transpose*. We already were introduced to transpose for vectors in chapter 3. For our purposes, the operation transpose interchanges rows and columns. Suppose we have a 3 by 3 matrix **A**.

$$\mathbf{A} = \begin{bmatrix} -3 & 1 & 4 \\ 2 & -2 & -1 \\ 5 & 0 & 3 \end{bmatrix}.$$

Then the transpose of **A** is denoted $\mathbf{A}^T$ and is equal to

$$\mathbf{A}^T = \begin{bmatrix} -3 & 2 & 5 \\ 1 & -2 & 0 \\ 4 & -1 & 3 \end{bmatrix}.$$

The diagonal elements remain in place during the transpose operation.

$$A[i,j] \to A[j,i]$$

and

$$A[j,i] \to A[i,j].$$

If

$$\mathbf{A}^T = \mathbf{A},$$

then the matrix is referred to as a *symmetric matrix* because

$$A[i,j] = A[j,i].$$

Symmetric matrices form an important special case found in some physical systems and assumed to exist in some neural models. Physical systems are often perfectly reciprocal; that is, the force of particle i on particle j is the same is the force of particle j on particle i. The same reciprocity is *not* true in the nervous system, although modelers often assume symmetry of connections because of a number of convenient and important mathematical properties of symmetric matrices. This assumption must be looked at with care if it is used. Sometimes useful results follow through if only statistical symmetry holds; that is, if the average strength of $A[i,j]$ is the same as the average strength of $A[j,i]$. However, each case must be examined carefully.

We never actually use the transpose of a matrix in the programs we will write, so no PROCEDUREs will be provided, although it should be easy to write one. However, transpose operations have a number of useful notational aspects. First let us consider what is meant by the transpose of a vector. We assumed (chapter 3) that all vectors are m by 1 column vectors. Then the transpose of a column vector must be a 1 by m row vector. As an example, suppose

$$\mathbf{f} = \begin{pmatrix} -1 \\ 2 \\ -2 \\ 0 \\ 1 \end{pmatrix},$$

then

$$\mathbf{f}^T = (-1 \quad 2 \quad -2 \quad 0 \quad 1).$$

Our rule for matrix multiplication, along the row of the first term and down the column of the second term, tells us what to do if we multiply two terms to form the transpose of a vector times a vector. That is, using our example vector as one vector, and defining a new vector, $\mathbf{g}$,

$$\mathbf{g} = \begin{pmatrix} 3 \\ -1 \\ 4 \\ 2 \\ -2 \end{pmatrix}$$

$$\mathbf{f}^T\mathbf{g} = (-1 \quad 2 \quad -2 \quad 0 \quad 1) \begin{pmatrix} 3 \\ -1 \\ 4 \\ 2 \\ -2 \end{pmatrix}$$

$$= (-1)(3) + (2)(-1) + (-2)(4) + (0)(2) + (1)(-2)$$

$$= [\mathbf{f}, \mathbf{g}]$$

$$= -15.$$

The important observation is that, given our rules for matrix multiplication, *the product of a row vector and a column vector is simply the inner product of the two vectors.*

What happens if the two terms are reversed, that is, we have a column vector times a row vector? The rows and columns are out of synch, that is, columns in the first term and rows in the second term are each only one element wide. So there should be many pairwise products of the terms in the two vectors, arranged in some kind of array. If the dimensionality of the system is n, then there are n^2 pairwise products in the product. The rules describing the multiplication of a vector by a matrix also suggest where these terms should go: they form a matrix of a particular kind called an *outer product matrix*. We can write down the outer product matrix for our test vectors as

$$\mathbf{gf}^T = \begin{bmatrix} 3 \\ -1 \\ 4 \\ 2 \\ -2 \end{bmatrix} (-1 \quad 2 \quad -2 \quad 0 \quad 1).$$

The way to arrange everything in the right order is to observe that the first row of the matrix will look like the first term in $\mathbf{g}$ times every term in $\mathbf{f}$, and so on. That is,

$$\mathbf{gf}^T = \begin{bmatrix} g[1] & * & (f[1] & f[2] & f[3] & f[4] & f[5]) \\ g[2] & * & (f[1] & f[2] & f[3] & f[4] & f[5]) \\ g[3] & * & (f[1] & f[2] & f[3] & f[4] & f[5]) \\ g[4] & * & (f[1] & f[2] & f[3] & f[4] & f[5]) \\ g[5] & * & (f[1] & f[2] & f[3] & f[4] & f[5]) \end{bmatrix}$$

$$= \begin{bmatrix} g[1]f[1] & g[1]f[2] & g[1]f[3] & g[1]f[4] & g[1]f[5] \\ g[2]f[1] & g[2]f[2] & g[2]f[3] & g[2]f[4] & g[2]f[5] \\ g[3]f[1] & g[3]f[2] & g[3]f[3] & g[3]f[4] & g[3]f[5] \\ g[4]f[1] & g[4]f[2] & g[4]f[3] & g[4]f[4] & g[4]f[5] \\ g[5]f[1] & g[5]f[2] & g[5]f[3] & g[5]f[4] & g[5]f[5] \end{bmatrix}$$

$$\begin{bmatrix} -3 & 6 & -6 & 0 & 3 \\ 1 & -2 & 2 & 0 & -1 \\ -4 & 8 & -8 & 0 & 4 \\ -2 & 4 & -4 & 0 & 4 \\ 2 & -4 & 4 & 0 & -2 \end{bmatrix}$$

Another name for the outer product is *Cartesian product*, that is, the set of pairwise products of elements.

The outer product matrix turns out to be extremely important for neural network models, as we will see in the next chapter. Let us write a PROCE-DURE Outer_product, given in program fragment 5.6, which will take two vectors as input and will generate the outer product matrix as the output.

```
                 Program Fragment 5.6.
               PROCEDURE Outer_Product;
PROCEDURE Outer_product (G,F: Vector; VAR A: Matrix);
    {                                                  }
    {We want to generate an outer product matrix A so that  }
    {             T                                     }
    {     A = g f                                       }
    {                                                  }
    VAR I,J: INTEGER;
    BEGIN
    FOR I:= 1 TO Dimensionality DO
        FOR J:= 1 TO Dimensionality DO
            A[I,J] := G[I] * F[J];
    END;
```

Eigenvectors and Eigenvalues

If we think of a vector as a geometric direction in a high-dimensional space, multiplication of a vector by a particular matrix, **A**, takes a vector pointing in one direction and usually transforms it into one pointing in another direction. However, some vectors have a special property: when they are multiplied by the matrix **A**, the product vector points in the *same direction* as the original vector. If we have a vector, say **x**, then a vector in the same direction as **x** is one that is a constant times **x**. The constant can be positive or negative. This means the new vector lies somewhere along the line defined by **x** and the origin. The vector can be pointing in the opposite direction as the original vector, but it must lie along the line.

Such vectors whose direction is the same when they are multiplied by a particular matrix, **A**, are called the *eigenvectors* of **A**, or, equivalently, the *characteristic vectors* of **A**. The constants are called the *eigenvalues* or *characteristic values*. If the matrix is **A**, **x** is an eigenvector and λ is the associated eigenvalue for **x**, we can write this relationship formally as

$$\mathbf{A}\mathbf{x} = \lambda\mathbf{x}.$$

The most obvious point about eigenvectors is that it is easy to calculate with them. Instead of performing the very large number of multiplications and summations required to multiply a matrix and a vector, we can save ourselves a lot of work and only perform one multiplication, a vector times a constant. Besides their intrinsic interest, eigenvectors perform a significant labor-saving function. Even more important, sometimes they have meaningful interpretations, and their behavior will tell us something important about the system the matrix describes.

Let us give an example with a clear geometric interpretation. Consider a matrix, which we will call **A**, that is defined as

$$\mathbf{A} = \begin{pmatrix} 0 & 1 \\ 1 & 0 \end{pmatrix}.$$

It is easy to see what this matrix does to a vector. Suppose we consider a general vector **f** with x component x and y component y, and see what happens when it is multiplied by A:

$$\mathbf{Af} = \begin{pmatrix} 0 & 1 \\ 1 & 0 \end{pmatrix} \begin{pmatrix} x \\ y \end{pmatrix} = \begin{pmatrix} y \\ x \end{pmatrix}.$$

This matrix simply interchanges the x and y coordinates. Figure 5.2 shows what is happening. Interchanging the x and y coordinates *reflects* the vector around the 45-degree line going through the origin.

Now using common sense and the picture (assuming they are different) we can see what the eigenvectors must be. The first is simply a vector that lies along the 45-degree line through the origin. Every point on this line has the two components equal. Let us call this vector $\mathbf{x}_1$. If we do the multiplication, we can see what the eigenvalue, λ_1, associated with this eigenvector must be:

Eigenvectors of a Simple Geometrical Transformation

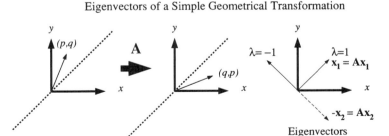

Reflection around the 45 degree axis

Figure 5.2 Eigenvector demonstration. A simple matrix performs reflection around the 45-degree axis. The eigenvectors are obvious from the geometry, and are shown in the righthand diagram.

$$\mathbf{Ax}_1 = \begin{pmatrix} 0 & 1 \\ 1 & 0 \end{pmatrix}\begin{pmatrix} x \\ x \end{pmatrix} = \begin{pmatrix} x \\ x \end{pmatrix} = (1)\begin{pmatrix} x \\ x \end{pmatrix} = \lambda_1 \mathbf{x}_1.$$

The eigenvalue, λ_1, is obviously 1. Note that the eigenvector can have *any* length as long as it is along this line, that is, it is only determined to within a constant. The convention is often made that the eigenvector is normalized, that is, of length 1.

The second eigenvector, $\mathbf{x}_2$, can also be inferred from geometry, but is a little less obvious. Is there another vector that points in the same direction after reflection? Such a vector would be one lying on the line 90 degrees from the first. When it is reflected, it is flipped over, but still lies along the same line. Points along this line are of the form $x_2[1] = x$, $x_2[2] = -x$. Doing the multiplication:

$$\mathbf{Ax}_2 = \begin{pmatrix} 0 & 1 \\ 1 & 0 \end{pmatrix}\begin{pmatrix} x \\ -x \end{pmatrix} = \begin{pmatrix} -x \\ x \end{pmatrix} = (-1)\begin{pmatrix} x \\ -x \end{pmatrix} = \lambda_2 \mathbf{x}_2.$$

In this case the eigenvalue is $\lambda_2 = -1$.

The theoretical and computational literature on eigenvectors and eigenvalues is enormous because many practical problems can be analyzed in terms of them. They can be computed with available numeric analysis programs, for example, EISPACK, a set of programs in the public domain. Some of the learning models we will discuss later in this book make statements about the behavior of the eigenvectors and eigenvalues of the connection matrix.

We must use two points. First, an n by n matrix has at most n distinct eigenvalues. In general, each distinct eigenvalue is associated with a distinct eigenvector. If, however, two or more eigenvalues are identical, some complexities are involved. We will consider a few of these additional problems when they arise.

Second, a symmetric matrix has real eigenvalues and orthogonal eigenvectors, as long as all the eigenvalues are different. In an n-dimensional system, any vector can be represented as the appropriate sum of n orthogonal eigenvectors. This is because there can be, at most, n linearly independent vectors in an n-dimensional system. If we have a new vector, the set of n eigenvectors and the new vector must be linearly dependent, otherwise the set of $n + 1$ vectors would be linearly independent. This means that a new vector must be representable as a weighted sum of the n eigenvectors.

Our example 2 by 2 matrix was symmetric and it had real distinct eigenvalues (1 and -1) and orthogonal eigenvectors.

Linear Systems

We now have enough background to see why linear systems are so convenient and often simple to analyze. It is important to understand the mathematical definition of a *linear system*.

A system in the sense we use it here is like the famous function machine that most of us were exposed to in high school or college. This pedagogic device, which acts like an abstract meat grinder, takes an input into the hopper and, when a crank is turned, corresponding to the lawful operations contained inside, generates a processed output. The input and output can be something as simple as a number, or they can be vectors or even functions. For example, if the operation inside the function machine was differentiation, we could put in a function and get out another function: the derivative.

A linear system is defined by the following property. Suppose we have a linear system, which we will call L, that when given arbitrary input f_1 produces output g_1. When given input f_2, L produces output g_2. A system is linear if the response of L to $(f_1 + f_2)$ is $(g_1 + g_2)$; that is, given any pair of inputs, the output of the system is the sum of the outputs to each input individually. This is referred to as the superposition principle. (One of the experiments described in chapter 4 was done specifically to show that the lateral inhibitory system obeyed superposition. See figure 4.15.) Figure 5.3 shows this behavior graphically.

The superposition principle is what makes linear systems so easy to work with. Consider the problem of obtaining the output of a linear system, L, from any possible input. In general, we will have an infinite number of complicated possible inputs to L. Suppose, however, we have a set of simple

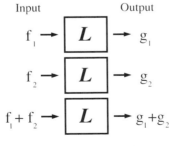

Principle of Superposition

Figure 5.3   A linear system, L, obeys the principle of superposition.

Analysis and Resynthesis

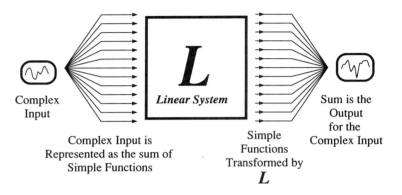

Figure 5.4   A linear system can be used to analyze complex systems. In an analysis phase, a complex input is expressed as the weighted sum of simpler components. The simple components are operated on by the linear system, and the outputs due to the simpler components are then summed. The summed output is the response to the complex input. These operations are possible only because linear systems obey the principle of superposition.

inputs that we have been able to characterize completely, so that we know what the response will be. Further suppose that any possible complicated input can be represented as a weighted sum of the simple inputs that we already know about. Then the response to the complicated input is simply the weighted sum of the response to the simple inputs, as diagrammed in figure 5.4

A particular example that will recur later in the book involves the eigenvectors of a symmetric matrix, $\mathbf{A}$. This is an orthogonal set of vectors, in general, with real eigenvalues. Let us assume that a complete set of n orthogonal eigenvectors exists. They are linearly independent. Since this is an n-dimensional space, every vector will be describable as some sum of these eigenvectors. Let a set of n orthogonal eigenvectors of $\mathbf{A}$ be $\{\mathbf{e}_1, \mathbf{e}_2, \ldots \mathbf{e}_n\}$.

Suppose we have a particular input vector, $\mathbf{x}$. We want to break $\mathbf{x}$ up into pieces, specifically to make it a weighted sum of the eigenvectors of $\mathbf{A}$. This is easy to do. We want a set of coefficients, $\{a_1, a_2, \ldots a_n\}$ so that

$$\mathbf{x} = \sum_{i=1}^{i=n} a_i \mathbf{e}_i.$$

To find a coefficient a_j take the inner product $[\mathbf{x}, \mathbf{e}_j]$. Then

$$[\mathbf{x}, \mathbf{e}_j] = \sum_{i=1}^{i=n} a_i [\mathbf{e}_i, \mathbf{e}_j].$$

But since the eigenvectors are orthogonal, all the inner products are zero except when $i = j$. So

$$[\mathbf{x}, \mathbf{e}_j] = a_i[\mathbf{e}_j, \mathbf{e}_j],$$

$$a_j = \frac{[\mathbf{x}, \mathbf{e}_j]}{[\mathbf{e}_j, \mathbf{e}_j]}.$$

We could have saved ourselves the bother of dividing by the inner product by normalizing the eigenvectors in the first place.

Now suppose we want to multiply the vector, $\mathbf{x}$, by the matrix, $\mathbf{A}$. First, we break up $\mathbf{x}$ into pieces so

$$\mathbf{x} = \sum_{i=1}^{i=n} a_i\mathbf{e}_i.$$

we already know that

$$\mathbf{A}\mathbf{e}_i = \lambda\mathbf{e}_i.$$

Then,

$$\mathbf{A}\mathbf{x} = \sum_{i=1}^{i=n} a_i\mathbf{A}\mathbf{e}_i$$

$$= \sum_{i=1}^{i=n} a_i\lambda_i\mathbf{e}_i.$$

One might wonder, with justice, why bother, since the numbers of multiplications involved in doing the decomposition and matrix multiplication are the same. However, there are good analytical reasons for looking at the coefficients of the eigenvectors and the eigenvectors themselves. The eigenvectors and associated eigenvalues may tell us something important about the behavior of the connection matrix, something that can be seen and understood best with this analysis. The eigenvectors and eigenvalues are special in many physical systems, and we will see that they are in some neural network models as well.

It is possible to do this kind of decomposition and resynthesis for many linear systems. The most familar case of what we just did is Fourier analysis. We can show that any reasonably smooth repetitive signal can be represented by a Fourier series of harmonically related sine and cosine functions, and any reasonable signal whatever by a generalization of the Fourier series, called the Fourier integral (Bracewell, 1986). The linear system, *L*, is then characterized by its response to sines and cosines of single frequencies. The response of *L* to all of the frequencies is called the *transfer function* and is a list or a graph of gains and phase shifts for the single frequencies. If we wish to know the response of L to a complicated signal, we first decompose the signal into a sum of single components (frequencies), compute the response of L to the components, and add up all the responses. By superposition, this will give the response of L to the complex input signal. Such convenient mechanical methods of solution (i.e., decomposition of the inputs and resynthesis of the outputs) are available for analyzing many linear systems of different kinds. More details about linear systems and linear algebra are available from myriad good textbooks at all levels.

References

R. Bracewell (1986), *The Fourier Transform and Its Applications*, 2nd ed. New York: McGraw-Hill.

6 *The Linear Associator: Background and Foundations*

This chapter presents a few of the elementary biological and psychological constraints on memory storage in humans. The Hebb rule *for synaptic learning is introduced. The Hebb rule requires that change occurs in the connection strength between cells when the presynaptic and postsynaptic cells are active together. Physiological evidence suggests the existence of Hebb synapses at various locations in the brain. The rule is often approximated in mathematical models by the* generalized Hebb rule *or* outer product rule, *where change in connection strength is given by the product of presynaptic and postsynaptic activity. Almost all variants of Hebbian learning rules give rise to* pattern associators, *where an input activity pattern is linked to an output activity pattern. A simple pattern associator, the* linear associator, *results from the use of the generalized Hebb rule in conjunction with a network of linear computing units.*

Synaptic Learning: Hebbian Rules

The subject of memory is of interest to almost everyone. Memory, loosely defined, is an adaptive change in the response of the organism based on experience. Current psychological theories of memory and learning usually involve the assumption of several different kinds of memory with different time courses. In the following discussion we simplify psychology as drastically as we simplified neurophysiology when we proposed the generic connectionist neuron.

It has been known, literally for millenia, that much, perhaps most, memory in humans is *associative*. That is, an event is linked to another event, so that presentation of the first event gives rise to the linked event. In the most simplistic version of association, a *stimulus* is linked to a *response*, so that later presentation of the stimulus will evoke the response. We will start by discussing and modeling this simple but inflexible form of association. As we discuss in chapter 16, more complex uses of association can form computing systems of power and flexibility. The simple form of associative memory we will develop in this chapter is a convenient but oversimplified approximation.

One form of memory, *short-term memory* (STM), applies to periods of seconds, minutes, and perhaps hours. It seems to have limited capacity in that it can retain simultaneously only a small number of items. This number is usually characterized by the phrase taken from a famous paper by George Miller (1956) entitled "The magic number seven, plus or minus two." Capacities as low as three or four have been suggested as more typical. The complexity of an individual item can vary over a far greater range than the number of separate items. For example, we can store about seven digits at a time, and also roughly the same number of words, which are more complex entities. By grouping items together in a process Miller called *chunking*, it is possible to enhance the effective size of STM, although the capacity for discrete items remains the same. An example from the Miller paper would be strings of ones and zeros. It is possible to repeat back a short string, roughly seven, of ones and zeros. If triples of the ones and zeros are converted on the fly into octal digits, it is possible to keep track of about 7 of them, or 21 binary digits. With proper coding techniques, very long strings of numbers can be maintained in memory. One subject was able to reproduce on the order of 100 slowly spoken numbers by chunking triples of digits as the times of track events.

Psychologists argue a good deal about the exact details of STM and whether it is a single process or, more likely, involves further subdivisions. What is important is that some fundamental differences exist between it and another form of memory, *long-term memory* (LTM).

Long-term memories are retained for durations of days or longer. Indications are that they are nearly permanent. The capacity of LTM is enormous and is subject to no such limitations on simultaneous number of stored items as afflict STM. For example, an educated adult has a vocabulary of perhaps 100,000 words, any one of which and its associated meaning can be retrieved from memory in roughly a second. Long-term memory for pictures and complex events is even more impressive: the average television viewer knows within seconds whether or not an episode of a favorite show is new or a rerun.

Memory for pictures has been shown to be extraordinarily large. Lionel Standing (1973) used large numbers of slides, most of which were routine vacation material volunteered by students and faculty at McMaster University. Even after seeing 10,000 slides for a few seconds apiece, with learning sessions spread over an entire week, subjects were still able to tell reliably if they had been seen the pictures previously when they were tested. If, in place of routine snapshots, more vivid and unusual images were used, recognition accuracy for 2000 pictures learned from 4-second presentations was over 98%.

Landauer (1986) was one of the few brave enough to estimate the capacity of long-term memory from experiments such as Standing's. He estimated capacity independently from three converging lines of evidence. His conclusion was that human LTM capacity is somewhere between 10^9 and 10^{10} bits. This is a large capacity, but it is not astronomical. Significantly, it is smaller than the number of neurons in the brain, and certainly much smaller than the number of connections.

At the present state of development of our techniques, we cannot model more than a few aspects of such a complex system. However, when we start to build models, we should leave enough flexibility to allow extensions toward greater realism.

Physical Basis of Memory

We have some practical experience with large information storage systems. Libraries, disk drives, and computers are examples. For these systems, the physical change corresponding to information is quite clear: marks on paper, changes in magnetization, and different states of an electronic switch.

Assumptions about physical storage strongly affect the way memory is organized. We know that a book is a discrete thing that is stored in a particular place. We look in the card catalog to find out where it is and then go get it. In well-organized libraries, books on similar subjects are stored nearby, allowing some ability to generalize by subject to be built into the physical storage scheme. This content coding is one of the most valuable functions of a library. But it is also tacitly assumed that a book is a self-contained bundle of information located in one physical place.

One can organize a library in lots of other ways. Suppose, for example, that the first pages (page 1s) of all the books in the library were stored on the first row of shelves, page 2s on the second, and so on. All the information would still be there, but getting it out would be a very different process than going to a particular shelf to retrieve all the pages at once.

Given the brief introduction to the brain in chapters 1 and 2, can we make any intelligent comments about the way memory might be organized in the brain?

One possibility would be that the physical memory-storage medium would be very small. One favorite suggestion for a medium with high information density is a coded macromolecule. Several candidates have been proposed over the past few decades, such as protein molecules and RNA molecules. Both are long strings of a number of possible components, 4 in the case of RNA and about 20 in the case of protein. Presumably, stored information would be coded in the sequence of base pairs or amino acids. Best of all, we already have highly developed mechanisms for reading sequences of RNA and DNA in place in the cell as part of the genetic machinery.

However, despite the initial attractiveness of this idea, there is no firm evidence for it. It is difficult to come up with efficient mechanisms for storage or quick retrieval. Experiments that supposedly showed molecular memory coding either are not replicable or have alternative explanations. For example, we are aware that large peptides can sometimes act as neurotransmitters. Changes in concentrations of these neuropeptides through stress or other general effects might mimic memory, if behavioral tests are not carefully designed.

Other observations about memory also constrain possible mechanisms. For example, as discussed in chapter 1, new neurons cannot routinely be constructed to store new memories, since the number of neurons is roughly fixed at birth, at least in mammals.

There also seems to be no particular indication that large blank areas of the brain are waiting to be filled with memories like empty shelves in a library. Such a strategy would seem to be unreasonable from a biological point of view. As we mentioned, neurons have a high energy consumption, are delicate, and require extensive and elaborate physical support. They must earn their keep at all times. The fitness of an organism would be greatly reduced if it had to carry around a significant number of neurons that had nothing to do until near the end of a long lifetime.

Synaptic Modification

Given the many ways memory is not likely to be stored, can we suggest a reasonable way that it that can be stored? Most neuroscientists assume, and have for a number of years, that the physical basis of memory lies in changes in the strengths of connection between neurons, that is, modification of synaptic strengths. This change could either be the formation of new synapses, or the elimination or modification of preexisting synapses.

Let us take as our working hypothesis the assumption that memory corresponds to changes in synaptic strength. The task of the neurotheoretican (or neurotheologian) is to see if it works. Discussions about these topics can be quite arcane. For example, learning and memory are sometimes held to be properties of the whole animal, where behavioral changes are observed to arise from particular events in the animal's history. Changes in strength of a synapse due to the neuron's past electrophysiologic activity only rarely can be linked directly to behavior. It has been possible to do this is a few cases, for example, the siphon withdrawal response of the gastropod mollusk *Aplysia*. Many papers have been written on *Aplysia* learning in particular and invertebrate learning in general. Two recent reviews of the neurobiology of synaptic plasticity that could serve as introductions are Carew and Sahley (1986) and Hawkins, Kandel, and Siegelbaum (1993). For a general discussion, Kandel's well-known book (1976) is still valuable.

In the best-studied vertebrate learning synapse, which we will discuss shortly, a direct link between a modifiable synapse and a modified behavior is not presently possible. It would be nice to be able to study a model synapse, quantify the synaptic change involved, and then study the behavior of the memory system that arises. However, synapses are small structures, and our techniques for studying them in vivo are poorly suited to learning how the details of small, long-term synaptic changes are related to behavior.

In the past decade some important neurophysiological evidence was obtained relevant to the kind of synaptic learning we want to model. A few physiological systems seem to show roughly the kinds of changes that a memory system should have. The details are still not fully understood. We will briefly describe this work in a later section, but first, it might be of interest to consider the problem from a slightly more abstract point of view: what kind of synaptic change *would* work; that is, how could we build a memory-storage system that is based on synaptic change?

The first idea anyone comes up with when thinking about general learning involves something along the lines of *strengthening by use*. That is, the more a synapse is used, the stronger it gets. (Metaphorically, the synapse, and by extension, the brain, is like a muscle.) One form of this learning rule would hold that the change in synaptic strength depends only on local presynaptic activity, and postsynaptic activity is irrelevant. The most serious problem with this approach is its lack of selectivity. If the function of the nervous system is to connect the appropriate input with the appropriate output, it seems necessary that the output must appear explicitly somewhere in the equation. However, simple models using strengthening-by-use synapses give rise to a model for recognition memory that has been applied to psychological data with some success (see chapter 7 and Anderson, 1968, 1973).

Associative Learning

For more complex kinds of learning, almost every learning model that has been proposed involves both the output activity and the input activity in the learning rule. The essential idea is that the amount of synaptic change is a function of both presynaptic and postsynaptic activity.

The first person who seems explicitly to have phrased a learning rule of this kind in terms of synaptic change was Donald Hebb (1949) in his influential book, *Organization of Behavior*. Hebb said, in a much-quoted sentence, that what is critical for synaptic learning is a coincidence between excitation in the presynaptic and postsynaptic cells:

When an axon of cell A is near enough to excite a cell B and repeatedly or persistently takes part in firing it, some growth process or metabolic change takes place in one or both cells such that A's efficiency as one of the cells firing B, is increased. (p. 62)

To show how ubiquitous the idea of learning by conjunction of input and output is, we can point out that a similar rule was formulated by William James, in his chapter on association from his classic text, *Psychology: Briefer Course*, first published in 1892:

When two elementary brain processes have been active together or in immediate succession, one of them on recurring, tends to propagate its excitement into the other. (p. 226)

James also has an activation equation (expressed in words) in the next paragraph, which starts, "*The amount of activity at any given point in the brain-cortex is the sum of the tendencies of all other points to discharge into it . . . .*" (p. 226). The simplest realization of this equation, if "point in the brain-cortex" is replaced by "neuron," is something like the *Limulus* equation.

James has a mechanistic and quite clear description of biological association based on discharges and tendencies, with a conjunctional learning rule that looks very much like the architecture of modern neural models for cognition. The model qualifies him, in my mind at least, for the title, first neural modeler or first connectionist, depending on the reader's preference. The architecture of James' "network," reprinted in figure 6.1, is quite similar to the network architecture for the simple linear associator shown in figure 6.7, and can serve as the schematic diagram of a pattern associator. (One wonders what James would have done with a computer.)

The Hebb rule (or James rule) for learning concerns itself only with conjunctions of excitation. Since we know inhibition of cells is as important as excitation in the actual function of the nervous system, we have three other conjunctions to worry about: presynaptic excitation with postsynaptic inhibition; presynaptic inhibition with postsynaptic excitation; and simultaneous presynaptic and postsynaptic inhibition. The neurobiological literature discusses the implications of the other conjunctions. See, for example, Levy,

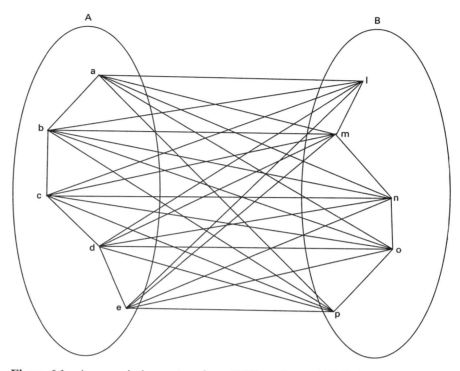

Figure 6.1 An associative system from William James (1892). Compare with the structure of a "modern" network associator in figure 6.7. James viewed a mental object or event as being composed of many pieces that were interassociated in complex ways. Presentation of part of the association could evoke the rest.

Colbert, and Desmond (1990), and Brown, Zador, Mainen, and Claiborne (1991).

In the mammalian nervous system, physiologic inhibition, in the sense of IPSP (inhibitory postsynaptic potential) generation, seems to be handled by distinct groups of inhibitory neurons. In the cerebral cortex, axons entering with information from the sensory systems or from other regions of cortex seem to all be excitatory, as do the connections among cortical pyramidal cells. Inhibitory neurons handle functions such as gain control and lateral inhibition. Therefore, it is not clear if a cell fires less because the inhibitory neurons driving it fire more, or if the excitatory neurons driving it fire less. In general, a system composed entirely of coupled excitatory units would be extremely unstable. When cortical inhibition is suppressed by a drug such as strychnine, spectacular seizures triggered by sensory inputs result. Since inhibition and excitation are handled by separate mechanisms that are biophysically and chemically distinct, we have no particular reason to expect that their learning rules are the same.

From both physiologic and modeling points of view, we can signal inhibition by other ways than using a strictly inhibitory synapse. There are two obvious possibilities. Many cells in the central nervous system (although not all) have a *spontaneous activity*, which means they discharge in the absence of obvious inputs. Sensory inputs and presumably other inputs modulate this spontaneous rate, in both a positive and a negative direction. This is evident in figure 4.6 of the eighth nerve discharges in the squirrel monkey, clearly showing transduction above and below a high spontaneous level.

One serious problem with using positive and negative deviations from the spontaneous rate as a way of coding negative connection strengths is that maximum allowable deviations in the rate are not symmetric. Most spontaneous rates are low compared with the maximum firing rate. There is much more allowable deviation in the positive (from spontaneous rate to maximum rate) than in the negative direction (from spontaneous rate to zero). Part of the reason for this must surely be that it takes a lot of biological energy to keep a cell firing all the time, and high spontaneous rates are wasteful.

Another slightly more complicated way of signaling both positive and negative activity would be by some mechanism involving pairs of cells; one signaling decrease and one increase. A number of sensory systems have examples of something like this. For instance, in vision in many mammals, classes of retinal ganglion cells seem to be organized into complementary groups; the centers of the receptive fields of one group are excited by increases in light intensity with annular regions surrounding the center where light causes a decrease in activity, and the centers of the other group are inhibited by increases in light intensity and surrounds excited by decreases.

Digression. It is hard to avoid comparison of the nervous system on this point to the well-known classification of electronic amplifiers (figure 6.2). Real-world signals, for example pressure waves in the air corresponding to sound, are both positive and negative. If the operating point of an active electronic element (transistor, vacuum tube, or FET (field effect transistor)) in an amplifier is set so the device is continuously biased on, and is not turned off by the largest possible input signal, the amplifier is called class A. (*Bias* means the device has a constant level of activation in the absence of input.) For one sign of input signal, the current (or voltage) in the output device decreases, for the other it increases. Since the active device must be biased to be active enough so the largest possible input signal will not turn it fully on or off, it continuously consumes large amounts of power. Serious audiophiles, those concerned with the absolute ultimate in fidelity of sound reproduction, agree that class A amplifiers are the most desirable from the point of view of faithful response, especially for small signals, but are flagrantly inefficient. They always consume about the same amount of power, no matter whether or not they are amplifying a signal.

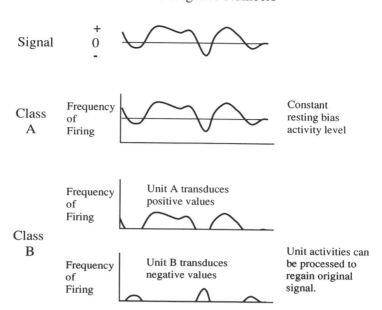

Figure 6.2   Speculation as to how neurons might handle negative numbers. Names are classification of amplifiers used in electronics. Class A amplifiers have a high resting rate that a signal moves up or down. (See figure 4.6 for a biological example.) Class B amplifiers use two elements, one to handle positive going signals and one to handle negative going signals. Class AB amplifiers also use two elements, but can transduce negative signals over a small range.

Class B amplifiers correspond to a pair of amplifying devices biased so they conduct no current with no input signal. When the signal becomes nonzero, the circuit is constructed so that one device amplifies the negative going portion of the signal and the other the positive going portion. The outputs from the two devices are then added, giving a complete replica of the input signal. Class B amplifiers are efficient because they are on only when they are amplifying a signal. However, they are subject to problems with small signals because amplifiers operating near zero input are often nonlinear, and the bias of the devices must be chosen to avoid a dead zone where a very small signal would turn on neither amplifier. This *crossover distortion* can be a severe problem with class B amplifiers, and produces a subjectively unpleasant corruption of an acoustic signal.

The most common kind of amplifier runs in what is called class AB. It uses pairs of devices, one primarily for positive and one primarily for negative signals like class B amplifiers, but biases both devices on at rest, like class A. The bias is sufficient to eliminate crossover problems with small signals. As soon as signals become large, the amplifiers split the load, with one device off and one amplifying. The output signals are combined to reconstruct the amplified input. This compromise amplifier uses more power at rest than class B but less than class A.

It is hard to avoid the conclusion that many neurons are biased so they operate metaphorically in class AB. They have a small but significant spontaneous rate allowing for accurate small signal transduction, but quite asymmetrical limits. When or if the elementary parts are later combined to represent both parts of a signal is unclear.

This may be a case where biological constraints from the real world have strong implications for the brain's information-processing strategy. The requirement for energetic efficiency may mean that the most faithful neural transduction scheme cannot be used, and a somewhat more complex compromise system is necessary. If we take this line of reasoning seriously, spontaneous activity also must earn its keep, just as individual neurons must, by providing a significant benefit. Perhaps this benefit is allowing accurate processing for small signals.

Real neurons use significantly more energy when they are active than when they are inactive, and ordinarily they are kept either off or firing at a low rate. When a brain region becomes active, this suggests that it will increase its metabolic rate. This increase in blood flow and oxygen consumption can be picked up with various modern imaging techniques such as positron emission tomography (PET) and functional magnetic resonance imaging (fMRI). It is assumed that local increase in blood flow means that the brain area involved is doing the processing. However, there might be other ways to run a nervous system!

This discussion has significant implications for the way we formulate models. The key representation question is whether we will allow the state vectors to have negative activities for element values. If we allow only positive numbers, we restrict ourselves to state vectors lying in one segment of state space. We know that many synaptic weights can be considered negative because inhibition is common. Models of brain function have been constructed using both assumptions.

Biological Basis of Hebbian Synaptic Modification

Even though the experiments are difficult, advancements in modern neurophysiological techniques have finally allowed us to see what appears to be Hebbian modification in several parts of the mammalian brain. A part of the cerebral cortex, the *hippocampus*, displays a number of characteristics that make it a popular candidate for an experimentally accessible learning system. The hippocampus plays an important although unclear role in memory. If it is damaged, the result is severe effects on memory.

A famous neurosurgical patient, H.M., had both hippocampi removed in an attempt to treat his epilepsy. He appeared normal after the operation, but was apparently unable to form long-term memories. It was possible for an experimenter to hold a long intelligent conversation with H.M., leave the room for a minute, and on returning, find H.M. had forgotten both the conversation and the experimenter. This led to a large experimental literature studying what H.M. could and could not learn. Although consciously accessible long-term memory was gone, for example, the name of the experimenter or the name of the current president, H.M. did show some kinds of learning. Practice on some largely motor skills such as following a moving dot with a stylus led to better performance over time, even though H.M. denied ever having seen the task each time he practiced it.

H.M. is described briefly in Gardner (1974) along with a number of other fascinating neurological cases. This book is still well worth reading.

Physiologically, the hippocampus shows an effect called *long-term potentiation*, in which its neurons can be induced to display long-term, apparently permanent, increases in activity with particular patterns of stimulation.

An early set of experiments by William Levy of the University of Virginia recorded from groups of cells in the hippocampus of the rat. Due to the neuroanatomical structure of the hippocampus, it is possible to excite the same group of cells by a weak path and by a separate strong path. The strong input provokes the cells to considerable activity; the weak input generates much less. If the two paths are excited independently, the amount of activity evoked does not change. If, however, the strong and the weak inputs are excited together, the picture changes. After pairing with the strong input, the weak input generates much more activity. This increase in response lasts for a long time. See Levy, Anderson, and Lehmkuhle (1985) for a review of this and related work.

A direct neurophysiological demonstration of Hebbian modification was provided by a group led by Thomas Brown. In an early study they used a slice of rat hippocampus maintained outside the animal (Kelso, Ganong, and Brown, 1986). The mechanical stability possible with a slice allowed long-term intracellular recording from single cells. Intracellular recording also allowed the investigators to control the activity in the postsynaptic cell by controlling the membrane potential. When the presynaptic cell was excited and the postsynaptic cell was inhibited, little or no change was seen in the efficacy of the synapse. If the postsynaptic cell was excited by raising the membrane potential (it was not necessary to fire action potentials) at the same time the presynaptic cell was active, the excitatory postsynaptic potential (EPSP) from the presynaptic cell was significantly enhanced. This enhancement was stable and long lasting (figure 6.3). Note that presynaptic activation alone, or postsynaptic activation alone, provoked by direct electrical stimulation did not cause a change in the size of the synaptic potential. But the same two stimuli, presynaptic and postsynaptic, applied together did cause a long-lasting increase in the size of the excitatory postsynaptic potential. This result was so like what Hebb had predicted that the authors entitled their paper "Hebbian synapses in hippocampus" without a question mark, just as a simple statement of fact.

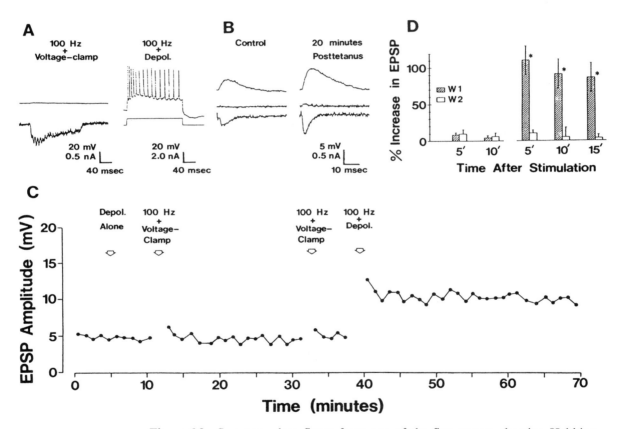

Figure 6.3 Summary data figure from one of the first papers showing Hebbian modification at the synaptic level in the mammalian hippocampus. The data were recorded from a slice of tissue from the hippocampus of the rat. Direct demonstration of the conjunctive mechanism. (A) (Left) Voltage-clamp record of inward synaptic currents (lower trace) and membrane potential control (upper trace) during the synaptic stimulation train. (Right) Current-clamp recording of postsynaptic spiking (upper trace) produced by an outward current step (lower trace) that is paired with the synaptic stimulation train. (B) Computer average of five single

We now have some idea of the biology of the leading candidate for a synapse showing Hebbian modification, the synapse displaying long-term potentiation. (For reviews see Brown, Kairiss, and Keenan, 1990; Brown, Chapman, Kairiss, and Keenan, 1990; Baudry and Davis, 1991.) The critical synapse uses the amino acid neurotransmitter glutamate. The cells involved have a second, more traditional synapse that also uses glutamate. The particular glutamate synapse that is believed to be Hebbian also responds to a glutamate analog called NMDA. As we discussed in chapter 1, neurons work by modulation of ionic flow through the neuron membrane. Ions pass through the membrane through *channels*, holes through the cell membrane that are formed by protein molecules. The action potential, the basic information transmission act of a neuron, reflects an enormous increase in the conductance to sodium ions of a *sodium channel*. The resulting electrochemical interactions make a large change in voltage inside the cell. Synaptic interactions also involve flows of ions through channels, which have the end result of changing the membrane potential of the cell.

The sodium channels are *voltage dependent*. As the membrane potential inside the cell becomes more positive, the sodium channels become more and more conductive, raising the membrane potential still more, creating a positive feedback system that terminates when the channels are fully open.

Many channels associated with synaptic strength are not strongly voltage dependent. Their conductivity is controlled by the presence or absence of particular molecules. An implication of this is that a single synaptic strength can be assumed to have a roughly constant value in the generic connectionist neuron.

◀ synaptic responses recorded under current-clamp (upper traces) or voltage-clamp condition (middle traces show potential control and lower traces show synaptic currents). (Left) Responses during initial control period. (Right) Responses 20 minutes after pairing synaptic stimulation with the outward current step. (C) EPSP amplitudes as a function of the time of occurrence (arrows) of three manipulations—an outward current step alone (Depol. Alone) or synaptic stimulation trains delivered while applying either a voltage clamp (100 Hz + Voltage-Clamp) or while applying an outward current step (100 Hz + Depol.). Each point is the average of five consecutive EPSP amplitudes. (D) Mean increases in the EPSPs produced by two synaptic inputs (W1 and W2) at the indicated times after stimulation. (Left) W1 and W2 were alternately presented stimulation trains while applying a voltage clamp to the postsynaptic cell. (Right) The W1 pathway was stimulated during application of an outward current step, and the W2 pathway was stimulated during application of a voltage clamp to the postsynaptic cell. Asterisks denote significant differences ($P < 0.05$) between the changes induced in the W1 and W2, responses (paired t test for dependent means). From Kelso, Ganong, and Brown (1986). Reprinted by permission.

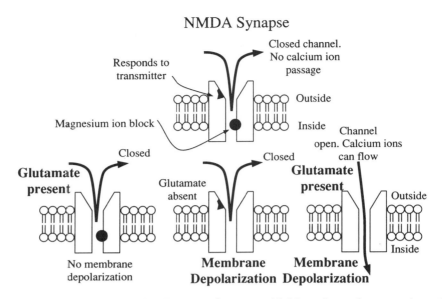

Figure 6.4 A schematic diagram of current thinking about the operation of the NMDA channel. The channel responds to both a chemical (glutamate) and the postsynaptic membrane potential. The receptor protein is associated with an ionic channel that, when open, is preferentially permeable to Ca^{++} rather than to Na^+. The channel is blocked by Mg^{++} until the postsynaptic cell is depolarized. A conjunction of presynaptic and postsynaptic activity is therefore necessary to open the channel. Somehow the immediate change is later made permanent. One suggestion for this process is that a retrograde transmitter could send a signal backward from the postsynaptic side of the synapse to the presynaptic side; that is, against the normal information flow of the cell, causing a presynaptic change in strength. A potential candidate for the retrograde transmitter is nitric oxide (NO).

The NMDA synapse combines aspects of both systems. Conductivity to ion flow is a function of both the presence of certain molecules and the membrane potential. The reason for this seems to be a site located in the channel that binds magnesium ions. Even if the channel is opened by a neurotransmitter, conductance through the channel can be blocked by the presence of magnesium ions. But this blockage is voltage dependent. Therefore if the postsynaptic cell becomes excited, the magnesium block is removed, and the channel can now conduct. Channel conductance is highest for calcium ions, which flow into the postsynaptic cell. Figure 6.4 schematizes some of these complex interactions.

The NMDA receptors are present in many parts of the brain and may play a role in variations in strengths of many synapses. In the hippocampus and probably some other structures, an additional effect occurs. For reasons not fully understood, activation of the NMDA channel activates a system that produces *permanent* changes in synaptic strength. It was not clear where the long-term changes in the NMDA synapse were located, on the presynaptic or the postsynaptic side of the junction. If the changes were on the presynaptic side, how did information flow backward from the presynaptic to postsynaptic cell? No one had seen a *retrograde transmitter* that would move information opposite to the traditional picture. Recently, evidence appeared suggesting that the small molecule nitric oxide (NO) is the retrograde transmitter. It now seems that besides being a component of smog, nitric oxide plays a major role in many parts of the nervous system. Nitric oxide was *Science* magazine's "Molecule of the Year" for 1992 for this and other important biological effects (Culotta and Koshland, 1992).

The use of a rapidly diffusing agent as a retrograde transmitter makes a number of remarkable predictions. For example, activity at one cell leading to release of NO, or any diffusible messenger, should cause long-term potentiation at other nearby cells, even though these cells were not themselves active. This striking result was found by Schuman and Madison (1994), where cells in hippocampus hundreds of microns away indeed showed LTP apparently communicated this way from an active cell. The consequences of this seeming loss of modification specificity have only started to be addressed by theorists.

The study of synaptic function is sure to produce many surprises in the future.

Modifiable synapses on hippocampal cells are located on structures called *dendritic spines*, which are thin processes a couple of microns long, with synaptic complexes at their end. We briefly described spines in chapter 2. As we mentioned, evidence accumulated over decades strongly suggests that dendritic spines are implicated in learning; for example, their shape changes in response to some kinds of environmental manipulation. By changes in spine geometry, it is possible to make changes in the influence of one cell on another. Figure 6.5 indicates how completely neurons in cortex are covered with dendritic spines. See Wallace, Hawrylak, and Greenough (1991) for a review of spine modification in the hippocampus in relation to long-term potential.

How does the experimentally observed conjunctional modification rule agree with the Hebb rule? The data are not yet clear enough to be sure. One major difference is the possibility that a *modification threshold* exists, so that no long-term change occurs unless activity exceeds some critical value. This means that small correlations between presynaptic and postsynaptic activity would not be learned.

Another point of interest is the similarity of the short-duration and the long-duration effects. Both the early current passed through the NMDA

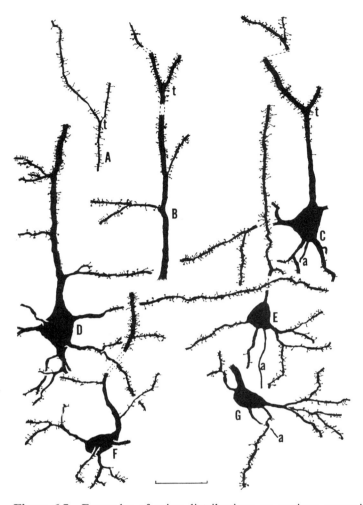

Figure 6.5 Examples of spine distributions on various pyramidal cell dendrites. Tracings from rat visual cortex, Golgi rapid impregnation. (A) Thin apical dendrite bifurcating into its terminal tuft (t). (B) Apical dendrite of deep layer II/III pyramid showing its initial spine-free portion connected to this segment by the terminal tuft bifurcation (t) of the same dendrite; this is connected by dotted lines to a tracing of the tip of one of the terminal tuft dendrites. (C) Upper layer II/III neuron; note that the apical dendrite forms very few spines before the terminal tuft bifurcation (t); the dotted line connects to a tracing of a more distal segment of the terminal tuft. (D) Layer V pyramidal neuron. (E) Layer IV pyramidal neuron. (F, G) Layer VI pyramidal neurons. The dotted lines on the apical dendrite of cell F connect to a tracing of a segment of the same apical dendrite in layer V. The ventrally directed dendrite from cell G projects onto the white matter. Calibration line 25 μ. From Feldman (1984). Reprinted by permission.

channel and the long-term change are in the same direction. This suggests that, as far as neural information representation is concerned, perhaps short- and long-term memory may be organized similarly. A very short-term Hebbian modification was suggested as a useful information processing technique by some modelers (von der Malsburg and Bienenstock, 1986).

Virtually all abstractions of the Hebb rule assume more than simple excitatory conjunction. Suppose the presynaptic cell is consistently excited when the postsynaptic cell is not firing or inhibited. Does this weaken the synapse? There is some evidence that anticorrelation can produce synaptic weakening, that is, *long-term depression* by a mechanism other than the NMDA synapse (Bear, Cooper, and Ebner, 1987; Kirkwood, Dudek, Gold, Aizenman, and Bear, 1993).

The Outer Product Hebbian Learning Rule

The Hebb rule in its generality implies that the change in a synaptic weight coupling two neurons is some function of presynaptic and postsynaptic activity. The actual rule in the brain almost certainly is not simple. Our approach in this chapter will be to investigate a simple generalization of the Hebb rule and see if it does anything interesting. We will see that it forms an effective associative system, which is an important result from the functional point of view. Virtually all variants of the rule form associative systems that work to some degree or other. Associativity seems to be a robust result of a conjunctional learning rule. Therefore we may be justified in looking at a particularly simple form of the rule that is easy to analyze. A more complex version of a basically Hebbian synapse with a sliding modification threshold was described by Bienenstock, Cooper, and Munro (1982).

Let us assume we want to make an association between an input and an output. Let us initially assume that making such an association is just something good to do, but we will see later (chapter 16) that evidence suggests that association is perhaps the primary memory operation in human cognition. Associativity means one thing to us, that *we wish to associate two state vectors*. We want to develop a system so that when an input state vector, say **f**, is an input, an output state vector, **g**, is the output. We will say the system has *learned* the association when, if **f** is the input, **g** is the output. Figure 6.6 shows a block diagram of what we are aiming for.

The simplest realization of such a system, capable of both learning and recall using the kind of parallel hardware we have to work with, assumes

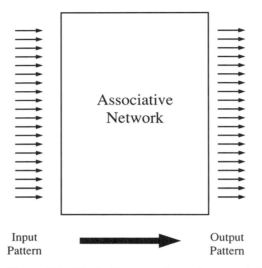

Figure 6.6 Block diagram of a vector associator. We want to develop a system in which presentation of an input vector gives rise to the associated output vector. Developing accurate vector associators with the right input-output characteristics is the goal of most current neural network learning algorithms.

that we have one set of units that projects to another set of units by way of modifiable connections, synapses (figure 6.7). (Compare figure 6.7 with figure 6.1 from 1892!) In this system the *ij*th connection, $A[i,j]$, connects the *j*th unit in the first set, $f[j]$, to the *i*th unit in the second set, $g[i]$. For learning we assume that a state vector, **f**, is present on the first set of units, and a state vector, **g**, on the second set of units. This assumes that learning involves having both the input and output simultaneously present.

   We assume that when the system learns, it will change the connection strengths, $A[i,j]$ by the rule,

$$\delta A[i,j] \propto f[j]g[i],$$

or

$$\delta A[i,j] = \eta f[j]g[i],$$

where η is a learning constant.

   This is often called the *generalized Hebb rule*. It is also called the *outer product learning rule* because the connection matrix **A** is formed from the outer product between **g** and $\mathbf{f}^T$. Values given by this rule are defined for both positive and negative values. Taken literally, the definition given by Hebb requires both values in the product to be positive. As an initial assumption about learning, however, the outer product rule is a natural place

Simple Network Associator

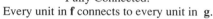

Input Pattern
f[j]
Modifiable Connections
A[i,j]
g[i]

f   **g**

Fully Connected:
Every unit in **f** connects to every unit in **g**.

Figure 6.7   Architecture of a specific type of vector associator: the linear associator. This very simple network has an input group of neurons connected by way of modifiable synapses to an output group of neurons.

to start, and it is easy to work with because it is concisely described by one of the fundamental matrix operations mentioned in chapter 5.

Now Print

Let us consider the mechanism for pattern association. Many years ago, Livingston (1967) hypothesized a "now-print" order, located separately from the learning synapses, that changed the biochemical weather in the brain sufficiently to cause the synapses to modify themselves according to whatever learning rule they were following. A number of conjectures have been put forward about what the now-print signal might be, and whether there is more than one. Some neural projection systems send out axons that spread widely over the brain and seem to release their transmitters "into space" from swellings in the axon, rather than as part of a synapse. One suggestion for some kinds of early learning is that cooperation exists between widely projecting systems using the neurotransmitter norepinepherine and one using the neurotransmitter acetylcholine (Bear and Singer, 1986).

One of the first things that is obvious from a consideration of both the mathematics and psychology of memory is that learning is dangerous. Most of life is boring, repetitive, and irrelevant, and it is not necessary to clutter up a memory system with its details, even a memory system with a large but still finite capacity. Many things are simply not worth learning. However, some events are critically important and have to be learned as quickly as possible. Often, biologically, these events are associated with danger, food, or sex, all of which have strong biochemical effects on the organism.

There is good evidence that we learn a little all of the time and a lot some of the time. The best example of the latter is what has been called the *flashbulb* memory. It usually is a traumatic, often unexpected event with a great deal of emotional involvement. For many years, "What were you doing when you heard about the assassination of John F. Kennedy?" was practically guaranteed to elicit a flashbulb memory. The explosion of the Challenger spacecraft was another more recent one. A car accident is a more individual example. The biochemical and psychologic upheavals involved in such events are unquestionable and perhaps correspond to the uncontrolled operation of the now-print system. The remarkable thing about flashbulb memories is their totality: everything is learned all at once, important things as well as unimportant things. It is as if something said, "Learn it all now and sort it out later." The ordinary mode of learning seems to have learning that is both more difficult and more selective.

The external control of learning has two important implications for modeling. First, it strongly suggests the learning parameter, η, is not constant, but under control of an external system. Second, and more important, an approximation to learning is the learning of associations of discrete pairs of events. That is, we assume learning occurs rarely, when conditions are appropriate. Then the magic now-print button is pressed, the now-print order is issued, and all of the synapses modify themselves. Then learning ceases until a now-print order is issued again. Learning of discrete pairs of events like this makes our modeling task very much easier and *may* not be a bad approximation to some of the structure of the real system. Learning discrete associations is so commonly assumed in artificial neural network theory that it is often not appreciated what a strange and unnatural assumption it actually is. It assumes that learning isolated flashbulb memories is the normal mode of operation. However, systems that learn continuously often have severe mathematical problems with stability, and this assumption is a good place to start.

Other practical problems may arise when an associator is realized in a neural structure. For example, some authors pointed out that a neural mechanism may be required to turn off old memory retrieval so that old memories do not interfere with learning new ones (Bower, 1991; Hasselmo, 1993). It was suggested that in olfactory cortex, acetylcholine can control the strength of entire groups of connections, allowing new learning and old retrieval to be separated to some degree.

Theory

Let us now make our discussion a little more formal. Given two sets of units, one projecting to the other, and connected by a matrix of connection strengths, **A**, we wish to associate the two activity patterns, **f** and **g** (figures 6.6 and 6.7). We assume **A** is composed of a set of modifiable connections.

We make two fundamental assumptions. First, the elements of **A** are changed during learning according to the generalized Hebb rule,

$$\Delta A[i,j] = \eta f[j]g[i].$$

Second, the unit linearly sums its inputs, as in figure 6.8, that is,

$$g[i] = \sum_j a[i,j]f[j],$$

allowing us to compute the output pattern, **g**, from the input pattern, **f**, by matrix multiplication,

$$\mathbf{g} = \mathbf{Af}.$$

Suppose that the matrix **A** starts with all values initially zero and a single association of **f** and **g** is formed, with the learning constant, η. Then **A** is given by

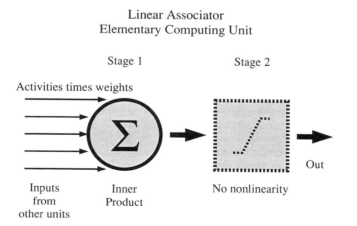

Linear Associator
Elementary Computing Unit

Stage 1   Stage 2

Activities times weights

Inputs   Inner   No nonlinearity
from   Product
other units

Figure 6.8 The elementary computing unit in the linear associator uses only the first box of the generic connectionist unit: it is linear and takes the inner product between input activity and synaptic weights.

$$\mathbf{A} = \eta \begin{bmatrix} f[1]g[1] & f[2]g[1] & f[3]g[1] & \cdots & f[N]g[1] \\ f[1]g[2] & f[2]g[2] & f[3]g[2] & \cdots & f[N]g[2] \\ f[1]g[3] & f[2]g[3] & f[3]g[3] & \cdots & f[N]g[3] \\ \cdots & & & & \\ f[1]g[N] & f[2]g[N] & f[3]g[N] & \cdots & f[N]g[N] \end{bmatrix}.$$

The elements are given by the product of the appropriate row and column elements of $\mathbf{f}^T$ and $\mathbf{g}$. This is the outer product matrix $\mathbf{A}$,

$$\mathbf{A} = \eta \mathbf{g} \mathbf{f}^T.$$

As we discussed in chapter 5, a transposed vector times an untransposed vector is an inner product,

$$[\mathbf{f}, \mathbf{g}] = \mathbf{f}^T \mathbf{g},$$

and a vector times the transpose of another is an outer product,

$$\mathbf{A} = \mathbf{g} \mathbf{f}^T.$$

Suppose after $\mathbf{A}$ is formed, pattern $\mathbf{f}$ is input to the system. Then since the matrix $\mathbf{A}$ has nonzero elements, a pattern, $\mathbf{g}'$ will be generated as the output to the system according to the simple matrix multiplication rule discussed before. This output, $\mathbf{g}'$, can be computed as

$$\mathbf{g}' = \mathbf{A}\mathbf{f},$$

$$= \mathbf{g}\mathbf{f}^T\mathbf{f},$$

$$\propto \mathbf{g},$$

since the square of the length, $\mathbf{f}^T\mathbf{f}$ or $[\mathbf{f}, \mathbf{f}]$, is simply a constant. Thus subject to a multiplicative constant, we have generated a vector in the same direction as $\mathbf{g}$. By normalizing the input vector, we can easily make this constant one for convenience, or we can specifically concern ourselves with the value of the constant. In either case, the cosine of the angle between $\mathbf{g}$ and $\mathbf{g}'$ will be 1, since they are in the same direction.

When more than one set of associations is formed and stored in the same matrix, the potential exists for interference between the associations. This observation leads to many of the immediately testable experimental predictions of the model.

Let us consider an important special case. Suppose we store a group of associations. We will assume that the input vectors, $\{\mathbf{f}\}$, are normalized, so we can assume the inner product, $[\mathbf{f}, \mathbf{f}]$, is 1. Suppose the set, $\{\mathbf{f}\}$, is composed of orthogonal vectors. We will assume that the learning constant $\eta = 1$, for convenience. Then, suppose each pair of associations

$$\mathbf{f}_i \rightarrow \mathbf{g}_i$$

generates an associative matrix

$$\mathbf{A}_i = \mathbf{g}_i \mathbf{f}_i^T,$$

and the overall connectivity matrix **A** is the sum of all the matrices of the individual associations,

$$\mathbf{A} = \sum_i \mathbf{A}_i.$$

It is easy to show that this associator works perfectly. Suppose a member of the set, $\{\mathbf{f}\}$, say, $\mathbf{f}_j$, with association, $\mathbf{g}_j$, and connection matrix, $\mathbf{A}_j$, one component of the sum, **A**, is the input to the system. Then the output is given by $\mathbf{A}\mathbf{f}_j$, or

$$\mathbf{A}\mathbf{f}_j = \sum_i \mathbf{A}_i \mathbf{f}_j$$

$$= \sum_{k \neq j} \mathbf{A}_k \mathbf{f}_j + \mathbf{A}_j \mathbf{f}_j$$

$$= \sum_{k \neq j} \mathbf{g}_k \mathbf{f}_k^T \mathbf{f}_j + \mathbf{g}_j,$$

since by assumption (or approximation!) the set $\{\mathbf{f}\}$ is orthogonal. That is, the inner product of $\mathbf{f}_j$ with any other vector $\mathbf{f}_k$ in the set is zero. Then all the terms in the sum are zero and we have shown

$$\mathbf{A}\mathbf{f}_j = \mathbf{g}_j.$$

If pairs whose inputs are orthogonal are stored, association is perfect and the system reconstructs the output pattern perfectly. Obviously the capacity in such a system has an upper limit. More than N orthogonal vector pairs cannot be stored, where N is the dimensionality of the vectors.

It can be seen from the structure of the system that the more orthogonal the input vectors to be learned, the better the system will work. If we take this approach to memory seriously, it means that there are considerable advantages to having mechanisms present that can orthogonalize patterns to some degree.

Numerical Example

Let us see what happens in a simple test system that we can compute by hand. Suppose we have four input elements and four output elements. Suppose we have a normalized input vector, $\mathbf{f}_1$, and an output vector, $\mathbf{g}_1$. The factor of 1/2 multiplying $\mathbf{f}_1$ is for normalization in a four-dimensional system. We will also define another pair of vectors, $\mathbf{f}_2$ and $\mathbf{g}_2$, for test vectors. Note that $\mathbf{f}_1$ and $\mathbf{f}_2$ are orthogonal.

$$\mathbf{f}_1 = \frac{1}{2}\begin{bmatrix} 1 \\ -1 \\ 1 \\ -1 \end{bmatrix} \qquad \mathbf{f}_2 = \frac{1}{2}\begin{bmatrix} 1 \\ -1 \\ -1 \\ 1 \end{bmatrix}$$

and

$$\mathbf{g}_1 = \begin{bmatrix} 1 \\ 1 \\ -1 \\ -1 \end{bmatrix} \qquad \mathbf{g}_2 = \begin{bmatrix} -1 \\ -1 \\ 1 \\ 1 \end{bmatrix}.$$

We can construct the connection matrix easily, according to our learning rule. Let us assume that the connection strengths are originally zero and we are associating $\mathbf{f}_1$ and $\mathbf{g}_1$. The learning constant, η, equals 1. Let us form a connection matrix $\mathbf{A}_1$.

$$\mathbf{A}_1 = \mathbf{g}_1\mathbf{f}_1^T = \frac{1}{2}\begin{bmatrix} 1 & -1 & 1 & -1 \\ 1 & -1 & 1 & -1 \\ -1 & 1 & -1 & 1 \\ -1 & 1 & -1 & 1 \end{bmatrix}.$$

The first thing we have to check is its associative function. According to our definition of association, if we put $\mathbf{f}_1$ as the input, then $\mathbf{g}_1$ should be the output. Checking this is easy,

$$\mathbf{A}_1\mathbf{f}_1 = \mathbf{g}_1\mathbf{f}_1^T\mathbf{f}_1 = \frac{1}{2}\begin{bmatrix} 1 & -1 & 1 & -1 \\ 1 & -1 & 1 & -1 \\ -1 & 1 & -1 & 1 \\ -1 & 1 & -1 & 1 \end{bmatrix}\frac{1}{2}\begin{bmatrix} 1 \\ -1 \\ 1 \\ -1 \end{bmatrix}$$

$$= \frac{1}{4}\begin{bmatrix} 4 \\ 4 \\ -4 \\ -4 \end{bmatrix} = \begin{bmatrix} 1 \\ 1 \\ -1 \\ -1 \end{bmatrix}$$

$$= \mathbf{g}_1,$$

just as we hoped it would be. We want the system to be selective, so that if we put in an input that it did not learn, the system will not respond to it. To check this, let us use $\mathbf{f}_2$, which was not learned, as an input and see what the output vector is.

$$\mathbf{A}_1\mathbf{f}_2 = \mathbf{g}_1\mathbf{f}_1^T\mathbf{f}_2 = \frac{1}{2}\begin{bmatrix} 1 & -1 & 1 & -1 \\ 1 & -1 & 1 & -1 \\ -1 & 1 & -1 & 1 \\ -1 & 1 & -1 & 1 \end{bmatrix}\frac{1}{2}\begin{bmatrix} 1 \\ -1 \\ -1 \\ 1 \end{bmatrix}$$

$$= \frac{1}{4}\begin{bmatrix} 0 \\ 0 \\ 0 \\ 0 \end{bmatrix}$$

$$= 0.$$

The connection matrix discriminates between $\mathbf{f}_1$ and $\mathbf{f}_2$.

Multiple Associations

Let us see if the matrix can learn more than one association. Suppose we form a matrix $\mathbf{A}_2$ by our learning rule:

$$\mathbf{A}_2 = \mathbf{g}_2\mathbf{f}_2^T = \frac{1}{2}\begin{bmatrix} -1 & 1 & 1 & -1 \\ -1 & 1 & 1 & -1 \\ 1 & -1 & -1 & 1 \\ 1 & -1 & -1 & 1 \end{bmatrix}.$$

It is important to realize that if we are learning by our rule, it does not matter what the preexisting connection strengths are, since we form a matrix of connection strengths based only on the input and output activity patterns, $\mathbf{f}$ and $\mathbf{g}$.

The most reasonable way to extend learning to multiple associations is simply to add the connection strength change matrices due to different associations. *This means that the strengths of individual connections are the sum of their history. Information from many associations is present in each connection strength.* Profound consequences follow from this assumption. Let us continue our numerical experiment and see if we can form a matrix capable of storing multiple associations.

By our earlier assumption, the complete connection matrix is the sum of the matrices from the two associations; that is,

$$A = A_1 + A_2 = \frac{1}{2} \begin{bmatrix} 0 & 0 & 2 & -2 \\ 0 & 0 & 2 & -2 \\ 0 & 0 & -2 & 2 \\ 0 & 0 & -2 & 2 \end{bmatrix}.$$

It is easy to show that

$$A\mathbf{f}_1 = \begin{bmatrix} 1 \\ 1 \\ -1 \\ -1 \end{bmatrix}$$

$$= \mathbf{g}_1$$

and

$$A\mathbf{f}_2 = \begin{bmatrix} -1 \\ -1 \\ 1 \\ 1 \end{bmatrix}$$

$$= \mathbf{g}_2.$$

So a matrix, A, which is a sum of matrices formed from individual associations, is capable of multiple associations. It can also discriminate against unlearned inputs, for example, random vectors, although it does not work very well because the dimensionality of the system is so small.

Heteroassociation and Autoassociation

We have described what is usually called *heteroassociation*, a term coined by Teuvo Kohonen. In heteroassociation, the input, $\mathbf{f}$, and the output pattern, $\mathbf{g}$, are not the same. If, however, $\mathbf{f} = \mathbf{g}$, the association of a vector with itself is referred to as an *autoassociative* system.

Suppose we are interested in looking at autoassociative systems storing outer products of the form

$$A = \eta \mathbf{f} \mathbf{f}^T,$$

where η is a learning constant (figure 6.9). It is easy to show that autoassociation will *reconstruct* a missing part of an input state vector. This is because the missing part is associated with the part that is present, and the

Autoassociative Network

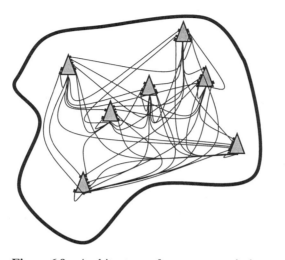

Figure 6.9   Architecture of an autoassociative network. In this case, there is only one set of units that connects to itself. Since different parts of an input vector are associated together, presentation of part of a learned pattern can give rise to the entire pattern, sometimes called the *reconstructive* aspect of neural network operation. This architecture is also used as the basis for some nonlinear networks. (See chapter 15.)

learned associative cross-connections built into the feedback matrix will fill in the missing element values.

Suppose we have a normalized state vector, $\mathbf{f}$, that is composed of two parts, say $\mathbf{f}'$ and $\mathbf{f}''$, that is, $\mathbf{f} = \mathbf{f}' + \mathbf{f}''$. Suppose $\mathbf{f}'$ and $\mathbf{f}''$ are orthogonal. The easiest way to accomplish this would be to have $\mathbf{f}'$ and $\mathbf{f}''$ be subvectors that occupy different sets of elements. Say $\mathbf{f}'$ is nonzero only for elements $[1 .. n]$, and $\mathbf{f}''$ is nonzero only for elements $[(n + 1) .. \text{Dimensionality}]$.

Then consider a matrix $\mathbf{A}$ storing only the autoassociation of $\mathbf{f}$, that is,

$$\mathbf{A} = \mathbf{f}\mathbf{f}^T$$

$$\mathbf{A} = (\mathbf{f}' + \mathbf{f}'')(\mathbf{f}' + \mathbf{f}'')^T.$$

Suppose at some future time part of the complete state vector, $\mathbf{f}$, say $\mathbf{f}'$, is presented at the input to the system. The output is then given by

$$(\text{output}) = \mathbf{A}\mathbf{f}'$$

$$= (\mathbf{f}'\mathbf{f}'^T + \mathbf{f}'\mathbf{f}''^T + \mathbf{f}''\mathbf{f}'^T + \mathbf{f}''\mathbf{f}''^T)\mathbf{f}'.$$

Since $\mathbf{f}'$ and $\mathbf{f}''$ are orthogonal, we can eliminate two terms and collect the other two, and (output) $= (\mathbf{f}' + \mathbf{f}'')[\mathbf{f}', \mathbf{f}']$,

$\qquad = \alpha \mathbf{f}$,

where α is a constant since the inner product, $[\mathbf{f}', \mathbf{f}']$, is simply a number. Therefore the system can reconstruct the missing part of the state vector in an autoassociative system. This is sometimes called the *reconstructive*, or the *holographic*, property of neural networks. The last name is based on a similar property, from similar causes, shown by optical holograms.

We will analyze this system in much more detail in chapter 15 when we discuss feedback models that are based on autoassociators. An obvious trick would be to reconstruct an autoassociated state vector from a given noisy or truncated example state vector, and pass the reconstructed state vector through the matrix again to see if the final result could be improved still more. It can be, and this process is the basis of *feedback* neural networks. We will defer analysis of feedback models until later chapters since these models contain significant nonlinearities in their dynamics, although they are basically autoassociative in the kinds of detailed computations they do.

Taylor Series

Digression. Two nice things about the particular form of connection strength modification we chose, the outer product of presynaptic and postsynaptic activity, are that it can generate an exact association and is easy to analyze. When the input vectors are orthogonal, recall is perfect. Is practical operation of the system critically dependent on the exactness of this learning rule?

Suppose we had a more general learning rule for connection strengths, so that

$$\delta a[i,j] = h(f[j], g[i]),$$

where $h(.)$ is some function. We can expand this function in a Taylor series.

The expansion contains an infinite sum of terms with all possible permutations of powers of $f[j]$ and $g[i]$. If we don't worry about the values of the constants of the Taylor series, the three lowest-order terms of the expansion will be

$$\delta a[i,j] = c_1 f[j] + c_2 g[i] + c_{11} f[j] g[i] + \cdots.$$

The first term is a traditional strengthening (or weakening) by use rule involving only the presynaptic term. We will discuss the implications of such a learning rule in chapter 7. It builds a simple recognition model. The second term is a different rule of the same kind involving only postsynaptic (output) activity. The third term is the rule we just analyzed: the Hebbian outer product linear associator that gives rise to the perfect association of orthogonal input vectors.

Higher-order terms containing only odd powers of $f[j]$ and $g[i]$ have similar qualitative behavior to the rule we analyzed, but they introduce distortions, since the output after learning is no longer an exact replica of the association learned. These higher-order terms can be incorporated to let the learning rule ignore small values of activity and enhance larger values, effectively constructing a learning threshold.

Higher-order terms containing even powers $f[j]$ and $g[i]$ no longer show the same qualitative pattern as the simple learning rule and must be analyzed more carefully. The linear associator may be a useful first approximation to a more complex learning rule.

History

As an historical note, many of the ideas that turned out to be fruitful in neural modeling, such as one or another variant of the Hebb synapse, distribution of activity, and parallel architecture, were invented by a number of workers independently and sometimes nearly simultaneously. For example, two early descriptions of the linear associator in the early neural network literature (Anderson, 1972; Kohonen, 1972, both reprinted in *Neurocomputing*) were published nearly simultaneously, but the authors were working independently and were not aware of each other until considerably after these papers appeared. A similar situation occurred more recently in the development of the backpropagation algorithm that was developed independently in several places.

The most complete analyses of this class of linear associative models are found in two books by Kohonen (1977, 1984) that give a great many extensions and examples. One of the predictions of the linear associator is the existence of "cross-talk" between nonorthogonal vectors. There should be strong interaction between separate memories, giving rise to a number of testable behavioral predictions. Chapter 11 provides a worked-out example of such a direct prediction. Gradient descent methods, such as the Widrow-Hoff least mean squares algorithm discussed in chapter 9, are effective extensions of the linear associator.

The major criticism of this class of models concerns their basic linearity. We will see why this is a problem when we discuss the perceptron. Neurons are not linear for large signals, and most later work building on the linear associator as a starting point included large signal nonlinearities. But the useful associative properties generated by the Hebb synapse carry over to more complicated systems: *Hebb synapses of virtually any type build associative systems.*

References

J.A. Anderson (1968), A memory storage model using spatial correlation functions, *Kybernetik, 5*, 113–119.

J.A. Anderson (1972), A simple neural network generating an interactive memory. *Mathematical Biosciences, 14*, 197–220.

J.A. Anderson (1973), A theory for the recognition of items from short memorized lists, *Psychological Review, 80*, 417–438.

M. Baudry and J.L. Davis (Eds.) (1991), *Long Term Potentiation: A Debate of Current Issues.* Cambridge: MIT Press.

M. Bear, L.N. Cooper, and F. Ebner (1987), A physiological basis for a theory of synaptic modification. *Science, 237*, 42–48.

M. Bear and W. Singer (1986), Modulation of visual cortical plasticity by acetylcholine and noradrenaline. *Nature, 320*, 172–176.

E.L. Bienenstock, L.N. Cooper, and P.W. Munro (1982), A theory for the acquisition and loss of neuron specificity in visual cortex. *Biological Cybernetics, 19*, 9–28.

J.M. Bower (1991), Piriform cortex and olfactory object recognition. In J. Davis and H. Eichenbaum (Eds.), *Olfaction as a Model System for Computational Neuroscience.* Cambridge: MIT Press.

T.H. Brown, E.W. Kairiss, and C.L. Keenan (1990), Hebbian synapses: Biophysical mechanisms and algorithms. *Annual Review of Neuroscience, 13*, 475–511.

T.H. Brown, P.F. Chapman, E.W. Kairiss, and Claude L. Keenan (1990), Long-term synaptic potentiation. *Science, 242*, 724–728.

T.H. Brown, A.M. Zador, Z.F. Mainen, and B.J. Claiborne (1991), Hebbian modifications in hippocampal neurons. In M. Baudry and J.L. Davis (Eds.), *Long Term Potentiation: A Debate of Current Issues.* Cambridge: MIT Press.

T.J. Carew and C.L. Sahley (1986), Invertebrate learning and memory: From behavior to molecules. *Annual Review of Neuroscience, 9*, 435–487.

E. Culotta and D.E. Koshland, Jr. (1992), NO news is good news. *Science, 258*, 1862–1865.

M.L. Feldman (1984), Morphology of the neocortical pyramidal neuron. In A. Peters and E.G. Jones (Eds.), *Cerebral Cortex*, Volume 1. *Cellular Components of the Cerebral Cortex*. New York: Plenum.

H. Gardner (1974), *The Shattered Mind*, New York: Vintage.

M.E. Hasselmo (1993), Acetylcholine and learning in a cortical associative memory. *Neural Computation, 5*, 32–44.

R.D. Hawkins, E.R. Kandel, and S.A. Siegelbaum (1993), Learning to modulate transmitter release: Themes and variations in synaptic plasticity. *Annual Review of Neuroscience, 16*, 625–665.

D.O. Hebb (1949), *The Organization of Behavior*. New York: Wiley.

W. James (1892/1984), *Briefer Psychology*. Cambridge: Harvard University Press.

E.R. Kandel (1976), *Cellular Basis of Behavior*. San Francisco: W.H. Freeman.

S.R. Kelso, A.H. Ganong, and T.H. Brown (1986), Hebbian synapses in hippocampus. *Proceedings of the National Academy of Sciences, 83*, 5326–5330.

A. Kirkwood, S.M. Dudek, J.T. Gold, C.D. Aizenman, and M.F. Bear (1993), Common forms of synaptic plasticity in the hippocampus and neocortex in vitro, *Science, 260*, 1518–1521.

T. Kohonen (1972), Correlation matrix memories. *IEEE Transactions on Computers, C-21*, 353–359.

T. Kohonen (1977), *Associative Memory—A System Theoretic Approach*. Berlin: Springer-Verlag.

T. Kohonen (1984), *Self-Organization and Associative Memory*. Berlin: Springer-Verlag.

T. Landauer (1986), How much do people remember? Some estimates of the quantity of information in long term memory. *Cognitive Science, 10*, 477–493.

W.B. Levy, J.A. Anderson, and S. Lehmkuhle (1985), *Synaptic Modification, Neuron Selectivity, and Nervous System Organization*. Hillsdale, NJ: Erlbaum.

W.B. Levy, C.M. Colbert, and N.L. Desmond (1990), Elemental adaptive processes of neurons and synapses: A statistical/computational perspective. In M. Gluck and D.E. Rumelhart (Eds.), *Neuroscience and Connectionist Theory*. Hillsdale, NJ: Erlbaum.

R.B. Livingston (1967), Reinforcement, In G.C. Quarton, T. Melnechuk, and F.O. Schmitt (Eds.), *The Neurosciences*, pp. 568–577, New York: Rockefeller University Press.

G.A. Miller (1956), The magic number seven plus or minus two: some limits on our capacity for processing information, *Psychological Review, 63*, 81–97.

R.G.M. Morris, S. Davis, and S.P. Butcher (1991), Hippocampal synaptic plasticity and N-methyl-D-aspartate receptors: A role in information storage. In M. Baudry and J.L. Davis (Eds.), *Long Term Potentiation: A Debate of Current Issues.* Cambridge: MIT Press.

E.M. Schuman and D.V. Madison (1994), Locally distributed synaptic potentiation in the hippocampus. *Science, 263,* 532–536.

L. Standing (1973), Learning 10,000 pictures. *Quarterly Journal of Experimental Psychology, 25,* 207–222.

C. von der Malsburg and E. Bienenstock (1986), Statistical coding and short term synaptic plasticity: A scheme for knowledge representation in the brain. In E. Bienenstock, F. Fogelman-Soulie, and G. Weisbuch (Eds.), *Disordered Systems and Biological Organization.* Berlin: Springer-Verlag.

C.S. Wallace, N. Hawrylak, and W.T. Greenough (1991), Studies of synaptic structural modifications after long-term potentiation and kindling: Context for a molecular morphology. In M. Baudry and J.L. Davis (Eds.), *Long-Term Potentiation: A Debate of Current Issues.* Cambridge: MIT Press.

The Linear Associator: Simulations

This chapter discusses some basic elements of neural network architecture. Behavior of any network depends on the statistics of what the network must learn and discriminate. A few important properties of random vectors are simulated. Supervised learning algorithms assume (unrealistically) that complete knowledge of input and output patterns is possible. This knowledge means that amount of error can be determined because the desired and the actual outputs can be compared. Several different measures of error are discussed. An elementary model of recognition memory shows that some general properties of networks are present even in very simple systems. Simulations of associative networks are described.

The linear associator we discussed in the last chapter illustrates the classic neural network architecture. When learning, we have a set of known input patterns that we want to associate with a set of known output patterns. We do this by modifying the weights of the connection matrix coupling two groups of neurons. We showed that for one simple case, when the input patterns were orthogonal to each other, a generalized Hebbian modification rule built a perfect associator.

We can proceed in many directions at this point. One common direction is to try to make the capacity of the network larger so it can store more pairs of associations, and to make the stored associations more accurate so that when a learned input pattern is presented, the network correctly reproduces the associated output. Besides giving a useful set of criteria to judge network performance, some measures of accuracy and capacity are easy and informative to compute. Moreover, sometimes use of these criteria allows network analysis to make direct contact with a large body of work in statistics and pattern recognition. Unfamiliar neural networks then become new versions of familiar problems. With respect to the linear associator and related models, the best and most complete analysis of this type is found in two books by Kohonen (1977, 1984).

One could claim, as has been done, that the function of neural networks and their learning algorithms is to produce architectures and sets of connection strengths that produce the "right" input-output relationships between

Supervised Learning

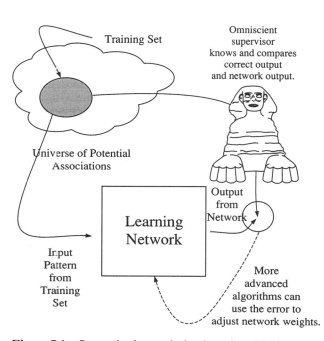

Figure 7.1 Supervised associative learning. We have a set of known associations, some or all of which form the *training set*. The network learns the training set using some rule for setting connection strengths. Supervised learning requires information about the correct answers from an omniscient supervisor.

patterns. In one sense this is a triviality, since what we want with any computing system is to put the data in one end and to get the right answer out the other. A peculiarity of neural networks is that they *memorize* the correct answers instead of computing them.

The most common approach to network learning is called *supervised learning in pattern recognition* (figure 7.1). Supervised learning takes the following form. We have a set of input patterns, $\{\mathbf{f}_i\}$, and a set of output patterns, $\{\mathbf{g}_i\}$. Either all the known input and output patterns, or a subset of them, form a *training set*. The network is shown the members of the training set, and the connection strengths are modified according to the learning rule that is being used. In the last chapter we used the outer product rule, that is, the generalized Hebb rule.

Next, we want to know how well our network works. We can use several classes of patterns to test it (figure 7.2). First, we can see how well the network does with members of the training set, that is, the patterns presented to the network during learning.

Testing After Learning

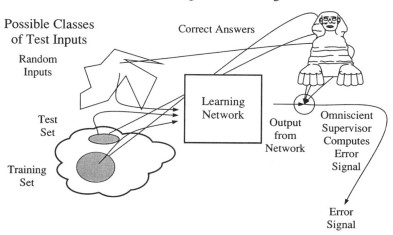

Figure 7.2 Once the network has learned the training set, its performance can be tested with various test inputs, that is, what it actually learned, new patterns of the same type it learned, to see if it can generalize, or, with other inputs to see if it can discriminate. The supervisor computes the error signal, since it knows the correct answer. Many kinds of error functions can be used.

Second, if we do not use all the pairs of input and output patterns that we have, we can test the network on pairs it did not learn. For example, we might have a number of samples of handwritten letters, say 100 As, 100 Bs, and so on. We train the network with 50 As, 50 Bs, and so on, and then test it both with the specific patterns it learned (training set) and with examples of letters that it has not seen (test set). Obviously, if we want to use the network to read hand-printed ZIP Codes from envelopes, it will have to give the correct output classification for digits it has never seen before. The ability to respond appropriately to novel patterns is called *generalization*, and it is supposedly one of the strengths of neural networks. We will see in later chapters (especially chapters 16 and 17), however, that correct generalization is an extremely subtle and difficult problem.

Third, and a test that is not used as often is it should be, we can give the network novel patterns, say random noise, or, a harder one, related patterns. For the letter example, we could present it Greek characters and see what happens. It is often extremely important to have a network decide that it does not know enough to make a decision. A practical example of this is asking a system to read handwritten numbers on checks. It is far better to return no answer at all than to read the wrong amount.

For all these reasons, we must have some way to measure accuracy of recall to see how well the network operates. We have assumed, since this is supervised learning, that we know exactly the desired output pattern. We can compute the output pattern that we actually obtain from the network and can compare it with the one we want. We next compute some kind of *error function* that tells us how well the network did. There are many possible error functions, and the one we choose has a profound effect on the kinds of systems we build, how we analyze them, and how we apply them. There is no "right" error measure.

Suppose the desired output pattern is **g**, which we know. Suppose the network produced **g'**. Perhaps the most commonly used error measure is computed from the difference vector, (**g** − **g'**), between the desired vector and what the network produced. This difference vector, (**g** − **g'**), is often called the *error vector* and has a length. Length is always equal to or greater than zero. Therefore, the closer the two patterns are to being identical, the smaller the length of (**g** − **g'**). Length is computed by summing the squares of the differences between elements, that is,

$$\text{Length}(\mathbf{g} - \mathbf{g'}) = \left[\sum_{i=1}^{\text{dim}} (g[i] - g'[i])^2 \right]^{1/2}.$$

If we have a lot of patterns, we can sum up the length of the error vector for each one and get the total error for the entire training set. When we discuss more complex learning rules than the Hebb rule, we will give rules that work by reducing this error measure to as small a value as possible by modifying network weights. The form of total error above is equivalent to the *least mean squares* criterion used in statistics for elementary operations such as regression; that is, finding the best-fitting straight line through a set of data points. "Best-fitting" generally means the line that reduces the sum of the distances between the line and the data points to a minimum. Unkind souls have commented that neural networks often solve linear or nonlinear regression problems by unusually complicated and time-consuming means.

One might also ask if there is any particular reason to form an error signal based on the sum of squares of differences between elements, rather than, for example, the sum of absolute values of differences between elements, or the sum of the fourth powers of differences, or any one of many other possibilities. The reason is largely psychological. Because of the structure of mathematics it is often possible to obtain simple closed-form expressions when a least mean squares criterion is used. Therefore an esthetic judgment is being pronounced: if it gives concise formulas, use it. This statement is more about human cognition than about abstract mathematics.

We can propose many other useful error criteria. For example, a vector has both a length and a direction. We noted in chapter 6 that a simple Hebb learning rule for orthogonal input patterns gives output patterns lying in the same direction as the correct association, but the output patterns may not be the right length because the length of the output depends on the learning constant, η.

In general, the length of the output vector is not well determined in simple Hebb learning. Consider our system that learns orthogonal vectors. If we learn the same vector *twice*, the associated output vector will be twice as long as a vector learned only once. If associations are learned many times, the length of the output patterns will keep increasing indefinitely. In a related problem, the size of the connection strengths will also increase indefinitely. This property may or may not be desirable.

A former favorite analogy was to compare the brain with a computer. However, a far better electronic analogy is to compare the brain with a radio. One can make an extended and relatively accurate comparison of many brain functions with different parts of an FM radio. Radios have many internal mechanisms to produce what is called automatic gain control, so that great differences in input signal strength do not affect the average amplitude of the output. A car radio, for example, may have the actual strength of the signal from the antenna rapidly vary by a factor of a thousand or even more as the car passes near a building, with no apparent change in volume from the loudspeaker.

Similarly, humans can understand speech and recognize objects visually when the average amplitude of the physical signal is varied over orders of magnitude. Many complex mechanisms working at all levels of the auditory and visual pathways collaborate to give us this ability. In FM radios the actual amplitude of the input signal is deliberately thrown away, and only changes in transmitted frequency are detected and used to reconstruct the audio signal. Automatic gain control (AGC) in AM or SSB radios usually involves feedback from later to earlier stages to control the overall gain of the receiver, so the average value of the output signal, as set by the volume control, remains constant.

Therefore, let us suggest two other reasonable criteria for error that we use in several places later in this book. First, suppose we have a desired output vector, **g**, and the output vector we actually obtained, **g'**. Instead of computing the length between **g** and **g'**, we could compute the cosine of the angle between them. The cosine is independent of the length of the two vectors, whereas the length of the error vector is not (figure 7.3). Perhaps we might reasonably think that reproducing the pattern is the hard part of the computation since it involves activities of many independent elements. Finding the best length, once the pattern is determined, involves only a single parameter, overall gain, that can be changed easily by many different

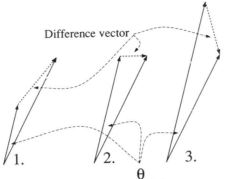

Difference vector

The angle, θ, is the same for all three cases.

The difference vector is different for all three cases.

Figure 7.3   Contrast between length of the error vector as a measure, and the cosine between actual output and correct output as an error measure.

means. Using the length of the error vector as the error criterion mixes amplitude and direction together.

Second, another possible error criterion is akin to how frequency is processed in an FM radio. Consider three sinusoidal waveforms of different amplitude (figure 7.4). These waveforms are very different from each other in terms of the mean square distance between them. However, they all cross the zero axis at the same points. These *zero crossings* give the frequency and are independent of the amplitude. Suppose one simply clipped the sine waves by using a limiter that cut off portions of the signal that are above or below upper and lower limits. The resulting signal—roughly a square wave—still contains the frequency information, but has discarded amplitude information (figure 7.5). Note the three sine waves look almost identical after *limiting*. Something like this trick is used in the limiters in FM radios. Therefore, we can form another error measure between the desired output, **g**, and actual output, **g′**, by checking to see if the sign of the activities are the same for each element. Again, we don't care about the amplitude of the activity. This error measure is hard to analyze mathematically, but easy to implement on a computer, and as we will see, lets us make easy comparisons of complex and high-dimensionality output patterns.

Some vision scientists have suggested that images can be best characterized by qualitative information related to zero crossings of various functions computed from the image, and not to detailed amplitude patterns (Marr, 1982). This cavalier approach to the length of the output vector may strike some readers as odd. For example, doesn't the size of the amplitude of the activity pattern give rise to perceived intensity, say, the loudness of a sound or the brightness of a light? Without going into details of biology, the an-

Three Sine Waves
of Different Amplitude

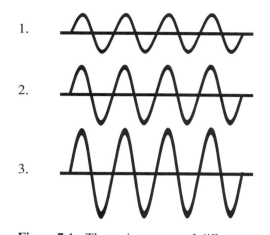

Figure 7.4 Three sine waves of different amplitudes. If we subtract two of these vectors there is a substantial error vector.

Same Sinewaves
Clipped

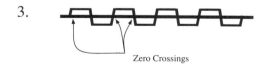

Zero Crossings

Figure 7.5 After clipping, the difference between the sine waves becomes very small. The zero crossings of the waveform remain the same.

swer is absolutely not. As only one example, anyone who has tried to take photographs without an exposure meter knows that subjective estimates of actual physical light intensity are extremely poor.

Given that we have an associator that links input patterns and output patterns, we will next do a series of computational experiments to see how well it works. First we will demonstrate some elementary properties of the inner product that underlie many of the important information-processing powers of neural networks. Then we will discuss an even simpler memory than the linear associator, but one whose mathematical properties actually do much of the work in the linear associator. Finally, we will look at the linear associator itself and see how well it works.

Properties of the Inner Product

We said earlier that much of the power of all neural networks, including the linear associator, arises from the selectivity of the inner product. The first stage of the generic neural network computing unit presented in chapter 2 takes an inner product between the input pattern and the connection strengths. We showed that the linear associator worked perfectly when it learned associations whose input patterns were orthogonal. Of course, this constraint is unreasonable. Let us consider a slightly more realistic case. Suppose the components of the input vectors are simply random, independent values taken from some distribution with mean zero. In that case, the inner product between pairs of such vectors is zero *on the average*, since the product of two independent random variables with zero mean also has mean zero. The larger the dimensionality, the better the approximation to average orthogonality. This observation follows from the definition of statistical independence. If the inner product is zero on the average, qualitatively, many of the same results follow as for orthogonality, although now when associations between nonorthognal vectors are formed and stored in the same matrix, there is interference from the other associations. This introduces noise into the system, and it is no longer a perfect associator. We will test this approximation in PROGRAM Inner_product_demo. To avoid cluttering up the text with program details, here and in the future we will only point out highlights of the programs, and give more complete listings in the appendixes. This program can be found in appendix D.

Although it was easy to analyze the orthogonal case in chapter 6, we are all rightfully suspicious of mathematical results, especially of idealizations (aren't we?). The structure of the test program is straightforward. PROCEDURE Make_normalized_random_vector generates a random vector whose elements are taken from a distribution with mean zero. The program then normalizes the resulting vector. A pair of such vectors is generated, and the inner product between them computed. This computation is done a number of times, and the resulting inner products are plotted as a simple histogram.

We have also included an elementary character-based histogram-plotting **PROCEDURE** Plot_histogram that plots the inner products. If a line of the histogram is going to run off the screen, that is, it contains over about 70 characters, a '+' is printed in the last screen column. The actual number of entries in a bin is also printed. The scale of the histogram can be expanded or contracted. This simple routine is useful for quick plots of data, and except for the esthetics involved, is as valuable as fancier routines and easier to modify.

Use of **PROGRAM** Inner_product_demo is straightforward. The program will prompt for the dimensionality and the number of dot products to compute. A dimensionality of 200 with 150 dot products is a good start. Figure 7.6 shows the relationship between width of the distribution of inner products and the number of elements in the vector.

Observe how the histogram contracts as the dimensionality increases. It becomes very narrow when the dimensionality grows large, indicating that our approximations are better for larger systems. The width of the histogram decreases as the square root of the dimensionality. We can see why this is so. We normalized the vectors in the program, which means that the sum of the squares of the elements is equal to 1, that is,

$$\text{mean squared element value} = \frac{1}{\text{dimensionality}}.$$

The mean of the elements is zero on the average. This relation is approximately equivalent to saying that the mean of the square of the elements is equal to the mean variance. (The actual calculation is slightly more complex, but the difference can be ignored if the dimensionality of the system is large.)

The terms in the inner product, which is what we are computing, will then be approximated by the sum of products of elements with zero mean whose variance is the product of the variances of the individual terms; that is, for a single element,

$$\text{variance of the mean squared product} = \frac{1}{(\text{dimensionality})^2},$$

and whose mean is zero. When we add all the terms together to form the inner product, the variance of the sum of the products is the sum of the variances of the elements, or

```
-0.20    0:
-0.18    0:
-0.16    1:*                              Inner Products of
-0.14    2:**                             Random Vectors
-0.12    3:***
-0.10    2:**
-0.08    7:*******
-0.06    5:*****
-0.04    5:*****
-0.02    9:*********
 0.00    9:*********
 0.02   11:***********
 0.04    7:*******
 0.06    6:******                          250 Dimensional
 0.08    7:*******
 0.10    5:*****
 0.12    1:*
 0.14    0:
 0.16    0:
 0.18    0:
 0.20    0:
```

```
-0.10    0:
-0.08    0:
-0.06    0:
-0.04    3:***
-0.02   20:********************
 0.00   36:************************************
 0.02   19:*******************
 0.04    1:*
 0.06    1:*                              2,500 Dimensional
 0.08    0:
 0.10    0:
```

```
-0.10    0:
-0.08    0:
-0.06    0:
-0.04    0:
-0.02    9:*********
 0.00   66:******************************************************************
 0.02    5:*****
 0.04    0:
 0.06    0:
 0.08    0:                               25,000 Dimensional
 0.10    0:
```

variance of the inner product $= \dfrac{1}{\text{dimensionality}}$.

The standard deviation, which is the customary measure of the width of the histogram generated by the program, is then proportional to the square root of the dimensionality. This relationship can be verified qualitatively by running a few simulations.

Element Distributions

We can experiment with the distribution that we used to derive the vector elements. The plots in figure 7.6 are what we obtained when the elements of the vectors were taken from a uniform distribution. Since the vectors are normalized, their tips all lie on what is called the *unit hypersphere*, that is, a high-dimensionality sphere centered at zero and of radius 1. Since we started with a uniform distribution of element values, it is tempting to say that the distribution of vectors is uniform on the unit hypersphere, but this is not the case.

Consider a two-dimensional vector, where the two elements are given by values of two uniformly distributed random components. Intuitively, a vector generated from uniformly distributed random components means that initial vectors are uniformly distributed on a *square*. The probability that a given vector will lie in a particular region of the square is only a function of the area of the region, and not where it is located on the square. If we normalize the vectors, we force their length to lie on the unit circle. But there is more area in the square near the diagonal than near edges, so it will be more probable that vectors lie near the diagonal than elsewhere. Figure 7.7 shows an intuitive way of seeing this. Regions of the square formed by different equal angle segments do not have equal area. Since the probability of a vector lying in a segment is strictly a function of the area of the segment, the angular distribution of the random vectors is not uniform.

It is easy and instructive to modify the PROCEDURE to change the distribution and see if the histograms change shape. Suppose Random is a uniformly distributed random variable between zero and 1. Then (Random − 0.5) is uniformly distributed with mean zero. Then, replacing the line generating the initial element values,

FOR I := 1 TO Dimensionality DO V[I] := Random − 0.5,

◀ **Figure 7.6** Distribution of values of inner product between 80 pairs of random vectors of zero mean for different dimensionalities. Note the distribution contracts as the dimensionality increases.

Random Points in the Square

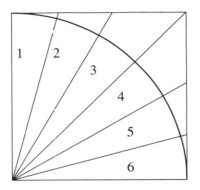

Figure 7.7   Choosing random independent components of a vector and then normalizing the vector does not produce a uniform distribution on the unit hypersphere. Choosing pairs of random, uniformly distributed components produces vectors equally likely to be anywhere on the square. If the vectors are then normalized, so they must lie on the circle, there are many more vectors in equiangle arcs near the diagonal, since the initial area is greater there.

with

R := Random − 0.5;

FOR I := 1 TO Dimensionality DO V[I] := R*R*R;

or

R := Random − 0.5;

FOR I := 1 TO Dimensionality DO V[I] := R*R*R*R*R;

in the PROCEDURE, and normalizing, will give different distributions of element values. Other distributions can be tried. Generating element values with a normal distribution requires a little more effort, but an efficient algorithm is given in Knuth's classic book, *The Art of Computer Programming,* Vol. II, (1981, p. 113). It is also easy to develop a technique that does generate random vectors uniformly distributed on the unit hypersphere (an exercise for the reader).

Experimenting with the distribution will display the important result that the exact distribution of the elements does not seem to matter very much for the inner product. As long as there are reasonable numbers of elements in the vectors, the resulting histograms look pretty much the same. In fact, they look like the normal distribution with mean zero. This simulation has become a demonstration of the famous central limit theorem, which states that the sum of a number of independent random variables from any reasonable distribution will approach the normal (Gaussian) distribution. The normal distribution of statistics is given by

$$n(x) = \frac{1}{\sigma(2\pi)^{1/2}} e^{-(1/2)/((x-\xi)^2/\sigma^2)}$$

where $n(x)$ is the probability density function at x, ξ is the mean, and σ is the standard deviation. Although the central limit theorem is frequently invoked both in models and to explain data, it is quite difficult to prove (Feller, 1966).

Since the normalization process in the simulation has equated the variances, and the means of the distributions are zero, we predict that, if the dimensionality is sufficiently large, the shapes for our histogram will look about the same no matter what distribution produces the element values. Figure 7.8 shows histograms of inner products for two different distributions of element values, the first a uniformly distributed random variable, and the other with elements given by the fifth power of a uniformly distributed random variable. For reference, the third histogram is generated from a normal distribution that is scaled to match the simulated histograms in mean and variance.

An important lesson to be drawn from this discussion is that sometimes the details assumed for a simulation really do not matter very much. The statistics involved erase the fine structure. On the positive side, this means that aspects of the behavior of large systems can be predicted with inadequate knowledge of the details. For example, it is likely that we will not have to know every last detail about brain structure to understand how the brain works. We have illustrated this point here with respect to the inner product. However, this observation is true as well for many nonlinear systems where wide classes of models will have almost identical behavior (see chapter 15). On the negative side, however, very different initial assumptions may become indistiguishable in their predictions. Therefore simulations can become much less valuable as a tool for discriminating between assumptions than one might hope.

Summed Vector Memories

At the end of chapter 6 we pointed out that the generalized Hebbian rule might be considered as the third term, the first term dependent on both presynaptic and postsynaptic activity in a series expansion of a more general modification function. Let us look at a model arising from a modification rule that contains only a single activity, with no interaction. We called it the

```
-0.130   130:*                                            First Power
-0.120   269:**
-0.110   461:***
-0.100   718:****
-0.090  1143:********
-0.080  1879:************
-0.070  2624:****************
-0.060  3641:***********************
-0.050  4690:******************************
-0.040  5891:*************************************
-0.030  6962:********************************************
-0.020  8179:***************************************************
-0.010  8717:*******************************************************
 0.000  9021:*********************************************************
 0.010  8565:******************************************************
 0.020  8066:***************************************************
 0.030  7149:*********************************************
 0.040  5977:**************************************
 0.050  4809:******************************
 0.060  3694:************************
 0.070  2554:****************
 0.080  1823:************
 0.090  1230:********
 0.100   737:****
 0.110   457:***
 0.120   251:**
 0.130   124:*
-------------------------------------------------------------------
-0.130   134:*                                            Fifth Power
-0.120   233:**
-0.110   441:***
-0.100   715:****
-0.090  1259:********
-0.080  1840:************
-0.070  2717:*****************
-0.060  3641:***********************
-0.050  4785:******************************
-0.040  6068:**************************************
-0.030  7085:*********************************************
-0.020  8025:***************************************************
-0.010  8689:*******************************************************
 0.000  8859:********************************************************
 0.010  8550:*****************************************************
 0.020  8111:***************************************************
 0.030  7105:********************************************
 0.040  5849:*************************************
 0.050  4741:*****************************
 0.060  3668:***********************
 0.070  2625:****************
 0.080  1848:************
 0.090  1185:********
 0.100   753:****
 0.110   420:***
 0.120   242:**
 0.130   137:*
```

Figure 7.8 The central limit theorem in action. The distribution of the histogram of 100,000 inner products of pairs of 500-dimensional random vectors. The elements of the vectors in the upper left histogram are drawn from a uniform random distribution. The elements of the vectors in the lower left histogram are computed from the fifth power of elements drawn from a uniform distribution. The right distribution is a normal distribution scaled to match the left distributions in mean and variance. The distributions are essentially identical.

```
-0.130  132:*                                          Normal Distribution
-0.120  246:**
-0.110  437:***
-0.100  737:*****
-0.090 1184:********
-0.080 1808:************
-0.070 2627:******************
-0.060 3632:************************
-0.050 4777:*******************************
-0.040 5978:*************************************
-0.030 7116:**********************************************
-0.020 8060:****************************************************
-0.010 8686:*********************************************************
 0.000 8905:*************************************************************
 0.010 8686:*********************************************************
 0.020 8060:****************************************************
 0.030 7116:**********************************************
 0.040 5978:*************************************
 0.050 4777:*******************************
 0.060 3632:************************
 0.070 2627:******************
 0.080 1808:************
 0.090 1184:********
 0.100  737:*****
 0.110  437:***
 0.120  246:**
 0.130  132:*
```

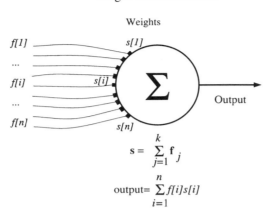

Summed Vector Memory
Single Unit Realization

Weights

$$\mathbf{s} = \sum_{j=1}^{k} \mathbf{f}_j$$

$$\text{output} = \sum_{i=1}^{n} f[i]s[i]$$

Figure 7.9 A single unit with a strengthening-by-use learning rule realizes a simple memory called a summed vector memory. The weights, $s[i]$, are the sum of their history. In this case, the members of the learned set are formed by summing the k vectors, $\mathbf{f}_j$, that are learned. The dimensionality is n.

summed vector model (Anderson, 1968). It is often not appreciated that non-Hebbian synapses can serve as useful, although limited, memories. A specific example is the system that arises when strengthening-by-use synapses are assumed. A crude description of this rule might be the synapse is a muscle; the more the synapse is activated the stronger it gets. (Or the weaker it gets. Some habituation phenomena seem to involve synaptic weakening with use.) A conjunction between presynaptic and postsynaptic activity is not necessary. We assume the overall synaptic strength is simply the sum of its history of presynaptic activation. Note, by the way, a converse observation: if indeed there are strengthening or weakening-by-use systems, they give rise to pattern recognizers of the type we discuss next.

Let us realize such a model with a single linear neuron (figure 7.9). This unit computes an inner product at its synapses, between the strengths of its connections and the input pattern. We will call **s** the vector giving the strength of the connections, so the value of the ith single connection is $s[i]$. We will assume a modification by use learning rule, so that when a pattern, **f**, is learned by the neuron, the change in the strength of the ith connection, $\Delta s[i]$, is given by

$$\Delta s[i] = f[i].$$

(We will assume that the learning constant is 1 for illustration, and that the mean of the distribution of element values is zero.)

Information Storage

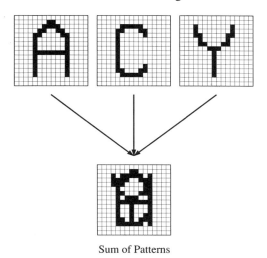

Sum of Patterns

Figure 7.10   Storage of information in the summed vector memory. The three letter patterns are summed to form the memory, *s*.

If we start from zero and if we have K memories, $\{\mathbf{f}_1, \mathbf{f}_2, \ldots \mathbf{f}_K\}$ that we want to store, the memory vector, $\mathbf{s}$, the connection strengths, is the sum of the K memories, that is,

$$\mathbf{s} = \sum_{i=1}^{K} \mathbf{f}_i.$$

This rule could easily be made Hebbian by saying the unit output is clamped at one when a pattern is to be learned and is zero otherwise. Figure 7.10 illustrates how a memory vector, $\mathbf{s}$, can be constructed from three input patterns. Notice that the sum of the input patterns looks rather random. In fact, when this model is implemented using the sum of random vectors, the central limit theorem suggests that the distribution of memory element values, $s[i]$, will look reasonably normally distributed. Information is still present, however, in what looks at first like random noise.

It is easy to store information in the system, but it is quite difficult to get information out, since any partcular stored vector is interfered with by the presence of the other stored information. This learning rule shifts the computational burden of memory from storage to retrieval. So much information has been lost that getting out what is left is challenging. In fact, this is true for all neural networks: they shift the burden of the system from learning to retrieval.

Learning is easy because it is dependent only on locally available information, simple unit activity for the summed vector model or a conjunction of local activity for the Hebb synapse. Because information is mixed together at the storage elements in all neural networks, however, unambiguous retrieval of stored information is not simple. For example, almost all network models assume that a network output is provoked by an input pattern. An input pattern contains a very large amount of information that determines the output pattern. It is not possible to "browse" through the values used to store a neural network memory, as can be done in a computer, a library, or any system in which information from different events is kept separate.

The summed vector model has strong similarities to more complex matrix models. The *rows* of the connection matrix in the linear associator form summed vector memories of this type, weighted by the associated output activities. This observation is used when we discuss prototype forming networks in chapter 11. As we said earlier, the ability of neural networks to discriminate between patterns is based on the selectivity of the inner product. Suppose we have k stored patterns, $\{\mathbf{f}_j\}$. Suppose one of the stored patterns, $\mathbf{f}_i$, is presented to the input of our single-neuron memory. The output of the neuron is given by

$$\text{output} = [\mathbf{s}, \mathbf{f}_i] = \sum_{j=1}^{k} [\mathbf{f}_j, \mathbf{f}_i] = [\mathbf{f}_i, \mathbf{f}_i] + \sum_{i \neq j} [\mathbf{f}_j, \mathbf{f}_i].$$

For illustration, let us assume that the stored patterns are orthogonal (or at least independent and of high dimensionality), normalized, random vectors of mean zero and length 1. The first term above is then 1, since it is the square of the length of $\mathbf{f}_i$. The second term is exactly zero if the stored patterns are orthogonal, or close to zero if the patterns are independent.

The input pattern, $\mathbf{f}_i$, acts as a *template*, as shown in figure 7.11. If this pattern is present, the size of the inner product is positive and large. If this pattern is not present in the memory vector, $\mathbf{s}$, the output of the inner product is small. The input pattern is used as a *filter* to detect its presence or absence in the memory vector. The size of the inner product provides some measure of how likely the input vector is to be present in the memory.

When the retrieval problem is formulated this way, it becomes equivalent to a classic problem in signal-detection theory: how best to determine the presence or absence of a signal of known form ($\mathbf{f}_i$) in the midst of noise (other stored information). The inner product template match realizes the spatial version of what is called the *matched filter*, which is the statistically optimum linear filter for detection. The matched filter is used in many signal-processing applications because of its simplicity and good performance (Davenport and Root, 1958).

Retrieval

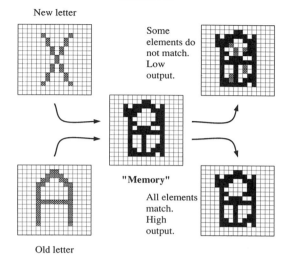

Figure 7.11   Retrieval of information in the summed vector memory. The old letter, A, is a perfect match and there is a high output. The new letter, X, does not match the memory in many places and there is a low output from the summer. Such a simple model is surprisingly effective as a recognition memory.

Detailed information has been lost in the act of storage, and we cannot reconstruct it as we could in the associative model. However, if we only want to perform something such as a computation of familiarity or recognition—have I seen this pattern before?—the model works nicely. Recognition involves knowing whether or not a new input has been presented before and is stored in the memory. Suppose we have an input, $\mathbf{f}_i$, that might or might not have been stored in the memory. We know exactly the signal we are looking for, because it is right there at the input to the system. Then the matched filter output is given by

matched filter output $= [\mathbf{s}, \mathbf{f}_i]$.

As we showed, this inner product gives a large positive output if the memory is present, and if it is not present, the output will be small on the average.

The matched filter in this system is effective in practice. This model also predicts a host of interference effects. If similar items are stored, a natural prototype forming system is produced that responds best to the average of the correlated cluster of state vectors. We will discuss this effect in chapter 11. This model and extensions of it have been used as the basis for several psychological models of recognition memory and simple concept formation. It was first proposed by Anderson in 1968, since then it has been used as a model of recognition memory for lists of words (Anderson, 1973), and as the basis of a theory of concept formation (Knapp and Anderson, 1984).

Program SVDEMO

It easy to construct a program to simulate this model. The program is a straightforward realization of the model. It starts by constructing random vectors with zero mean and length 1 using PROCEDURE Make_random_vector. These are added together to form a memory. A number of other test random vectors are constructed the same way. Then the inner product of the learned and the unlearned vectors are computed, and the results plotted in a rudimentary histogram display. If not too many vectors are learned compared with dimensionality, good separation is possible between the learned and unlearned vectors when the filter output is computed. The bulk of the PROCEDUREs and FUNCTIONs used in SVDEMO are identical to those that we presented in earlier chapters; the program itself can be found in appendix E.

Because of its simplicity, it is easy to analyze how well this model works for recognition. A commonly used measure of system performance in signal detection is the *signal to noise* ratio. One definition of average signal to noise ratio for our system is the ratio of the average squared output when a vector has been stored (signal present), over the filter output when the vector has not been stored (signal absent). For example,

$$(S/N) = \frac{\text{(average squared output with signal present)}}{\text{(average squared output with signal absent)}}.$$

The outputs are squared, because, of course, the average filter output with no signal is zero. Another common definition of signal to noise ratio uses the square root of the average squared output, the *root mean square* (*rms*) value.

The computation of signal to noise ratio is not difficult and can be found in Anderson (1972, 1973). Similar computations have been performed for a number of simple network memories, and the same functional relationship is common, although sometimes the value below is multiplied by a constant (Willshaw, Buneman, and Longuet-Higgins, 1969; Palm, 1982). If the dimensionality of the system is n and there are k stored independent vectors, the signal to noise ratio is roughly

$$(S/N) = \frac{n}{k}.$$

This expression, or the square root of it, arising from the alternative definition of signal to noise ratio, indicates that system operation improves as the dimensionality (*n*) increases, and becomes poorer as the number of stored items (*k*) increases, both perfectly intuitive results. Thus for this entire class of models, bigger memories might work better. Although this result seems trivial, many artificial information-retrieval system show the opposite behavior: the bigger they are the worse they work. Large libraries and data bases are good examples.

The computer program SVDEMO will repay a little experimentation. It shows that a system can work quite nicely if *k* is less than a few percent or so of dimensionality. Figure 7.12 shows several 2000-dimensional simulations as the number of stored vectors is increased. The histograms display the output of the system when the stored items are used as input ('*'), and when new random vectors are used ('+'). The histograms for new and old items are well separated until a large number (100) of items is stored.

Associative Memory

Next we will simulate the properties of the linear associator. We immediately run into the problem we mentioned earlier—how do we know the system is working? Is it possible to look at a large state vector generated as an output of the associator and say that the output is correct? First, we will compute statistical properties of the output vector, specifically its direction and its length. Second, we will represent inputs and outputs as strings of characters, so we can use our human abilities to read strings of letters to evaluate the accuracy of output patterns. These two measures correspond to the two error measures we discussed earlier.

Program Amdemo

Program AMDEMO, to test the simple linear associator, is given in appendix F. First, we generate a number of zero mean normalized random vector pairs using the PROCEDUREs that we used previously. Pairs of vectors are called F and G. Then we form the outer product matrix, Delta_A, between the pair using PROCEDURE Outer_Product. Delta_A is added to the connection matrix called A.

As the system learns, it is tested. We will look first at the direction of the output, and measure the cosine between the stored output, and what we actually obtained when we used the associator. After each change in the connection matrix, the input, **f**, of the pair that was just learned is used as a test input. If learning was noise free, presentation of **f** should lead to exactly **f** at the output. The cosine between **Af** and **g** (i.e., between actual output and desired output) should then be exactly 1. Since the input set is not composed

```
-0.2                                          10 Stored Vectors
-0.1 +++++++-+++++++++++++++++
 0.0 +++++++-++++++++++++++++++++++++++++++++++++++++++++++++
 0.1 +++++++-++++++++++++
 0.2 ++
 0.3
 0.4
 0.5
 0.6
 0.7
 0.8
 0.9
 1.0 *******
 1.1 *
 1.2 **
 1.3
-----------------------------------------------------------------------
-0.4                                          25 Stored Vectors
-0.3 +
-0.2 +++++++++
-0.1 +++++++++++++++++++++
 0.0 ++++++++++++++++++++++++++++++
 0.1 +++++++++++++++++++++++++++++
 0.2 +++++++++++
 0.3 +
 0.4
 0.5
 0.6
 0.7
 0.8
 0.9 *****
 1.0 *******
 1.1 ******
 1.2 *****
 1.3 **
 1.4
-----------------------------------------------------------------------
-0.5 +                                       100 Stored Vectors
-0.4 +++++
-0.3 +++++
-0.2 +++++++++++++
-0.1 ++++++++
 0.0 ++++++++++++++
 0.1 +++++++++++++++++++++
 0.2 +++++++++++++++
 0.3 ++++++++
 0.4 ++++++
 0.5 +++
 0.6 ***
 0.7 ********
 0.8 *******
 0.9 ***************
 1.0 *********************
 1.1 *****************
 1.2 ************
 1.3 *****
 1.4 ****
```

Figure 7.12 Simple demonstration of the summed vector memory, using zero mean random vectors. This figure is composed of an edited version of the output of SVDEMO. The vertical axis gives the inner product between the memory and the test vectors. The '*' signs are the inner product values of learned vectors and the '+' are inner product values of random test vectors. All the simulations are 2000 dimen-

```
Test of associative memory.
250 Dimensions

Pairs learned:    1.  Cosine between G[ 1] and actual output: 1.00000
Pairs learned:   11.  Cosine between G[11] and actual output: 0.98099
Pairs learned:   21.  Cosine between G[21] and actual output: 0.95377
Pairs learned:   31.  Cosine between G[31] and actual output: 0.93498
Pairs learned:   41.  Cosine between G[41] and actual output: 0.94238
Pairs learned:   51.  Cosine between G[51] and actual output: 0.91490
Pairs learned:   61.  Cosine between G[61] and actual output: 0.94540
Pairs learned:   71.  Cosine between G[71] and actual output: 0.88384
Pairs learned:   81.  Cosine between G[81] and actual output: 0.89217
Pairs learned:   91.  Cosine between G[91] and actual output: 0.84158
Pairs learned:  101.  Cosine between G[101] and actual output: 0.85240
```

Figure 7.13 Simple demonstration of the linear associator. This 250-dimensional system learned to associate pairs of random vectors. The output gives the cosine between the desired output vector and the actual output vector. Only every tenth pair is reproduced. Reproduction is perfect when only 1 pair is learned and deteriorates as 100 pairs are learned. This figure is edited output from program AMDEMO.

of orthogonal vectors, but statistically independent vectors, this will not happen, and recall will not be exact. The cosine between actual and stored output is computed and displayed. The cosine drops steadily as more and more associations are learned, but stays reasonably high if the dimensionality of the system is large compared with the number of stored associations.

After learning is complete, the final connection matrix, A, is tested with all the learned vectors. The cosine between the actual output and the desired output is computed and displayed. Note that the first association learned has a cosine of 1.000 when only one association was present, but after several are learned, the cosine drops considerably. The program gives as output a histogram of the cosines for learned patterns. Figure 7.13 gives a typical example of the output of this program.

Demonstration Amdemol

The linear associator discriminates against novel random vectors, just as the summed vector memory did, and for the same reasons. We can see this if we look at the length of the output vector, as can be seen in the next demonstration program, AMDEMOL. This program is similar to AMDEMO (details are in appendix G). The learning process is identical.

Instead of the cosine between actual and desired output vectors, the length of the output vector is computed. Since the system is learning nor-

◄   sional. The top histogram was from a system that learned 10 vectors, the middle from a system that learned 25 vectors, and the bottom from a system that learned 100 vectors.

```
0.0                                                      10 Learned Pairs
0.1  ++++
0.2  +++++++++++++++++++++++++++++++++++++++++
0.3  ++++++++++
0.4
0.5
0.6
0.7
0.8
0.9
1.0  **********
1.1
1.2
1.3
1.4


------------------------------------------------------------------

0.0                                                      25 Learned Pairs
0.1
0.2  +++
0.3  ++++++++++++++++++++++++++++++++++++
0.4  +++++++++++++++++
0.5
0.6
0.7
0.8
0.9
1.0  **************
1.1  **********
1.2
1.3
1.4


------------------------------------------------------------------

0.0                                                      50 Learned Pairs
0.1
0.2
0.3
0.4  ++++++++++++++++++++++++
0.5  ++++++++++++++++++++++
0.6  ++++
0.7
0.8
0.9
1.0  ****
1.1  ****************************************************
1.2  ****
1.3
1.4
```

Figure 7.14 The length of the output of the linear associator can be used to discriminate between learned and unlearned input vectors. This 250-dimensional system plots the length of the output vector for 10 learned pairs (upper) and 50 unlearned vectors, and for 25 learned pairs (middle) and 50 unlearned vectors. This histogram is an edited version of the output of program AMDEMOL.

malized random vectors with length of 1, the output length should also be 1. Because of the noise introduced by the learning of nonorthogonal input vectors, the length is not exactly 1, but close to it, when small numbers of associations are stored. The length of the output vector for new, unlearned, normalized random vectors is small. This program uses some screen control procedures and now plots a histogram of output lengths to make the length descriminations more graphic (figure 7.14).

The rows of the matrix, **A**, are summed vector memories. When a random vector is input to the system, the length of the output vector will be quite small on the average because it does not match any of the stored vectors, as we showed in SVDEMO. The system is acting like an amplitude filter for learned vectors, and will only give a high-amplitude output if the input vector is a stored item.

If we were to base a recognition decision—that is, has the input item been seen before—strictly on the basis of length of output, it would be possible to do quite well even with large numbers of stored pairs. In fact, a little experimentation would show that the system could respond accurately to familiarity, based on length of output vector, even with poor retrieval performance as measured by cosines. This is at least qualitatively consistent with much human psychological data, which generally find that recognition performance is more accurate than detailed retrieval of previously encountered information.

A number of attempts have been made to model the details of human memory performance with very simple network models. Of particular note are those by Pike (1984), Murdock (1982), Metcalfe (1990), and others to make detailed fits of experimental psychological data to models. Some of the papers in a recent collection of papers edited by Hockley and Lewandowsky (1991) provide a good entry to the literature coupling classic data-oriented experimental psychology with network modeling. This is currently an active area. In later chapters we will discuss attempts to model human behavior in other domains with neural networks.

Character Simulations

It is useful to perform simulations with random vectors, as we did in AMDEMO and AMDEMOL, but the artificiality of the stimuli make the simulations unsatisfying. It is also difficult to construct state vectors with the kind of internal structure that makes the real world so interesting.

Therefore we next take an arbitrary way to construct stimuli that is useful for high-dimensional simulations of inputs with structure. We will construct state vectors composed of strings of characters (humans are very good at working with strings of characters), which is easy to do in most computer languages. For example, Pascal compilers contain predefined functions, CHR( ) and ORD( ), that are based on the ASCII character list. Given a number, ORD( ) returns the character in that location in the ASCII list, and given a character, CHR( ) returns the number of that character on the list.

The ASCII list that we are concerned with contains 128 characters, including all lower- and upper-case letters, a number of symbols, and some nonprinting characters. If we were to use these programs for anything practical, it would be wise to have a less arbitrary relationship between letters and numbers.

Our strategy will be to take the numerical position of a character on the ASCII list and represent it as a byte of eight $+1$s and -1s. We will include a parity bit, so we require eight elements to represent a character. Even though we really only need seven elements, using eight allows us to construct orthogonal bytes with pairs of characters. This can be a convenience. The parity element requires that a vector always contain an even number of $+1$ elements. We can concatenate characters to make higher-dimensionality strings.

Suppose we have a four-character string, "test." Figure 7.15 shows how this string is converted into a 32-element state vector composed of 1s and -1s. This is a distributed coding. We cannot look at any one vector element and identify the character; that requires many elements.

Because we will make use later of state vectors that contain many zeros, we must have a character that is interpreted as eight zeros. The underline

```
Character          Position              Byte representing
                   on ASCII              Character
                   list                  (+ = +1, - = -1)

   'T'    →        ASCII  84     →        + + - + - + - -
   'e'    →        ASCII 101     →        - + + - - + - +
   's'    →        ASCII 115     →        + + + + - - + +
   't'    →        ASCII 116     →        - + + + - + - -
```

The resulting state vector for the string 'Test' is a 32 dimensional string of +1 and -1's:

```
        'Test'  →  ++-+-+---++--+-+++++--++-+++-+--
```

Figure 7.15 Arbitrary character based coding used in high-dimensionality simulations. Don't use this coding for anything important!

character, '_', is used, so that

'_' → (0, 0, 0, 0, 0, 0, 0, 0).

Since the output of the linear associator contains noise, when we convert a state vector back into characters it sometimes helps to use an interpretation threshold. Elements above the threshold or below the negative of the threshold are given the values 1 or −1. Elements in the band between the threshold and minus threshold are given the value zero. We want the system to give a character as output only when it is sure of the character, which we define to mean that all the elements composing the character (except the parity element) are above threshold. If any of them is below threshold, we print the underline character, '_'. Suppose we have a state vector with the following elements.

0.4   0.6   −0.1   0.1   −0.8   0.2   −0.7   −1.0.

If the threshold is 0.15, this state vector is interpreted as:

1   1   0   0   −1   1   −1   −1

and printed as '_'. If the threshold is 0.05, this state vector is interpreted as:

1   1   −1   1   −1   1   −1   −1

and printed as T. The other conventions are that nonprinting characters, which can and do arise in noisy outputs, are printed as '#'. Boring details about the PROCEDUREs are provided in appendix H.

Program Amchardemo

We can now use our demonstration program to see how well the linear associator works with less optimal material. We are using meaningful character strings that will be far from orthogonal in practice (the detailed list is given in appendix I). It requires inputs to be provided in the form of character strings. These strings are provided by a separate text file that can be created with an editor. The text file is composed of pairs of lines, corresponding to input and output strings, separated by lines that are ignored when the file is read and that can be used for notes, a template so characters can be lined up, or sequence numbers. An example is given below.

Using characters is a rigorous test of retrieval accuracy, since a single bit with the wrong sign will change the output character. As a demonstration of the kind of results that are given by character strings, let us consider the

```
123456789012345678901234 5   (A template for writing characters.)
  Abbott and Costello
    Laurel and Hardy
2
  Ludwig von Beethoven
 Wolfgang Amadeus Mozart
3
2 4 6 8 10 12 14 16 18 20
A list of ten even numbers
4
qwertyuiopasdfghjklzxcvb
Keys in keyboard order
5
Capital of Rhode Island
Providence,thelargestcity
6
ASDF1234QWER5678ZXCV6789/?
A bunch of random characters
```

Figure 7.16 Input file for the character-based association program. Every third line—first, fourth, seventh, and so on—is not used by the program and can be used for notes, order information, or templates. The other lines are converted into 200-dimensional state vectors using the scheme in figure 7.15. Blanks are coded as ASCII 32. The first association is (input) "Abbott and Costello" → (output) "Laurel and Hardy".

strings in figure 7.16. This file is the input to the program, which uses 25 character strings corresponding to 200-dimensional state vectors. The first string (Abbott and Costello) gives rise to the input pattern, **f**, and the second string (Laurel and Hardy) gives rise to the output, **g**.

AMCHARDEMO learns one association at a time, starting from the first on the list until all are learned. It will then test for the entire set. It will also print out the cosine between desired and actual output state vectors and the length of the output. Below is an edited sample output. The interpretation threshold is zero for the example, so the interpretation responds to the pattern of signs in the output state vector. When only one pair of associations is learned,

Abbott and Costello → Laurel and Hardy,

the output to all inputs will be the pattern of the single learned output. This is because the connection matrix is given by

$$\mathbf{A} = \mathbf{g}\mathbf{f}^{T}.$$

When a new input pattern, **f′**, is presented, it can be decomposed into its projection onto **f** and an orthogonal component. The orthogonal component gives zero output; the projection onto **f** gives the output pattern **g**, with length proportional to the cosine of the angle between the learned and the test vectors. The outputs for the inputs from pairs 2, 3, 4, 5, and 6

```
Correct length of output vectors: 14.14214
After  1 learned associations.

1.   Cosine: 1.00000.    Length: 14.14213
Input            ==>     Abbott and Costello
Desired Output ==>       Laurel and Hardy
Actual Output  ==>       Laurel and Hardy

2.   Cosine: 0.18000.    Length: 6.78823
Input            ==>     Ludwig von Beethoven
Desired Output ==>   Wolfgang Amadeus Mozart
Actual Output  ==>       Laurel and Hardy

3.   Cosine: 0.16000.    Length: 1.97990
Input            ==> 2 4 6 8 10 12 14 16 18 20
Desired Output ==> A list of ten even number
Actual Output  ==>       Laurel and Hardy

4.   Cosine: 0.22000.    Length: 2.54559
Input            ==> qwertyuiopasdfghjklzxcvb
Desired Output ==> Keys in keyboard order
Actual Output  ==>       Laurel and Hardy

5.   Cosine: 0.16000.    Length: 2.26274
Input            ==> Capital of Rhode Island
Desired Output ==> Providence,thelargestcity
Actual Output  ==>       Laurel and Hardy

6.   Cosine: -0.30000.   Length: 0.28284
Input            ==> ASDF1234QWER5678ZXCV6789/
Desired Output ==> A bunch of random charact
Actual Output  ==> _____3#####_###_7#####_____
```

Figure 7.17 After "Abbott and Costello" → "Laurel and Hardy" is learned, the only output pattern observed is "Laurel and Hardy" with its length dependent on the cosine between the input vectors and "Abbott and Costello". Notice that association 6 has a negative cosine, which is interpreted as a different set of characters because all the signs are inverted. The character "#" is used for a nonprinting character.

have positive cosines, and are interpreted as the string "Laurel and Hardy." The input from pair 6 has a negative cosine, interpreted as the negative of "Laurel and Hardy." The negative pattern is interpreted as a string with a number of nonprinting characters, represented as '#.' This result suggests that our "meaningful" character strings are not very orthogonal (figure 7.17).

Figure 7.18 shows the results after five pairs have been learned. There are some mistakes in the associations. The sixth association, since it has not been learned, is a nonsense string, although it reproduces a couple of fragments of other output associations. Accuracy of string reconstruction is poor even though the output cosines for learned vectors are quite high, all around 0.9 or above.

It would be hard to use this model for a practical retrieval system that required high accuracy. However, first, it is possible to improve accuracy by

```
After  5 learned associations.

1.   Cosine: 0.88366.   Length: 18.78875
Input          ==>    Abbott and Costello
Desired Output ==>    Laurel and Hardy
Actual Output  ==>    Laurel and Hardy

2.   Cosine: 0.87126.   Length: 19.64052
Input          ==>    Ludwig von Beethoven
Desired Output ==>    Wolfgang Amadeus Mozart
Actual Output  ==>    wolfeang aeideusbeopabt

3.   Cosine: 0.95265.   Length: 16.49581
Input          ==> 2 4 6 8 10 12 14 16 18 20
Desired Output ==> A list of ten even number
Actual Output  ==> A list of ten even number

4.   Cosine: 0.92223.   Length: 16.82530
Input          ==> qwertyuiopasdfghjklzxcvb
Desired Output ==> Keys in keyboard order
Actual Output  ==> Keys in keyboard order

5.   Cosine: 0.90857.   Length: 18.89617
Input          ==> Capital of Rhode Island
Desired Output ==> Providence,thelargestcity
Actual Output  ==> Providence,thelargestcity

6.   Cosine: 0.21813.   Length: 1.50413
Input          ==> ASDF1234QWER5678ZXCV6789/
Desired Output ==> A bunch of random charact
Actual Output  ==> KEys kn keYjgard -Zdor0
```

Figure 7.18 After five pairs of associations are learned, the network gave reasonable recall, but made some mistakes on association 2. Association 6 was not learned, but the input was nearest to the input pattern of association 4, and the output resembles a noisy version of association 4.

simple means and, second, more than strict accuracy of recall may be involved in a neural network computation. This last observation will be discussed in later chapters.

Autoassociation

Let next simulate autoassociation. We want to see if an autoassociator can be used to reconstruct the missing parts of a state vector, as our calculation in chapter 6 suggested. Let us construct a small data base to show this. We will use 25-character strings.

A set of 15 states and their capitals is given in figure 7.19. The 25-character strings in the figure generate state vectors for use in our association programs. Since we will be testing autoassociation, the two input lines, corresponding to the input, **f**, and the output, **g**, will be identical in the text files.

```
          States and Their Capitals
             25 Character Strings


1234567890123456789012345

Boston       Massachusetts
Providence   Rhode Island
Sacramento      California
Bismarck     North Dakota
Oklahoma City    Oklahoma
Harrisburg   Pennsylvania
Charleston   West Virginia
Baton Rouge     Louisiana
Concord      New Hampshire
Columbia     South Carolina
Albany           New York
Trenton      New Jersey
Juneau           Alaska
Talahassee       Florida
Austin               Texas
```

Figure 7.19   Edited data set of state names and capitals used for an autoassociative data-retrieval system.


```
     Inputs and Outputs for Outer Product Autoassociation
                   Five and Ten Associations

                   5 Learned Associations

         Input String                   Output String

Boston       Massachusetts  →  Bkstina'      i'csachesedta
Providence   Rhode Island   →  Rcoviden'e    Bhcde Mslela
Sacramento      California  →  Sacramen'g       calidcbnma
Bismarck     North Dakota   →  Biseafak       Bkrdh Dakota
Oklahoma City    Oklahoma   →  Oklahoma City     Oklahoma

                  10   Learned Associations

         Input String                   Output String

Boston       Massachusetts  →  Boctmf''      masb'citsimta
Providence   Rhode Island   →  Bcgviden'e      ''bdeahshina
Sacramento      California  →  Cacriean'g      ''b'oidghnia
Bismarck     North Dakota   →  Biseibak      dgbdaadciola
Oklahoma City    Oklahoma   →  Cklaigea c    i''' Oilchoma
Harrisburg   Pennsylvania   →  Barricce'g     'abdcilwinia
Charleston   West Virginia  →  Cicrmecd'f    eas''gipginia
Baton Rouge     Louisiana   →  Batgm'bo'g      '''doahsiina
Concord      New Hampshire  →  Concmb''      lew''cipshiva
Columbia     South Carolina →  Columbaa      met''cipchina
```

Figure 7.20   An autoassociator should reproduce a learned input exactly. The network makes many errors for both 5 and 10 learned associations.

Results are not encouraging at first sight. Figure 7.20 shows the input and output vectors for the first 5 and the first 10 states when they are learned by outer product autoassociation. Although retrieval is not accurate, it is interesting that the network seems to have concluded from the learned examples that all state names end in the letter a, and state capitals begin with either B or C. We will discuss this observation further in chapter 11 when we tackle prototype learning.

Information Retrieval

Suppose we want to make this network serve as a fact retrieval system. One way to do this would be to use autoassociation to construct the missing information, that is, put in a state, and let the network complete the string with the capital, and vice versa. We can test this idea by putting in partial state vectors. We will use the same matrix as before, with 5 or 10 learned examples. We will probe the 5 learned examples with the state name and the 10 learned examples with the capital name.

```
               Five Learned Associations
                  Probe With Capital

        Input String                    Output String

Boston_____  →  Bkstina'     i'csachesedta
Providence_____  →  Providence   Rhode Island
Sacramento_____  →  Sacramen'g       calidcbnma
Bismarck_____  →  Biseafak     Bkrdh Dakota
Oklahoma City_____  →  Oklahoma City     Oklahoma

               Ten Learned Associations
                   Probe with State

        Input String                    Output String

_____Massachusetts  →  Boctof''     Masb'citsilta
_____Rhode Island  →  Bcovidanae    p'jdgaislana
_____California   →  Cicriead'g   ''b'nifghnia
_____North Dakota  →  Biseirck     Dgrdi'dciota
_____Oklahoma  →  Cidcigaa c   ''''''oidchoma
_____Pennsylvania  →  Barricbe'g   'ebdcilwinia
_____West Virginia →  Ciarmgad'f   ges' Girginia
_____Louisiana  →  Bcfwm''e'g    '''dgahsiina
_____New Hampshire  →  Concob''     New''cipshiva
_____South Carolina  →  Columbaa     oet' Cipghina
```

Figure 7.21 The autoassociator can be used to answer questions. For example, given the state name, the capital can be retrieved due to autoassociative reconstruction, although not very accurately. A simulation in chapter 8 uses the same information as data for an error-correcting network and performs much more accurately.

Figure 7.21 shows results when partial state vectors are used as input to the autoassociative system. Seeing such results, very few would be tempted to rush out and buy a neural network data base. The linear associator is poor at forming accurate input-output relationships. In the rest of this book we will discuss ways of improving performance, and ponder the critically important question, what should we expect a neural network to do, anyway? When even a model as simple as the linear associator is applied to the right problems, it can do quite a bit of interesting computation.

References

J.A. Anderson (1968), A memory storage model using spatial correlation functions. *Kybernetik, 5,* 113–119.

J.A. Anderson (1972), A simple neural network generating an interactive memory. *Mathematical Biosciences, 14,* 197–220.

J.A. Anderson (1973), A theory for the recognition of items from short memorized lists. *Psychological Review, 80,* 417–438.

W.B. Davenport and W.L. Root (1958), *An Introduction to the Theory of Random Signals and Noise.* New York: McGraw-Hill.

W. Feller (1966), *An Introduction to Probability Theory and Its Applications*, Vol. II. New York: Wiley.

W.E. Hockley and S. Lewandowsky (1991), *Relating Theory and Data: Essays on Human Memory in Honor of Bennet B. Murdock.* Hillsdale, NJ: Erlbaum.

A.G. Knapp and J.A. Anderson (1984), Theory of categorization based on distributed memory storage. *Journal of Experimental Psychology: Learning Memory and Cognition, 10,* 616–637.

D.E. Knuth (1981), *The Art of Computer Programming: Volume II, Seminumerical Algorithms. 2nd Edition*, Reading, MA: Addison Wesley.

T. Kohonen (1977), *Associative Memory—A System Theoretic Approach.* Berlin: Springer-Verlag.

T. Kohonen (1984), *Self-Organization and Associative Memory.* Berlin: Springer-Verlag.

D. Marr (1982), *Vision*, San Francisco, CA: W.H. Freeman.

J. Metcalfe (1990), Composite holographic associative recall model (CHARM) and blended memories in eyewitness testimony. *Journal of Experimental Psychology: General, 119,* 145–160.

B.B. Murdock (1982), A theory for the storage and retrieval of item and associative memory. *Psychological Review, 89,* 609–626.

G. Palm (1982), *Neural Assemblies,* Berlin: Springer Verlag.

R. Pike (1984), A comparison of convolution and matrix distributed memory systems. *Psychological Review, 91,* 281–294.

D.J. Willshaw, O.P. Buneman, and H.C. Longuet-Higgins (1969), Non-holographic associative memory, *Nature, 222,* 960–962.

Early Network Models: The Perceptron

This chapter is devoted to the perceptron, *an influential early neural network. The classic perceptron had an input sensory layer, often called a* retina. *It was partly connected to an* association *layer, which then projected to a* response *layer by way of a single layer of modifiable synapses. The perceptron usually functioned as a pattern classifier. If two pattern classes were* linearly separable, *that is, they could have a hyperplane drawn between them, the* perceptron learning theorem *showed that a simple learning rule would eventually learn the classification. The theorem will be proved in this chapter. Although this was a powerful theorem, the perceptron was severely limited with respect to what it was capable of learning.*

In the last two chapters we analyzed and simulated a particularly simple neural network, the linear associator. As we saw, the basic network had a limited capacity and made errors. Now we will discuss other early networks and show how intrinsic theoretical limitations in their performance lead to important extensions of network ideas. Two learning algorithms we will discuss in this chapter and the next—the perceptron and early gradient-descent algorithm, the ADALINE—are the best-known early examples of neural network algorithms. Like many pioneering efforts, besides producing networks that worked, they did something else that was probably more important in the long run. They framed the network learning problem in a way that was consciously or unconsciously accepted and used by future models. Work on the perceptron and the ADALINE accepted most of the assumptions and techniques used in the related field of pattern recognition. Although these assumptions are reasonable and are often useful for practical applications, it is not clear if they are always appropriate for brainlike computation. We discuss this point in later chapters.

Supervised and Unsupervised Learning

Many common neural network system have a similar learning process. Suppose we are using a supervised learning algorithm. We have a set of data, a *training set*. It is assumed that we know the appropriate output from the system, for example, whether it is correct or incorrect. Or, given a member of the training set as input to the network, we might know exactly what the output is supposed to be and how it differs from what we actually get. If we can use it effectively, the more information we have about the training set and the network, the better we can learn.

There is also a class of algorithms in which an omniscient supervisor does not exist. *Unsupervised* algorithms are of great interest because they look at the input data and somehow organize the world in an interesting and appropriate way. This almost seems like magic: structure and organization arise from the raw data. It is not magic at all, of course, but because they have little initial information to work with, unsupervised algorithms are difficult to construct and use, and generally make strong assumptions about what they are trying to do. We will discuss a couple of unsupervised algorithms later in this book. The adaptive mapping algorithms (chapter 14) are one example; the use of a neural network for radar clustering (chapter 15) is another.

Pattern Recognition

The structure assumed for supervised network learning is fundamentally associative. We have a given set of input data, the training set. We know what the output is for a given input. The output could be another state vector of any form, as was assumed for the linear associator. Another possible output pattern would correspond to a *classification*, which is what is usually assumed for pattern recognition. One approach to pattern recognition assumes that the world is divided up into a number of discrete categories. For example, a set of black and white pixels might correspond to 1 of 26 letters, or 1 of 10 digits. Or a furry quadruped might be a dog, a cat, a marmoset, or a lion. Given a set of input data, the pattern-recognition device must decide what category is most appropriate.

Suppose we have a set of input data. In the neural network context, this would correspond to a set of input patterns, that is, state vectors. There are a number of output categories. The assumption is often made that the state space of possible input patterns contains discrete regions, where all the patterns within a region are given the same categorization. Figure 8.1 suggests this architecture.

This structure is based on observations about real world categories. Things that are very close in state space are likely to be in the same category. For example, if we have two birds that differ by only a single feather, they

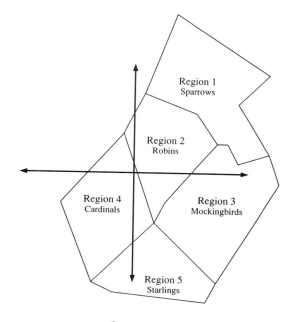

State Space

Figure 8.1   A common form of pattern recognition. A state space representing possible input patterns is divided into regions, each of which corresponds to a particular classification.

are almost certainly the same species. This observation about natural categories is made the explicit basis of *nearest neighbor techniques,* discussed in chapter 13.

A Cognitive Aside

The experiment that most resembles classic pattern recognition is called *confrontation naming.* An object, drawing, description, or whatever is presented to a subject and the subject is supposed to name it. If you try this experiment yourself or on acquaintances, the results are likely to be interesting. Consider the pattern shown in figure 8.2. What might responses to it be in a confrontation-naming experiment? Perfectly correct possibilities could be:

"ay."

"The letter A."

"A capital letter."

A

Figure 8.2 A confrontation naming experiment. What would a human response be to this pattern?

"An article."

"A capital letter A."

"First letter in the alphabet."

"A Times Roman bold-face cap."

"A grade."

"ASCII 65."

... and so on.

Clearly, human pattern recognition does a lot more than simple categorization. It couples a rich associative structure with flexibility. The human system can be programmed for many different responses, responds appropriately to context, and can generate a considerable amount of associated output. Of course, if properly instructed, a human can act as a very good simple character recognizer, although most humans rapidly get bored doing so and start making mistakes. When they do act as letter recognizers, it is often in the service of such complex tasks as reading, where the letter-recognition step is unconscious, and detecting higher-level structures (words, meaning) is the real task.

Learning Networks

Given the basic structure of the problem, simple categorization, let us consider the problem of finding the best set of connection strengths for a network so it can determine the proper category for an input pattern. Most simple categorizers assume a grandmother cell output representation, where one unit is on, representing the category, and the others should be off. Hebbian learning, as used for the linear associator, can act as a categorizer (figure 8.3). We have an input pattern and an output category, represented as a pattern on a set of output units. We know all the details of the input and output association we are trying to learn. In general, each category will have many examples, instead of single pairs of independent state vectors.

We discuss in chapter 11 how a network of this type can function as a categorizer and act as reasonable model of human concept formation. In

Simple Associator

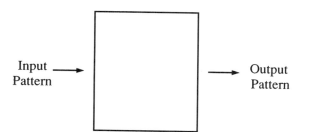

Figure 8.3 A simple learning associator.

Error Correction

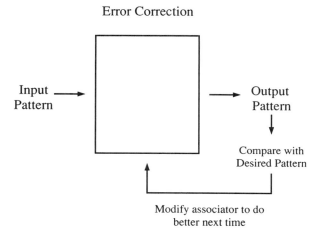

Figure 8.4 Most modern neural networks use feedback during learning. If the output pattern is not exactly what it is supposed to be, the associator is modified to do better next time.

many cases, however, the basic linear associator does not give very good results. As we saw in chapter 7, output associations can be noisy, and storage capacity is limited. One way to improve accuracy would be to provide feedback as to how well the network is doing and arrange it so the network can change its connection strengths so as to perform better. Both the perceptron and the contemporary model, ADALINE (discussed in chapter 9), use knowledge of past results to improve the performance of the network (figure 8.4). Both take the reasonable point of view, beloved of coaches everywhere, "keep at it until you get it right."

The information used to determine whether the network is acting correctly is quite different between the ADALINE and the perceptron, and this difference has profound effects on how they behave. In the perceptron the feedback provided is only as to *correctness*; that is, whether or not the network made the appropriate classification. In the ADALINE, a continous measure of error is minimized; that is, the length of the vector representing an *error signal* is to be made as small as possible by the learning procedure. The error signal is defined as the difference between the desired output and the actual output produced by the first part of the network, as discussed in chapter 7. This means that the perceptron only learns when it obtains wrong answers, whereas the ADALINE learns all the time. We will first discuss the perceptron and in chapter 9, the ADALINE.

The Perceptron

The perceptron was the first neural network that made an impact on a large community, particularly engineers, who became excited and intrigued with the possible practical implications of a real learning machine. Many later and more complex learning networks are really only modest extensions of the perceptron. The perceptron was proposed by Frank Rosenblatt in 1958 in a widely distributed technical report from the Cornell Aeronautical Laboratory, with a somewhat more formal presentation in an article in *Psychological Review*. Rosenblatt was a psychologist and knew a great deal about psychological learning theories.

Rosenblatt's high public visibility during the late 1950s and early 1960s led to a problem that has haunted the field of neural networks for decades. He and others doing perceptron research in this era often claimed that perceptrons could do too much, almost anything, in fact, eventually. As time progressed, it turned out to be much harder to build useful devices than anyone had realized. Perceptrons also had some serious theoretical limitations, which were pointed out by Minsky and Papert (1969) in their well-known book *Perceptrons*. We will discuss these theoretical problems later.

The 1970s were supposedly the dark ages for neural networks, and interest in the engineering community at large did not revive until the early 1980s. Some accounts to the contrary, however, people interested in networks in this era did not slink around in the shadows, concealing their identities and hiding their faces from the the light. A good deal of interest continued in the psychological community, because simple neural networks turned out to have considerable value as psychological models. One reason for this is that the theoretical limitations of perceptrons—that is, the limitations that made them less effective as computing devices than seemed desirable—overlapped those of humans. Perceptron-like models for human psychology turned out to have considerable predictive and explanatory value and interest in network models was kept alive, and indeed flourished among psychologists and cognitive scientists. Eventually, continued development reignited interest among a wider community. This led to many proclaiming the rebirth and rediscovery of what never actually went away.

There was no single perceptron model. Numerous variants were proposed and studied over a decade: multilayer machines, learning rules of varying complexity, and complex feedback architectures. Some of these more complex architectures contained ideas that are only now being rediscovered. We will discuss first the perceptron suggested by Rosenblatt in his early papers, and show how it was simplified into what has become known as the perceptron.

Threshold Logic Units

In chapters 6 and 7 we discussed simple networks of linear units. The perceptron used highly nonlinear elements. The basic computing element in the perceptron is a device known in the early literature as a *threshold logic unit* (TLU) (Nilsson, 1965).

The TLU is one form of the generic connectionist neuron that we have used throughout this book (figure 8.5). It has n input weights, with strengths $w[i]$. The TLU forms the sum of the input pattern activities times the weights; that is, the familiar inner product between the input pattern and the connection strengths. This version of the generic neuron has a nonlinear second stage in which the output of the TLU can take on only one of two values. Output values are usually zero and $+1$ in the early papers, but $+1$ and -1 often make the system easier to analyze. The TLUs have a threshold, θ, suggested by the thresholds of real neurons. The basic TLU output rule, given an input pattern vector, $\mathbf{x}$, is

TLU output $= +1$ if $\sum w[i]x[i] > \theta,$

TLU output $= -1$ if $\sum w[i]x[i] \leq \theta.$

It is ugly and confusing to carry around a threshold in the equations and programs describing perceptrons. We will follow a common convention that allows us to make the threshold one of the weights. This makes for more readable programs, as long as the unique nature of the weight is clear. This extra weight is enumerated either as $w[0]$ or $w[n + 1]$, where n is the dimension. We assume that this weight is connected to a special input, $x[0]$ (or

Threshold Logic Unit
(TLU)

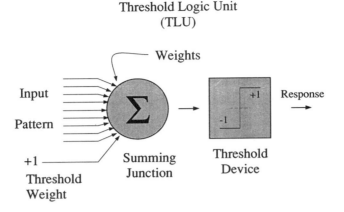

Figure 8.5   A threshold logic unit.

$x[n + 1]$), which always has the value 1. Therefore the contribution of this weight to the sum formed by the synapses of the TLU always has value $w[0]$ (or $w[n + 1]$). That is, the sum is always offset by this amount. Therefore this weight behaves as the negative of the threshold. To provide the fixed value, $+1$, as the input to this threshold weight, the input patterns are *augmented* by having the value 1 placed in the appropriate position of the vector (0 or $n + 1$), increasing the dimensionality of the system by 1.

With augmentation of the vectors, we can form a simpler expression for the behavior of the TLU,

TLU output $= +1$ if $\sum w[i]x[i] > 0,$

TLU output $= -1$ if $\sum w[i]x[i] \leq 0.$

This rule is used for all the units in the perceptron.

Perceptron Architecture

Rosenblatt and co-workers assumed several layers of TLUs, with complex sets of interconnections between them (figure 8.6). The perceptron started with a sensory surface, usually called a *retina* in the literature, that connected to a second layer, the *association* layer (A layer). The connections from the model retina to the A layer could be random, or could have localized connectivity. It is important to realize that each unit in the A layer only saw a *subset* of the cells in the retina. The pathway from the retina to the A layer was not *fully connected*, where every A layer unit would see every retinal unit, but *partially connected*, where an A layer unit saw only some of the retina. Therefore each A layer unit was computing a *different* function of

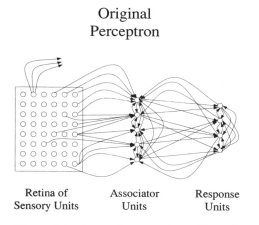

Retina of   Associator   Response
Sensory Units   Units   Units

Figure 8.6 An early perceptron. Note the extensive lateral and feedback connections. Most of these connections function as an extensive winner-take-all network, to ensure a grandmother cell output representation in the top layer of response units (R units). From Block (1962, figure 2).

the image on the model retina. Rosenblatt placed considerable theoretical weight on this point, as did Minsky and Papert in their analysis of perceptrons. Many, probably most, modern network architectures start by assuming full connectivity, which means that all units in a fully connected layer will see the same input pattern.

The A layer was reciprocally connected to a third layer, the *response layer* (R layer). Units in the A layer connected to units in the R layer and vice versa. The modifiable connections were those from the A layer to the R layer. The activation of a single appropriate R layer unit for a given input pattern or class of input patterns was the goal of operation of the perceptron. It was thought to be undesirable to have more than one R layer unit be on at a time. To prevent this, a set of reciprocal inhibitory connections was used, so that an R layer unit inhibited all the A layer units that did *not* connect to it. Therefore, when an R layer unit was activated, it suppressed the response of competitors. The obvious biological analogy to this behavior was lateral inhibition in the *Limulus*. Networks that show this behavior are winner-take-all (WTA) networks, as discussed in chapter 4. Using grandmother cells for R layer units was an important representational assumption. It was not essential for the perceptron architecture, but it led to analysis of the device primarily as a pattern classifier and not as a pattern associator.

Rosenblatt and associates also considered systems with two layers of association units and other complex variants. One of the most complex perceptrons analyzed in detail is found in Block, Knight, and Rosenblatt (1962).

Simple Perceptron

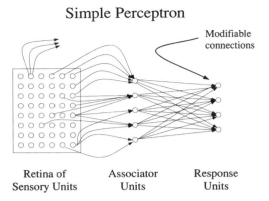

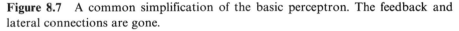

| Retina of | Associator | Response |
| Sensory Units | Units | Units |

Figure 8.7 A common simplification of the basic perceptron. The feedback and lateral connections are gone.

They showed that a complex perceptron was capable of some impressive feats of sequence learning and self-organization. Nilsson's 1965 book is still the most lucid review of the early work in neural network adaptive systems.

Simplified Perceptron

A perceptron with many reciprocal feedback connections is too complicated to analyze easily. Therefore almost all mathematical analysis of the perceptron concerned itself with a simplified device with only feedforward connections, as shown in figure 8.7. Since the perceptron was then analyzed as a pattern classifier with grandmother cells at the R layer, the output layer, the classification power of the system could be no greater than the classification power of a single model neuron, an R layer unit. The simplified perceptron is almost always assumed to be fully connected between A and R layers, so that every unit in the A layer connects to every unit in the R layer.

The R layer units start with random weights for their connections from the A layer units. We want an R layer unit to start to respond when and only when members of a particular class of events are presented to the retina. An example of desired behavior mentioned in the literature is R layer units that turn on when the retina "sees" a triangle as opposed to units that turn on only when it "sees" a square.

The major problems of the perceptron came in this area. What exactly could it learn to discriminate? All the early workers were aware of at least one severe limitation on what a simple perceptron could compute, and it arose directly from the functioning of the TLU. When the perceptron was used as a classifier, it classified inputs into categories based on the discrete output of the TLU, that is, is the input pattern a member of class 1 or class

Linear Separability

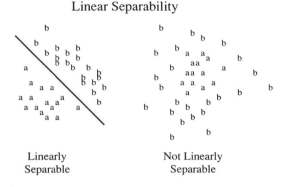

Linearly   Not Linearly
Separable   Separable

Figure 8.8   Linear separability.

2? One way this would be done for a single model neuron would be to
identify a +1 output with class 1 and a −1 output with class 2.

Consider a point **x** in state space. The weights of the TLU form a vector,
w. The TLU forms the sum of pairwise products between weight elements
and pattern elements, that is, the inner product, [**x**, **w**]. Consider all the
points where the inner product [**x**, **w**] = 0, that is,

$$[\mathbf{x}, \mathbf{w}] = \sum_i x[i]w[i] = 0.$$

The resulting expression is the equation of a hyperplane in state space. All
points on one side of the hyperplane will be given output value +1, and all
points on the other side −1. Therefore the space is dichotomized by this
hyperplane. The hyperplane forms a *decision surface* separating one category from another.

Suppose we have two distinct groups of patterns, {**a**} and {**b**}. We want
our TLU to respond +1 if an item from group {**a**} appears and −1 if an
item from group {**b**} appears. This is only possible for a threshold logic unit
if the two groups of items are *linearly separable*; That is, the two classes can
be separated by a hyperplane, the only decision surface that the TLU can
form. Linear separability is intuitive (figure 8.8). Unfortunately, many interesting classifications are not linearly separable and require more complex
decision surfaces to separate the classes.

Note how important the assumption of grandmother cells becomes. If we
categorize based on the responses of groups of R layer units, the regions
associated with a particular classification could have more complex shapes
formed from several hyperplanes. Presumably another layer could look
at this pattern, forming a *multilayer perceptron* that would be much more
powerful than one with only one layer. Successful learning rules for multi-

layer perceptrons were not developed until much later, but their power was obvious from the beginning.

If two classifications are linearly separable, it is possible to give rules for changing the connection strengths of the TLU so that eventually a set of weights will be generated that will correctly make the classification. Moreover, given a finite set of linearly separable patterns to classify, the learning rules will produce a set of weights giving a solution in a finite time.

This result is the famous *perceptron convergence theorem* or *perceptron learning theorem*. In his first paper (1958) Rosenblatt did not have proofs of this theorem, but empirically observed that some pattern classes with structure were easy to learn and the perceptron could classify many new examples correctly, whereas memory for random patterns was more limited. We discuss this theorem below.

Perceptron Learning

There are many variations on perceptron learning rules, but the ones that can be shown to converge are simple. Only the connections from the A layer to the R layer are modifiable; those from the retina to the A layer are fixed. Again, based on the assumption about output layer representation, we only have to look at the connections of the one R layer unit that does the classification.

We assume we have two linearly separable pattern classes, which we will call $\{\mathbf{a}\}$ and $\{\mathbf{b}\}$. We assume we have a certain number of examples of each class. We present the examples to the perceptron in some kind of sequence, referred to as a training sequence, $\{\mathbf{y}\}$. The kth member of the training sequence is referred to as $\{\mathbf{y}_k\}$. The supervisor knows the correct classification for every pattern and can adjust all the weights.

Let us give the system a pattern. If the system makes the *correct* response, we do nothing. Suppose, however, the system makes an *incorrect* response. The TLUs in our perceptron only signals -1 or 1. Suppose the R layer unit that should have been $+1$ is actually -1. Then, we add a positive amount, c, to the connection strengths from A layer TLU cells with activity $+1$ to the incorrect R layer unit. This will increase the positive input to the R layer unit so it will be more likely to exceed threshold next time. The connections of the incorrectly responding R layer unit to A layer cells with activity -1 have an amount $-c$ subtracted from their weights. Again, this will make the incorrect R layer unit more likely to exceed threshold next time, since it will receive less negative input. If the R layer unit gave output $+1$ and should have responded -1, the same rule is applied with the opposite sign: connections from activity $+1$ A units are weakened by an amount c and from A units with activity -1 are strengthened by c. The amount of weakening in the simpler rules was simply a constant, c, although other rules where c was a function of things such as amount of error, number of learning trials, and so on were also used.

When put in mathematical form, this is actually a simple operation involving inner products. Let us consider the weight vector, $\mathbf{w}$. The rule corresponds to taking the presented pattern, multiplying it by c, and adding or subtracting it from the weights, depending on whether or not the R layer unit had activity $+1$ or -1.

Suppose $\mathbf{y}_k$ is the kth member of the training sequence. Suppose $\mathbf{w}_k$ is the set of weights on the R layer unit we are studying at the kth learning trial. Then, if the classification is correct:

$$\mathbf{w}_{k+1} = \mathbf{w}_k.$$

If the classification is incorrect:

$$\mathbf{w}_{k+1} = \mathbf{w}_k + c\mathbf{y}_k$$

if the R layer unit activity should have been $+1$ and

$$\mathbf{w}_{k+1} = \mathbf{w}_k - c\mathbf{y}_k$$

if the R layer unit activity should have been -1.

Consider the weights after the correction was made. Suppose the R layer unit activity should have been $+1$ but was wrong. That means that the inner product,

$$[\mathbf{w}_k, \mathbf{y}_k] \leq 0.$$

We added c times $\mathbf{y}_k$ to the weights. If the pattern $\mathbf{y}_k$ occurred again, the perceptron is more likely to give it the correct classification because the new inner product in response to y_k is

$$
\begin{aligned}
[\mathbf{w}_{k+1}, \mathbf{y}_k] &= [(\mathbf{w}_k + c\mathbf{y}_k), \mathbf{y}_k] \\
&= [\mathbf{w}_k, \mathbf{y}_k] + c[\mathbf{y}_k, \mathbf{y}_k] \\
&> [\mathbf{w}_k, \mathbf{y}_k].
\end{aligned}
$$

This quantity is more positive than the result before learning, since the inner product of $\mathbf{y}_k$ with itself is always positive, the length squared. Therefore, the weights are being changed in the right direction so as to make a correct response next time. The second rule will tend to *decrease* the output reaching the R layer unit by the same mechanism. Both these rules will make the answer more likely to be correct the next time this pattern is presented.

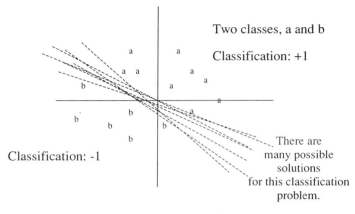

Two classes, a and b

Classification: +1

Classification: -1

There are many possible solutions for this classification problem.

Figure 8.9   This decision hyperplane discriminates pattern groups **a** and **b**. There may be many such hyperplanes successfully discriminating the two groups.

This rule is a modified form of Hebb learning where learning occurs only when an error is made. When learning takes place, connections are incremented by an amount that is the product of the desired state of the output unit times input pattern.

The Perceptron Convergence Theorem

Since, by the usual assumption, only a finite number of training patterns exists, and performance tends to get better and better, it can be proved that *if a solution is possible*, that is, if the pattern classes are linearly separable, the solution will be found. There are many ways of proving the result.

Let us start by assuming that the training set is indeed linearly separable, and is composed of two classes of patterns, {**a**} and {**b**}, with desired outputs $+1$ and -1, respectively. That means that at least one hyperplane exists, so that the inner products of all the examples of set {**a**} with the weights, **w**, that is, $[\mathbf{a}_i, \mathbf{w}]$, are positive, and the inner products of all the examples of set {**b**} with the weights, $[\mathbf{b}_i, \mathbf{w}]$, are negative. Usually, many possible decision planes will solve the problem (figure 8.9). We have to show that we can find one of them.

Let us introduce the useful notion of *weight space*, which we will use a number of times in this book. The weights of the TLU form a point in a space of the dimensionality of the number of weights. As figure 8.10 shows, there is a *solution region* of weights where all weights in the region will be correct solutions. In general, this region can be seen to take the form of a cone passing through zero. That is because, if a set of weights **w** is a

Weight Space

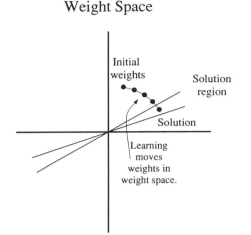

Figure 8.10   Weight space. Every set of weights corresponds to a point in weight space with the same dimensionality as the patterns to be discriminated. As the perceptron learns, its weights move in weight space. For the simplest perception learning rule, each correction step will be of a fixed size. If the pattern classes are linearly separable, there will be a cone-shaped *solution region* in weight space. To prove that the perceptron can find a solution, it is necessary to show that eventually the weights will enter the solution region.

solution, so is any nonzero multiple of **w**. That is because the classification decision is made based only on the sign of the inner product.

To simplify bookkeeping, we can use a trick. We have two pattern classes in the training sequence, one with a negative inner product with the weights and one with a positive inner product. It will be a lot simpler if we replace all the examples of set $\{\mathbf{b}\}$ with their negatives. This means that the negative inner product of $\{\mathbf{b}\}$ with the weights now is positive because if

$$[\mathbf{b}_i, \mathbf{w}] < 0,$$

then

$$[-\mathbf{b}_i, \mathbf{w}] > 0.$$

We will call this *new* training set, which now contains only one class, $\{\mathbf{y}'\}$. Examples of $\{\mathbf{y}'\}$ will be written as $\mathbf{y}'_1$, $\mathbf{y}'_2$, and so on. We have assumed that our TLU definition requires the inner product between weights and input activity to be strictly greater than zero. All the examples of the training set, from both classes, are now on one side of the hyperplane, say the positive side (figure 8.11), so that if a set of weights, **w**, is a solution, $[\mathbf{y}'_i, \mathbf{w}] > 0$ for all the members of $\{\mathbf{y}'\}$.

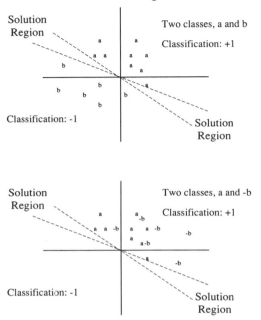

Replacement of the -1
Class with Its Negative

Figure 8.11 Illustration of one trick used to prove the perceptron convergence theorem. The pattern class **b** is replaced by its negative, moving both classes to the same side of the solution region. This means that for a set of weights to be a solution, all the pattern points will have a positive inner product between weights and pattern.

We want to show that learning ceases after a finite number of learning trials, that is, the weight vector now correctly classifies the training sequence. The proof will follow that in Nilsson's proof 2 from 1965 and is largely geometrical. We know what the solution region for $\{\mathbf{y}'\}$ looks like: it is the set of points where $[\mathbf{y}'_i, \mathbf{w}] > 0$ for all the members of $\{\mathbf{y}'\}$.

We know that to be in the solution region requires that the inner product between weights and patterns be greater than zero. But this means that we can choose a subregion of the solution region where the inner product is greater than any value we want. This is because strictly speaking, greater than zero means that the inner product for the training set must have a minimum value. By multiplying the weights by a constant, we can scale the smallest inner product to any value we want. This means that we can create what Nilsson calls an *insulated region* where all the inner products are greater than a constant value. Let m be the largest value of the length squared in the training set, that is, the maximum of $[\mathbf{y}'_i, \mathbf{y}'_i]$. Let b be an

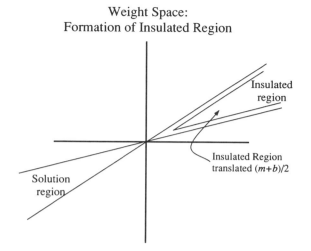

Figure 8.12 Formation of an *insulated region* in weight space. Weight vectors in the insulated region have an inner product with the members of the training set greater than a certain value.

arbitrary constant. We can choose the insulated region so that

$$[\mathbf{y}_i', \mathbf{w}] > \frac{(m + b)}{2}.$$

The edges of this new cone will be parallel to the old ones since they have simply been translated a distance $(m + b)/2$ (figure 8.12).

We would have our proof if we could show that every time the perceptron learned, the distance between a weight vector in the insulated region, which is a solution, and the new set of weights produced by learning always *decreased* by an amount greater than a fixed amount. Since the distance decreased is larger than a fixed and nonzero value, learning must stop after a finite number of learning trials. That is what we will try to do.

Suppose after a number, say k, of learning trials we have a set of weights $\mathbf{w}_k$. Suppose $\mathbf{w}$ is a weight vector in the insulated solution region. We can compute the distance squared between the two weight vectors easily:

$$|\mathbf{w} - \mathbf{w}_k|^2 = [\mathbf{w} - \mathbf{w}_k, \mathbf{w} - \mathbf{w}_k] = [\mathbf{w}, \mathbf{w}] - 2[\mathbf{w}, \mathbf{w}_k] + [\mathbf{w}_k, \mathbf{w}_k].$$

When the perceptron learns, we add the misclassified vector, $\mathbf{y}_i'$, to the weights, times a constant, c. For convenience we will assume $c = 1$. We do not lose generality in doing this, because we can rescale everything if necessary. Therefore the new weight vector, $\mathbf{w}_{k+1}$, since it has had $k + 1$ learning trials, is given by

$$\mathbf{w}_{k+1} = \mathbf{w}_k + \mathbf{y}_i'.$$

Consider the squared distance between the solution vector, $\mathbf{w}$, and the two set of weights, $\mathbf{w}_k$ and $\mathbf{w}_{k+1}$, the new one and the old one. We can form an expression for the distance between the two squared distances, which we will call d_{k+1}:

$$d_{k+1} = |\mathbf{w} - \mathbf{w}_k| - |\mathbf{w} - \mathbf{w}_{k+1}|$$

$$= |\mathbf{w} - \mathbf{w}_k| - |\mathbf{w} - (\mathbf{w}_k + \mathbf{y}'_i)|.$$

This quantity is the quantity that we want to show is greater than some minimum amount after every learning trial; that is, the old distance to a solution vector is greater than the new distance by a fixed value or larger. Expanding these expressions and doing a little algebra lead us to

$$d_{k+1} = -2[\mathbf{w}_k, \mathbf{y}'_i] + 2[\mathbf{w}, \mathbf{y}'_i] - [\mathbf{y}'_i, \mathbf{y}'_i].$$

We know that the inner product in the first term must be equal to or less than zero, because this pattern was misclassified. Therefore, the negative value means this first term must always be positive or zero. We can take it out and set up an inequality, so that

$$d_{k+1} \geq 2[\mathbf{w}_k, \mathbf{y}'_i] - [\mathbf{y}'_i, \mathbf{y}'_i].$$

But we know that, by assumption, patterns in the training set have a maximum value of length, which we called m. Therefore the maximum value for the last term in the expression is m, and we know that

$$d_{k+1} \geq 2[\mathbf{w}, \mathbf{y}'_i] - m.$$

But by the inequality that defined the insulated region, we know that the inner product between the solution, $\mathbf{w}$, and any of the members of the training set, $\mathbf{y}'_i$, is,

$$[\mathbf{w}, \mathbf{y}'_i] > \frac{(m + b)}{2}$$

and a little algebra has

$$2[\mathbf{w}, \mathbf{y}'_i] - m > b.$$

We can substitute this in our final expression for d_{k+1} and

$$d_{k+1} > b > 0.$$

Therefore, our desired result is true; subject to the assumptions we made, each time the perceptron learns, it reduces the distance between a solution weight vector and the current weight vector by a finite amount.

This result shows that if a solution exists, the perceptron learning rule will find it. That is an elegant and powerful result. It gives a mechanical algorithm for finding correct weights. The algorithm is also *local*, so that an individual connection only need know the A layer unit activity, the R layer unit activity, and the strength of the connection between them. It is a parallel algorithm because the requisite vector operations can be done in parallel.

As is typical of most neural network algorithms, it lends itself well to special-purpose hardware. Several special-purpose devices were constructed around 1960 to realize the perceptron and related algorithms. One functioning example was the Mark I perceptron made at Cornell Aeronautical Laboratories, which had a 40 by 40-unit retina connected to a video camera, and contained 512 A layer units and 8 R layer units. Synaptic weights were set with motor-driven potentiometers (Block, 1962).

One practical problem with perceptrons is that developing a completely correct classifier with a number of known test examples can be extremely slow. This is partly a function of the way learning is done: corrections are made only when the system is wrong. Therefore system learning slows down as the network becomes more accurate. Another problem is the difficulty the perceptron has dealing with a common problem, noisy categories. Very often the items to be discriminated are corrupted with noise. The same point may have different classifications in different samples. In this case, the perceptron will never stabilize, but may make large changes in weights forever.

Another conceptual problem is the nature of the solution. A region of the weight space contains correct solutions if they exist, that is, sets of weights that correctly discriminate the training sets, as we saw in figure 8.9. Any solution lying in the solution region is equally correct because it correctly classifies the training set. What part of this weight space of correct solutions is actually reachable during learning depends in part on the learning constant, c, since the final set of learned weights that correctly classifies the training set cannot penetrate the region with correct solutions deeper than c allows. For some purposes, parts of this correct solution region may be better than others, for example, areas that are somehow most distant from members of the training sets. This might allow better generalization to new examples that are not members of the training set. The only criterion the perceptron cares about is correct classification; nothing else is optimized about the solution discovered. (Variants of the perceptron act slightly differently in this respect. See the reviews in Nilsson, 1965, or Hertz, Krogh, and Palmer, 1991, for details.)

More than perceptron learning is going on in humans. We continue to learn a task after we finish performing it correctly. For example, in a classification task the response time continues to drop even when performance is perfect. The beneficial effects of practicing even well-learned tasks are known to all.

The Attack on Perceptrons by Minsky and Papert

The perceptron was, and is, powerful and easy to implement. Unfortunately, all during the 1960s it promised great commercial potential that never materialized: there just were not many practical problems that really required a perceptron. By the end of the decade, little hard analysis and few practical results emerged from perceptrons after years of work and a lot of government and industry money. Minsky and Papert were irritated for most of that time, with some justification, by the claims and overclaims of perceptron enthusiasts.

Minsky was one of the leading proponents of *symbol-processing artificial intelligence* (AI). Symbol processing claims that intelligence primarily involves logical or rule-based operations on symbols. The belief was that, first, this was the way all intelligent organisms worked because of the fundamental nature of intelligence, and, second, properly done, symbol processing was the only way to get truly intelligent machines. Although current thinking in AI has retreated somewhat from the stronger form of these claims, an early statement of this position was given in a progress report from the MIT AI Laboratory in 1972 (pp. 1–2):

Thinking is based on the use of SYMBOLIC DESCRIPTIONS and description-manipulating processes to represent a variety of kinds of KNOWLEDGE—about facts, about processes, about problem-solving, and about computation itself, in ways that are subject to HETERARCHICAL CONTROL STRUCTURES—systems in which control of the problem-solving programs is affected by heuristics that depend on the meaning of events. The ability to solve new problems ultimately requires the intelligent agent to conceive, debug, and execute new procedures. Such an agent must know to a greater or lesser extent how to plan, produce, test, modify, and adapt procedures; in short, it must know a lot about computational processes. We are not saying that an intelligent machine, or person, must have such knowledge available at the level of overt statements or consciousness, but we maintain that the equivalent of such knowledge must be represented in an effective way somewhere in the system.

Of course, the perceptron did not use symbols and did not have much in the way of complex control structures. It derived any higher-level structure that appeared in its responses from the learning of special cases. This is a general statement about neural networks that is true today. Explaining the apparent use of complex rule-governed behavior in language, or even the explicit use of high-level rules, that humans can indeed use on occasion is still a major difficulty for neural networks. No satisfactory way of approximating symbolic descriptions and explicit rule-governed behavior using low-level neural networks exists at present.

The thrust of Minsky and Papert's book was that perceptrons *theoretically* had some very important limitations and *could not* compute some things that it was important to compute. Even a third edition in 1988 did not change the authors' generally negative view of neural networks.

Their argument was brilliant, clear, and incisive, and is a classic in the theory of computation. It is also true. That is, computational limitations of the type that they suggest are typical of neural nets. They can be overcome in special cases, but they reflect intrinsic properties. Neural net enthusiasts have spent much time denying these limitations, when their efforts might have been better applied investigating the strengths of the architecture.

Minsky and Papert's analysis used the fact that the TLUs from which perceptrons are built are essentially McCulloch-Pitts neurons. That means that they could be considered to be doing inner products between weights and input patterns, which is the customary interpretation used today, or they could be considered to be computing logical predicates. Minksy and Papert started by defining a TLU, but they cast it in terms of computing logical predicates. Since the TLU gives an all-or-none output, this corresponded to logical true or false.

The researchers considered the operation of a perceptron as a two-stage logical process. First, a number of simple properties or features are computed; they are simple because they are easy to compute. This first step is really a data-representation step. The early perceptron made a strong assumption about data representation; that the retina was *partially* connected to the A layer units. The implicit claim was that this representation was adequate to solve many problems. The authors showed that it was not.

Second, the simple features are combined to form the output. Minsky and Papert, like almost everyone else, assumed a simple weighted linear combination of the features, the familar inner product between weights and input activities. The inner product gives rise to a TRUE output if it is above a threshold, or a FALSE if it is equal to or below it. The computational problem now becomes, what logical predicates could perceptrons compute and what could they not compute? As we mentioned, and as was well known, only linearly separable pattern classes can be separated by the perceptron.

Consider one very simple case: the two logical predicates, Inclusive-OR (INC-OR) and Exclusive-OR (X-OR). Suppose the logic unit has two input lines, as we discussed for McCulloch-Pitts neurons. This system forms a two-dimensional state space (figure 8.13). For Inclusive-OR it is easy to

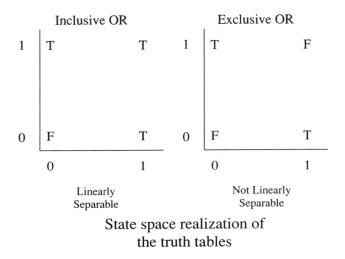

State space realization of
the truth tables

Figure 8.13 State space realization of the simple logic functions Inclusive-OR and Exclusive-OR.

Expanded Representation
Solution for Exclusive Or

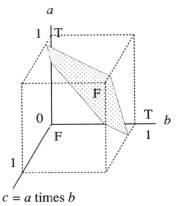

$c = a$ times b

Figure 8.14 By expanding the input state space representation, Exclusive-OR can be solved. One possible separating plane is shaded.

draw a straight line so it dichotomizes the state space into the correct true and false regions. However, it is not possible to find a single straight line that correctly segments the space for X-OR. Therefore it is not possible to compute X-OR with a perceptron using this representation.

Another way to solve X-OR in practice is by changing the data representation. Suppose we allowed interaction between the two inputs, so we generated a new input to the cell, c, that was the logical product of a and b. This new representation can solve the problem because the resulting three-dimensional state space for X-OR is now linearly separable (figure 8.14). However the product requires another unit to do the expanded representation. This additional unit is neither an input unit nor an output unit. In later work it was called a *hidden unit* (see chapter 9) because it was not directly accessible at the input or output.

The moral is that, with enhanced representation, it is often easy to compute what you want to compute. But this effectively involves multiple-layer systems. The assumption of the classic perceptron is that the data representation formed from taking a partial retinal input is adequate. For every network, when an input representation is presented to an input layer, strong assumptions about the data representation determine whether or not an explicit hidden layer interaction is necessary to solve the problem.

Connectedness

The heart of the analysis of perceptrons by Minsky and Papert was in what would seem to be their home ground, geometry and, effectively, visual perception. It was not realized until it was studied for a while just how difficult visual perception is. Over half the primate brain is concerned to some degree with visual information processing. A multitude of specialized modules handle different aspects of visual analysis. The idea that a retina could feed directly into a simple one-layer learning device and compute a complex visual discrimination is ludicrous. But in the early days of perceptrons, and in the early days of AI as well, this difficulty was not appreciated. The claim that a retina could feed into a perceptron, which could then compute the difference between a square and a triangle, seemed reasonable.

Minsky and Papert considered how to compute the geometric predicate *connectedness*, an intuitive predicate that people can see immediately in simple figures (figure 8.15). We cannot do justice to their full analysis, but it is possible to show that a perceptron with reasonable limitations *cannot* compute connectedness for some simple cases.

Connected Figures

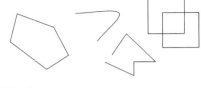

Not Connected

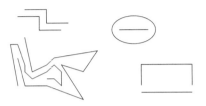

Figure 8.15   Examples of connected and disconnected figures.

Assume, as Rosenblatt did, that the perceptrons are looking at a retina, which is composed of discrete points (pixels). The retina is like ours in that an image is projected on it, and we will use line drawings so that pixels are only on or off. The usefulness of the perceptron is that the low-level feature computations are somehow simple. Minsky and Papert suggested that there were two different kinds of simplicity (figure 8.16). First, connectivity could be *order limited*. This meant that a unit had only a small number of actual connections, compared with the number of potential connections. Second, connectivity could be *diameter limited*. This meant that the unit looked at a restricted region of the retina. That is, all the retinal locations of connections from the retina of an A layer unit could be contained within a circle of some fixed diameter. Experimental observation finds that visual units in the earlier parts of the mammalian brain seem to be both order and diameter limited.

It is essential to assume some limitations on connectivity of the connections from the retina. Otherwise a simple perceptron *could* compute connectivity, or any other predicate based on retinal activity, for that matter, although in a trivial way. Suppose we had a fully connected retina, where every A layer unit was connected to every retinal unit. Suppose we had a total of n units in the retina. Then, if we had 2^n units in the A layer, we could compute connectedness just by having each A layer unit respond to one and only one pattern of retinal activity and connect to the appropriate R layer grandmother cell. This "solution" is shown in figure 8.17.

It is moderately difficult to show that order-limited perceptrons cannot compute the predicate connectedness. But it is easy to show that a diameter-

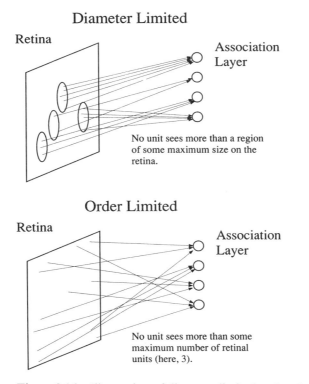

Figure 8.16   Illustration of diameter-limited and order-limited perceptrons.

limited perceptron cannot compute connectedness. The proof in *Perceptrons* is so simple it is worth showing.

Because perceptrons are diameter limited, let us consider very long figures, larger than the diameter, such as the pattern in figure 8.18. Since diameters are small relative to the size of the pattern, there are only three possibilities for the units that see the pattern. The first group (group 1) sees only the left side of the pattern, the second group (group 2) sees only the right side, and the third group (group 3) sees the middle but not the ends. Each input computes simple, necessarily local, functions on its input. It is critical to the proof that no units see both ends of the figure at once. This means that for the predicate connectedness to be computed, the inner product between weights and input pattern that feeds the connectedness unit has three parts, 1, 2, and 3, corresponding to the three different groups of cells.

But look at the four patterns in figure 8.19. Let us start with the connected figure on the top. Because this is a connected figure, the inner product between the pattern and the weights must be greater than zero.

Suppose we move a line segment from the bottom of the left side of the figure to the top of the left side, as shown in the second line of figure 8.19.

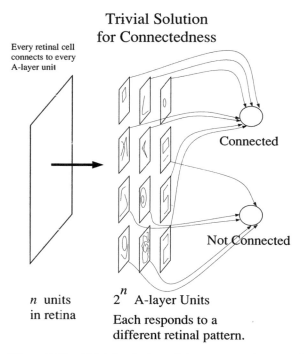

Figure 8.17   Trivial solution of the connectedness problem by a simple perception. Give me enough units and I will categorize the world …

The resulting pattern is now not connected. Therefore the inner product between pattern and weights must now be *less* than zero to signal "not connected." But the only units that saw the change were the units (group 1) that saw the left end. Therefore contribution to the value of the inner product from the group 1 must have dropped sufficiently to force the overall sum to be less than zero.

We do the same manipulation to the right side of the starting figure, moving a line segment from the top of the right side to the bottom. Again the resulting figure is not connected. In this case, only the units in group 2 saw the change. Therefore the contribution from the group 2 units must have dropped so that the overall sum is less than zero.

In the final step of the proof, shown on the last line of figure 8.19, we move both the line segments at once. The resulting figure is connected, so the inner product must be greater than zero. But the left end of the figure is identical to the "not-connected" figure in the second line, and the right end of the figure is identical to the not-connected figure in the third line. To signal not-connected, the outputs from both the group 1 (left end) and group 2 (right end) units had to drop so as to force the overall inner product below zero. Therefore the new connected figure must have an inner product less

Diameter Limited Perceptron

(Redrawn from Minsky and Papert)

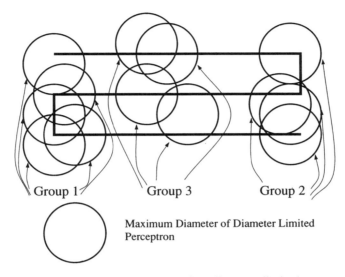

Maximum Diameter of Diameter Limited
Perceptron

Figure 8.18 The figure shows that diameter-limited perceptrons cannot compute connectedness. Group 1 units only see the left end of the figure and group 2 units see only the right end of the figure. No units see both ends at once.

Proof by Contradiction

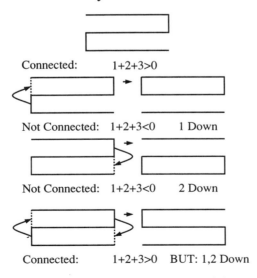

Figure 8.19 Visual demonstration of the connectedness argument.

than zero. But we claimed that this diameter-limited perceptron could correctly compute connectedness. It cannot, because such an assumption leads to a contradiction.

This result shows that a perceptron has fundamental computational limitations. The argument is essentially similar to that used for X-OR. Both cases involve a violation of monotonicity, that is, more of something makes the answer the same as having none of the something.

We must now investigate our own motives. Are we interested in building computational devices, or in understanding cognition and perception? If we are interested in understanding human behavior, we should now ask if humans or animals show the same computational weaknesses in some areas as perceptrons. If this is so, the perceptron might be a useful model of the mind, and its limitations would be reflected in human performance.

People and even animals can cope with X-OR, although with a little difficulty. If one tries to teach a discrimination that is a realization of logical X-OR to rats, what happens is that the rats pick up Inclusive-OR almost immediately, which gives 75% accuracy. Then, with difficulty, they increase their accuracy somewhat, but only over many training sessions (Griffith, Davis, and Kause, 1968).

Connectedness shows a similar pattern of difficulty. Connectedness or lack of it in the simple figures in figure 8.15 can be seen immediately. But figure 8.20 shows two figures that appear on the most recent edition of *Perceptrons*. (The first edition had the same figure, using curved lines.) One of the figures is connected and the other is not. There is no immediate perception of connectedness. It is necessary to trace out the contours, that is, to become a serial machine, to determine connectedness. Perhaps perceptrons cannot compute connectedness in general, but then perhaps humans cannot either, at least in the initial stages of perception.

So, during the dark ages of the field, neural networks were used to investigate human information processing. It turned out that a number of interesting psychological models could, indeed, be based on the simplest neural networks. Even with their severe theoretical processing limitations, networks made interesting and plausible models of human behavior. Although interest might have been lost in the larger engineering and computer science communities, neural networks were kept alive and well among cognitive psychologists.

Probably the Minsky and Papert book had its most damaging effect on granting agencies. The money stopped, hitting the area where it hurt most and causing the most significant harm. The problem was compounded by some speculations made by the authors. After a brilliant book of analysis

Figure 8.20 Cover of the third edition (1988) of Minsky and Papert's *Perceptrons*. Which figure is connected and which is not connected? Reprinted by permission.

they concluded that even multilayer perceptrons could not do any better than simple ones. In a much-quoted passage they said (pp. 231–232):

The problem of extension (to many layers) is not merely technical. It is also strategic. The perceptron has shown itself worthy of study despite (and even because of!) its severe limitations. It has many features to attract attention: its linearity; its intriguing learning theorem; its clear paradigmatic simplicity as a kind of parallel computation. There is no reason to suppose that any of these virtues carry over to the many-layered version. Nevertheless, we consider it to be an important research problem to elucidate (or reject) our intuitive judgment that the extension is sterile.

Coming after their virtuoso analysis, this judgment carried great weight. The granting agencies effectively said, if simple perceptrons are too limited to be of any value, and complex ones share the same limitations, why support them? The agencies were not willing to continue support when even the critics themselves said that perceptrons had some important research problems that were still worth studying. Unfortunately, without research funds, this "important research problem" was not investigated for a number of years. Now, many techniques can teach multilayered systems of perceptron-like elements some of the things that cannot be learned by a single-layer system. Also, people have a much more realistic acceptance of what networks can and cannot do, as well as some appreciation of the virtues of simple neural networks for explaining human cognition.

References

H.D. Block (1962), The perceptron: A model for brain functioning. I. *Reviews of Modern Physics, 34*, 123–135.

H.D. Block, B.W. Knight, Jr., and F. Rosenblatt (1962), Analysis of a four-layer series coupled perceptron. II. *Reviews of Modern Physics, 34*, 136–142.

V.V. Griffith, J.A. Davis, and R.H. Kause (1968), Learning of the exclusive-OR logic function in rats. In H.L. Oestreicher and D.R. Moore (Eds.), *Cybernetic Problems in Bionics*. New York. Gordon Breach.

J. Hertz, A. Krogh and R.G. Palmer (1991), *Introduction to the Theory of Neural Computation*. Redwood City, CA: Addison-Wesley.

M. Minsky and S. Papert (1969), *Perceptrons*, 1st ed. Cambridge: MIT Press. (Third edition, 1988).

M. Minsky and S. Papert (1972), *Artificial Intelligence: Progress Report*, Artificial Intelligence Memo No. 252. Cambridge: MIT AI Laboratory.

N. Nilsson (1965), *Learning Machines*. New York: McGraw-Hill. (Reprinted by Morgan-Kauffman Publishers, Los Altos, CA.)

F. Rosenblatt (1958), The perceptron: a probabilistic model for information storage and organization in the brain. *Psychological Review, 65*, 386–408.

9 *Gradient Descent Algorithms*

... feed my brain with better things.
G.K. Chesterton

Striving to better, oft we mar what's well.
Shakespeare
King Lear, I, iv, 371

This chapter describes the basic ideas behind the gradient descent *learning algorithms that minimize network error by modifying weights using supervised learning. The gradient is reviewed briefly. The early ADALINE model of Widrow and Hoff is discussed in detail. This technique provides a simple and effective way to minimize network mean square error. An important generalization of gradient descent to a nonlinear, multilayer feedforward network, called* backpropagation, *is derived. Backpropagation is a supervised algorithm that learns by first computing an error signal and then propagating the error backward through the network by assuming the network weights are the same in both forward and backward directions. Three well-known applications of backpropagation are described: NETtalk, image compression, and digit recognition. Some of the strengths and weaknesses of gradient descent are discussed.*

In the last chapter we saw our first iterative learning algorithm, one where a rule is applied over and over again. The perceptron simply kept on learning until it got it right, if that was possible. The perceptron learning theorem guarantees that, if a solution is possible, it will be found eventually. At present, the vast majority of neural network systems use some kind of iterative learning procedure. The majority of the vast majority use one or another of several *gradient descent* procedures discussed in this chapter. The best known and most widely used, *backpropagation*, is an iterative gradient descent procedure.

Widrow-Hoff Gradient Descent Procedure

Let us look at a useful extension of the linear associator, an error-correcting technique sometimes called the Widrow-Hoff procedure, the LMS algorithm (LMS stands for least mean squares, the error quantity the technique tries to minimize), or the delta method.

This technique was originally proposed in 1960 by Widrow and Hoff, who were active in the first wave of enthusiasm for neural networks. The error-correcting technique was proposed in the context of improving the behavior of a device called the ADALINE, a simple neural model. ADALINE originally stood for the initial letters of the phrase, ADaptive LInear NEuron. When neural networks became less popular, ADALINE became an ADaptive LINear Element. However, it did exactly the same thing as before, and was used for a number of practical applications in filters, telecommunications, and adaptive antennas.

Based on the correctness or incorrectness of its immediate past classification, the perceptron changed its weights. The learning rules sometimes took a long time to converge to a set of weights that classified correctly. Even if successful weights were found, they were rarely unique and it was not clear if they were optimum in any sense. The ADALINE learned quickly and accurately, and more to the point, the way it worked was clear.

Widrow and Hoff assumed an input pattern. The adaptive neuron then produced an output classification of the input pattern. Initially, the researchers assumed allowable input and output activity patterns were binary, that is, elements could only take values of $+1$ or -1. They used a *supervised* algorithm where the correct answer was known for a particular input pattern.

In their initial analysis, Widrow and Hoff considered the behavior of a *single neuron*. So far in this book we have emphasized the behavior of groups of neurons. However, we will follow tradition and analyze the behavior of a single computing element in this chapter. Most of the ideas generalize immediately to groups of neurons. The computing unit Widrow and Hoff used they called an adaptive neuron and was a straightforward modification of the generic connectionist neuron, discussed at the end of chapter 2 (figure 9.1).

The first stage of the neuron computed the inner product between the input pattern and the set of synaptic weights. To avoid mathematical problems with thresholds, a bias weight was added, that is, a weight always connected to a value of $+1$. This was the same trick that was used to simplify the analysis of the perceptron in chapter 8. The output of the first

Widrow-Hoff Computational Unit

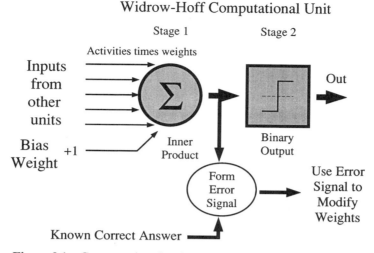

Figure 9.1 Computational unit used in the Widrow-Hoff gradient descent algorithm. It is the threshold logic unit used in the perceptron with the additional ability to compare the output of the first stage (inner product) and the correct output. An *error signal* is formed that measures the difference between what the value of the inner product from the first box is and what it should have been.

stage was fed to the nonlinear second stage that was a threshold device, with only two possible output states, $+1$ and -1. The output value was determined by whether the output of the first stage was equal to or above zero $(+1)$ or below zero (-1). The only difference between this adaptive neuron and the generic connectionist neuron is that it is assumed that the supervisor has access to the output of the first box, the inner product, as well as to the final classification.

When we looked at the threshold logic unit in the perceptron we only knew the output state, $+1$ or -1. The perceptron learned only when the final state was wrong. Who knew or cared what was happening inside the computing unit? With the assumption that the output of the first stage was available, it meant that the teacher or supervisor could compute an *error signal* between what the output was supposed to be $(+1, -1)$ and what the summer actually computed. Figure 9.1 indicates where the error signal is computed. This assumption meant that the network could have a nonzero error, even though it got the right classification.

In the original ADALINE it was assumed that the correct output of the first, linear stage should be the same as the output of the second, nonlinear stage, that is, $+1$ or -1. Change of synaptic weights could continue even if the classification was correct, because the output of the first stage, the inner product, was rarely exactly $+1$ or -1. One reason perceptrons took so long to learn was that synapses were changed only when the output classification was incorrect.

Analysis: Weight Space

We discussed error signals in chapters 7 and 8. Widrow and Hoff assumed that we are interested in reducing the square of the error signal (a scalar) to its smallest possible value. The square of the error signal is used, of course, because a scalar error signal could be positive or negative. If the error signal is a vector, as it was in chapter 7, the error term would represent a distance.

Consider the weights of the adaptive neuron. Suppose there are n of them. Let us consider this set of weights as being a point in an n-dimensional space, which we will call a *weight space*. (We first introduced this idea in chapter 8.) Suppose we have a set of patterns that we want our ADALINE to discriminate, that is, to classify the patterns into two groups, one with a $+1$ output and one with a -1 output. Every time the ADALINE sees a pattern it generates an error signal. This means that every set of weights has associated with it a scalar error value; if we change the weights we get a new error value. The error values for every set of weights define a *surface* in the weight space.

What is the function of learning in this network? One reasonable way to view the task of learning is to require learning to *reduce the error* to the greatest degree possible. An astonishing number of important problems can be put in the form of minimizations (or maximizations) of some function. In the learning rule, we want to reduce the error function as much as we can. In business, we might want to minimize cost or maximize profit. We might want to minimize stress in our lives, or maximize happiness, which may or may not be the same thing. There are a great many ways to do minimization; however, this problem is often extremely difficult both in principle and practice. In the neural network field, one of the first to point out the close connection between the operation of a neural network and a minimization (or maximization) problem was Oliver Selfridge (1958) with his Pandemonium model.

One obvious way to find a minimum would be to look at very many points, that is, brute force. In general, this approach will not work because the problem will rapidly outrun the capabilities of even the largest computer in the world. Consider an ADALINE. Suppose it has one weight. Let us assume that if we divide up a weight into 10 pieces, that will be good enough for an adequate approximation to the minimum. This assumption means

Error Landscape
in Weight Space

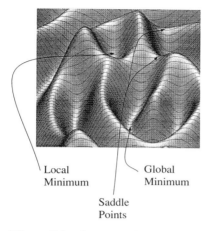

Local   Global
Minimum   Minimum

Saddle
Points

Figure 9.2   An error landscape in *weight space*. Total error is a function of the values of weights, so every set of weights has an associated error. When the errors are plotted, they form a surface in a weight space. These surfaces potentially can have *local minima*, for example, mountain valleys, and *global minima*, the lowest overall error. *Saddle points* are sometimes found where surfaces go up in one direction and down in others. Surface courtesy of F. Reichel.

that we only have to make computations, and choose the value of weight that gives the lowest error. However, if we have two weights in our ADALINE, we must compute 10^2 values and choose the smallest. If we have 100 weights (one network we demonstrated in chapter 7 had 200 weights per unit) we must do 10^{100} computations to find the minimum. This is orders of magnitude more than the number of particles estimated to be in the universe.

That is not the way to proceed. We must be more clever in how we perform our search. Our problem is that of a hiker in the fog, we cannot see where the valleys are; however, we can see the local topography around our feet. One thing to do would be always to take steps downhill. Eventually we will come to a point at which we cannot go any farther, because every direction is up. To a hiker, this would be the lowest point in a valley. Such a point it is called a *local minimum.* There may be many local minima, and some local minima may be higher than others. The lowest minimum is called the *global minimum* (figure 9.2). Unfortunately, complex error surfaces have many local minima. An algorithm that always goes downhill is *not* guaranteed to find the global minimum except in the glorious situation where there is only one minimum and that is the global minimum. We will see that this is the case for the Widrow-Hoff rule.

Directional Derivative

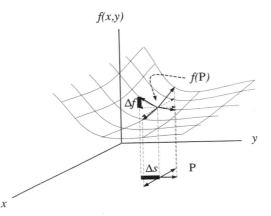

Figure 9.3   A directional derivative. The rate of change of the function at a point P is a determined by the direction moved from P.

Going downhill will eventually get us to a minimum. However, suppose we are impatient because (computer) time is money. We want to get to the bottom as fast as possible. Let us return to our error surface. If we can only move a fixed amount, we want to know what direction to move that will make the *largest* reduction in error.

The Gradient

The direction of steepest descent is straightforward to compute and it is called the *gradient*. The following discussion is roughly the same as that in many elementary calculus books (see Thomas and Finney, 1988, for a clear description).

We all know that the derivative of a function of one variable, $f(x)$, involves evaluating the function at $f(x)$ and $f(x + \Delta x)$, subtracting the two values, forming the quotient, and taking the limit

$$\underset{\Delta x \to 0}{\text{limit}} \left(\frac{f(x + \Delta x) - f(x)}{\Delta x} \right) = \frac{dy}{dx}.$$

The derivative is a measure of how rapidly the function is changing. When the function is of higher dimensionality than 1, the deriviative is not so straightforward. For example, in figure 9.3 the two-dimensional function, $f(x, y)$, gives different values for $\Delta f(x, y)$ depending on the direction moved. We can define a directional derivative, dw/ds, by analogy with the one-dimensional derivative. Now, however, the function is a little more compli-

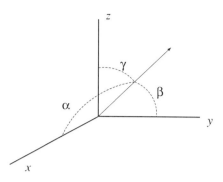

Figure 9.4   Direction cosines for a three-dimensional space. The angles α, β, and γ determine direction in space.

cated: Δw is the change in the function $f(x, y)$, and Δs is the amount moved, in the direction given by angle ϕ.

$$\underset{\Delta s \to 0}{\text{limit}} \left(\frac{f(w_1) - f(w_0)}{(\Delta x^2 + \Delta y^2)^{1/2}} \right) = \frac{dw}{ds}.$$

In this two-dimensional case we can take limits and show that

$$\frac{dw}{ds} = \frac{\partial f}{\partial x} \cos \phi + \frac{\partial f}{\partial y} \sin \phi.$$

A similar expression can be derived for the directional derivative in three dimensions, using the direction cosines, and by extension for higher dimensions. In three dimensions, suppose the angles with the x, y, and z coordinate axes are α, β, and γ (figure 9.4). Then, it can be shown that

$$\frac{dw}{ds} = \frac{\partial f}{\partial x} \cos \alpha + \frac{\partial f}{\partial y} \cos \beta + \frac{\partial f}{\partial z} \cos \gamma.$$

The directional derivative is not a vector quantity, but its computation contains components that are vector quantities. A good strategy for computing a directional derivative might be to break the computation into its natural vector parts. The first part is the direction of the directional derivative. A standard notation in engineering and physics is to define unit vectors **i**, **j**, and **k** in the directions of the x, y, and z axes. Then we can define a unit vector, **u**, in the direction we take the directional derivative, as

$$\mathbf{u} = \mathbf{i} \cos \alpha + \mathbf{j} \cos \beta + \mathbf{k} \cos \gamma + \cdots$$

We just write three terms of the vectors, but the expressions can be extended to higher dimensions. At this point we *define* a vector of partial derivatives,

$$\mathbf{v} = \mathbf{i}\frac{\partial f}{\partial x} + \mathbf{j}\frac{\partial f}{\partial y} + \mathbf{k}\frac{\partial f}{\partial z}\cdots$$

The reason that we went to all this effort is because we can now compute the directional derivative as an inner product between two vectors. Consider the inner product, $[\mathbf{u}, \mathbf{v}]$. When multiplied out, we get

$$[\mathbf{u}, \mathbf{v}] = \frac{\partial f}{\partial x}\cos\alpha + \frac{\partial f}{\partial y}\cos\beta + \frac{\partial f}{\partial z}\cos\gamma\cdots = \frac{dw}{ds}.$$

The vector $\mathbf{v}$ has several names:

$\mathbf{v}$ = the gradient of $\mathbf{f}$ at point p_0,

$\quad$ = grad $\mathbf{f}$,

$\quad$ = $\nabla\mathbf{f}$ = "del" $\mathbf{f}$.

Let us return to the original problem: how to find the direction that will move us downhill the fastest, based only on local information. This means that we want to find the largest value of dw/ds, or of $-dw/ds$, depending on whether we want to climb the hill or descend into the valley. This computation is easy if we view it in terms of a dot product. At the point p_0, we can write the inner product as

$$[\mathbf{u}, \mathbf{v}] = \frac{df}{ds} = |\nabla\mathbf{f}||\mathbf{u}|\cos\theta,$$

where θ is the angle between the direction and the gradient. This expression has a maximum when the cosine of θ is $+1$ and a minimum when the cosine of θ is -1. So the function *increases* most rapidly when the direction of movement is *in the direction of the gradient*. The function *decreases* most rapidly when the direction of movement is *in the direction of the negative of the gradient*.

The Widrow-Hoff Algorithm

The gradient tells what direction to move to get downhill fastest. Therefore, we want to adjust the weights so that the change moves the system down the error surface in the direction of the locally steepest descent, given by the negative of the gradient. To do this means we must be able to compute the gradient. There is no particular reason why this should be easy, since it involves computing a (very large) number of partial derivatives. However, for the problem as formulated by Widrow and Hoff, computing the gradient is amazingly simple.

Gradient Calculation

Widrow and Hoff showed how to compute the gradient in their 1960 technical report. It is worth going through this calculation because the strategy they followed was so influental on later neural network research. More details about the algorithm, more detailed calculations, variants, extensions, and applications to signal processing can be found in Widrow and Stearns (1985). The overall goal of learning is to minimize the total error for the entire set of patterns to be learned. However, learning proceeds in practice by correcting the error on a single pattern, then correcting it on another pattern, and so forth. Correcting the error for one pattern may increase it for another, but on the average, the network will be following the gradient.

Let us consider a computing unit with n weights $(w[1],\ldots,w[n])$, and a bias weight, $w[0]$, always connected to the value $+1$. We will assume we are learning an input pattern, **f**. Suppose the correct output has the scalar value, t, for target value. This value is known to the supervisor. (Remember, we are analyzing a single unit.) Let us follow the original derivation and assume that we only have two weights, $w[1]$ and $w[2]$, and a bias weight, $w[0]$. Generalizing to any number of weights is straightforward. The error, e, for pattern, **f**, is given by

$$\text{error} = e = t - w[0] - w[1]f[1] - w[2]f[2].$$

We are interested in minimizing the *square* of the error. Therefore, we must square the error term, so

$$e^2 = t^2 + w[0]^2 + (w[1]f[1])^2 + (w[2]f[2])^2$$
$$- 2tw[0] - 2tw[1]f[1] - 2tw[2]f[2]$$
$$+ 2f[1]w[0]w[1] + 2f[2]w[0]w[2] + 2f[1]f[2]w[1]w[2].$$

To compute the gradient, we have to compute the partial derivatives of the square of the error for each weight, that is,

$$\frac{\partial(e^2)}{\partial w[0]}, \frac{\partial(e^2)}{\partial w[1]}, \quad \text{and} \quad \frac{\partial(e^2)}{\partial w[2]}.$$

As a demonstration, let us compute one of these derivatives.

$$\frac{\partial(e^2)}{\partial w[0]} = f[1](-2t + 2w[0] + 2w[1]f[1] + 2w[2]f[2]).$$

Sharp eyes will notice that the term in parentheses is the error, e, term given above, multiplied by -2.

In a supervised learning system, we assume that we can actually *measure* the error term, by forming the difference between actual and desired output. If that is the case, we can compute all our partial derivatives from the error term, *e*, by

$$\frac{\partial(e^2)}{\partial w[j]} = -2f[j]e.$$

Since we are minimizing the error, we want to use the negative of the gradient. *Therefore, we want to change the weight by an amount proportional to the error times the input weight*, that is,

$$\Delta w[j] = \eta f[j]e,$$

where η is a learning constant. This expression forms a Hebbian learning rule, with the desired output, *t*, replaced by the error in the output. Because we are learning the difference (Δ) between desired and actual outputs, this technique is sometimes called the *delta rule*. We have only computed a weight change due to a single pattern. If we present many patterns, however, changing weights after each one according to this rule, the learning will follow the gradient on the average.

The best possible situation for finding the global minimum would be to do gradient descent in a space where there was only a single minimum. That is the case here. Our computation of error squared shows us that all we can ever have in our expression for error is constants, various combinations of first powers, and squares of weights. This means that the error surface in a weight space of any dimension is a *paraboloid*, that is, a high dimensional parabola, a function that has only a single minimum. Therefore, gradient descent starting from anywhere will find the global minimum. Figure 9.5 sketches this geometry.

Another way of deriving this rule is the one used in the best-known derivation of the delta rule and its generalization, backpropagation (Rumelhart, Hinton, and Williams, 1986a). It uses the chain rule for partial derivatives. Ths calculation is simple:

$$\frac{\partial(e^2)}{\partial w[j]} = \frac{\partial(e^2)}{\partial(e)} \frac{\partial(e)}{\partial w[j]}.$$

We know the error, *e*, is

$$e = \left(t - \sum_j w[j]f[j] \right),$$

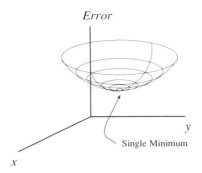

Figure 9.5   Widrow-Hoff learning in a simple network has the agreeable property that simple gradient descent always finds the global minimum, since there is only a single minimum error in the weight space. All downward paths take you there.

so

$$\frac{\partial(e)}{\partial w[j]} = -f[j],$$

and

$$\frac{\partial(e^2)}{\partial w[j]} = 2e\frac{\partial(e)}{\partial w[j]} = 2e(-f[j]) = -2ef[j].$$

Widrow and Hoff's paper was greatly concerned with practical implementations and applications for these learning rules. Concrete suggestions for devices to realize the rules were made, and some pilot hardware was constructed for demonstrations. At recent neural network meetings, Widrow has given demonstrations of one or another of the original ADALINEs, which have worked perfectly for 25 years, much to the delight of the audience. It is only recently that a similar flowering of hardware designs has occurred as that which flourished in the first golden age of neural networks.

Vector Widrow-Hoff

It is easy to generalize this rule, derived for a single unit, to the pattern learning that we have emphasized in this book. The learning rule is *local*. To change a weight, all that is required is information that is present at the single adaptive element: the input, the output, and what the output should have been, allowing computation of the error signal. The adaptive unit does not care what other neurons are doing in a single-layer, multiunit system.

A simple change in the programs discussed in chapters 6 and 7 will allow us to implement a system that will correct errors. In the linear associator, suppose we have an input, **f**, and a desired output, **t** (for target pattern). If we want to associate **f** and **t**, we construct the outer product matrix according to the learning rule:

$$\mathbf{A} = \eta \mathbf{t} \mathbf{f}^T.$$

There is no feedback as to correctness of association. We saw that **t** is unlikely to be reproduced exactly if more than a few associations are stored.

Let us use the Widrow-Hoff algorithm to improve accuracy. Suppose we want to reproduce the association, **f** → **t**, that we have learned, but because other information is stored, when **f** is the input, our matrix **A** gives rise to

$$\mathbf{A}\mathbf{f} = \mathbf{g}.$$

Unfortunately, the actual output **g** is not exactly the correct association, **t**. We form the vector difference (**t** − **g**). This difference is the *error signal* we saw before, although now it is a vector.

Each individual term can be used to implement the Widrow-Hoff procedure for its unit. We made the point before that because the Widrow-Hoff is learning the product of the error and the input, it is implementing the outer product generalized Hebbian learning rule, with the error vector in place of the output vector. This makes intuitive sense. The final matrix will eventually be the sum of the desired associations plus a number of correction terms, all of which tend to reduce the error.

Let us describe a single iteration. Let us call the old matrix **A** and the correction matrix we are going to add to **A**, Δ**A**. Let us assume η is a learning constant. Now, suppose we learn

$$\Delta \mathbf{A} = \eta (\mathbf{t} - \mathbf{g}) \mathbf{f}^T$$

or, equivalently,

$$\Delta \mathbf{A} = \eta (\mathbf{t} - \mathbf{A}\mathbf{f}) \mathbf{f}^T,$$

and add this to the **A** that already exists.

Error correction will not be perfect. Therefore we will have to keep learning corrections until the network gets the right answer to some required degree of accuracy; it is an iterative algorithm like the perceptron. In summary,

1. Choose a random input-output pair (**f** → **t**) from the training set.

2. Let the network operate on the input, **f**, to generate an output, **g**.

3. Compute the error vector, $(\mathbf{t} - \mathbf{g})$.

4. Add a matrix $\Delta\mathbf{A}$ to the connection matrix, $\mathbf{A}$, which is proportional to the outer product, $(\mathbf{t} - \mathbf{g})\mathbf{f}^T$, between the error vector and the input.

5. Choose another random pair and do the correction again.

6. Continue until the error is small enough.

If the matrix $\mathbf{A}$ is initially zero, the output pattern $\mathbf{g}$ will be zero and the error correcting learning rule will produce the same outer product Hebbian rule used for the linear associator. If the matrix gives perfect responses, that is, $\mathbf{g} = \mathbf{t}$, there will be no learning at all since $\Delta\mathbf{A}$ will be zero. If the entire set of inputs gives the correct associated outputs, learning ceases. Error correction as we have described it heads toward a system where the actual output vector is of the same length and direction as the desired output vector. The decoupling of length and direction discussed in chapter 7 is not so natural for this algorithm.

If $\eta = 1/[\mathbf{f}, \mathbf{f}]$, the reciprocal of the square of the length of $\mathbf{f}$, error correction can be made perfect for a single pair of associations, $\mathbf{f} \to \mathbf{t}$. Suppose we are working with the associated pair of vectors, $\mathbf{f} \to \mathbf{t}$ and

$$\mathbf{A}\mathbf{f} = \mathbf{g}.$$

We increment the matrix $\mathbf{A}$ according to the rule so that the new $\mathbf{A}'$ after the matrix has learned is given by

$$\mathbf{A}' = \mathbf{A} + \frac{(\mathbf{t} - \mathbf{g})\mathbf{f}^T}{[\mathbf{f}, \mathbf{f}]}.$$

Then, if $\mathbf{f}$ is the input to the system,

$$\mathbf{A}'\mathbf{f} = \mathbf{g} + \frac{(\mathbf{t} - \mathbf{g})[\mathbf{f}, \mathbf{f}]}{[\mathbf{f}, \mathbf{f}]}$$

$$= \mathbf{t},$$

which is the correct association.

Some care has to be taken with the value of η. If it is too large, the system will oscillate. That is because the correction will be too large and will overshoot the target. If η is very large, the overcorrection will be larger in magnitude than the original error. Then the size of the correction will increase in magnitude indefinitely as the output gyrates around the correct answer, and the network will "blow up." This can happen suddenly if parameters in simulations are changed, and is a common sources of malfunctioning simulations. However, this error is so dramatic it is easy to find and fix. If the learning constant is too small, the system will take a long time to converge to a correct answer. If η is a constant, **A** may never converge to an unchanging value but will oscillate around the best solution.

If η gradually drops toward zero as learning progresses, eventually the matrix will cease to change, even if errors are not completely corrected. It can be shown that the matrix in that case is the best one in the sense of linear regression; that is, the matrix will minimize the mean square error between the actual and desired outputs over the input set. A trick sometimes used is to let

$$\eta = \text{constant}/n,$$

where n is the number of learning trials. Overall, the Widrow-Hoff rule is a stable, reliable technique and insensitive to choice of parameters. It is one of those algorithms that "wants to work."

Short-Term Memory

Suppose we have a set of associations, $\{\mathbf{f}_i \rightarrow \mathbf{g}_i\}$. Suppose we present pairs from this set to the learning system at random. If η is relatively large, on the order of $1/[\mathbf{f}, \mathbf{f}]$, the association will be perfect or nearly so (as shown above) for the immediately past pair. It will be less good if a pair has not been presented recently, because the corrections to later pairs will have corrupted the association between earlier pairs. This error correction system displays what psychologists call a recency effect in that it gives the best accuracy to the immediate past associations, and becomes progressively less and less accurate for more temporally distant associations. Serial position effects are inherent in most error-correction algorithms.

The critical procedure for implementing the Widrow-Hoff algorithm on the computer is given in program fragment 9.1.

```
                  Program Fragment 9.1
PROCEDURE Widrow_Hoff (Correct_g, Actual_g, F: Vector;
                   VAR Delta_A: Matrix);
  VAR Difference_vector,
      Weighted_vector  : Vector;
      Learning_constant: REAL;      {Chose appropriate value.}
  PROCEDURE Scalar_times_vector (S: REAL; A: Vector;
      VAR B: Vector);
    {Multiplies A times S, Product is B.}
    VAR I: INTEGER;
```

```
    BEGIN
    FOR I:= 1 TO Dimensionality DO B [I]:= S * A [I];
    END;
  PROCEDURE Subtract_vectors (A, B: Vector;
     VAR Difference: Vector);
    {Subtracts B from A. The Difference Vector is Difference.}
    VAR I: INTEGER;
    BEGIN
    FOR I:= 1 TO Dimensionality DO Difference [I]:= A [I] - B [I];
    END;
  PROCEDURE Outer_product (A, B: Vector; VAR C: Matrix);
    VAR I,J: INTEGER;
    { Note:  We want the outerproduct procedure call to be in the
      form gf transpose format.  Then C = ab transpose. }
    BEGIN
    FOR I:= 1 TO Dimensionality DO
        FOR J:= 1 To Dimensionality DO C[I,J]:= B[J] * A [I];
    END;
BEGIN {Procedure Widrow_Hoff.}
Subtract_vectors (Correct_g, Actual_g, Difference_vector);
Scalar_times_vector(Learning_constant,Difference_vector,
  weighted_vector);
Outer_product (Weighted_vector, F, Delta_A);
END;  {Procedure Widrow_Hoff.}
```

Demonstration of Widrow-Hoff Learning

In chapter 7 we learned a small data base of states and state capital names using the linear autoassociator (figure 7.19). Recall was not very good. The output strings were almost unreadable if more than five associations were learned in a 200-dimensional system. Using the same set of associations, we would expect better results if we used the Widrow-Hoff error-correction technique. That is indeed the case. Using the 10 patterns given in figure 9.6, with Widrow-Hoff learning we get a far more accurate system. We will consider the ability to reconstruct missing information (i.e., answer questions) as the most important indicator of success in the system. When 10 vectors are learned, recall is nearly perfect. Compare these results with those in figure 7.21, where the linear associator produced unreadable associations when 10 vectors were learned. (More details about this demonstration are in appendix J.)

Widrow-Hoff Error Correction

Autoassociative State Capital Data Base

Input Pattern		Output Pattern	
Boston	Massachusetts →	Boston	Massachusetts

(Output Patterns)
Boston	Massachusetts
Providence	Rhode Island
Sacramento	California
Bismarck	North Dakota
Oklahoma City	Oklahoma
Harrisburg	Pennsylvania
Charleston	West Virginia
Baton Rouge	Louisiana
Concord	New Hampshire
Columbia	South Carolina

Reconstruction with Capital Probe

Input Pattern		Output Pattern	
Boston_____ →	Boston	Massachusetts

Boston	Massachusetts
Providence	Rhode Island
Sacramento	California
Bismarck	North Dakota
Oklahoma City	Oklahoma
Harrisburg	Pennsylvania
Charleston	West Virginia
Baton Rouge	Louisiana
Concord	New Hampshire
Columbia	South Carolina

Reconstruction with State Probe

Input Pattern		Output Pattern	
_____Massachusetts →	Boston	Massachusetts

Boston	Massachusetts
Providence	Rhode Island
Sacramento	California
Bismarck	North Dakota
Oklahooa City	Oklahoma
Harrisburg	Pennsylvania
Charleston	West Virginia
Bc|on Roege	Lguisiana
Concord	New Hampshire
Columbia	South Carolina

Figure 9.6 Demonstration of accurate recall using the Widrow-Hoff rule. These simulations use 10 patterns from the state capital data base used in chapter 7 (figure 7.19). Nothing was changed in the simulation except that the Widrow-Hoff algorithm was used instead of the plain linear autoassociator. Recall is close to perfect; there are five character errors. Compare with the accuracy shown in figure 7.21.

Backpropagation

Many people think that neural networks *are* backpropagation. *Backpropagation*, also called *backward error propagation* or *backprop*, is by far the most popular and widely used network-learning algorithm. It is a more complex gradient descent algorithm than the Widrow-Hoff learning rule, but it does essentially the same thing, it builds an adaptive system that minimizes an error signal by using gradient descent. We will discuss its general structure next but not give as many details as can be found elsewhere. Haykin (1994), in particular, provides a long and thorough technical review that is highly recommended.

As is often the case in science, backpropagation was discovered independently by several groups. The best-known early description is in Rumelhart, Hinton, and Williams (1986a), a chapter in the so-called PDP books. This chapter provided a number of impressive examples of its operation, reasonably clear derivations of the algorithm, and showed that backprop was able to solve some difficult problems in interesting ways. At about the same time, back-prop was also discovered by David Parker (1985), at that time at MIT, now an independent consultant, and by Yann Le Cun (1986), who is now at AT&T Bell Laboratories. Any potential priority disputes were avoided by the discovery that Paul Werbos, now at the National Science Foundation, had proposed and analyzed back-prop in his Harvard Ph.D. thesis in 1974.

Why Backprop Is Necessary

We saw in our discussion of the perceptron that simple neural networks have a severe limitation: they cannot compute certain classes of functions. Up to this point, we have only considered neural networks with a single set of modifiable connections. This can take the form of a network with an input set of units and an output set of neurons, or one with a single set of units connecting to itself by a recurrent set of modifiable connections (figure 9.7). A more general architecture (figure 9.7c) is a set of neurons with two or more modifiable sets of weights. We still have an input and an output set of units. The middle groups of units are called the *hidden layers* since they are neither input nor output, and their activities are not accessible from outside the network. It is not hard to show that *if the network contains nonlinearities*, it can have more computing power than one without a hidden layer. For example, it solves problems such as Exclusive OR.

The network must be nonlinear because a many-layer linear network reduces to a single-layer network. Consider a three-layer linear network

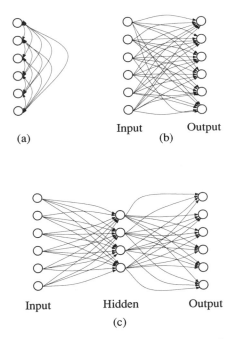

Figure 9.7   Figures (a) and (b) show the network architectures we have used up to now. Figure (a) shows an autoassociative architecture; (b) is the standard network associator with an input and an output layer. Both (a) and (b) have a single layer of modifiable connections. Figure (c) shows a multilayer architecture. In a multilayer architecture, in addition to the input and output layers, there are also one or more layers of *hidden* units that are neither input nor output. Figure (c) has two layers of modifiable connections. A generalization of gradient descent called *back-propagation* provides one way of changing the weight in the hidden layers.

(figure 9.8). Assume the input pattern is **f**. The connection matrix, **A**, connects input pattern, **f**, with hidden layer pattern, **g**. If this is a linear network, then

Af = g.

A connection matrix, **B**, connects the hidden layer pattern, **g**, with output layer pattern, **h**. Therefore,

Bg = h.

But since the network is linear,

Bg = h = B(Af) = h

BAf = h.

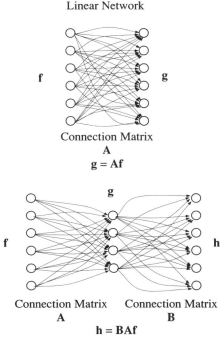

Linear Network

Connection Matrix
A

$$g = Af$$

Connection Matrix   Connection Matrix
A   **B**

$$h = BAf$$

Figure 9.8   A linear network with two layers of modifiable synapses. This architecture exhibits no gain in processing power over a network with a single layer of modifiable synapses.

We can replace a linear network with a hidden layer with a network that has a single layer of weights with connection strengths given by the matrix product, **BA**. Therefore we gain no processing power.

The major technical problem in making an adaptive multilayer network is providing a rule for modifying the weights at the hidden layers. If we have only a single layer of weights, we have two effective error-correcting algorithms, the perceptron and the Widrow-Hoff rule. They both depend on being able to measure the error and on the fact that only the weights on a single unit contribute to the error for that unit. But consider a more complex architecture, for example, figure 9.9. The weights on a unit contribute to the error of the unit. But so do the weights from the input layer to all the hidden layer units. If there is an error, which of these many possible contributors caused it? If the network gets the answer right, who did the right thing? This is called the *credit-assignment problem* and it is a difficulty for all complex systems, including neural networks, governments, armies, and business organizations.

Backpropagation gives a rule that solves the credit-assignment problem for a feedforward network using nonlinear units. The network is called feed-

Credit Assignment Problem

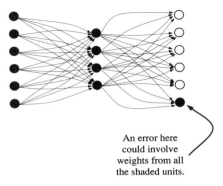

An error here
could involve
weights from all
the shaded units.

Figure 9.9 The *credit-assignment problem*. If there is an error at an output unit,
which of many possible weights caused it?

Backpropagation Computing Unit

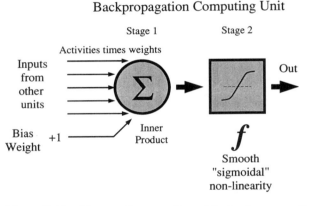

Figure 9.10 The nonlinear unit used in backpropagation. This is the generic connectionist unit. The nonlinearity is usually a smooth, monotonically increasing function, *f*.

forward because information goes only from input to output; there are no
recurrent or backward connections. The need for nonlinearity requires us to
use the general version of the generic neural network neuron (figure 9.10).
The nonlinear second stage is almost always considered to contain a monotonic sigmoid function. Rumelhart, Hinton, and Williams (1986a) referred to
this as a *semilinear* function, which means it is monotonic and differentiable.
The output of the first box is the inner product between the weights, **w**, and
the input activity, which we will call **a**. The nonlinear function in the second
stage we will call *f* and its derivative *f'*. The actual output activity of the unit,
is given by

unit output = $f([\mathbf{w}, \mathbf{a}])$.

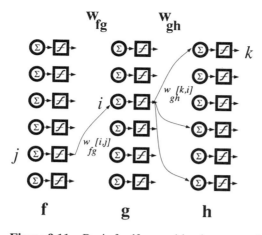

Figure 9.11   Basic feedforward backpropagation architecture.

We will sketch the derivation of backpropagation here. The basic strategy is like that we used for the second derivation of the Widrow-Hoff rule, that is, application of the chain rule for partial derivatives. Unfortunately, the derivation is somewhat unintuitive. This discussion follows that in the PDP book (Rumelhart, Hinton, and Williams, 1986a). The notation we will use is to label the layers from input to output $\mathbf{a}, \mathbf{b}, \mathbf{c}, \dots, \mathbf{f}, \mathbf{g}, \mathbf{h}$. There may be numerous hidden layers. Figure 9.11 shows the labeling conventions. Subscripts represent layers, that is, $\mathbf{w_{gh}}$ are the weights connecting particular units in the $\mathbf{g}$ and $\mathbf{h}$ layers. We use our usual computer convention to represent a single element of the weight, that is, $w_{gh}[i,j]$ is the weight connecting the element j in $\mathbf{g}$, $g[j]$, with element i in $\mathbf{h}$, $h[i]$.

Suppose layer $\mathbf{f}$ projects to layer $\mathbf{g}$. Suppose we know what the output activity, $\mathbf{g}$, is when layer $\mathbf{f}$, that drives layer $\mathbf{g}$, has a particular input. We are computing the square of the error magnitude, as we did with the Widrow-Hoff calculation. For a single unit, say the ith unit in layer $\mathbf{g}$, with $t[i]$ as the unit's desired activity and $g[i]$ as the actual value, the error e is

$$e = (t[i] - g[i]).$$

The error we are trying to minimize is the square of this quantity,

$$e^2 = (t[i] - g[i])^2.$$

We want to implement gradient descent for this pattern, so that the next time we see the pattern, the network will have less error. We will do this for all the patterns in the training set eventually, but right now we are just concerned with the error for a particular pattern, say, $\mathbf{f}$. We must change the weights of the unit in the direction of the gradient of the square of the error.

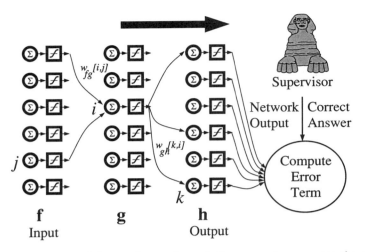

Figure 9.12 A forward pass through the network computes the output for a given pattern. The supervisor has provided the correct answer. The error vector can then be computed for the output layer.

This means we must compute the partial derivatives of the square of the error with respect to the weights. We are concerned with the expression

$$\Delta w[i,j] = -\eta \frac{\partial (e^2)}{\partial w[i,j]},$$

where η determines the step size of the gradient descent. The minus sign is used because we are going downhill on the error surface. If we look at the picture of the generic neuron in figure 9.10 and the network, in figures 9.11 and 9.12, we see that we can break up the derivative computation into two parts by using the chain rule. Suppose we are looking at output units, $g[i]$. The weight, $w_{fg}[i,j]$, couples the jth unit in the f layer to the ith unit in the g layer. Then,

$$\frac{\partial (e^2)}{\partial w[i,j]} = \frac{\partial (e^2)}{\partial ([\mathbf{w_{fg}},\mathbf{f}])} \frac{\partial ([\mathbf{w_{fg}},\mathbf{f}])}{\partial (w[i,j])}.$$

The inner product for the ith unit, $[\mathbf{w_{fg}},\mathbf{f}]$, is the familiar inner product between the weights coupling the **f** layer to the **g** layer. The second term is quite easy to calculate. Remember that the inner product for the ith unit is simply

$$[\mathbf{w_{fg}},\mathbf{f}] = w_{fg}[i,1]f[1] + w_{fg}[i,2]f[2] + \cdots + w_{fg}[i,j]f[j] + \cdots$$

The partial derivative of this expression with respect to the single weight, $w_{fg}[i,j]$, is going to be $f[j]$, that is, the output of the unit $f[j]$.

Let us use the delta notation from the PDP book to maintain consistency with a frequent notation in the field. We *define* an expression that refers to units in the g layer. For a unit i in the **g** layer, with activity, $g[i]$, we define a delta term (which we will see is a generalized error term), $\delta[i]$,

$$\delta[i] = -\frac{\partial(e^2)}{\partial[\mathbf{w_{fg}}, \mathbf{f}]}.$$

The minus sign arises from the desire to go downhill. With this *definition* we go back to our original gradient descent formula and now have,

$$\Delta w[i,j] = \eta\delta[i]f[j].$$

This definition allows us to compute $\delta[i]$ for more complex nonlinear networks as well and come out with the same formula. For example, yet another application of the chain rule gives

$$\delta[i] = -\frac{\partial(e^2)}{\partial[\mathbf{w_{fg}}, \mathbf{f}]} = -\frac{\partial(e^2)}{\partial g[i]}\frac{\partial g[i]}{\partial[\mathbf{w_{fg}}, \mathbf{f}]}.$$

In the linear case, the output of the unit, $g[i]$, equals the inner product, $[\mathbf{w_{fg}}, \mathbf{f}]$ for this ith unit, and this step accomplishes nothing. But for the nonlinear, two-stage neuron where

$$g[i] = f[\mathbf{w_{fg}}, \mathbf{f}],$$

the second term will equal the derivative of the function f, that is, f'.

We have all the pieces in place to finish the calculation. We have two distinct cases to worry about.

First Case: Output Layer

Suppose we are in the output layer of the network. Then

$$\frac{\partial(e^2)}{\partial g[i]} = 2e\frac{\partial e}{\partial g[i]}.$$

Suppose the desired output from the supervisor is $t[i]$. Then the error

$$e = (t[i] - g[i])$$

and

$$\frac{\partial(e^2)}{\partial g[i]} = -2e.$$

We now have the raw material for computation of $\delta[i]$:

$$\delta[i] = 2(t[i] - g[i])f'([w_{fg}, f])$$

and

$$\Delta w_{fg}[i,j] = \eta \delta[i] f[j].$$

The form of this expression reproduces the form that we found earlier for single-layer Widrow-Hoff learning with linear neurons, that is, the product of the input activity, $f[j]$, and an error term, $\delta[i]$. The function f' in that case will be 1, since $g[i] = [\mathbf{w_{fg}}, \mathbf{f}]$. We can immediately compute $\delta[i]$ for the output layer since the knowledge present in the supervisor lets us compute the output error pattern.

Second Case: Earlier Layer

What can we do for earlier layers? The basic idea of back-propagation is to take this error expression for all the units in one layer and *propagate it backward to earlier layers*. This, of course, assumes that the network can run backward, which is a very large assumption. It assumes that the forward weight, $w[i,j]$, connecting the jth unit in $\mathbf{f}$ with the ith unit in $\mathbf{g}$ *is exactly the same as the backward weight connecting the two units*. It is this assumption that abandons claims of biological plausibility for this class of neural networks.

We will assume that we know the error in the $\mathbf{h}$ layer. We want to compute the error at the $\mathbf{g}$ layer. We are going to compute the error term, $\delta[i]$, at the $\mathbf{g}$ layer as a weighted sum of error terms coming from the $\mathbf{h}$ layer. Let us apply the chain rule again. We know the connection strengths between layers $\mathbf{h}$ and $\mathbf{g}$, because they are just the same as the connection strengths between $\mathbf{g}$ and $\mathbf{h}$. Using the chain rule gives us

$$\delta[i] = -\frac{\partial(e^2)}{\partial[\mathbf{w_{fg}}, \mathbf{f}]} = -\frac{\partial(e^2)}{\partial g[i]} \frac{\partial g[i]}{\partial[\mathbf{w_{fg}}, \mathbf{f}]}.$$

Let us consider the first term in the expression, since the second term, which turned out to be $f'(g[i])$, will not be affected. The value of $g[i]$ can affect the activity of every unit in the $\mathbf{h}$ layer. To find the error due to $g[i]$, we must sum the errors from all the units in $\mathbf{h}$, the layer beyond $\mathbf{g}$, because the value of $g[i]$ can affect the error at all these units. We denote the units in $\mathbf{h}$ as $h[k]$. Let us assume that the error is known at layer $\mathbf{h}$, so that

$$e^2 = \sum_k (t[k] - f[\mathbf{w_{gh}}, \mathbf{g}])^2.$$

The inner product $[\mathbf{w_{gh}}, \mathbf{g}]$, will, of course, differ for each of the $\mathbf{k}$ units in the $\mathbf{h}$ layer since the weights $\mathbf{w_{gh}}$ will differ. We can differentiate this expression with respect to $g[i]$ and get

$$\frac{\partial(e^2)}{\partial g[i]} = \sum_k \frac{\partial(e^2)}{\partial[\mathbf{w_{gh}}, \mathbf{g}]} \frac{\partial[\mathbf{w_{gh}}, \mathbf{g}]}{\partial g[i]}.$$

Looking at the expression immediately above,

$$\sum_k \frac{\partial(e^2)}{\partial[\mathbf{w_{gh}}, \mathbf{g}]} \frac{\partial}{\partial g[i]}(w[k,1]g[1] + w[k,2]g[2] + \cdots + w[k,i]g[i] + \cdots)$$

$$= \sum_k \frac{\partial(e^2)}{\partial[\mathbf{w_{gh}}, \mathbf{g}]} w[k,i].$$

Notice that the first partial derivative term is the definition of $\delta[k]$, the error of the kth unit in the $\mathbf{h}$ layer. Therefore, we can write the final form of $\delta[i]$, the error in the $\mathbf{g}$ layer, as the sum of the $\mathbf{k}$ layer error terms, or

$$\delta[i] = f'[\mathbf{w_{fg}}, \mathbf{f}] \sum_k \delta[k]w[k,i],$$

and

$$\Delta w[i,j] = \eta \delta[i] f[j].$$

Once we have our expression for the error term, we can compute the weight changes to implement gradient descent in the layer one layer in from the output. Once we have computed the error at this layer, we can repeat the process and compute the weight changes at earlier layers, using the values for $\delta[i]$ iteratively. Figure 9.13 suggests the nature of the backward pass.

We have described how to compute the weight changes due to a single pattern. If weight changes are made for every pattern in the training set, the total error for all patterns will approximate gradient descent. There are two ways to compute weight changes due to many patterns. One is to change the network weights immediately after each pattern is presented. The other is to accumulate a total error for the entire set of patterns, and then change the weights once after all patterns in the training set have been presented to the network. (Sometimes an entire cycle through the training set is called an *epoch*.)

The second technique involves less computation. Computing a summed error before changing weights is also easy to do with some kinds of parallel computers. Many copies of the network can be run in parallel, each computing error for a different member of the training set. Then the total error

264   Chapter 9

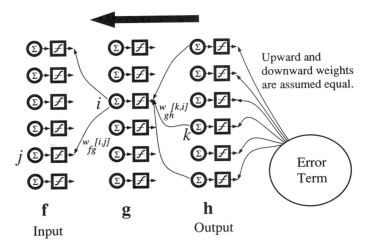

Figure 9.13 A backward pass through the network using the error vector, and running the network units "in reverse" allows computation of the correct error to implement gradient descent at the earlier layers.

for the entire set of patterns can be computed in a single global summing operation.

It is not immediately obvious which technique is best; however, extensive discussion on e-mail networks seem to have reached consensus that weight correction after each pattern presentation seems to be most satisfactory. Although summing corrections and doing weight changes only once per training set cycle (epoch) is superficially attractive, learning seems to be less effective and takes longer. The total amount of computer time used for the two techniques becomes roughly equivalent.

What we have described is the plainest form of backpropagation. The literature describes numerous variations. Like most algorithms, it can always be made to work better when it is modified to conform to the demands of a particular problem. For the past few years, any neural network meeting had dozens of papers discussing ways of making backprop work better, which almost always means work faster. Haykin (1994) discusses some of the better modifications. One problem with many commercial neural network simulators, by the way, is that it may not be possible to alter the code if that is desirable.

One modification, *momentum*, which was mentioned by Rumelhart, Hinton, and Williams (1986a), is a straightforward example of how to adapt an algorithm to the details of a problem. Momentum is perhaps the most commonly used modification to basic backprop. The gradient descent derivation assumes that small steps are taken. Real gradient descent uses a finite step size to navigate the error surface. The bigger the step, the faster the minimum will be reached, but if the step is too large, various computational problems may occur. One problem that seems quite frequent in practice is to have a region of the error surface shaped like a ravine with fairly steep sides. A large step size can oscillate back and forth across the ravine. If, however, we take a step that is something like the sum of previous step sizes, the oscillation will average out, and the component heading down the ravine will dominate the direction of the step. To construct a momentum term, keep track of previous weight changes, and add them to the current weight change, for example, after n learning trials, the weight change, $\Delta w_n[i,j]$, is given by

$$\Delta w_n[i,j] = \eta \delta[i] f[j] + \alpha \Delta w_{n-1}[i,j].$$

Applications of Backpropagation

We will discuss three examples of backpropagation out of many. First is NETtalk, a neural network that learned to translate written English text into spoken text. Second, we will discuss the interesting statistics that appear when a network is applied to the data compression of a visual image. Third is the current state-of-the-art of neural network character recognition.

NETtalk

NETtalk (Sejnowski and Rosenberg, 1987) is a program that learns to pronounce English words. English spelling is notoriously inconsistent. However, it does follow rules, although weak ones with exceptions and qualifications. As an example, consider the sentence, "This is a test." The two letters "is" occur in both the first and second words, each time followed by a space, so the immediate context of s is identical. However, in the word "this" the s is unvoiced, that is, pronounced /s/. In the word "is," the s is voiced, that is, pronounced /z/.

NETtalk takes a strings of characters forming English text and converts them into strings of phonemes that can serve as input to a speech synthesizer. Digital Equipment Corporation sells a commercial product named DECtalk that also turns text into speech, hence the similarity of names. DECtalk is a complex, rule-based, expert system developed over a number of years.

Sejnowski and Rosenberg used a three-layer feedforward network with input and output layers of units, as well as a layer of hidden units (figure 9.14). The input layer looked at a seven-character window of text. The net-

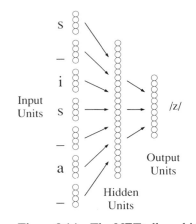

Figure 9.14   The NETtalk architecture. A seven-character window gives rise to the pronunciation of the central character in the window. The seven input characters are taken from the string in the text, "This is a test." The s in the word "is" is voiced, that is, pronounced like a /z/. The s three positions up is the last letter in "this" is unvoiced, that is, prounounced like an /s/. Redrawn from Sejnowski and Rosenberg (1987).

work generated an output corresponding to the center character in the window. A window of this size seems to be enough to pronounce most words; however, some pronunciation decisions require detailed knowledge of English, for example, to differentiate the verb "to lead" from the metal "lead."

The representation of characters at the input layer is localized, with 1 active unit representing a character, space, or punctuation, for a total of 29 units, times 7 character positions, for a total of 201 input units. The number of hidden units varied from simulation to simulation, but for continuous speech 80 hidden units were used. Twenty-six features (23 articulatory features, 3 for stress) were represented in the output layer. Phonemes are represented by many simultaneously active output units. Simulations used about 300 units total with around 20,000 connections.

Sejnowski and Rosenberg taught the system both isolated words and continuous speech. Backpropagation was used to train the network to give the correct output phoneme when presented with input character strings. The authors found samples of transcribed speech in the child language literature and used this as the training set.

After training, NETtalk was producing phonemes from the training set correctly 95% of the time. New samples of similar text showed a drop in accuracy, but still around the 80% level. Sejnowski has played a developmental tape of NETtalk's output at meetings. The simulation passes through what sounds like a babbling phase, then starts approximating speech more and more closely. By the time learning stops, it is generally understandable, although it still makes errors.

Sejnowski and Rosenberg were interested in what was going on in the hidden layers, because of the theoretical reasons for believing that the hidden layer units develop efficient representations. Units in the hidden layers seem to respond to several input characters, and have no obvious interpretations. Later work has pursued this point. Some units seem to respond more strongly to some classes of characters, for example, vowels, than others, but nowhere is there strict localization. Representations in the hidden layers are distributed to some degree.

Although it provided a fascinating demonstration, NETtalk did not lead to practical applications. Competitors did better. As an example, cheap computer memory made it feasible to use memory-based techniques, that is, essentially storing the dictionary. The pronounciation of unlisted words could be guessed by using the pronunciation of the closest stored strings, an example of the nearest neighbor techniques discussed in chapter 13.

Data Compression and Principal Components

One of the important claims for multilayer networks is that units in the hidden layers "discover" effective ways of representing the input information. The most frequently cited paper on backpropagation is by Rumelhart, Hinton, and Williams (1986a). Several astonishing examples of the development of such representations are known; for example, the hidden layer units developed in an early genealogic simulation developed hidden layer units that seemed to respond to such complex properties of the input as nationality, generation, and so on (Rumelhart, Hinton, and Williams, 1986b).

Neural networks are statistics for amateurs. Statistics is a way of extracting meaning from an environment. The deep connections between neural network algorithms and statistics are clear. As one example, a good deal of recent evidence suggests that many neural networks, including some variants of backpropagation, are doing something like *principal component analysis*. It is worth describing this briefly because the connection to the simpler neural net algorithms is powerful and obvious. A good introduction can be found in Jolliffe (1986). The expansion used to find principal components is sometimes referred to as a *Karhunen-Loeve expansion* in engineering (Young and Calvert, 1974).

Principal components are easy to compute. Mathematically, suppose we have a set of data vectors, $\{\mathbf{x}_1, \mathbf{x}_2 \ldots, \mathbf{x}_n\}$, and we want to describe them. The first thing we do is form an outer product matrix, $\mathbf{A}$, where

$$\mathbf{A} = \sum_{i=1}^{n} (\mathbf{x}_i \mathbf{x}_i^T).$$

In statistics, when points of experimental data are used and with appropriate normalization, this is sometimes called the *sample covariance matrix*.

The principal components are the *eigenvectors* of this matrix. Because the matrix is symmetric, the eigenvectors are orthogonal to each other. The *eigenvalues* are a measure of how much variance of the data set that eigenvector accounts for. The larger the eigenvalue, the better job that eigenvector does describing the members of the data set.

Suppose we have a set of examples of data that we wish to work with, examples from a training set, for instance. They form a cloud of points in a very high-dimensional state space. To describe the location of a particular point takes a great many numbers. Suppose we want to describe a point in a more compact way, that is, we want to find the particular set of vectors that does the most accurate job of descibing members of the data set as the sum of a small number of components. We are going to *throw away* information, but we would like to throw away as little important information as possible. The principal components are the vectors that minimize the *mean square error* between the actual points in the data set and the points that can be described with a given smaller number of components. Sometimes the principal components are described as the best set of linear features. Given some fixed number of descriptors, that number of the principal components, ranked by eigenvalue, will give the most accurate description of the data set in terms of mean square error. That is, the principal component with the largest eigenvalue will account for the largest amount of variance, the next largest eigenvalue the next largest amount, and so on (figure 9.15).

The relation with simple neural networks is now obvious. The outer product matrix is what the connection matrix looks like with a Hebbian learning rule in an autoassociative system. Because the outer product form of Hebb synapse is essentially a measure of covariation between presynaptic and postsynaptic activity, it is not surprising to find this connection (see Oja, 1982). Linsker (1988) described a particularly interesting kind of self-organization that may be relevant to early vision with the analysis based on Hebb learning and principal component analysis.

Principal components arise from a linear analysis. Many interesting features are nonlinear. The representations in the hidden layer in backpropagation are derived from the operation of a nonlinear system. And, in a general nonlinear system, where an arbitrary set of inputs is to be associated with arbitrary outputs, the connection to the covariance matrix is not clear. One might argue that the nonlinear systems used in neural networks are often

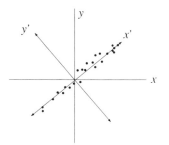

Figure 9.15 A sample data set suggests the meaning of principal components. The *x* and *y* axes do not provide a very compact description of a data point; values on both axes are required to specify accurately the location of a data point in the space. A single value along the *x* or *y* axis would be far away from the actual location of the data points. However, a value along the axis *x'* would provide a single number that accurately approximates most data points. Principal component analysis provides a way of discovering in which directions most of the variance in the data lies. The first principal component for this example lies along *x'*, the second is orthogonal to it and lies along *y'*. If you only could use a single number to describe a data point, would you prefer to have the value along the *x* axis or the *x'* axis?

not *that* nonlinear. The sigmoid used in the standard connectionist neuron is sometimes called semilinear. The hidden layer is doing something much like efficient feature extraction with dimensionality reduction, because it must represent the input information in, usually, a smaller number of hidden units than input units.

Cottrell, Munro, and Zipser (1989) were interested in an application of neural networks to the general and very practical problem of *image compression*. When visual data are stored in a computer they take up a very large amount of disk space and take a long time to transmit over expensive communications channels. If some more efficient way to store them could be found, it would be of great practical value. The network used is called an *encoder* network (figure 9.16). It has three layers, but the input and output are constrained to be the same pattern; that is, given an input, the network must reproduce it. But there are fewer hidden units than input units; in the case of this paper, there are 64 input and output units and 16 hidden units. The input pattern is described by a small number of hidden units; the reconstruction at the output layer is based on the activity of the hidden units. The practical importance of the problem is that it can be used to save time on a communications channel, where time is literally money. If you can describe a pattern on 64 units of the input layer satisfactorily with 16 activities, why not just send the 16 hidden-layer activities and speed up transmission by a factor of 4?

Image Compression
Encoder Network

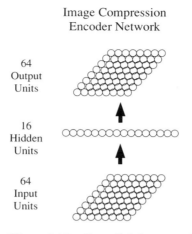

64
Output
Units

16
Hidden
Units

64
Input
Units

Figure 9.16   Cottrell, Munro, and Zipser used backpropagation for compression of visual images. This is what is called an *encoder* network. The input and the output patterns are identical in this class of networks. The function of the network is to reproduce the input pattern at the output. They used 8 by 8 samples of pixels (64 units) from natural images and 16 hidden units. Compression can arise because, instead of sending all 64 pixels, satisfactory reconstruction can be attained by only sending the values of the 16 hidden units, that is, there is a 4 to 1 compression ratio. Information is lost using this technique, but the resulting image quality is good. Redrawn from Cottrell, Munro, and Zipser (1987).

The data used by the authors were raw pixel image data from several photographs of natural scenes. Backpropagation was used to develop the best data representation at the hidden layer. The best (mean square) solution to this problem in the linear case we just discussed is the 16 principal components with the largest eigenvalues. The network result is that processing by the hidden units was very close to what would have been obtained by principal component analysis, although nothing so simple as a one hidden unit–one principal component identity occurred. The space spanned by activation of the 16 hidden units was roughly the same as that generated by the first 16 principal components. That is, the hidden units and the principal components were describing the same region of state space. The major difference was that the activities of the hidden units were such as to equalize the information transmitted across the hidden units; that is, each hidden unit was roughly of the same importance in the information it transmitted about the image. The amount of variance accounted for by the principal components is a function of the eigenvalue and can differ a great deal between principal components.

A second paper, by Baldi and Hornik (1988), put this experimental result on a firmer theoretical foundation. *If we assume that the hidden units are*

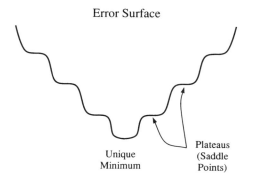

Figure 9.17 Baldi and Hornik (1988) were able to get some insight into what the hidden units in simple backpropagation were doing using a three-layer network with *linear* hidden layer units. The weight space had a single global minimum but also contained a series of plateaus where the gradient is very small. It can take a long time to traverse the flat spots on the error surface in weight space. Redrawn from Baldi and Hornik (1988).

linear, Baldi and Hornik showed that the solution that the network will find is indeed to have the hidden units span the principal component space. The hidden units, as found by Cottrell, Munro, and Zipser, will have roughly equal importance to information transmission. Baldi and Hornik made another interesting observation. It is known that backpropagation, which adjusts weights to minimize error by using gradient descent, is very slow. These authors suggested that one reason for this is that the error surface that the network is descending has many flat spots on it, although only a single global minimum (figure 9.17). The gradient is small on a plateau, and the network changes weights slowly there. For working with real-world data such as natural images, the nonlinear system seems to behave much like the straightforward and easily computed linear solution. From a practical point of view, it might be much more efficient to compute the principal components directly than to use backpropagation to come up with the same solution!

Character Recognition

One of the first sets of test data that a new neural network is trained on is usually crude representations of characters. Recognizing characters is one of those problems that at first seems simple but in fact is extremely difficult. The current state of the art in the use of a neural network for letter recognition is from a group at AT&T Bell Laboratories (Jackel, Graf, Hubbard, Denker, Henderson, and Guyon, 1988; LeCun, Boser, Denker, Henderson, Howard, Hubbard, and Jackel, 1990).

The task they set themselves was to use neural network techniques to recognize hand-printed digits. The data set used in the first paper consisted of two sets of digits produced by their own group, one carefully printed and one done more hastily. The second paper used a more challenging data set: the U.S. Post Office's ZIP Code data base, which consists of over 9000 hand-printed digits taken from envelopes that passed through the Buffalo, NY, post office. (The source of this data set also suggests one of the potential practical applications of digit recognition.) Figure 9.18 shows some digits from this set *after* the digits have been isolated. Some of them are so badly written that one of the Bell Labs authors commented that perhaps some writers don't actually want their letters to get to where they are supposed to be going.

A psychological problem with neural networks is that they work "pretty well" almost immediately. With little effort it is possible to obtain results in character recognition that are encouraging. In networks as in anything, doing things right is hard. To get 75% correct on a character classification task may take a week, and can be done with a commercial simulator package right out of the box, working on nearly unprocessed data. To get over 95% may take years of work. Unfortunately, 95% accuracy is the barely acceptable minimum for most digit recognition applications.

The most difficult problem in character recognition in many cases is not even recognition but in discovering where the characters are. *Segmenting* a number with many digits, that is, picking out single digits from a ZIP Code or an amount on a check, is a subtle and difficult problem. But let us suppose we can start with an isolated digit. The digit is first scaled so all examples become roughly the same size on a 16 by 16-pixel block. This can involve expansions, contractions, and shearing transformations of the image. Simply applying a powerful learning algorithm to the raw pixel array and expecting the system to learn to classify digits correctly does not work. This is referred to in the Bell Labs group's 1988 paper as the "no brains required" strategy.

The major differences between the first paper (1988) and the second (1990) fall in the way they handled the early stages of analysis of the pixel array. The problem is choosing the right features. The first paper used a set of features chosen by the experimenters, based on their intuition and experience. The features were spatially localized, that is, only involved patterns on a small contiguous neighborhood of pixels, 5 by 5 or 7 by 7 pixels. Features at a particular location were detected by templates that were constructed to respond strongly to oriented line segments, the ends of lines, and arcs. This particular set of features was suggested by the kind of processing known

Figure 9.18   Entire digitized ZIP Codes (top) and single digits used by the Bell
Laboratories group to test a backpropagation digit-recognition system. These digits
were taken from ZIP Codes on envelopes passing through the Buffalo post office.
The bottom set has been scaled so the characters are of roughly constant size. From
LeCun, Boser, Denker, Henderson, Howard, Hubbard, and Jackel (1990). Reprinted
by permission.

to exist in the mammalian visual system. The feature vectors were then concatenated to make up a 180-dimensional vector, which was the input actually used for the recognition computation. Once the large vector representing the digit was constructed, the network went to work.

What are the best features to use for the early stages of analysis of the image? If one knew how to do it, the best strategy would be for the network to learn to choose its own features, based on the data it must classify. The second paper used backpropagation with a constrained architecture designed largely to allow discovery and use of a good set of low-level features by the network. Instead of everything connecting to everything—the usual fully connected back-propagation network—the first two hidden layers were severely limited in their connectivity. The first layer was designed to pick up local features, so that only a small number of pixels in a local region projected to a single unit in the layer. Moreover, the first hidden layer was assumed to be composed of a number of modules. Each module was formed from a group of units, each unit looking at different local regions on the pixel array. During learning, each unit in a module was forced to have *identical* weights (with the exception of the bias or threshold weight). Therefore every unit in a module looked for the same feature, but in different locations in the array. This is sometimes called *weight sharing*. The first hidden layer had 12 such independent *feature maps*. The features that were developed during learning by the first layer turned out, in some cases, to be similar to the intuitively chosen feature set and to biologically observed visual features. The second hidden layer looked at features of the feature maps. The same technique was used as in the first layer, but its units could pick up higher-order correlations in the input, over larger spatial extents (figure 9.19).

This final network was run using standard digital signal-processing (DSP) chips in conjunction with a personal computer. Learning was somewhat slow, requiring 167,693 pattern presentations. The final network misclassified 0.14% of the training set patterns and 5.0% on a test set. A summary comment in the 1990 paper is, "Our results seem to be at the state of the art in digit recognition" (Le Cun et al., p. 549).

A number of important lessons are to be learned from these attempts at digit recognition. Prominent among them is that there is no magical neural network substitute for hard work and engineering insight. A good deal of the structure of the problem was designed into the network. The authors knew they were not dealing with random vectors, but with spatial patterns. A great deal of experience with vision problems, and the example of biological vision, suggested that one should start with local feature detection, and then look for features over wider and wider areas. A great deal of pre-

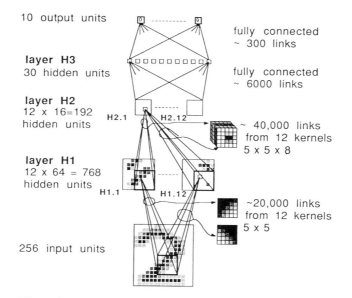

Figure 9.19   The network used by the Bell Laboratories group. The hidden layer units have constrained weights. Each hidden layer unit has the same weights, that is, responds to the same feature, but looks at a different point in space. From LeCun, Boser, Denker, Henderson, Howard, Hubbard, and Jackel (1990). Reprinted by permission.

processing was involved in segmenting, normalizing, and orienting the digits. A different problem might have suggested a very different network. The power of the learning algorithms involved should not obscure the fact that the realization of a network for a problem is likely to be very problem specific if it is to work well.

What's Wrong With Gradient Descent?

If used with care, nothing at all. It is appealing to think of descending rapidly from the heights of error to the valleys of truth. In many pattern-recognition tasks, as we have seen in our examples, gradient-descent techniques are practical and robust. It would be fair to say that most of the successes of neural networks have come through the use of backpropagation in one form or another. For many important problems a crude representation of the input and output information will allow backpropagation to produce good solutions, with good generalization.

However, unmodified gradient-descent techniques are not the ultimate solution to neural network computation. They have some problems.

First is the presence of local minima in the error surface in backpropagation. The best solution may not be found. Although backpropagation often finds good solutions, it is not guaranteed. For example, it almost always finds solutions to the exclusive-OR problem that caused the perceptron such difficulty. However, with some starting points, solutions are not found.

Second, the technique depends on the assumption that minimizing error for a training set is the right thing to do. This depends as much on the definition of error, the choice of training set, and the function of the network as on the technique itself. When error is defined strictly as the length of the error vector, implementing gradient descent gives rise to simple equations. Other definitions of error may not be so agreeable. The proper definition of error is as task dependent as the data representation, and is tied to it in subtle ways.

Third, if learning goes on too long, generalization often suffers. In statistics, this is sometimes called *overfitting* the data. It means that the peculiarities of the training set are accurately modeled, but the individual peculiarities may not be what you want the network to learn. This also means that the performance of the network may vary a great deal from training set to training set. Practically, therefore, for best generalization, learning must be stopped well before the minimum training set error is reached. Geman, Bienenstock, and Doursat (1992) discussed this problem, the importance of which is sometimes not fully appreciated.

Fourth, backpropagation learning times are extremely long. Thousands or tens of thousands of learning trials are common for even simple problems. Worse, the number of trials required goes up rapidly with the size of the network; backpropagation scales poorly.

Fifth, as many have pointed out, backpropagation is unbiological. No evidence whatsoever exists for "weights" running backward. There is substantial evidence for backward connections from higher to lower levels in the nervous system. In fact, many systems, including cortex, have more backward than forward connections; feedforward networks in the brain are plausible but mythical. These connections certainly give rise to recurrent networks with complex temporal dynamics, but are unlikely to be part of error propagation in feedforward learning. We do, however, have evidence that simple Widrow-Hoff error correction has close ties to some kinds of biological learning. Widrow-Hoff learning could be viewed as a kind of learning of "surprise," that is, if what happened was not what was expected, then learn it. This is the basis of a well-known theory of classic conditioning called the Rescorla-Wagner model. Sutton and Barto (1981) pointed out that the Rescorla-Wagner model describing rat learning was equivalent to the Widrow-Hoff algorithm. Others also used Widrow-Hoff learning to explain data from both human and animal experiments (Gluck and Bower, 1988; Gluck and Thompson, 1987).

Sixth, some backpropagation networks can show what has been called *catastrophic unlearning*. A network is taught a set of training patterns. After the network has learned, another training pattern is added to the training set. Sometimes learning the changed training set correctly can take a very long time because all the old connections are completely changed to accommodate the new information. This problem can be alleviated with proper network design and data representation.

Seventh, and perhaps the biggest problem, is neural network mysticism. A neural network may solve a practical problem, but it can be difficult to understand how it solved it. In some cases what is going on is clear, for example, the relation between image compression and principal component analysis or the genealogy network. For many problems, however, the hidden layer is not doing an obvious analysis. Sometimes this is acceptable, but it can be dangerous. If you don't know what was done, it can be hard to improve it. A strong tendency is to say, "Who cares? The network works." This approach is rarely the road to either progress or wisdom.

Let Minsky and Papert have the last word. In 1988 they wrote an epilog to *Perceptrons*, the book that had such a devastating effect on neural networks at the end of the 1960s. They commented on gradient descent (p. 260–261):

In the early years of cybernetics, everyone understood that hill-climbing was always available for working easy problems, but that it almost always became impractical for problems of larger sizes and complexities.... The situation seems not to have changed much—we have seen no contemporary connectionist publication that casts much new theoretical light on the situation. Then why has [backpropagation] become so popular in recent years? In part this is because it is so widely applicable and because it does indeed yield new results (at least on problems of rather small scale). Its reputation also gains, we think, from its being presented in forms that shares, albeit to a lesser degree, the biological plausibility of [the perceptron]. But we fear that its reputation also stems from unfamiliarity with the manner in which hill-climbing methods deteriorate when confronted with larger-scale problems.

References

P. Baldi and K. Hornik (1988), Neural networks and principal component analysis: Learning from examples without local minima. *Neural Networks*, 2, 53–58.

G.W. Cottrell, P, Munro, and D. Zipser (1989), Image compression by back propagation: An example of extensional programming. In N.E. Sharkey (Ed.), *Advances in Cognitive Science*, Vol. 3. Norwood, NJ: Ablex.

S. Geman, E. Bienenstock, and R. Doursat (1992), Neural networks and the bias-variance dilemma. *Neural Computation, 4,* 1–58.

M.A. Gluck and G.H. Bower (1988), Evaluating an adaptive network model of human learning. *Journal of Memory and Language, 27,* 166–195.

M.A. Gluck and R.F. Thompson (1987), Modelling the neural substrates of associative learning and memory: A computational approach. *Psychological Review, 94,* 176–292.

S. Haykin (1994), *Neural Networks.* New York. Macmillan.

L. D. Jackel, H. P. Graf, W. Hubbard, J. S. Denker, D. Henderson, and I. Guyon (1988), An application of neural net chips: Handwritten digit recognition. *Proceedings of the 1988 International Conference on Neural Networks, San Diego, 1988.* San Diego, CA: IEEE San Diego Section.

I.T. Jolliffe (1986), *Principal Component Analysis.* New York. Springer.

Y. LeCun (1986), Learning processes in an asymmetic threshold network, In E. Bienenstock, F. Fogelman-Soulie, and G. Weisbuch (Eds), *Disordered Systems and Biological Organization.* Berlin: Springer-Verlag.

Y. LeCun, B. Boser, J. S. Denker, D. Henderson, R. E. Howard, W. Hubbard, and L. D. Jackel (1990), Backpropagation applied to handwritten zip code recognition. *Neural Computation, 1,* 541–551.

R. Linsker (1988), Self-organization in a perceptual network. *Computer Magazine, 21,* 105–117.

M. Minsky and S. Papert (1988), *Perceptrons, 3rd ed.,* Cambridge: MIT Press.

E. Oja (1982), A simplified neuron model as a principal component analyzer. *Journal of Mathematical Biology, 15,* 267–273.

D. Parker (1985), Learning Logic, Technical report TR-87. Center for Computational Research in Economics and Management Science, MIT, Cambridge, MA.

D.E. Rumelhart, G.E. Hinton, and R.J. Williams (1986a), Learning internal representations by error propagation. In D.E. Rumelhart and J.L. McClelland (Eds.), *Parallel, Distributed Processing,* Vol. I. Cambridge: MIT Press.

D.E. Rumelhart, G.E. Hinton, and R.J. Williams (1986b), Learning representations by back propagating errors. *Nature, 323,* 533–536.

O.G. Selfridge (1958), Pandemonium: A paradigm for learning. *Mechanisation of Thought Processes: Proceedings of a Symposium Held at the National Physical Laboratory, November 1958.* London: Her Majesty's Stationery Office.

T.J. Sejnowski and C.R. Rosenberg (1987), Parallel networks that learn to pronounce English text. *Complex Systems, 1,* 145–168.

R.S. Sutton and A.G. Barto (1981), Toward a modern theory of adaptive networks: Expectation and prediction. *Psychological Review, 88*, 135–170.

G.B. Thomas, Jr. and R.L. Finney (1988), *Calculus and Analytic Geometry*. Reading, MA: Addison-Wesley.

P.J. Werbos (1974), Beyond regression: New tools for prediction and analysis in the behavioral sciences, Ph.D. Thesis, Harvard University, Cambridge, MA.

B. Widrow and M. Hoff (1960), Adaptive switching circuits. *1960 WESCON Convention Record*, 96–104.

B. Widrow and S.D. Stearns (1985), *Adaptive Signal Processing*. New York: Prentice-Hall.

T.Z. Young and T.W. Calvert (1974), *Classification, Estimation and Pattern Recognition*. New York: Elsevier.

You can observe a lot by watching.
Yogi Berra

This chapter discusses perhaps the most important aspect of a neural network com-
putation: what information to represent by the unit activity patterns and how to repre-
sent it. A number of biological information-representation techniques are described
briefly. Topographic maps of important aspects of the stimulus are given for three
sensory systems: somatosensory, vision, and audition. The sonar system of the bat is
discussed as an example of several topographic maps for information processing. Rep-
resentations used for cognitive information are less well understood, and some exam-
ples are provided. Finally, a few general rules for information representation in a
neural network are proposed.

Up to this point we have concerned ourselves with networks that learn to
connect things together: activity patterns with activity patterns or activity
patterns with classifications. It would be possible to devote an entire book
on neural networks only to the problems involved in making better and
more accurate connections between arbitrary activity patterns. This approach, in fact, is taken by most current books. Networks can be discussed
at their highest level of generality by placing few constraints on the details
of what is being processed.

To be useful, the activity patterns that are associated or used for classification must describe aspects of the real world. The data *representation* tells
how this important information is coded in the state vector. In general,
the data representation is the single most important decision that a neural
network designer makes. With a good input data representation, almost any
network will function well. With a poor one, even powerful learning algorithms will fail. Despite claims, an effective internal representation usually
cannot be developed without an appropriate initial coding of the input data.
In general, the more we know about the details of the problem, the less
learning the network will have to do and the better the generalization will
be.

In this chapter we will discuss something about the nature of information representation in the biological nervous system. Are neurons the basic computing elements, or something else? In a brain with large numbers of neurons, do many neurons respond to a particular property, thing, or event, or does only one? We will present some information about the systems and data representations that seem to be used by animals. Some of these ideas are valuable for artificial networks and for understanding natural ones.

It is necessary to make one comment on the many figures and their captions presented in this chapter. It is part of the culture of neuroscience to produce complex, multipart figures and to write elaborate, detailed captions describing them. These captions are often packed with information and sometimes are small informative essays all by themselves. We have reproduced many published figures and, often, most or all of their captions. Sometimes the captions use biological terminology that may be unfamiliar to some readers; however, others will find it useful. Reading original papers in this area is highly recommended. Often the real beauty and elegance of biological systems are found in the details. Those with an appreciation of marvelous engineering will find many aspects of neurobiology quite wonderful.

Neurons, or What?

For the formal models we discussed starting in chapter 6 we were concerned with operations on state vectors. Considering all the information we presented about neurons in chapters 1, 2, and 4, it would be hard not to infer a connection between single vector elements and the activities of single neurons in the nervous system. The generic connectionist neuron, discussed in chapter 2, is a highly connected integrator based on greatly simplified models of biological neurons. Although this plausible identification is accepted by most who work with neural networks, it is not necessarily true. A few alternatives have been proposed.

First, a single real neuron might contain several neural network neurons. A real neuron is complex. Just because it is a single biological unit does not necessarily mean that it is also a single functional unit. For example, a possibility is that the dendrites of a single neuron can each act as an independent, nonlinear, integrating device. The output of numerous dendritic integrators could be summed by the cell body. Evidence suggests that dendrites in some large cells can fire action potentials. Cerebellar Purkinje cells are the best-known example. Some invertebrates are known to carry on

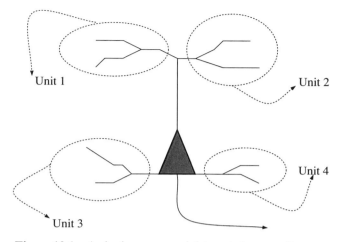

Figure 10.1   A single neuron might contain more than one region of integration. A single biological neuron may contain several functional elements, each of which could act like a neural network computing unit.

independent operations in different axonal branches. It has also been suggested, more speculatively, that smaller structures such as localized groups of synapses might act as neuron-like integrators. A practical consequence of this line of thinking is that the nervous system might contain many times more elementary units than neurons (figure 10.1).

Another possibility is that neurons are connected in functional groups, and multineuron groups are elementary entities. One example of a functional grouping that some modelers said might serve this function is the *cortical column.* In mammalian cerebral cortex, cells in a small area often respond to many of the same aspects of a stimulus. When electrical recordings were first made from the visual regions of cerebral cortex it was discovered that if an electrode penetrated perpendicular to the surface of cortex, all the cells encountered would respond to the same orientation. There is also evidence for regional anatomical groupings. Dendrites are often physically very close to each other and run together, and cell bodies are often lined up in columns perpendicular to the surface of cortex.

We have several ways to model such structure. The most popular is to assume that these multineuron units are indeed groups, but that the important group output parameter is average firing frequency. Therefore, the entire column acts formally like the generic connectionist computing unit. The only difference between a network made of such groups of neurons and one made of neurons is that the former has a lot *fewer* elementary computing units than neurons.

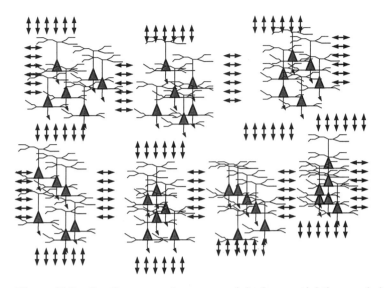

Figure 10.2 Small groups of neurons might form a tightly coupled local element. Each group does a local computation and passes the results to neighboring groups. The output might take the form of an average output value, or a pattern of output activity. A single neural network computing element might be composed of a number of physical neurons.

Another starting point for modeling might be to take the group idea seriously. If each column is a group, perhaps it signals its output to other groups not as a scalar, a single value such as average activity, but as an activity pattern, a vector. This line of thinking leads to viewing a neural computation as the operation of a network of networks. A group is a functional single unit held together by rich neuronal interconnectivity, but with equally complex patterned input and output. This idea abandons an important feature of the generic connectionist unit, the single valued output. The resulting network of networks is complex, but may not be difficult to analyze in some cases. As an intermediate level of organization, this approach has interesting and largely unexplored consequences (figure 10.2).

A related idea is the notion of *cell assemblies*, which was first suggested by Donald Hebb (1949) in the same book in which the Hebb synapse was postulated. Hebb's idea was that, perhaps due to correlational synaptic modification, neurons would become tightly connected: when one fired, another one would be strongly induced to fire. As he put it, "It is proposed ... that a repeated stimulation of specific receptors will lead slowly to the formation of an 'assembly' of association-area cells which can act briefly as a closed system after stimulation has ceased; this prolongs the time during which the structural changes of learning can occur and constitutes the

Cell Assembly

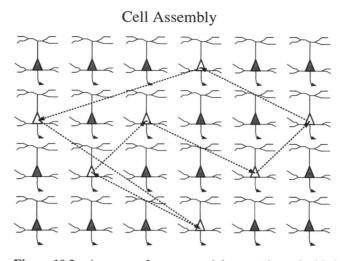

Figure 10.3 A group of neurons might contain embedded *cell assemblies* (Hebb, 1949). These units would be coupled together, and the entire cell assembly might be activated in sequence (as here) or simultaneously to represent the presence of an idea or perception.

simplest instance of a representative process (image or idea)" (p. 60). This suggests that complex objects can become represented by activation of one or more of these coupled groups. The neurons composing the assembly are embedded in the midst of many cells that are not part of the group. Firing a few units in the group could lead to activation of the entire group. Activation could eventually spread to other cell assemblies, coupled to the first group, as a way of forming associative links (figure 10.3).

Attempts to test this appealing idea led to one of the first computer simulations of a neural network (Rochester, Holland, Haibt, and Duda, 1956). The idea is attractive, but it is difficult to make it work properly without the use of powerful inhibitory and control mechanisms; the heavy use of recurrent excitation can give rise to unstable networks. Group behavior with a resemblance to cell assemblies is inherent in some of the energy-minimizing networks (see chapter 12) and in some applications of nonlinear dynamic systems theory, for example, work on olfactory cortex (Skarda and Freeman, 1987). Although cell assemblies are composed of neurons, their functional coherence makes it possible to move analysis to a level at which one might most appropriately talk about interactions between cell assemblies and not the neurons comprising them. Such mechanisms could provide a much-needed bridge between the structures found in high-level cognition and the nervous system.

Part of the historical reason for the focus on single neurons as elementary entities is that those are the only microscopic structures in the brain that can be recorded from easily because of the large electrical currents associated with action potentials. It is difficult with current technology to test the notion that many units exist within a single neuron membrane for more than a few large, favorably arranged cells. It is even more difficult to test the idea that many cells work together to form an elementary processing unit such as a cell assembly or a local network. Even if it is possible to record from more than one neuron at a time, it is technically difficult, and it is hard to show that the neurons are parts of a functional assembly. Statistical analysis and interpretation of multiple spike train data is difficult, not well understood, and involves massive amounts of computation. A little evidence now exists for neural interactions suggestive of assemblies of some type. However, there are almost no data on the details of what the assemblies might be doing in a computational sense (Gerstein, 1988, 1990; Aertsen, Gerstein, Habib, and Palm, 1989).

Distribution versus Specificity

How many active cells respond when a specific object or event is represented? We mentioned this issue once before, in chapter 6, and now we return to it. We could suppose that each cell in a particular region of the nervous system has a label on it, and that when this cell fires, the label is "perceived." This position is sometimes called the grandmother cell theory of the nervous system. It is an extreme version of *localized* coding. It assumes that one unit represents the idea of grandmother. This unit is activated when grandmother appears, no matter what the details of her appearance that day, or even whether she exists physically there at all. She might be imagined, hallucinated, or remembered. The discharge of the cell *represents* grandmother in a direct and fundamental way. This clear, simple, and understandable representation is a great favorite and is unfortunately used for the bulk of practical neural networks. (For some influential ideas about biological representations and degree of distribution found in them, see Barlow, 1972.)

Grandmother cells are particularly common as the output representation in categorization systems. It is natural to have one and only one output unit activated as the output of categorization. Most engineering applications of networks use this representation. For example, a neural network letter-recognition system might have the output correspond to a single active unit representing the letter. A speech-recognition system might have single output layer units corresponding to a particular word, syllable, or phoneme. The perceptron (chapter 8) contains a layer of output units (R units) that *in practice* corresponded to grandmother cells, that is, one and only one R unit is active at a time.

One particularly useful aspect of grandmother cell representations is that the state vectors representing different objects are orthogonal. This is because only one cell, different for each different object, is active. Therefore the inner product between representations of different recognizable objects is zero. One important application of winner-take-all (WTA) networks, mentioned in chapter 4, is to provide a mechanism to form grandmother cells.

However, serious problems are associated with a grandmother cell representation. First, a single unit is vulnerable to damage or loss. This criticism is usually handled by assuming a degree of redundancy, so that several identical cells represent grandmother. Therefore, the loss of one cell will not produce selective memory loss for grandmother. It can be argued that we can do this because we have lots of neurons.

Second, considerable difficulties are involved in constructing a biologically plausible learning system that can give rise to such an extreme degree of specificity, although such a system could, in theory, be built. In an artificial system we can do whatever is convenient, of course.

Third, and probably most serious, is the difficulty grandmother cell systems have in generalizing appropriately. Grandmother cell representations of different objects, as we mentioned, are orthogonal. One of the most useful claimed aspects of neural networks is their ability to generalize to new situations, but this is hard to do when there is great specificity of representation. As an example, let us consider another formulation of the grandmother cell, this one due to Naomi Weisstein, called the yellow Volkswagen detector. For figure 10.4 we will assume we have a classic car detector, in this case, a Mercedes-Benz SSK, one of the most magnificent cars ever produced.

Let us assume these cars come in several colors, so we have a grandmother cell activated by a white car, and a nearby cell activated by a black one. But suppose a gray SSK is sighted. Either a new cell is suddenly formed to handle the new shade or, more likely, both the white and the black cells are activated. It is clearly the SSK, and similar to both the black and white cars. But does the activation of two cells in a grandmother cell system signal an intermediate percept? And, clearly, all colors of SSK have similarities in shape, size, classic status, use and so on. But if they are represented by separate cells, how can we get desirable similarity relations between them to be computed? Perhaps we could generate a new grandmother cell very rapidly, but how could we have the new cell make all the interconnections

Problems with Grandmother Cells

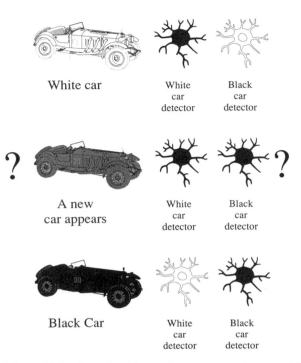

Figure 10.4   Grandmother cell systems may have problems with generalization. We assume that there is a grandmother cell for a black classic car and another one for a white classic car. What happens when a gray classic car is seen? Do both units become active?

with other stored information that would have to be made—number of cylinders, era of manufacture, historical anecdotes about the designer (Dr. Porsche), problems with the suspension, and so on? A way out would be to say that we have car cells and color cells, and both become active to represent the gray car. However, this natural first step toward a distributed code defeats the purpose of grandmother cells.

Fourth, combinatorial problems exist with the number of cells that are required to realize these systems. Suppose a localized visual perceptual model requires analysis of visual information into 1 of 5 possible shapes (triangle, square, circle, etc.), 1 of 5 possible colors (red, green, etc.), and 1 of 5 possible sizes (very large to very small) at 1 of 20 possible x locations and 20 possible y locations. This simple analysis requires 50,000 output units to represent all possible combinations of properties. If we suddenly decide that it would be nice to respond to motion as well, and real cells in the visual system generally do, then we have say, 5 possible directions of motion and

5 possible speeds, increasing the number of cells required from 50,000 to 1,250,000. A more detailed analysis of the visual scene than this minimal one requires even more cells. Selective representation rapidly is involved in a combinatorial explosion. Again, an obvious and plausible way out assumes separate structures to recognize color, form, location, and so on. But this also introduces distribution into the coding.

A fifth problem is what might be called amplification. When a neuron fires an action potential, energy on the order of magnitude of 10^{-7} joules is released. If firing of a single critical neuron routinely provoked a large animal into furious action because prey or a predator was detected, the result is effectively a huge amplification of the energy released by that neuron, perhaps 10 or more orders of magnitude of energy amplification. This enormous gain places extraordinary demands on the reliability and accuracy of a single neuron and its connections. Single bits in a computer are reliable enough to serve such a function, but this reliability is unlikely to be found in a small biological structure.

Distributed Representations

An extremely localized representation of information in the brain is unlikely, but this has not prevented grandmother cells from being employed as a useful simplification in many specific problems. We have a wide range of alternatives to localization, however. Let us consider a system that has to represent information about four items, a rock, a dog, a wolf, and a sheep. We used the coding of ASCII characters as an another example in chapters 3 and 7. Figure 10.5 contrasts the two representations applied to this small set of items. Each unit in a distributed representation is ambiguous and merely restricts the possibilities that the character can be. Different items can have many active units in common.

Binding Problems

As soon as we get into even moderately distributed systems we run into what are sometimes known as *binding problems*. How do we connect together all the physically separated fragments of a complex object so they can be processed as a whole? A red triangle is obviously a different object than a red square or a green triangle, yet some of the same cells will be active in representing them. Cells in distributed systems can be ambiguous. Notice that a grandmother cell representation completely finesses the binding problem by concentrating meaning into a single, active, basic entity.

Grandmother Cells

```
Rock:   1 0 0 0 0 0 0 0 0 ...

Wolf:   0 1 0 0 0 0 0 0 0 ...

Dog:    0 0 1 0 0 0 0 0 0 ...

Sheep:  0 0 0 1 0 0 0 0 0 ...
```

Distributed Coding

```
Rock:   0 1 0 1 0 0 1 1 0 0 0 0 0 1 1 0 0 1 0 0 0 0 0 1
Wolf:   0 1 1 0 1 0 0 0 1 0 0 0 0 1 1 0 0 1 0 0 0 0 0 1 0
Dog:    0 1 1 0 1 0 0 0 1 0 0 0 0 1 0 1 0 1 0 0 0 0 0 1 0
Sheep:  0 1 1 0 0 1 0 0 1 0 0 0 0 1 0 1 1 0 0 0 0 0 0 1 0
```

Identical Values

Figure 10.5 Here several items are represented by a grandmother cell representation, where a single active position corresponds to a particular item, and by a distributed representation. In the distributed representation a single element may have the same activity for many different items. In a grandmother cell representation, cell activity is not ambiguous. The distributed coding used is explained in figure 10.34.

In this book we simply accept the ambiguity and work with state vectors as elementary entities, that is, patterns of unit activity. Patterns can be just as selective as grandmother cells. In general, however, binding problems are difficult, and simply assuming patterns are primitives is only part of the answer. We touch on these issues in chapter 16 on associative computation, and chapter 11, where a network is demonstrated that uses ambiguous units to compute large-scale object motion.

Binding problems become particularly subtle and difficult in cognitive science or linguistics, where systems seem to obey rule-based behavior, for example, the complex but lawful syntactic relations in language. How can we operate at a high level of generality using state vectors, such as asserting that every sentence is composed of noun phrase and a verb phrase, when the specific examples of noun phrase and verb phrase will be different in every sentence? We surely cannot learn every possible sentence ahead of time. How can we bind together unit activity patterns to form higher-level temporary structures? There are currently no satisfactory answers to these questions.

Neuroscientists have displayed a degree of enthusiasm for the idea that 40-Hz oscillations of electrical activity in the cerebral cortex are an electrophysiological correlate of binding. The idea is that units in different regions of cortex oscillate in synchrony when a higher-level entity is bound together (see Crick and Koch, 1992, for further discussion). Unfortunately, this attractive idea is not the entire story, as more recent experimental results have shown, but it suggests a direction in which to look.

Motor Output

Briefly consider the data representation at the output side of the nervous system. The ultimate output is not information but motor activity. The motor neurons, as pointed out by Sherrington, a famous turn-of-the-century British physiologist, and, after him, every first-year lecturer in a neurobiology course, are the "final common path" for the nervous system: all parts ultimately lead to the same place. This obtains unless, of course, we believe that the nervous system is similar to the fabled write-only memory (WOM) of the semiconductor world that stores data but cannot retrieve it. There are several different kinds of muscles; however, in the mammalian nervous system, the motor output that concerns us is the one that drives the so-called voluntary muscles. What laymen think of as muscle is in fact this voluntary muscle, and it is under our more or less conscious control. (Other kinds of muscles—smooth and cardiac—are autonomous to some degree and do not have to be driven from the nervous system to function, although they do receive inputs from the central nervous system.)

Muscle fibers are similar to nerve cells in that they fire action potentials, show electrical summation of inputs, have thresholds, and have many of the same ionic currents involved in their function. However, in muscle, the action potential is accompanied by a mechanical twitch, an all-or-none event. Controlled and graded motor movements are produced by the mechanical summation of a great many of these all-or-none twitches in a large structure. The fact that many muscle fibers are present in a given muscle allows mechanical smoothing of the jerky individual fibers.

Muscles only contract. They have to move around a jointed framework, the skeleton. If a structure is to move forcefully in both directions, it must be actively driven in both directions. To go one way, one muscle contracts and the other relaxes; to go the other way, the other muscle contracts and the first relaxes. When pairs of muscles are connected in this reciprocal fashion, they are called agonists and antagonists. Since they directly oppose each other, special circuitry in the spinal cord is devoted to making sure that they don't fight each other, but that one is inhibited when the other is active.

The human body has several hundred muscles. It is fairly easy to show that virtually every motor act involves the activation of many muscles. For example, if the arm is raised, it changes the center of gravity of the body, causing compensatory contractions or relaxations of many postural muscles. The geometry of real joints is sufficiently complex that many of the muscles connected to any one joint respond in some way when it moves.

The point of this digression is that it justifies an extremely important statement about the output of the mammalian nervous system: it is a *pattern* of motor neuron discharges. It is never, for any significant movement, a single-output fiber. Humans have something on the order of 2.5 million motor neurons, that is, cells that directly drive muscles. Therefore the output of the nervous system is generally the excitation of some significant fraction of these cells, which must be precisely orchestrated so muscles do not fight each other, and the brain's will is executed in the most efficient and harmonious way possible. As has been put by many neuroanatomy teachers over the years, "The task of the nervous system is that facing a person playing a piano with two and a half million keys."

Layered Structures

If the output of the nervous system is a *pattern* of activation, and there do not seem to be grandmother cells, what facts do we know about the details of biological data representation? One of the most common arrangements of neurons in the vertebrate nervous system is a layered two-dimensional structure organized with a topographic arrangement of unit responses. Perhaps the best-known example of this is the mammalian cerebral cortex, although there are numerous others.

This is not a book on neuroscience, but it is worth sketching the outlines of cortical data representation because it says something about how the brain computes things, and may suggest practical ideas for data representation in artificial systems. An information-processing system with half a billion years of design experience must be taken seriously as a source of good ideas.

The cerebral cortex is the outside surface of the brain. In German it is called the *Hirnrinde*; that is, the rind of the brain, like the skin of an orange. Cortex is a two-dimensional structure, although in many larger and/or more intelligent animals it is extensively folded with fissures (*sulci*) and hills (*gyri*). The pattern of folds shows some differences from individual to individual, but the larger folds are relatively stable and can be used as landmarks. In humans about two-thirds of the cortex is buried in the folds. The cell layer (*gray matter*) is on the outside of the brain. The total surface area of the human cortex is perhaps 2300 cm^2, roughly the size of a dish towel (Prothero and Sundsten, 1984). Bundles of axons, the cabling of cortex, take up a large fraction of the volume of the brain and are called the *white matter*. When a brain is fixed for anatomical study, the texture of the white matter is very much like that of string cheese, and bundles of axons can be separated from each other.

Two different kinds of cortex exist: an older form with three sublayers, called *paleocortex*, and a newer form that is most prominent in animals with more complex behavior such as ourselves, a structure with six or more sublayers called *neocortex*. It has been noted for generations that the more intelligent an animal seems to be, the larger is its neocortex compared with other brain structures. (Comparative size is a key variable, because there is a strong relation between body size and brain size. The cortex of a porpoise is considerably larger in area and more convoluted than that of a human.) One of the things that characterizes human evolution is the increase in size of neocortex compared with our nearest primate relatives, the chimpanzee and the gorilla. It is hard to avoid the conclusion that the function of cortex is somehow connected with complex information processing.

Virtually every neuroanatomy textbook contains a picture showing a sequence that illustrates the increase in cortical size (the stippled area in figure 10.6) as animals increase in behavioral complexity from a codfish through frogs, pigeons, and cats to humans. Behavioral complexity and adaptability are probably better terms than "intelligence." One universal truth about biology is that "simple" animals turn out to be very complex when studied in detail. The hatched structure in the figure is the optic tectum (superior colliculus in mammals), which is a brain stem structure that is involved in spatial localization and does not increase very much in size with behavioral complexity, whereas neocortex becomes relatively much larger. (In humans the superior colliculus seems to be involved in moving the eyes to fixate interesting parts of visual space; in frogs and toads it is essentially the entire visual system.)

One can see visible differences between different regions of cerebral cortex. The two-dimensional sheet is divided into many regions, each presumably corresponding to a processing module. For example, most incoming visual information is sent to a cortical region in the back of the human head, referred to as V1, whereas acoustic information is sent to a region several centimeters away. The anatomy of two regions differs in detail, but the overall architecture is remarkably similar. Current estimates are that there are perhaps 40 functionally distinct areas. Boundaries between the areas are based on "the pattern of thalamic connections, along with slight local variations in cellular composition and connectivity" (Shatz, 1992, p. 237). Figure 10.7 shows an early parcellation of the human cerebral cortex by Brodmann, based on these small differences in neuroanatomy. The impression that one

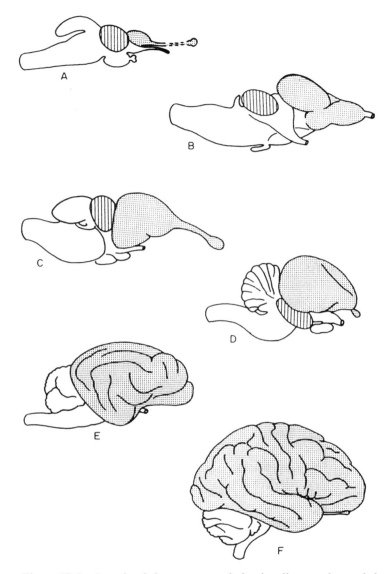

Figure 10.6   As animals become more behaviorally complex and show greater intelligence (whatever that is), the neocortex becomes bigger relative to other brain structures. A series of side views of brains of various vertebrate orders illustrates the progressive increase in relative size of the cerebral hemisphere (stippled). A, Codfish; B, frog; C, alligator; D, pigeon; E, cat; F, human. The optic tectum, indicated by vertical shading, is obscured from view in the two mammalian forms by the overlying cerebral hemisphere. From Nauta and Karten (1970) by copyright permission of the Rockefeller University Press.

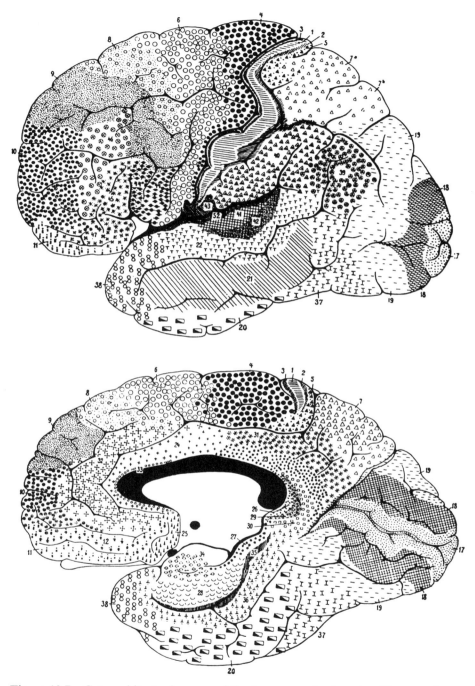

Figure 10.7 Cytoarchitectonic areas of the human cerebral cortex. These regions are often referred to as Brodmann's areas. This much reprinted figure is from Brodmann (1909).

receives is of a remarkably uniform structure that can expand and contract with the demands placed on it.

One insight of the early anatomists was that these cortical anatomical regions were also functional regions. Each region is composed of a large number of cells with similar statistical connectivities and performing a specific computation using sensory inputs and inputs from other regions of cortex. Figure 10.8 shows a map of the different visual regions in the macaque monkey (the human visual system is very similar to the macaque). Figure 10.9, from the same paper, shows a map of the connectivity of the modules in the mammalian visual system as it is currently known (Van Essen, Anderson, and Felleman, 1992).

The more detail we know about the connectivity between cortical regions, the more complex the overall connectivity of the visual system becomes. In no sense is this a picture of a simple feedforward network. It should be realized that the lines, indicating massive projections, go in both directions. That is, for every upward projection to a higher level of processing, there is a downward projection of equal or sometimes greater size. In addition, substantial sideways projections exist between layers at the same level.

This system shows distinct signs of a hierarchy of processing, although the idea of "level" in a hierarchy may not always be the best way to think of such a tightly coupled system. Detailed functions for many regions are not known, but even in the visual system, our best-studied cortical system, function is clearly separated in some cases.

For example, the projection from V1 → MT → MST seems to be a pathway concerned primarily with computations involving motion. (See the simulation and discussion in chapter 11.) Crudely, area V1 contains units that respond to local motion, and units in area MT respond to the motion of large moving objects, a complex computation that requires integrating many ambiguous local motion signals. Area MST seems to respond to even more complex rotations and dilations. It has been suggested that there are parallel visual pathways involving different aspects of vision. One theory, for example, holds that visual perception may be distinct from the visual control of actions. Telling that an object is "out there" uses a "ventral stream of projections from the striate cortex to the inferotemporal cortex," whereas reaching out to grasp the object uses a "dorsal stream projecting from the striate cortex to the posterial parietal region" (Goodale and Milner, 1992, p. 20; DeYoe and Van Essen, 1988).

A point worth making is that the depth of cortical processing is not very great. The richly interconnected weak hierarchy suggested in figure 10.9 does not contain very many layers. Once sensory information enters the cortical sensory receiving areas, it may not pass through many synapses

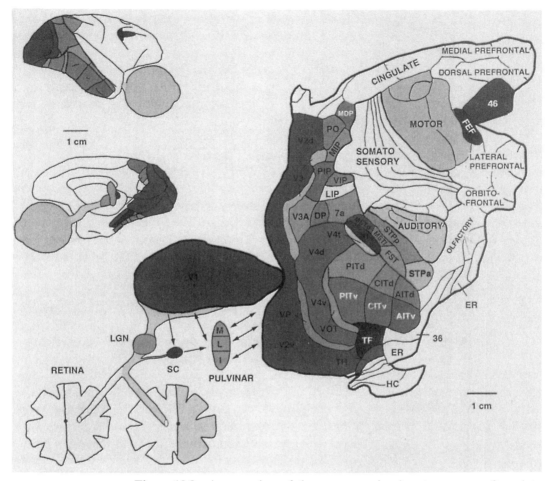

Figure 10.8 An overview of the macaque visual system as seen from lateral and medial views of the right hemisphere, and from unfolded representations of the entire cerebral cortex and major subcortical visual structures. The cortical map contains several artificial discontinuities, for example, between V1 and V2. Minor retinal outputs (∼10% of ganglion cells) go to the superior colliculus (SC), which projects to the pulvinar complex, a cluster of nucleii having reciprocal connections with many cortical visual ares. All structures except the much thinner retina are ∼1 to 3 mm thick. From Van Essen, Anderson, and Felleman (1992). Reprinted by permission. Copyright 1992 by A.A.A.S.

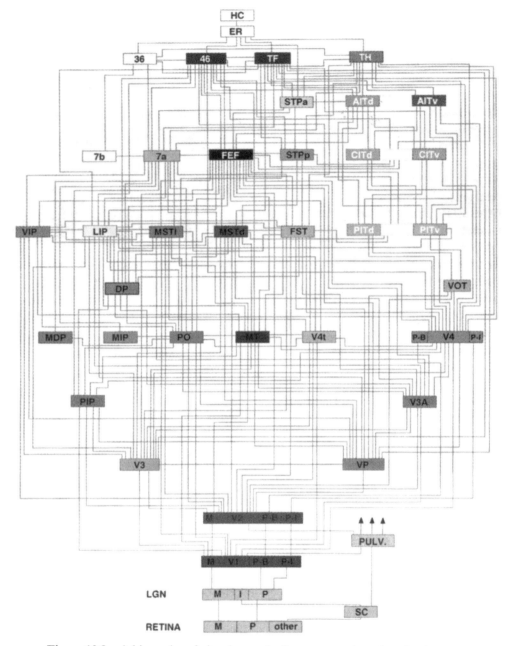

Figure 10.9   A hierarchy of visual areas in the macaque based on laminar patterns of anatomical connections. About 90% of the known pathways are consistent with this hierarchical scheme; the exceptions may reflect either inaccuracies in the reported anatomical data or genuine deviations from a rigid hierarchical scheme. From Van Essen, Anderson, and Felleman (1992). Reprinted by permission. Copyright 1992 by A.A.A.S.

before processing is complete and it emerges as a motor signal. A guess based on connectivity patterns might be on the order of a dozen synapses from input to output, and often fewer. There is no sign of extended chains of more and more complex serial computations where hundreds or thousands of steps might be involved in obtaining a result. This is a place where neural computation diverges fundamentally from digital computation. Elementary brain operations may be very powerful, but there are not many of them in series. The answers must be obtained in a small number of processing steps. The later chapters of this book (in particular chapters 11, 16, and 17) suggest some of the consequences of this observation.

Arrangement of Units Within Cortical Regions

A great deal is known about the neurons in cerebral cortex and their arrangement. Cerebral cortex does not have a great many cell types. The most common are *pyramidal cells*, so called because of their pyramid shape (figures 10.10 and 10.11). An apical dendrite comes from the apex of the pyramid, oriented perpendicular to the cortical surface. Pyramidal cells are present in most of the layers of cortex, but the apical dendrite always runs to layer 1, where it branches extensively. Pyramidal cells are the long-range cells, and their axons project outside of cortex to other cortical regions. Their dendrites are highly branched and impressive to see. Local anatomical groupings of pyramidal cells have suggested the existence of anatomical columns in cortex. The dendrites of several pyramidal cells may run together to the surface.

The pyramidal cells receive synaptic inputs to their dendrites at the tips of structures called dendritic spines, discussed in chapters 2 (figure 2.4), 6 (figure 6.5), and 10 (figure 10.12). The small spine process gives the neuron a fuzzy appearance under the light microscope. The size and appearance of spines respond to environmental manipulation, and it is generally believed that spines have something to do with the synaptic changes underlying learning.

Local Circuitry

Dendritic synapses on pyramidal cells are excitatory. Incoming sensory axons and contacts between pyramidal cells are excitatory. Inhibitory synapses seem to be localized to the cell body. The other classes of cells in cortex, several different kinds of what are called *stellate cells*, mediate inhibi-

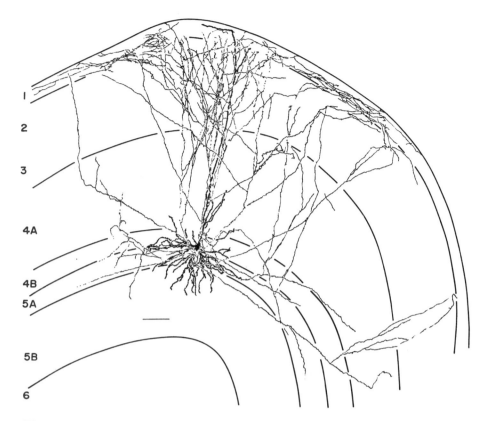

Figure 10.10   Pyramidal cell in visual cortex of the cat. (Horseradish peroxidase stain.) The cortical sublayers are numbered. The pyramid-shaped cell body is located in layer 5A, and the thick apical dendrite projects perpendicular to the cortical surface up to layer 1. Note the extensive collateral branches (lighter lines). Pyramidal cell in layer 5A with projections predominantly to supragranular layers. Complex receptive field is 0.9 by 2.0 degrees. Driven polysynaptically, length summation properties not tested. The axon extends 1050 μ anteriorly and 550 μ posteriorly. Scale bar: 100 μ. From Martin (1984). Reprinted by permission.

tion. Stellate cells are local in that they do not appear to project outside of local regions in cortex. It is suggested that detailed modifiability of the kind we have assumed for our simple Hebbian network models is found in the excitatory synapses. Whether learning or modifiability is involved in the inhibitory systems in cortex is not known.

A characteristic of pyramidal cells is the presence of large numbers of *recurrent collaterals*. When the pyramidal cell axon leaves cortex, it branches extensively and sends recurrent collaterals back into the local region of cortex around the cell (figure 10.13). A cell's collaterals might contact a number of other pyramidal cells within a few millimeters of the cell. The

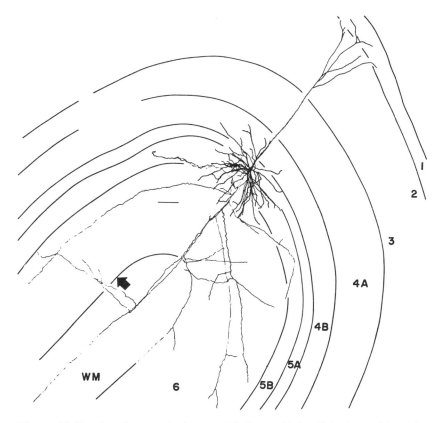

Figure 10.11   Another cortical pyramid. Pyramidal cell in layer 5A with axonal projections mostly to layers 5 and 6. Complex receptive field (standard complex cell) is 1.5 by 1.4 degrees. It is monosynaptically driven by Y-like afferents. This neuron could also be antidromically activated from the two electrodes in the optic radiations, giving an approximate conduction velocity of 17 m/sec. The arrow indicates the area 17/area 18 (V1/V2) border. The intracortical axon extends 1050 μ anteriorly and 2150 μ posteriorly. Scale bar: 100 μ. From Martin (1984). Reprinted by permission.

recurrent collaterals make contact on the dendrites of pyramidal cells and are excitatory. There are huge numbers of them, and they may be the most common single class of fibers in neocortex. Their functional role is unclear, but they may form the basis of local computation and provide the substrate for powerful local excitatory feedback (see chapter 15).

If modifiable connections in cortex are largely excitatory and also recurrent, the potential exists for disastrous instability caused by local positive feedback. It is surprising that the brain is so stable and displays normal operation over a lifetime. Powerful local inhibition presumably provides control and stability. If local inhibition is disabled by drugs such as strych-

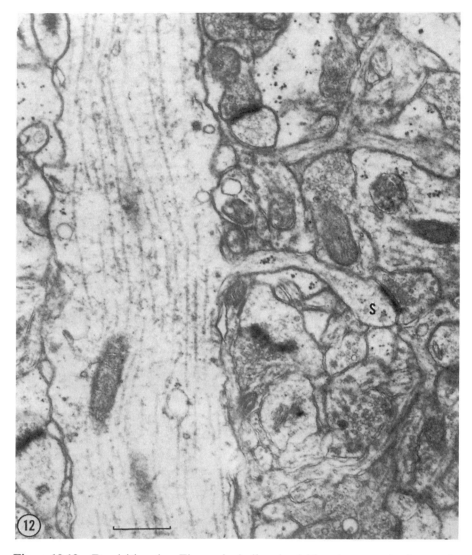

Figure 10.12   Dendritic spine. Figures including dendritic spines can also be seen in figures 1.3 and 6.5. Longitudinal section of an apical dendrite in rat visual cortex. A spine emerges from the dendritic shaft and forms an asymmetrical synapse at its terminal enlargement (S). The prominent cytoplasmic constituents of the dendritic shaft are the axially oriented microtubules. Calibration line: 0.5 μ. From Feldman (1984). Reprinted by permission.

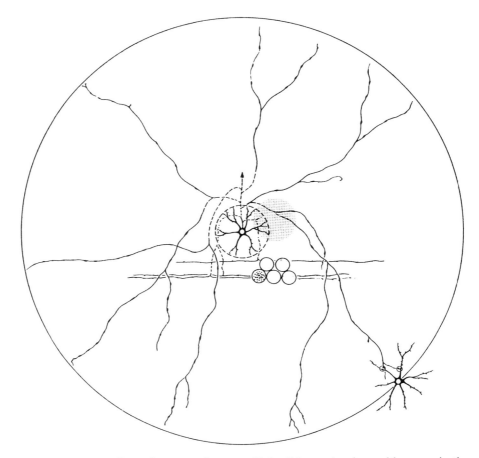

Figure 10.13 Collateral system of a pyramidal cell (center) as it would appear in the view from the surface (semidiagrammatic reconstruction). The large circle having a diameter of 3 mm indicates the potential territory reached by the collaterals. The pyramidal cell below right on the perimeter would be a potential recipient of synaptic contacts from the central cell. The small circles (100 μ diameter) correspond to the apical dendrite bundles containing 20 to 30 apical dendrites of a pyramid cell cluster. The round stippled area at right of the central pyramidal cell (300 μ diameter) indicates the width of an arborization column of corticocortical afferents. From Szentagothai (1978). Reprinted by permission.

nine, seizures quickly result, suggesting a considerable potential for instability that is kept under tight inhibitory control.

Connectivity

The abstract models we discussed in previous chapters are usually *fully connected* in that every unit connects to every unit. This is clearly untrue for the brain. According to Nelson and Bower (1990), a rough calculation shows that if cortical neurons were arranged on the surface of a sphere and fully connected by connections 0.1 μ in diameter, the sphere would have to be 20 km in diameter. Estimates are that pyramidal cells have thousands of synapses, for example, a range of 7000 in visual cortex to 50,000 for large pyramidal cells in motor cortex. This connectivity is very much less than the number of cells in a cortical region. If we assume that primate primary visual cortex has on the order of 100 million cells (estimates range from 70 million to 150 million), a single pyramidal cell contacts less than a hundredth of 1% of the other cells in that region. Other cortical regions have similar low connectivities.

The usual neural network assumption of full connectivity is grievously in error. This major difference alone casts serious doubt on many attempts to apply ideas from artificial neural networks to the nervous system. If an artificial network depends critically on complete connectivity to function, and many do, it cannot be a satisfactory model of the biological nervous system.

Maps

A good deal has been learned about the spatial arrangement of the responses of neurons within cortical areas. A common pattern in cortex is a topographic map of one or more sensory properties, which was discovered as soon as scientists began to look carefully at the responses of a cortical region. One of the earliest and still the most famous is the sensory motor homunculus described by Penfield and Boldrey (1937) (figure 10.14). By electrically stimulating the cortex of conscious patients undergoing surgery for epilepsy, the neurosurgical team headed by Penfield discovered that the surface of cortex contained a map of the body surface. Most interesting, the map displayed systematic distortions; that is, equal areas on the body surface did not map into equal areas on the cortical map. The homunculus had

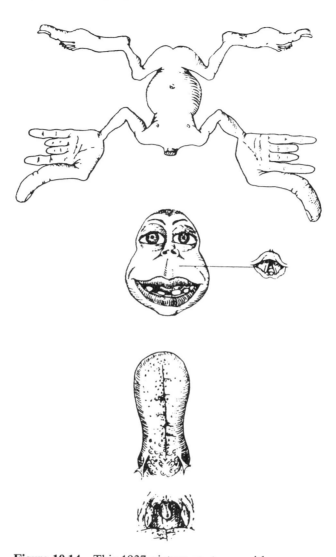

Figure 10.14   This 1937 picture captures, with some agreeable artistic license, the relative sizes of cortical regions associated with different parts of the body. They were sketched out by electrical stimulation of regions of somatosensory and motor cortex in the brains of conscious patients during brain operations. Sensory and motor homunculi were prepared as a visualization of the order and comparative size of the parts of the body as they appear from above down on the rolandic cortex. The larynx represents vocalization, the pharynx swallowing. The comparatively large size of thumb, lips and tongue indicates that these members occupy comparatively long vertical segments of the rolandic cortex as shown by measurements in individual cases. Sensations in the genitalia and rectum lie above and posterior to the lower extremity but are not shown. From Penfield and Boldrey (1937). Reprinted by permission of Oxford University Press.

very large hands and face, and small feet. Also, an unusually large area involved the tongue and mouth. The most obvious interpretation of this distorted map is that the size of the representation reflects to some degree the behavioral importance of the structures and the amount of peripheral innervation of the structure. The hands, tongue, and mouth are richly innervated. Humans have a great variety of important behaviors involving hands and the speech areas, and relatively few involving the toes and feet.

A more modern and accurate, but less entertaining, map of the somatosensory system is shown in figure 10.15. This segment of monkey cortex is labeled with the names of the regions with which they are concerned. Again, notice the rather small region associated with the foot, trunk, and arms, and the large regions responding to the mouth, tongue, and digits.

Regions associated with different parts of the body can have quite sharp boundaries. Figure 10.16 shows some maps of monkey cortex that have clearly separated regions corresponding to the different digits (Kaas, Nelson, Sur, Lin, and Merzenich, 1979). Adjacent cortex also has maps of the digits. In this case the two maps are mirror images of each other.

The precision of the maps and the separation of the digits is a matter of great interest. Did this separation arise from precise wiring during development, or was it learned? One of the most important and significant findings in cortical physiology in recent years is that maps such as this are plastic to some extent even in adult mammals. It has been known for many years that it is easy to disrupt and change cortical organization in young organisms. For example, even brief periods of patching an eye can cause permanent changes in stereopsis (depth perception) in kittens, young monkeys, and children. Recordings show that when this is done, cortical cells may no longer respond to both eyes and lose binocular interactions. The changes appear to be permanent.

Therefore, it was surprising when it was found that in other cortical areas in adults, for example, those devoted to the skin senses, the maps are plastic. For example, if a finger is repeatedly stimulated, especially if a reward is involved, the cortical region associated with that finger will grow. In normal monkeys the regions associated with the digits are cleanly separated. Units respond to one finger or the other, but not both. In a clever experiment workers fused two fingers in a monkey and showed that, after several weeks, the previous separation between the digits vanished, and some cells now responded to stimulation of skin from both digits (Clark, Allard, Jenkins, and Merzenich, 1988). Many experiments along these lines have been done. With electrical stimulation and proper techniques, changes in cell responses can been seen in as little as hours, suggesting far more cortical plasticity in adults than was expected.

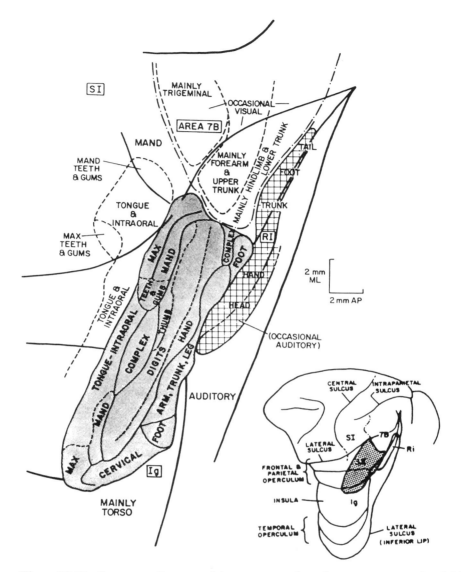

Figure 10.15 Summary diagram of the representation of cutaneous receptive fields along the upper bank of the lateral sulcus in a macaque monkey. The sulcus has been unfolded orthoplanimetrically, with the dorsal surface of the corpus callosum serving as the reference point for the reconstructions from serially aligned frontal sections. An unfolded view of the sulcus has been drawn attached to the parietal and frontal lobes in the lower right inset. The shaded area in this inset corresponds to the similarly shaded SII area on the left. Somatosensory areas surrounding SII include Si medially, area 7b posteriorly, retroinsular area (Ri) posteriorly, and granular insula (Ig) laterally. Scale, which applies to the figure on the left, marks the ML and AP planes of the brain; note that the long axis of SII runs anterolateral to posteromedial with respect to the frontal plane. From Burton (1986). Reprinted by permission.

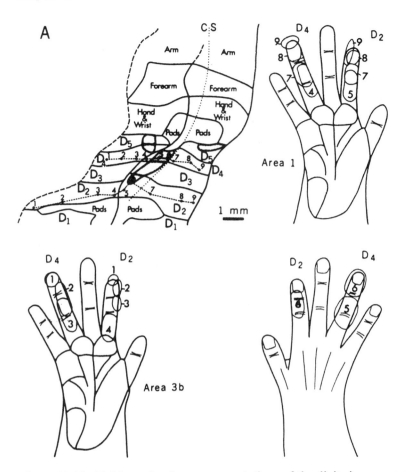

Figure 10.16 Evidence for three representations of the digits in macaque monkeys. For purposes of illustration, receptive fields are shown for only a few recording sites from much more extensive mapping experiments. A, Receptive fields for rows of recording sites across the representations of digits two and four in areas 3b and 1 in monkey 77–52. Cortex to the left of the dotted line is buried in the central sulcus (CS). Shaded areas indicate the representation of dorsal hairy surfaces of digits. Each architectonic field separately represents the digits, the representations are joined together along the bases of the fingers (rather than the palm as in the owl monkey), and the fingertips point in opposite directions. B, Receptive fields for rows of recording sites across the representations of the first three digits in areas 1 and 2 of monkey 77–39. The digits are represented separately in each architectonic field, and the two representations are joined at the fingertips so that they are approximately mirror images of each other. From Kaas, Nelson, Sur, Lin, and Merzenich (1979, figure 2). Reprinted by permission. Copyright 1979 by A.A.A.S.

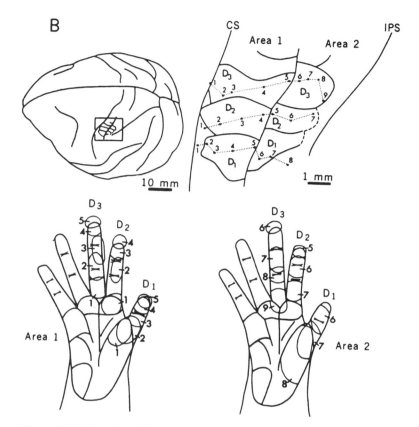

Figure 10.16 (continued)

At the same time, the range of plasticity seems to have limitations. It would probably not be possible to develop receptive fields that responded to both a finger and a toe, for example. What seems to be occurring is that cortical cells are connected to a much wider region of sensory surface than they actually respond to. If they lose their strongest input, or if the correlational statistics of the input change, these silent connections can become active and the cell will respond to them. These ineffective connections can also be unmasked with some drugs. Plasticity in this system has become a favorite subject system for modelers (Pearson, Finkel, and Edelman, 1987). The basic effect can be demonstrated in a variety of simulations. What seems to be required is some variant of a Hebbian learning rule combined with a normalization rule, so the totality of synaptic activation cannot get too large.

The homunculus in figure 10.14 is symmetrical around the mid-line, half of the figure representing the somatosensory map and, in a mirror image across the fissure, a motor map. This might suggest, and, in fact, it was long believed, that the motor cortex displayed a similar precise topographic map to that found in sensory cortex. However, recent work (see, for example, Scheiber and Hibbard, 1994) has found a far less strict mapping in the cortical motor regions. One cortical region is largely involved with, for example, movements of the hand and fingers. However, units found throughout this region are associated with activation of more than one digit or the wrist, in different proportions. Instead of the clean and obvious separation of regions found in the somatosensory maps, motor response is much more widely distributed.

Visual System

In vertebrates, the lens of the eye projects an image onto the retina where it is transduced into electrical events by photoreceptors. This image intrinsically forms a well-ordered topographic map onto a neural structure. After considerable complex retinal processing, in mammals the processed image is sent to the lateral geniculate nucleus (LGN) and from there to visual cortex. (A significant fraction of the visual output of the retina also goes to the superior colliculus.) The optics of the eye are such that the only part of the image with good spatial resolution is near the optic axis. The distribution of photoreceptors is extremely nonuniform. Specialized regions of the retina with high receptor and ganglion (output) cell density are present near the optic axis in structures called the *fovea* in primates and the *area centralis* in cats. Therefore the initial data representation of the image has far more cells associated with central vision than with peripheral vision. This pattern continues up to cortex.

Figure 10.17 shows the regions of visual space where the best responses are found for cells in cat primary visual cortex (V1) (Tusa, Palmer, and Rosenquist, 1977). A relatively good topographic map of the visual field is on the surface of the cortex. In the figure, parts of the cat brain are dissected away to show how the map follows the folds of the cortex. The map also contains significant distortions. The area near the optic axis of the eye, where the VM (vertical meridian) and HM (horizontal meridian) lines cross, located in the *area centralis*, has far more cortical area associated with it than areas in the peripheral retina. Since the density of cells in most cortical regions is rather uniform, differences in area directly reflect differences in cell number.

It must not be thought that these maps are extremely precise. At a particular location cells will respond to roughly the same region of space, but the

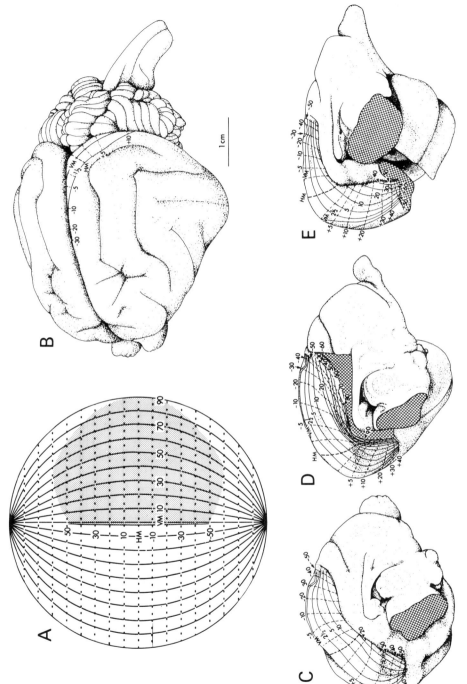

Figure 10.17 Summary diagram of the representation of the visual field in area 17 (now called V1). A, A perimeter chart shows the extent of the visual field represented in area 17. It is based on a world coordinate scheme in which the azimuths are represented as solid lines and the elevations as dashed lines. The location of the visual field in area 17 is illustrated in the four sketches of the cat brain shown in B through E. From Tusa, Palmer, and Rosenquist (1978). Copyright © 1978. Reprinted by permission of John Wiley and Sons, Inc.

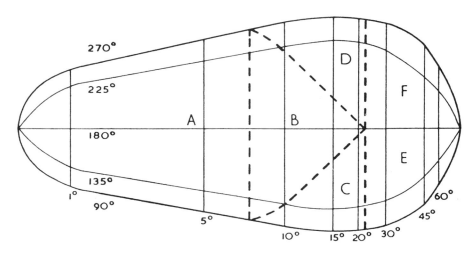

Figure 10.18 A map of the visual field onto V1 when the cortex is flattened out. Note the very large foveal representation and the much smaller representation of the peripheral visual field. A projection onto a plane of the reconstructed surface for the left hemisphere. This surface is folded along the heavy dotted lines so that F touches E, D and C touch B, and A folds round so that it touches and overlaps the deep surface of B. From Whitteridge and Daniel (1961). Reprinted by permission.

areas of best response shift around to some degree. These areas of best response can be quite large, but on average, the mapping is reasonably accurate in the early visual system.

Figure 10.18 shows the distorted map of visual space, similar to that in the cat, in the primary visual cortex of the monkey when the cortex is "laid flat," that is, the folds are removed and the cortex is repesented in the form of a flat sheet. Again, the huge amount of cortex devoted to the area around the area of high resolution, the fovea, is clear.

The minimum angle of resolution (the smallest angle where two black lines can be separated) is plotted against angular distance from the fovea in figure 10.19. Also plotted is the reciprocal of the *cortical magnification factor* derived from figure 10.18. Cortical magnification factor is the amount of cortex (in millimeters) devoted to a degree in visual space. If a great deal of cortex is devoted to a small number of degrees, as is the case near the fovea, the visual system has high angular resolution. The relationship is quantitative, as shown in the figure. High performance requires more cells to do the analysis.

Because the peripheral retina has relatively few cells, we would expect that cells there would respond to a much larger region of visual space than cells near the optic axis. This is true, and physiological data from cats and monkeys and psychophysical data from humans suggest that receptive field

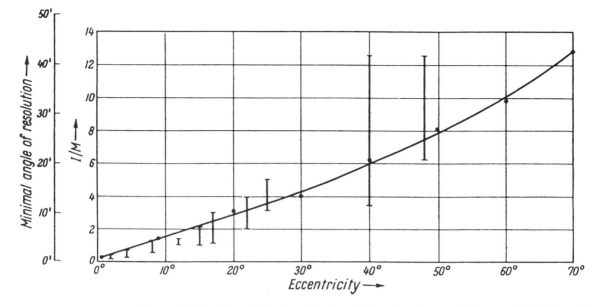

Figure 10.19 The minimal angle of resolution for a human (Weymouth, 1958), and the reciprocal of the cortical magnification factor for the monkey, both plotted against eccentricity; that is, angular distance from the fovea. From Whitteridge and Daniel (1961). Reprinted by permission.

size increases *linearly* with angular distance from the center of the visual field, as does the minimum angle of resolution plotted in figure 10.19.

Fischer (1973) did some simple calculations based on this observation and also on the known maps of retinal cell densities. He pointed out, first, that the numbers of cells interacted with receptive field sizes to suggest that *the same number of units looked at every point in visual space.* In the periphery there are few cells but large receptive fields, in the regions of high acuity there are many cells but small receptive fields. Fischer next observed that that this geometry determined a logarithmic transformation. Cavanagh (1978) and Schwartz (1984) noted that this transformation itself might perform useful information processing. For example, natural transformations such as magnification around the optic axis (which could be produced by moving toward or away from a fixated object) or small rotations around the optic axis tended to be transformed into small translations on the cortical surface by the logarithmic mapping function.

A great deal is known about the details of visual processing; this is by far our best-understood part of cortex. We will not review the huge literature here. However, significant fine structure exists in the topographic maps as well. Units in primary visual cortex respond best to moving oriented bars or

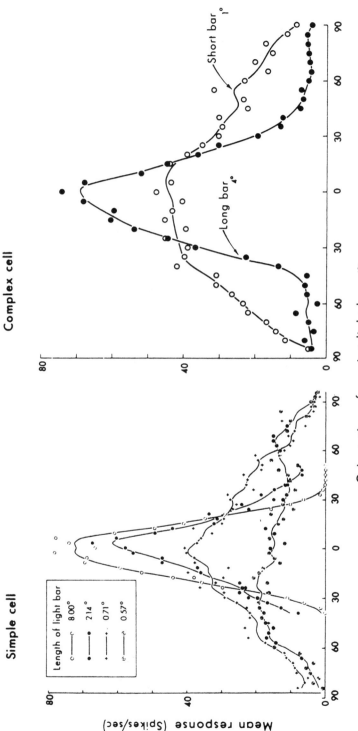

Figure 10.20 Responses of single cells in primary visual cortex of cat. Note cells respond best to moving bars at a particular orientation. *Simple* and *complex* cells are two different cell types in V1, a classification going back to Hubel and Wiesel (1962). Simple cells tend to respond best to a bar at a particular location, and complex cells tend to respond to a bar over a more extended region of the visual field. Orientation tuning curves for bars of different length in a simple (S) and complex (C) cell. In both the curve broadens for shorter bars even though no inhibitory flanks were observed in the receptive field of the cell. For both cells the shortest bar was smaller than the discharge region. From Henry (1985). Reprinted by permission.

edges, not to spots of light. (To the computer scientists: cortical units are not coding pixels.) Figure 10.20 gives orientation tuning curves of some V1 units. These could be described as relatively low-Q orientation sensors. The half-width of the orientation response is several tens of degrees. This pattern is typical of cell responses in sensory cortex, where even when a unit is responsive to a single parameter, it is rarely sharply tuned. We will see some other examples of this in chapter 11.

The orientation to which cells in primary visual cortex respond best changes in an orderly fashion across cortex. Nearby cells respond best to similar orientations, slightly rotated, with a constant number of degrees of rotation per millimeter of cortical surface. It is now possible to get some idea of what the *orientation columns*, that is, regions of cortex responding best to one angle, look like with neuroanatomical imaging techniques (figure 10.21) (Blasdel, 1992). These spectacular pictures were made with an optical imaging technique, using a camera looking directly down on exposed monkey cortex. It can be seen that orientation shifts smoothly in some regions of cortex, but there are also some discontinuities. These take the form of loops and triradii based on the topology. The regions may be of special importance for information processing, because all orientations are present in close proximity and could be integrated by the dendritic field of a single cortical pyramidal cell. Modern imaging techniques, of which these two figures are only one example, will transform our understanding of cortical physiology and anatomy because for the first time we can move away from emphasis on single units toward the properties of large groups of interacting neurons. The degree of orderliness in all the cortical systems we have seen is remarkable.

It might be thought that, as analysis became more and more precise at the higher levels of visual processing, the resulting maps also might become finer grained and more precise. This does not seem to be the case. Often areas of the image giving the best response become very large, whereas the stimulus configuration giving the response becomes more selective. At the highest levels of the visual system, the organization of cell responses in still unclear. Some cells have extremely selective responses. For example, cells in inferotemporal cortex of the monkey have been found that respond best to particular faces. This might indicate a grandmother cell representation; however, these cells also respond, but less strongly, to other faces. Evidence suggests that these "face" cells, taken together, form a distributed representation where they are graded in their responses to behavioral relevant cues such as physical similarity, familiarity, or even, conceivably, position in the laboratory social hierarchy (Young and Yamane, 1992).

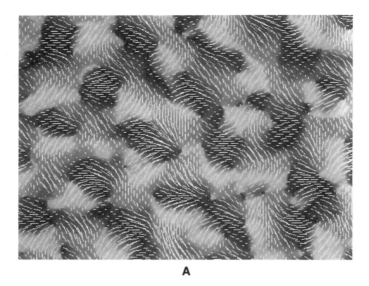

A

Figure 10.21 Orientation preferences are indicated by short, oriented lines. A, The correspondence of orientation with color (shades of gray) can be seen in this figure, where it also should be apparent that orientation selectivity can be indicated as well by the length of each line. B, Overlay of short oriented lines (indicating orientation preference and selectivity) with an inverse contrast image of the gradient (dark values indicate regions of discontinuous change). Note that lines tend to be longer (indicating high selectivity) in the continuous regions, and that more orientations per unit area are represented near regions of discontinuous change (indicated by dark spots and lines). Singularities are apparent from the gradient as dark spots. As one can see from the short oriented lines, these are regions where all orientation preferences converge. Because of the continuity outside singularities, the preferred

Other cells in primate inferotemporal cortex respond best to rather complex shapes and combinations of features, for example, a bar plus a disk, a combination of a dark disk and a light disk, or a star. The receptive field sizes of these elaborate cells were quite large, averaging over 10 degrees in inferotemporal cortex. Areas of the cortex associated with object vision (inferotemporal cortex, for example) have a much less strict topographic organization than those we saw at lower levels (Tanaka, Saito, Fukada, and Moriya, 1991). However, the organization is still columnar, so a single column will contain cells that respond to similar visual features, although adjacent columns may be quite different in their responses (Fujita, Tanaka, Ito, and Cheng, 1992).

It is not efficient to have cells doing the same thing. One way of using maps and decorrelating at the same time is to use several maps simultaneously. The primary visual cortex, the best-studied example, has maps of

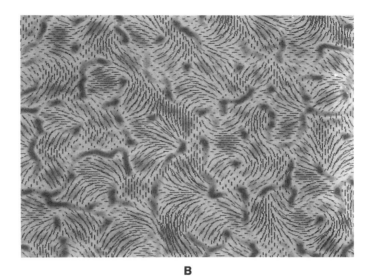

B

orientation around each one cycles through 180 degrees along any closed path. It does so in either a positive or a negative direction, which, for a counterclockwise path, means a counterclockwise or a clockwise rotation. These are apparent in the contours formed by short line segments as loops and triradii, which, as one can see, are roughly balanced. In addition to bringing all posssible orientation preferences together within the smallest conceivable radii, both organizations ensure that perpendicular orientations are always represented on opposite sides. These are the only places in cortex where all possible orientation preferences are accessible within the dendritic fields of typical upper-layer cells, that is, within radii of less than 250 to 350 μ. From Blasdell (1992). Reprinted by permission.

orientation preference, eye dominance, location in the visual field, and perhaps other parameters. A cell is under the influence of several maps and can develop an individual response pattern. This may be even more true of higher visual processing areas such as inferotemporal cortex. Cells may respond to several parameters, causing each cell to have an individual response pattern all its own.

Maps seemed to be used to develop cells with commonalities of response, but not necessarily with highly correlated responses to particular stimuli. Although most neural network mapping models, for example, Kohonen maps described in chapter 14, are based on formation of local correlations, in fact the nervous system seems to try to avoid these because they are not efficient.

Hubel and Wiesel (1977) proposed the notion of the cortical *hypercolumn*, which captures some of this idea. In the visual cortex a hypercolumn is a group of cells about 1 mm^2 in area. Because units respond to a region of the visual field, all the units in the hypercolumn respond to roughly the same area of visual space. The hypercolumn contains all orientations, all degrees of binocularity, and so on. All the cells in a hypercolumn see the same region in space, but all respond to different aspects of the pattern in that region. Commonality of response properties does not necessarily mean high correlation of unit response over all patterns.

Auditory System

The mammalian auditory cortex displays a degree of *tonotopic* organization in which cells responding best to different frequencies are arranged in orderly fashion. We will discuss the auditory system of the mustached bat, which provides some remarkable and illustrative examples of biological signal processing and topographic maps. Figure 10.22 gives some idea of the many tonotopic maps present in the cat auditory cortex (Imig and Reale, 1980).

It is worth making one point about single unit properties in the auditory system because the way cortex codes information can sometimes be unintuitive. Figure 10.23 displays responses of auditory units in auditory cortex (AI) as the intensity of a stimulus is increased (Phillips, Orman, Musicant, Wilson, and Huang, 1985). The monotonic units recorded in AI look very much like eighth nerve units, which are presumably very close to the auditory transducer outputs. The upper part of figure 10.23 shows the responses

Figure 10.22 Tonotopic relations in auditory cortex. CF stands for characteristic ▶ frequency. This refers to the frequency of the pure tone at which the unit first starts responding as the intensity of the tone is increased. The primary (AI), anterior (AO), posterior (P), and ventroposterior auditory fields contain a "complete and orderly" best-frequency representation. A ventral area (V) of the peripheral auditory belt with a second auditory area (AII) is also shown. The organization of this area is less clear. Map of CFs of neurons or neuron clusters in auditory cortex of the cat. The small inset shows a lateral view of the cat brain with the rostral end pointed upward and the ventral part to the left. The dashed lines outline the mapped area shown in the exploded view. Shaded areas represent the depths of the posterior ectosylvian sulcus, which has been opened up in this drawing to show portions of the fields normally buried there. Area in AI circled is the site of a small injection of tritiated proline. Stippled areas in A, P, and VP are the sites of terminal labeling that resulted from the injection (Brugge and Real, 1985). From Imig and Reale (1980). Reprinted by permission. Copyright © 1980 by Wiley-Liss, a division of John Wiley and Sons, Inc.

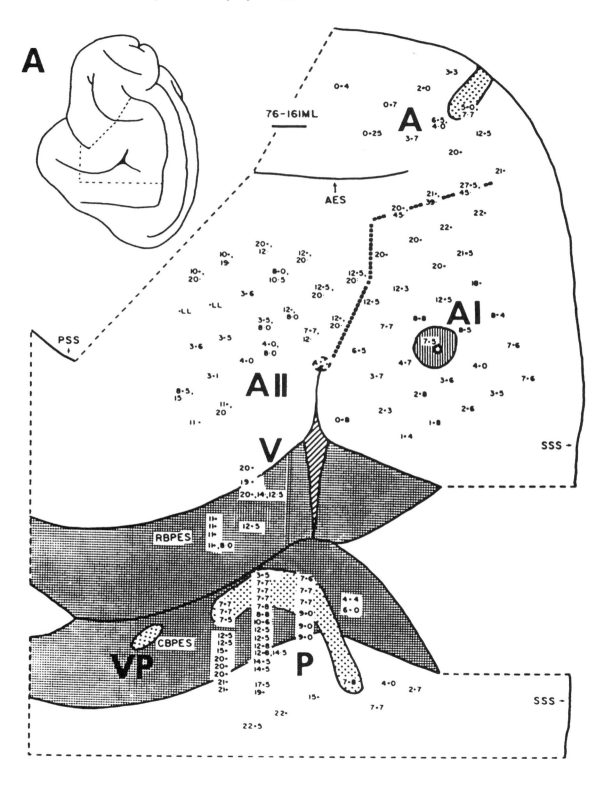

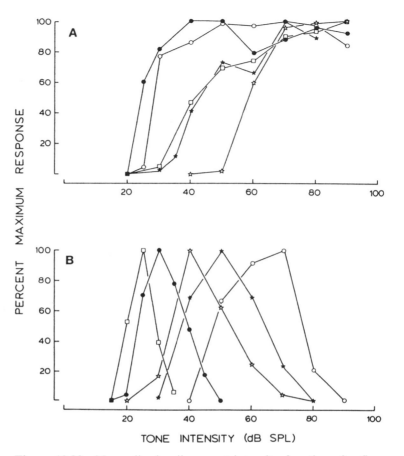

Figure 10.23 Normalized spike count-intensity functions for five representative monotonic (A) and five nonmonotonic (B) cells. Data obtained using 50 presentations of CF tones delivered either monaurally to the contralateral ear or to both ears simultaneously. From Phillips, Orman, Musicant, Wilson, and Huang (1985). Reprinted by permission.

of monotonic units. It is often claimed that the auditory system does a crude spectral analysis of the incoming signal. In many eighth nerve auditory units, and in the monotonic units in figure 10.23A, response begins at a low intensity to a particular frequency (CF, or characteristic frequency) and as intensity is increased, they respond over a somewhat broader range of frequencies. The output pattern looks somewhat like the response of a band-pass filter with a Q of 10 or 12. (The actual picture is more complex with significant contribution from precise timing relations as well.) As the amplitude of the input increases, the cell responds more and more strongly, until it reaches its maximum firing rate. This pattern corresponds to the intuition that as we turn up the gain on the stereo, cells in our auditory system should fire faster, although, as can be seen in the figure, the dynamic range of the cell is quite small, perhaps 10 to 15 db.

However, the picture is more complex in the nonmonotonic units in auditory cortex (figure 10.23B). A number of cells respond to limited ranges of frequencies, as before, but they also respond to limited ranges of amplitudes. As the sound intensity increases, the cell starts to respond, fires faster, and fires at its maximum rate. As intensity increases further, the firing rate of the cell *decreases* and ultimately drops to zero. A nonmonotonic and highly nonlinear relationship exists between sound intensity and cell activity. These cells respond best to a *region* in a two-dimensional *intensity-frequency space*. Such a nonmonotonic response causes no conceptual difficulties when a cell is assumed to be selective to something like the orientation of a bar. But such coding seems unnatural as a way to code acoustic intensity information in a neural network, say, for a neural network speech system. In fact, when a variant of this representation was used in a speech system, it worked well (Anderson, Rossen, Viscuso, and Sereno, 1990).

Motor Output Distribution

We observed previously that the structure of the motor system, which had to coordinate millions of motor neurons, did not look at all like a grandmother cell system. The same seems to be true earlier in the motor system as well. Work in primate motor cortex suggests a remarkably interesting averaging system for movements controlled by motor cortex (Schwartz, Kettern, and Georgopoulos, 1988; Georgopoulos, Kettner, and Schwartz, 1988; Georgopoulos, Lurito, Petrides, Schwartz, and Massey, 1989).

When they recorded from cells in motor cortex during movements of the monkey's arm to a target, they found that "the discharge rate of ... 83.6 percent (of the cells) varied in an orderly fashion with direction of movement: discharge rate was highest with movements in a certain direction ... and decreased progressively with movements in other directions" (Schwartz et al., 1988, p. 2913). A single cell would respond over quite a wide range of directions of movement; that is, it was not very selective. However, when this area of motor cortex is considered as a population, the *average* direction of movement of the cells in the population "points in the direction of movement in space well before the movement begins" (Georgopoulos et al., 1988, p. 2926). This kind of averaging is completely consistent with a distributed representation at the motor side, a point we discussed earlier in much more general terms. The suggestion that averaging of responses of broadly tuned

Figure 10.24 Individual neurons in the motor cortex of the monkey are *broadly* tuned with respect to the direction of arm movement toward visual targets in three-dimensional space.... many cells will be active for any particular movement ... the generation of movement in a particular direction depends upon the activity in the neuronal ensemble. We tested this "population coding" hypothesis by assuming that the contributions from individual neurons add vectorially to yield a neuronal population vector. We found this population vector accurately predicted the direction of movement in space before the onset of movement. An example of population coding of movement direction. The lines represent the vectorial contribution of individual cells in the population (N = 475). The movement direction is marked with the large white arrow. The direction of the population vector is marked with the small white arrow. The original figure was color coded. From Georgopoulos, Kettner, and Schwartz (1988). Reprinted by permission.

units is involved is significant because averaging of this kind is perhaps the most natural neural network computation. It also indicates how a population can be more accurate in performance than the elements that go to make up the population (figure 10.24).

Theory of Computation

Nelson and Bower (1990) provided an interesting analysis of some of the computational issues involved in the arrangement of computing units in parallel computing systems. Because connections have costs reckoned in size, energy consumption, and potential for damage, the connections should be as short as possible. This requirement means that the physical arrangement of units is of great practical importance. If everything is simply connected to everything, it does not much matter how the units are arranged, but this is not the case in the brain. According to Nelson and Bower, programmers of parallel computers point to two sources of inefficiency, *load imbalance* and *communications overhead*.

In a parallel computer, load imbalance means that some processors are working harder than others. Because "working harder" in a computer tends to translate into "takes more time," this means that many processors may have to wait for one processor to finish. Because neurons are analog processors, time considerations are not so clear. In mammalian cortex, where neurons tend to be members of large groups of what appear to be equivalent neurons, it is suggested that all neurons are equally important. It certainly seems that this is the way topographic maps are constructed in cortex. For example, the foveal region has many more cells devoted to it than peripheral regions (see figures 10.17 and 10.18). Presumably, if cells are equally important, having more of them in a region balances the load. The details of the topographic map then provides a direct measure of average processing load, measured, for example, in square millimeters of cortex.

Grandmother cells are the extreme case of load imbalance, where the entire organism depends on a computation to be performed by a single cell. However, in some invertebrates, particularly important responses are controlled by large specialized cells, such as the squid giant axon escape reflex. These specialized units can be fast, dedicated, and reliable. Some nervous systems may have reasons to use them for specific functions; load imbalance is not necessarily bad.

Communications overhead can be controlled to a large extent by the physical arrangement of the units. Overhead depends critically on the computation being performed. Nelson and Bower described three common types of physical arrangement for sensory computation in cortex. These classes are idealizations of what is actually seen.

The first arrangement is *continuous maps*. Most computation in early vision is local. For example, responding to the presence of an oriented edge in an image is largely local since only nearby points on the image are involved. Response to local movement is another example. If the geometrical relations of the image are maintained in the projection to cortex (as they are), many computations can be carried out with few short connections. Topographic maps are characteristic, as we have seen, of the early stages of sensory systems with obvious spatial metrics: seeing, hearing, and touching.

A second arrangement is *scattered maps*, where there is no obvious locality of computation. The best example is the olfactory system. No one has managed to suggest a useful metric for olfaction; it appears to be a somewhat arbitrary function of the shape of an odorant molecule interacting with possibly thousands of receptor types. Olfactory cortex indeed shows a scattered interconnection pattern in which a cell seems to be about as likely to connect to nearby units as to distant ones.

An interesting speculation arises here. Anatomical evidence suggests that olfactory cortex is the original cortex. The organization of the rest of cerebral cortex is based on modifications of the pattern first set by olfactory cortex. If olfactory discrimination involves operations with essentially arbitrary activity patterns, perhaps the rest of cortex has inherited this ability. Local topographic relationships are imposed on a structure that is designed to deal with arbitrary state vectors. Perhaps that is why mammalian cerebral cortex is so flexible and so good at generalizing: by accident of evolution it is particularly good at dealing effectively with distributed codings.

The third arrangement is *patchy maps*, which have a local spatial order—a "patch"—but the patches are scattered. This would be characteristic of computations intermediate between the first two arrangements, in which a degree of local order is combined with global interactions. Nelson and Bower suggest that the cerebellum is a place where this anatomical organization is found. Another speculative possibility would be the highest levels of the visual system, where both local and global information is necessary to identify an object.

The number of potential interconnections in both artificial and natural neural networks grows much faster than the number of units. Therefore, even computer simulations of very large networks cannot be fully interconnected because soon there would be too many connections to fit in memory. For large artificial networks, as well as large natural ones, both the spatial arrangement of data and the data representation have a major impact on the efficiency and accuracy of the computation.

Other Structures with Topographic Organization

Not only cortex uses maps. The superior colliculus also does, and forms a particularly interesting example of a distributed computing system. The superior colliculus of mammals is a midbrain structure that serves as a primarily visual, though actually multisensory, integrating center for what has been called the *visual grasp reflex*. The visual grasp reflex centers the image of an interesting or novel object or event on the retina. Moving the eyes to look at the location of an unexpected noise is a typical example. Because humans and other mammals have high-acuity vision only in the area of the retina near the optic axis (the fovea, or area centralis), the eyes, heads, and/or bodies must be moved so that the part of the retinal image to be analyzed is placed on the high-resolution portion of the retina.

The superior colliculus has an important role in controlling eye movements. It is a layered structure. Crudely, the input from the retina enters at the top layer, and the motor output leaves from the bottom layer. As we found for the cortex, a topographic map of the visual field is present on the surface of the colliculus. This was first discovered by Julia Apter in the 1940s, and the map was influential on the thinking of Pitts and McCulloch in their 1947 paper on geometrical transformations in the nervous system (figure 10.25). The map is highly distorted, with a much larger area corresponding to the high-resolution parts of the retina. The Pitts and McCulloch paper is also notable for its elegant use of the spatial layout of the neural elements to do difficult network computations such as those related to size invariance. Many of the ideas about nervous system computation first proposed by these authors keep being rediscovered.

Eye movement computation is closely bound to the map. If the surface of the superior colliculus is electrically stimulated, the eyes will move to bring the high-resolution part of the retina to the location on the map corresponding to the stimulation. It is as if the stimulation is perceived as something important, and the eyes move to take a closer look at it.

One way to do this computation would be to have a fine-grained map of the visual field. A point of maximum activation is precisely localized on the map, and the eyes are directed to that point. However, this is *not* what is done. The visual receptive fields (the area of visual space that affects the discharge of a cell) at the input layer to the colliculus are quite small and

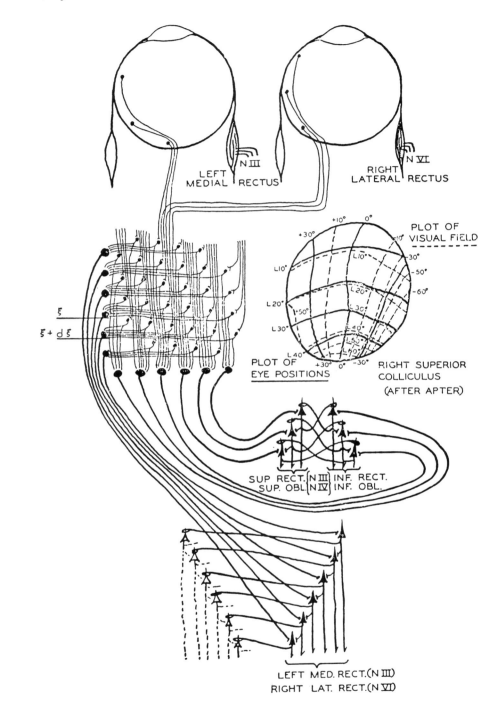

LEFT
MEDIAL RECTUS

RIGHT
LATERAL RECTUS

N III

N VI

PLOT OF
VISUAL FIELD

ξ

$\xi + d\xi$

PLOT OF
EYE POSITIONS

RIGHT SUPERIOR
COLLICULUS
(AFTER APTER)

SUP RECT. N III INF. RECT.
SUP. OBL. N IV INF. OBL.

LEFT MED. RECT.(N III)
RIGHT LAT. RECT.(N VI)

localized. However, receptive fields in the colliculus markedly *increase* in size moving toward the output, being many times larger in the lower layers. It is as if the high spatial precision in the input is being deliberately thrown away. As McIlwain (1976) stated, however, information is not lost, merely rearranged. From a spatially localized code the information has been re-represented as a distributed code, a notion that should be familiar by now.

McIlwain (1975) introduced the important concept of the *point image*, which is the set of units that see a point of light in the visual field. In a widely distributed representation the number of cells in the point image can be very large. This set of units takes up a certain amount of space in the neural structure. Both colliculus and cerebral cortex have a homogeneous neural substrate with sensory inputs impressed on it. In the case of the colliculus it is a distorted two-dimensional map similar to what we saw in cerebral cortex.

If we have spread out information about precise localization in space onto many neurons, how can we get spatial precision back again so we can make the correct eye movement? That is, what is the appropriate *output* representation? We saw in the discussion of motor cortex above that broadly tuned units seem to be associated with particular directions of movement, with the resulting output reflecting the sum of unselective output units.

Something similar seems to be going on in the organization of the output of the colliculus. McIlwain (1976) described a particularly simple way to use a distributed output representation. The mechanics of moving the eyeball around are quite straightforward. The extraocular muscles for an eye movement either move the eyes up and down (one set) or left and right (another set). Suppose the output layer of the colliculus is connected to these extraocular muscles. A region of the colliculus that receives inputs from a part of the visual field that is up and to the right from the center of gaze will connect to motor units that *primarily* are connected to the muscle groups that drive the eyes up and to the right. If the visual field location is primarily to the right of the center of gaze, the cells in the output layer connect primarily to the muscle system driving the eyes to the right (figure 10.26).

◄ **Figure 10.25** Model and map of the superior colliculus. A simplified diagram shows occular afferents to left superior colliculus, where they are integrated antero-posteriorly and laterally, and relayed to the motor nucleus of the eyes. A figure of the right superior colliculus mapping for visual and motor response by Apter is inserted. An inhibiting synapses is indicated as a loop about the apical dendrite. The threshold of all cells is taken to be 1. From Pitts and McCulloch (1947). Copyright 1947. Reprinted with kind permission of Pergamon Press Ltd.

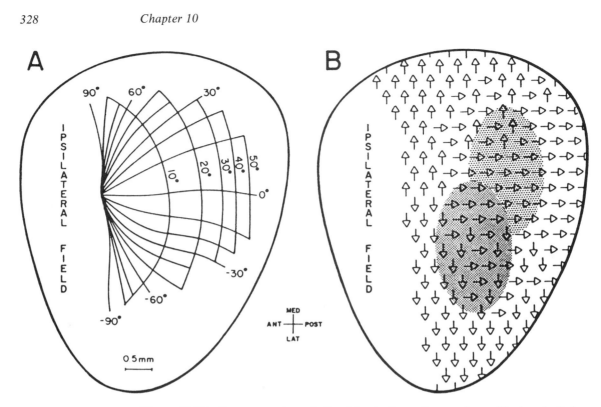

Figure 10.26 Sensory motor relationships in the cat's superior colliculus. A, Polar representation of visual projection to the cat's left superior colliculus. Zero degrees is horizontal meridian. B, Cells are symbolized by downward- and upward-directed arrows, connected to systems moving the eyes downward and upward, respectively. Arrows pointing to right signify cells are connected to systems that deflect eyes laterally and to the right. Stippled ellipses refer to two point images in the superficial gray layer involving the class of cells having the largest receptive fields. From McIlwain (1976). Reprinted by permission.

Suppose large numbers of cells are weakly connected to the motor system. Then suppose a point of light appears in the visual field, and the animal wants to move its eyes to that point. Because of the distribution of information about the point, many cells in the output of the colliculus are activated, each one (or small group of them) having a slightly different proportion of connections to the up-down and right-left systems, tending to move the eyes in a different direction appropriate to their location on the map. The overall motor output will be the sum of all these directions given by many cells. This average output direction can be highly accurate, even though most of the cells that have been activated do not by themselves move the eyes in the right direction.

Some experiments can test this idea. Suppose that the computation merely looked for peak activation on the colliculus. If part of the colliculus away from the peak was inactivated, the peak would not be moved. But in a distributed output code, if part of the colliculus was inactivated, it would make a difference in the resulting eye movements, because a large group of cells would not contribute to the eventual movement. Moreover, we could predict in which direction the error would be made, depending on where the inactivation occurred (Lee, Rohrer, and Sparks, 1988). The results of this experiment strongly support the idea of a distributed output code, summing up movement contributions from many units, although the details of the code are still not clear. Another prediction of most averaging systems is that if not one but two spots of light were close together, the eye movement would tend to fall between them, a result that is also seen experimentally.

Maps and Bats

Bats have advanced sonar systems that provide a good example of biological signal processing. Many of the ideas on data representation above can be seen in operation in the bat.

The mustached bat emits an extremely loud, short, ultrasonic signal and detects the return echo from flying insects. Once a target is detected, the bat flies toward it, intercepts it in the air, catches the insect in its tail membrane, and flips the insect into its mouth. This whole process takes under a second. As the bat closes on the target, its cries become shorter and more frequent. The bat can reliably detect insects up to perhaps 10 feet away. It can also detect stationary objects using sonar, an essential ability, because most bats hunt at night.

Two distinct kinds of sonar or radar exist, and the mustached bat uses both. (See Skolnik, 1980 for more details on radar.) In *continuous-wave* (CW) sonar or radar, the transmitter emits a continuous signal and listens for the return. No information is given about the position of the object; however, if the target is moving relative to the source, the Doppler effect causes a frequency shift in the echo (figure 10.27). In *pulsed radar* (figure 10.28), a brief pulsed signal is emitted. The return echo is detected, and the time delay between the two gives the distance to the target.

As in any radar or sonar system, the initial problem for the bat is to tailor the emitted signal so that the return signal can be best used to determine necessary information. Thus it uses both Doppler and pulsed sonar, each of which yields different information about the target (figure 10.29). A long *constant-frequency* (CF) portion at the beginning of the cry acts like a Doppler sonar, and a short chirp at the end of the cry acts like a pulsed sonar.

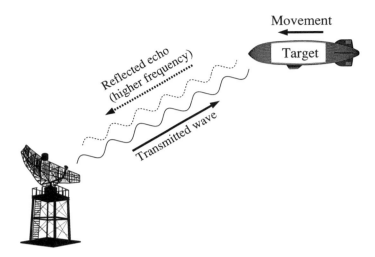

Figure 10.27   One form of bat sonar (or radar) involves emission of a CF (constant frequency) signal. When the signal is reflected off a moving target, the frequency is shifted because of the Doppler effect. By analyzing the amount of shift, the relative velocity of emitter and target can be determined. Careful analysis of the fine structure of the signal allows determination of such important information as the frequency of the target's wingbeat.

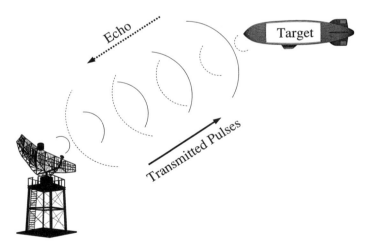

Figure 10.28   The other form of bat sonar (or radar) involves the emission of brief, high-intensity pulses. The time required for the echo to return from the target gives the distance to the target. Mustached bats do not emit extremely brief pulses, but a longer and more complex type of FM signal called a *chirp*, a rapid downward sweep in frequency. When the returning echo from the chirp is detected by a *correlation detector*, the time resolution of the detector output is much shorter than the length of the chirp. In radar this technique is sometimes called *pulse compression*. Bats used it long before humans invented it.

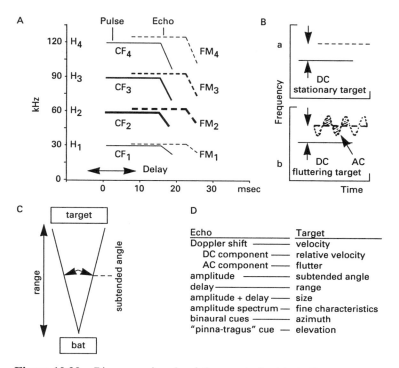

Figure 10.29 Biosonar signals of the mustached bat, *Pteronotus parnelli rubiginosus*, and the information carried by its signals. A, Schematized sonogram of the mustached bat orientation sound (solid lines) and the Doppler-shifted echo (dashed lines). The four harmonics (H_{1-4}) of both the orientation sound and the echo each contain a long CF component and an FM component (FM_{1-4}). The thickness of the lines indicates the relative amplitude of each harmonic in the orientation sound: H_2 is the strongest, followed by H_3 (6–12 dB weaker than H_2), H_4 (12–24 dB weaker than H_2), and H_1 (18–36 dB weaker than H_2). B, When the mustached bat flies toward or near a stationary object, the frequency of the echo becomes higher than the emitted sound by the Doppler effect (graph a). This steady shift is called the DC component of the Doppler shift. When the bat flies toward a fluttering target, for example, a flying moth, the Doppler shift of the echo consists of a DC component proportional to the relative velocity and the periodic FM proportional to the speed of the wing beat (graph b). This periodic FM is called the AC component of the Doppler shift. The AC component is complicated because the insect's four wings are moving in complex patterns and in different phase relationships relative to the bat. C, Target size is determined from both target range and subtended angle. D, Relationship between echo properties and target properties. From Suga (1985). Reprinted by permission.

The chirp at the end, a rapid drop in frequency, is actually an "advanced" radar technique, although the bat has been using it for millions of years. When a chirped signal is combined with a properly designed detector (a *correlation* detector, related to the matched filters we saw in chapters 6 and 7), the time accuracy to which the echo return can be measured is many times smaller than the duration of the chirp. In human engineering, this technique is called *pulse compression.* Instead of a very short pulse with very high peak power, it allows use of a lower power emitter transmitting for a longer time. Suga (1985) called the chirp the *frequency-modulated* (FM) part of the cry.

Because signal processing requirements are so different for the two types of radar, different analyses of the signal seems to be handled by different, specialized parallel pathways in the mustached bat. In the visual system, for example, as we mentioned, there is evidence for anatomically and physiologically distinct processing pathways depending on the computation being performed. Figure 10.30 indicates some of the various subregions involved in sonar processing. Many of these regions are involved in quite specific computations on the sonar signal. The mustached bat separates distance and velocity both in its cry and in its cortex. Other bats use different kinds of signal analysis for perception of the world through sonar. For example, the big brown bat, common in New England, has a cry composed of only a shaped descending frequency sweep. This bat is able to perform feats of perceptual signal processing that would be considered beyond state-

Figure 10.30   Map of best frequency on the mustached bat's brain. Note the extraordinary amount of cortex devoted to analyzing the region around 61 kHz, presumably corresponding to the Doppler return to echo from the strong second harmonic of the CF portion of the bat cry. Dorsolateral view (A) and the distribution of best frequencies along the tangential plane of the auditory cortex (B). A, The branching lines represent the branches of the median cerebral artery. The long branch is on the sulcus. The area surrounded by the rectangle is shown in B. B, Numbers and lines represent iso-best-frequency contours. Orderly tonotopic representation is clear in the areas with solid contour lines, but not so systematic in the areas with dashed contour lines. In the areas where contour lines overlap, neurons are tuned to different frequencies. In the areas where contour lines are not drawn, the tonotopic representation, if present, is obscure. Some of the best frequencies obtained in these obscure areas are shown. The area surrounded by the dotted line is the primary auditory cortex (AI), which contains the DSCF area at the center. The FM-FM and CF/CF areas are probably a part of the AI as well. In each of the subdivisions of the FM-FM, CF/CF, and DF areas, and a tiny area anteroventral to the DSCF, two different frequency bands are represented. The areas dorsal or ventral to AI are nonprimary auditory cortex. From Suga (1985). Reprinted by permission.

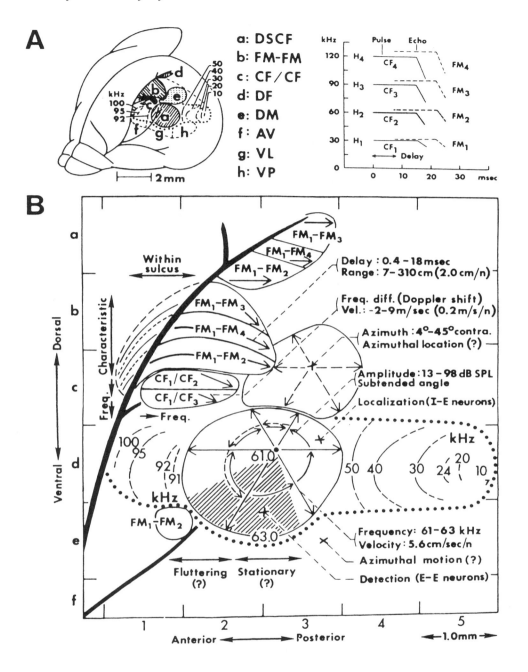

of-the-art if a human-engineered sonar system performed them (Simmons, 1989; Dear, Simmons and Fritz, 1993).

One problem with Doppler radar signal processing is that a powerful transmitter is only a centimeter away from a sensitive receiver. A weak return close in frequency to the transmitted signal must be detected. In the mustached bat's cry, maximum signal energy is around 60 to 63 kHz. One way to handle interference is to have a set of very narrow filters to analyze the return. In common with good engineering design, the bat uses filters in series: a narrow bandwidth from the peripheral auditory system is further sharpened and shaped in the cortex. The observed Qs (a measure of bandwidth) of the tuning curves in the bat peripheral auditory system can be as high as 200 in the region around 61 kHz. It is not completely clear how such narrow bandwidths are achieved, but they may be due to some mechanically resonant structures in the peripheral auditory apparatus. Further improvement of the filter occurs in the auditory cortex. The narrowest-bandwidth cell in cortex has a best frequency of 61.5 kHz and a bandwidth of 0.3 kHz, with essentially infinite skirt slopes, giving extremely good narrow band pass filter characteristics.

The region of cortex concerned with the constant-frequency portion has an extremely large number of cells, about 30% of the cells in that cortical region, responsive to the frequency range around 61 kHz, to process the Doppler return. (See the huge area of bat cortex devoted to this region in figure 10.30.) This large group of cells seems to have a local mapping topography imposed on the more global tonotopic map, in that neurons respon-

Figure 10.31 Processing of the Doppler return in the bat cortex. Single units in the CF region respond to precise frequency offsets between different harmonics of the CF portion of the call. Note the extraordinarily high Q (narrow response frequency band) and extremely steep skirts of these extraordinary biological filters. Excitatory frequency tuning (dashed lines) and facilitation frequency tuning (solid lines) of four neurons (A–D) denoting excitatory and facilitation areas. An excitatory area was measured by delivering a 34-msec-long single CF tone, and a facilitation area was measured by delivering a 34-msec-long CF tone (conditioning tone) simultaneously with another 34-msec-long CF tone (test tone) at a fixed frequency and amplitude. A, CF_1/CF_2 facilitation neuron. (CF_1 and CF_2 refer to the regions of the first and second harmonic components of the CF portion of the call.) The test sound used to measure the CF_1 facilitation area was 59.39 kHz and 60 dB SPL; for the CF_2 facilitation area it was 29.75 kHz and 70 dB SPL. These fixed values are indicated by the x marks in A. B and C, Two CF_1/CF_3 facilitation neurons. D, $CF_1/CF_{2,3}$ facilitation neurons. The test sounds used for the data in B, C, and D are indicated by x marks like those in A. Note the extremely sharp facilitation areas for CF_2 and CF_3. From Suga (1985). Reprinted by permission.

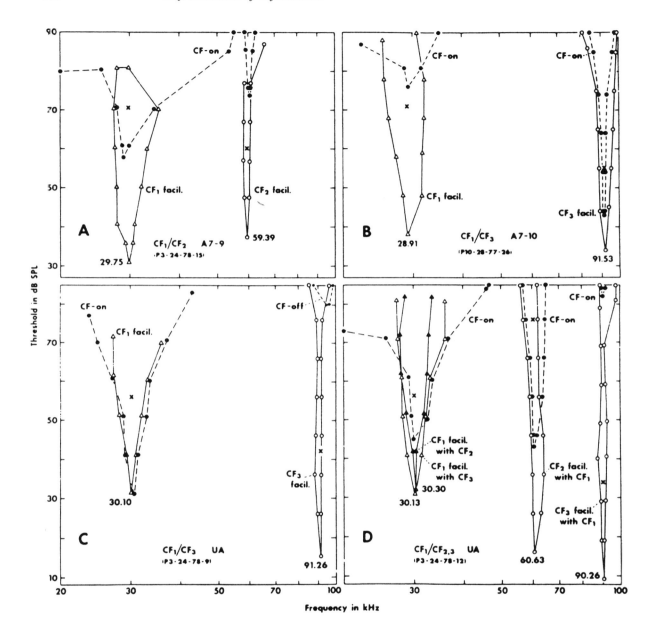

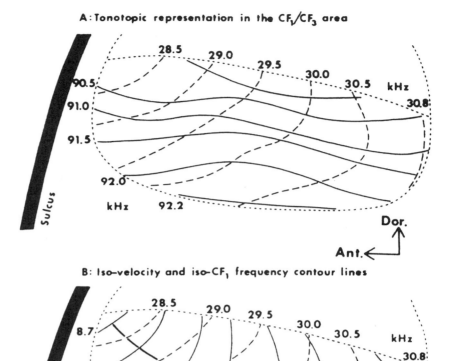

A: Tonotopic representation in the CF₁/CF₃ area

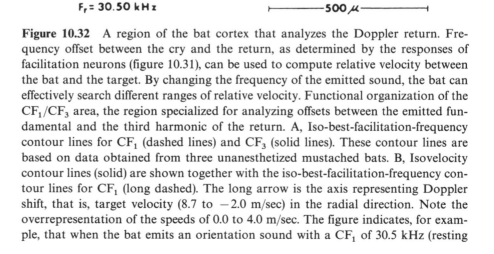

B: Iso-velocity and iso-CF₁ frequency contour lines

Figure 10.32 A region of the bat cortex that analyzes the Doppler return. Frequency offset between the cry and the return, as determined by the responses of facilitation neurons (figure 10.31), can be used to compute relative velocity between the bat and the target. By changing the frequency of the emitted sound, the bat can effectively search different ranges of relative velocity. Functional organization of the CF_1/CF_3 area, the region specialized for analyzing offsets between the emitted fundamental and the third harmonic of the return. A, Iso-best-facilitation-frequency contour lines for CF_1 (dashed lines) and CF_3 (solid lines). These contour lines are based on data obtained from three unanesthetized mustached bats. B, Isovelocity contour lines (solid) are shown together with the iso-best-facilitation-frequency contour lines for CF_1 (long dashed). The long arrow is the axis representing Doppler shift, that is, target velocity (8.7 to -2.0 m/sec) in the radial direction. Note the overrepresentation of the speeds of 0.0 to 4.0 m/sec. The figure indicates, for example, that when the bat emits an orientation sound with a CF_1 of 30.5 kHz (resting

sive to the center of the frequency band (61.5 kHz) are central to the region, and neurons away from the center (63 kHz) are on the circumference of the region. Some of the behavioral significance of the Doppler sonar involves the velocity modulation of the return due to the wingbeats of a flying insect. Wingbeat frequency and amplitude can serve as a means of identifying insects.

To compute the Doppler return requires knowledge of both the emitted frequency and the echo frequency, that is, it is necessary to determine the difference between two frequencies to find the relative velocity. Cells are found in the CF/CF cortical region (figure 10.31) that respond to two frequencies. These units respond to a low frequency, around 30 kHz, presumably the emitted fundamental (CF_1), and also to a very narrow band of frequencies in the range of the second harmonic (CF_2) around 60 kHz or third harmonic (CF_3) around 90 kHz. (Remember, considerably less energy is present in the emitted fundamental than in the harmonics.) Cells become very active when both frequencies are present. The very narrow band frequency band facilitation areas are not in strict harmonic relationships with the fundamental. If the emitted frequency is known, the relative frequency difference between the emitted and the received signal can be determined by noting which facilitation frequency cells respond. If the emitted frequency is CF_1 and the cell is facilitated best by a frequency CF_3, response of the cell signals an offset between an emitted frequency and echo frequency of

$$3CF_1 - CF_3,$$

referred to the third harmonic.

Remarkably, these cells responding to two frequencies are mapped in the two CF/CF regions, one with facilitation regions concerned with the second harmonic (CF_1/CF_2), and the other with facilitation regions in third harmonic (CF_1/CF_3). There is a dual tonotopic relationship on the map (figure 10.32); the facilitation region for the harmonic runs roughly up and down on the map, and the facilitation region responding to the fundamental runs roughly back and forth. The plots in figure 10.33 show the isovelocity contour lines, where the offset between the two components of the cell response signals a constant relative velocity.

◄ frequency), neurons in the CF_1/CF_3 are best activated by targets moving with relative velocities of -1.2 to 2.0 m/sec. When the frequency of CF_1 is reduced to 29.5 kHz, they are stimulated best by targets moving with relative velocities of 3.4 to 6.1 m/sec. Isovelocity contour lines similar to the above have also been found in the CF_1/CF_2 area. From Suga (1985). Reprinted by permission.

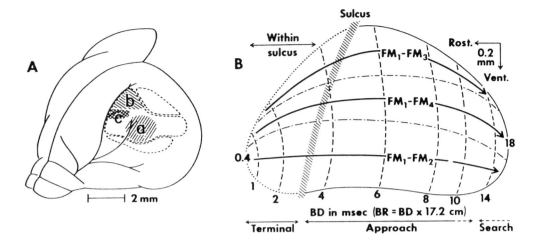

Figure 10.33 Map of best distance, that is, target range on mustached bat cortex. Odotopic (distance based map) in the FM-FM area. A, The left cerebral hemisphere of the mustached bat. The large auditory cortex (dotted lines) contains at least three areas specialized for processing biosonar information: a, DSCF; b, FM-FM; and c, CF/CF. The branched lines are arteries. There is a sulcus below the largest branch. B, The FM-FM area consists of three major clusters of delay-sensitive neurons: FM_1-FM_2; FM_1-FM_3; and FM_1-FM_4 facilitation neurons. Each cluster shows odotopic representation. Iso-best-delay contours and range axes are schematically shown by dashed lines and solid arrows, respectively. Best delays (BDs) of 0.4 and 18 msec correspond to best ranges (BRs) of 7 and 310 cm, respectively. Range information in the search, approach, and terminal phases of echolocation is represented by neural activity at different locations on the cerebral hemisphere. C, Relation between BD and distance along the cortical surface. The data were obtained from six cerebral hemispheres, indicated by six different symbols. The regression line

The bat has a degree of control over the emitted frequency, and the facilitation regions, particularly the CF_1 region, are not infinitely narrow. Therefore a particular cell with its two facilitation regions does not signal an abolute relative velocity. As the cry changes, the map may signal different relative velocities. With an emitted frequency of 30.5 kHz, the CF_1/CF_3 region signals a range of relative velocities from 8.7 to −2.0 m/sec with an expanded region of higher resolution around 0.0 to 4.0 m/sec.

Another portion of the bat cortex is concerned with analyzing the chirp, the FM part of the cry. In pulsed radar or sonar the feature of interest is the time it takes for the signal to reach the target and for the echo to return. This time delay determines target distance. In chirped signals, however, computing the time delay is not just a matter of starting and stopping a clock. A complex correlational filter must be used to determine an accurate time delay between cry and echo. The bat is able to do this calculation and forms

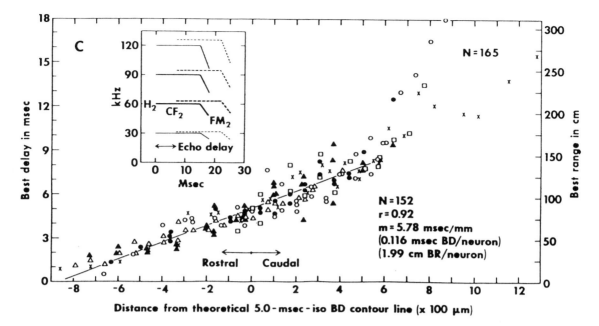

Distance from theoretical 5.0 - msec - iso BD contour line (x 100 μm)

represents an average change in BD with distance along the cortical surface. Since the 5-msec iso-BD contour line always crosses the central part of the FM-FM area along the exposed surface of the cortex, the 5-msec BD on the regression line is used as a reference line to express distance. The slope of the regression line corresponds to 0.12 msec BD per neuron, that is, 1.99 cm BR per neuron. Note that the time axis of the FM-FM area is used to represent the time interval between two acoustic events. The inset in C is a schematized sonogram of an orientation sound and a Doppler-shifted echo in the approach phase of echolocation. All the data were obtained from unanesthetized bats. From Suga (1985). Reprinted by permission.

a topographic cortical map of target range. Many neurons in this cortical region respond to a best delay in milliseconds between cry and echo. The cortical surface has a linear map of best delay. As the bat moves toward its target, an area of activity will sweep from one end of the cortical region to the other.

Some countermeasures to bat sonar are taken by prey. Some insects listen for bat cries and then, depending on the amplitude of the cry, either turn and go away from the bat, or, if the bat is close, start evasive maneuvers, such as power diving toward the ground (Roeder, 1962).

The effectiveness of the bat system is due to careful attention to detail at all levels, as well as many special-purpose systems, each suited to do part of the job. For example, selective filters in the peripheral auditory system are further sharpened centrally. One simple step does not do the job; several stages are used, with distinct parallel pathways, each optimized for one aspect of the computation. Sometimes a global organization is present over a large region, the range map, and sometimes global organization is subservient to the need for very large numbers of analyzers for one aspect of a signal, analyzing the Doppler return.

Maps may be quite difficult to construct. Both the range map and the relative velocity map involve complex multistage neural processing to set them up. Clearly, the advantage of constructing the map is worth the trouble. This animal shows a complex interplay between distribution and localization. The simple use of topographic maps shows a degree of localization. Yet many cells in a local region will respond to a signal. The degree of distribution and localization of cells responding to a given stimulus is surely the result of some kind of optimization, of what exactly we are not sure. For more examples and information about biological maps, see Knudsen, du Lac, and Esterly (1987).

Cognitive Representations

Up to this point we have discussed data representation in the context of neuroscience. Most of the examples are closely tied to the physical stimulus. Cognitive scientists and psychologists are also extremely concerned about representation. In fact, large parts of experimental psychology have to do with representation in the sense that the point of the experiment is to find out what is important for the mental computation. Often the important aspects of a stimulus are far removed from physical stimulus.

The most popular model of representation among cognitive scientists who do neural networks—the connectionists—is based on *features*. It is usually realized in a neural network by putting the value of the feature, for example, either present or absent or a continuous value, in an appropriate location in a state vector.

Features in cognitive science almost always refer to highly complex entities. Neuroscientists sometimes refer to selective neurons, for example, cells that respond to oriented line segments, as feature detectors. However, the word "feature" as used by a neuroscientist and a cognitive scientist is usually constructed by several levels of complexity, causing frequent failures of communication.

One of the best-known examples of feature analysis is in linguistics. Phonemes, the elementary units of speech, have long been considered to be composed of 15 or so independent *distinctive features*. Virtually any elementary textbook in linguistics, psycholinguistics, or cognitive science will contain or mention distinctive features somewhere. (Random examples from three decades: Clark and Clark, 1977; Anderson, 1985; Gleason and Ratner, 1993). Experimental evidence supports the idea that phonemes are represented this way, for example, many misperceptions are due to errors in detecting a single feature. However, features are complex, temporally varying patterns of frequencies.

An example might be the feature *continuant*, which refers to sounds made with continuous air flow such as vowels or fricatives (/f/, /s/, etc.), as opposed to sounds made with interrupted air flow such as stop consonants (/p/, /t/). Another example might be the feature *voicing*, which detects whether or not the vocal cords are vibrating during the phoneme. Although this might be considered an example of a simple physically based feature that detects the presence or absence of a band of frequencies, in fact it is not that simple. The telephone, for example, removes the fundamental frequency of the vocal cord vibrations quite nicely, since it only transmits frequencies from about 300 to 3000 Hz. It is still possible to discriminate voiced from unvoiced consonants over the phone.

Other connectionist models use elementary features that are even farther removed from the physical stimulus. As a typical example, consider the feature representation proposed in a well-known paper by McClelland and Kawamoto (1986). Figure 10.5 showed the feature decomposition seen in figure 10.34, but now we have added the names of the features that were used to give the representation. Although these seem like perfectly reasonable high-level descriptors, much effort is required to connect them to a sensory input. Most of the effort involved in constructing such a feature set is involved in reflecting the subjective similarity judgments of the experimenters in the coding, or to match experimentally derived similarity measures.

In low-level neural systems we often have a reasonable idea about what the neurons are responding to. As problems become more and more cognitive, we know less and less about the details of the data representation. We know very little about how complex information should appropriately be represented in a state vector.

Similarity

Let us suppose that we have two state vectors that represent complex cognitive constructions. Perhaps their elements reflect degree of activity of high-level features. We would like to know how *similar* the two entities are. Much neural network computation is based on operations that look a lot like

Grandmother Cells

```
Rock:   1 0 0 0 0 0 0 0 0 0 ...

Wolf:   0 1 0 0 0 0 0 0 0 0 ...

Dog:    0 0 1 0 0 0 0 0 0 0 ...

Sheep:  0 0 0 1 0 0 0 0 0 0 ...
```

Distributed Coding

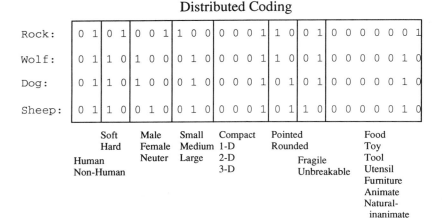

Figure 10.34 The feature decomposition. A 1 corresponds to the presence and a 0 to the absence of that feature. The features are listed below the appropriate blocks. From McClelland and Kawamoto (1986).

similarity. For example, the inner product computation of the generic connectionist neuron is such a measure. But the inner product reflects the data representation. In the example given in figure 10.34, the distributed representation suggests that dogs and wolves are similar to each other, less similar to sheep, and not similar at all to a rock. The grandmother cell representation suggests that they are all equally dissimilar, that is, orthogonal.

It is difficult to know what kind of information is used to determine if two stimuli are similar. Many cognitive scientists have observed that similarity is a very tricky concept (Tversky, 1977; Medin, Goldstone, and Genter, 1993). This last group said, "Similarity has its own mysteries which we are only beginning to understand" (p. 60).

We have a simple example in figure 10.35. Consider the top pair of figures, then decide which of the bottom two pairs of figures is most similar to the top two. Note that one pair (the left) is closer by measures of *physical* similarity, and this physical similarity will be surely be reflected in the low-level

Which pair is most similar?

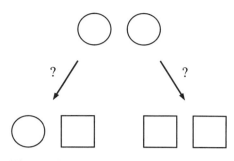

Figure 10.35 Judging similarity is not so simple as it seems. Which of the bottom pair of figures is most similar to the top pair? (Figure suggested by Goldstone, Medin, and Gentner, 1991.)

neural representation. Children frequently choose this pair as most similar. However, with neutral instructions, adults chose the rightmost pair about 70% of the time. This pair is *relationally similar*. Both the top pair and the rightmost pair are *identical*. The match is made based on a complex relationship between two objects. Judgments of similarity in adults often turn out to be based on high-level relations rather than simple physical similarity. Another example is provided in figure 10.36 (Genter and Markman, 1993). Which circle in the bottom triplet is most similar to the circle marked with an arrow in the top triplet? Again, most adults make a judgment based on relational similarity (the end circle) rather than on physical identity (the middle circle).

One significant part of this example is that it is easy to bias subjects to make judgments one way or the other. With neutral instructions, adults will choose the rightmost pair in figure 10.35 about 70% of the time. With instructions that bias them toward looking at relationships, this can be raised to about 90%. With instructions that bias subjects to look for matching attributes, it can be dropped to about 20%. The conclusion from this and many other experiments is that human similarity judgments are extremely flexible, even for what appear to be straightforward perceptual matches. The word *attention* is sometimes used to refer to certain kinds of perceptual and cognitive flexibility. Unfortunately, it covers a large number of different phenomena from the lowest to the highest levels of cognition. However, incorporating mechanisms to accomplish flexible computation into neural networks may be the most significant future advance in artificial neural networks.

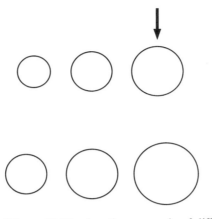

Figure 10.36   Another example of difficulty in judging similarity. Consider the top set of circles. One is marked with an arrow. Which circle in the bottom set of circles is most similar to this circle? (Example suggested from a figure in Genter and Markman, 1993.)

Reciprocity

Virtually all mathematical similarity measures we have seen such as distance and inner product are symmetrical, so the inner product $[f, g]$ is equal to $[g, f]$. One firm conclusion of studies of psychological similarity is that estimated similarities between complex cognitive entities may not be symmetrical. A number of examples demonstrate this, many due to Tversky (1977). For example, subjects consistently rated the similarity of Red China to North Korea to be less than the similarity of North Korea to Red China. Tversky proposed a model to explain these asymmetries based on the fact that we know less about North Korea than China. So judgments in one direction were matching a different set of attributes than judgments in the other direction. This approach could be modeled in a neural network by using different parts of the state vector to do the comparison, depending on the order of presentation of the items compared.

A very large part of human cognition is based on reasoning by analogy and metaphor. Similarity judgments are obviously central to such reasoning, but it is difficult to see why what is being compared is chosen, since generally only a subset of attributes is involved in the comparison. As we mentioned, these comparisons are not symmetrical. To use an example from Medin, Goldstone, and Genter (1993), "To say that surgeons are like butchers means something different than to say butchers are like surgeons. The former criticises surgeons and the latter compliments butchers" (p. 17). We can be sure that the comparisons are not based primarily on the fact that both surgeons and butchers wear white coats.

Natural Data Representations

Because of the evolutionary pressure on nervous systems to be efficient, watching how animals do important computations might suggest good rules to use in artificial systems, at least as a place to start. We will finish this chapter by suggesting a few general rules for data representation that seem to be followed to some degree in the vertebrate nervous system.

First, similar events should give rise to similar representations. In general, if a bird looks like a robin and acts like a robin, it probably is.

Second, things that have to be separated should be given different representations. It would be best if separate categories had orthogonal representations. One effective way to accomplish this would be to separate important distinct events or features on a map. The limited dynamic range of neurons makes it difficult to represent orthogonality faithfully if all or nearly all units are active, because nonlinear interactions are likely to occur among representations of different items.

Third, if something—a sensory property or feature of the input—is important, lots of elements should be used to represent it. This coding strategy is used in every neurobiological map we have seen. It is as if each neuron contributes a roughly equal amount of processing power to the computation. If you want more processing power, use more neurons. This strategy is very unlike the one used in a digital computer, where some bits may be far more important than others in their implications for the course of the computation. This strategy may also be a consequence of the unreliability and limited dynamic range of neurons. One cannot help be reminded of a comment from the 19th-century German materialist Fichte, "The brain secretes thoughts and consciousness as the liver secretes bile." If you want more bile, make a bigger liver; if you want better thoughts about something, devote more neurons to it.

Fourth, do as much preprocessing as possible, so the adaptive parts of the network has to do as little work as possible. Build invariances and abstractions specific to the problem into the hardware, preferably in the initial wiring, and do not require the system to learn them.

Fifth, make the representation flexible and easy to program. We know very little about why mental operation is so flexible. But surely at least part of it is because we can use input data in various combinations and ignore data if necessary. Mechanisms such as attention are usually ignored in artificial neural network theory, but are critical to real mental life. Artificial neural networks are usually configured to solve only one class of problems; real neural networks do not have this luxury. If there is a strong spatial component to brain computation in terms of detailed maps of sensory properties and in terms of modular organization, as seems to be the case, it would be easy to use rather crude spatial means—say, spatially organized excitation or inhibition—to emphasize or deemphasize one or another aspect of the computation. These control structures may not be very sophisticated, but may be effective and reliable if the data representation allows them to be used.

References

A.M.H.J. Aertsen, G.L. Gerstein, M.K. Habib, and G. Palm (1989), Dynamics of neuronal firing correlation: Modulation of "effective connectivity." *Journal of Neurophysiology, 61*, 900–917.

J.A. Anderson, M.L. Rossen, S.R. Viscuso, and M.E. Sereno (1990), Experiments with representation in neural networks: Object motion, speech, and arithmetic. In H. Haken and M. Stadler (Eds.), *Synergetics of Cognition*. Berlin: Springer.

J.R. Anderson (1985), *Cognitive Psychology and Its Implications*. New York: Freeman.

H.B. Barlow (1972), Single units and sensation: A neuron doctrine for perceptual psychology? *Perception, 1*, 371–394.

G.B. Blasdel (1992), Orientation selectivity, preference, and continuity in monkey striate cortex. *Journal of Neuroscience, 12*, 3139–3161.

K. Brodmann (1909), *Vergliechene Localisations lehre der Grosshimrinde in ihren Prinzipien dargestellt auf Grund des Zellenbaues*. Leipzig: J.A. Barth.

J.F. Brugge and R.A. Reale (1985), Auditory cortex. In A. Peters and E.G. Jones, (Eds.), *Cerebral Cortex: Association and Auditory Cortices*. New York: Plenum Press.

H. Burton (1985), Second somatosensory cortex and related areas. In E.G. Jones and A. Peters (Eds.), *Cerebral Cortex*. Vol. 5. New York: Plenum Press.

P. Cavanagh (1978), Size and position invariance in the visual system. *Perception, 7*, 167–177.

H.H. Clark and E.V. Clark (1977), *Psychology and Language*. New York: Harcourt Brace.

S.A. Clark, T. Allard, W.M. Jenkins, and M.M. Merzenich (1988), Receptive fields in the body surface map in adult cortex defined by temporally correlated inputs. *Nature, 332*, 444–445.

F. Crick and C. Koch (1992), The problem of consciousness. *Scientific American, 267*(3), 152–159.

S.P. Dear, J.A. Simmons, and J. Fritz (1993), A possible neuronal basis for representation of acoustic scenes in auditory cortex of the big brown bat. *Nature, 364,* 620–623.

E.A. DeYoe and D.C. Van Essen (1988), Concurrent processing streams in monkey visual cortex. *Trends in the Neurosciences, 11,* 219–226.

M.L. Feldman (1984), Morphology of the neocortical pyramidal neuron. In A. Peters and E.G. Jones (Eds.), *Cerebral Cortex.* Vol. 1. *Cellular Components of the Cerebral Cortex.* New York: Plenum Press.

B. Fischer (1973), Overlap of receptive field centers and representation of the visual field in the cat's optic tract. *Vision Research, 13,* 2113–2120.

I. Fujita, K. Tanaka, M. Ito, and K. Cheng (1992), Columns for visual features in monkey inferotemporal cortex. *Nature, 360,* 343–346.

D. Gentner and A.B. Markman (1993), Analogy—watershed or Waterloo? Structural alignment and the development of connectionist models of analogy. In S.J. Hanson, J.D. Cowan and C.L. Giles (Eds.), *Advances in Neural Information Processing Systems 5,* 855–862.

A.P. Georgopoulos, R.E. Kettner, and A.B. Schwartz (1988), Primate motor cortex and free arm movements to visual targets in three-dimensional space. II. Coding of the direction of movement by a neuronal population. *Journal of Neuroscience, 8,* 2928–2937.

A.P. Georgopoulos, J.T. Lurito, M. Petrides, A.B. Schwartz, and J.T. Massey (1989), Mental rotation of the neuronal population vector. *Science, 243,* 234–236.

G.L. Gerstein (1988), Information flow and state in cortical networks: Interpreting multi-neuron experiments. In W.V. Seelen, G. Shaw, and R. Leinhos (Eds.), *Organization of Neural Networks.* Wenheim: VCH Verlagsgesellschaft.

G.L. Gerstein (1990), Interactions within neuronal assemblies: Theory and experiment. In J.L. McGaugh, N.M. Weinberger, and G. Lynch (Eds.), *Brain Organization and Memory: Cells, Systems and Circuits.* Oxford: Oxford University Press.

J.B. Gleason and N.B. Ratner (1993), *Psycholinguistics.* Fort Worth, TX: Harcourt Brace Jovanovich.

R.L. Goldstone, D.L. Medin, and D. Genter (1991), Relational similarity and the nonindependence of features in similarity judgments. *Cognitive Psychology, 23,* 222–262.

M.A. Goodale and A.D. Milner (1992), Separate visual pathways for perception and action. *Trends in the Neurosciences, 15,* 20–25.

D.O. Hebb (1949), *The Organization of Behavior.* New York: Wiley.

G.H. Henry (1985), Physiology of cat striate cortex. In A. Peters and E.G. Jones (Eds.), *Cerebral Cortex.* Vol. 3. *Visual Cortex.* New York: Plenum Press.

D.H. Hubel and T.N. Wiesel (1962), Receptive fields, binocular interaction, and functional architecture in the cat's visual cortex. *Journal of Physiology, 160,* 106–156.

D.H. Hubel and T.N. Wiesel (1977), Functional architecture of macaque monkey visual cortex. *Proceedings of the Royal Society of London, Series B, 198,* 1–59.

T.J. Imig and R.A. Reale (1980), Patterns of cortico-cortical connections related to tonotopic maps in cat auditory cortex. *Journal of Comparative Neurology, 192,* 293–332.

J.H. Kaas, R.J. Nelson, M. Sur, C.-S. Lin, and M.M. Merzenich (1979), Multiple representations of the body within the primary somatosensory cortex of primates. *Science, 204,* 521–523.

E.I. Knudsen, S. du Lac, and S.D. Esterly (1987), Computational maps in the brain. *Annual Review of Neuroscience, 10,* 41–65.

C. Lee, W.H. Rohrer, and D.L. Sparks (1988), Population coding of saccadic eye movements by neurons in the superior colliculus. *Nature, 332,* 357–360.

J.L. McClelland and A.H. Kawamoto (1986), Mechanisms of sentence processing: Assigning roles to constituents of sentences. In J.L. McClelland and D.E. Rumelhart (Eds.), *Parallel Distributed Processing,* Vol. 2. Cambridge, MA: MIT Press.

J.T. McIlwain (1975), Visual receptive fields and their images in the superior colliculus of the cat. *Journal of Neurophysiology, 38,* 219–230.

J.T. McIlwain (1976), Large receptive fields and spatial transformations in the visual system. *International Review of Physiology, 10,* 223–246.

K.A.G. Martin (1984), Neuronal circuits in cat striate cortex. In E.G. Jones and A. Peters (Eds.), *Cerebral Cortex.* Vol. 2. *Functional Properties of Cortical Cells.* New York: Plenum Press.

D.L. Medin, R.L. Goldstone, and D. Genter (1993), Respects for similarity. *Psychological Review, 100,* 254–278.

W.J.H. Nauta and H.J. Karten (1970), A general profile of the vertebrate brain with sidelights on the ancestry of cerebral cortex. In F.O. Schmitt (Ed.), *The Neurosciences: Second Study Program.* New York: Rockefeller University Press.

M.E. Nelson and J.M. Bower (1990), Brain maps and parallel computers. *Trends in the Neurosciences, 13,* 403–408.

J.C. Pearson, L.H. Finkel, and G.M. Edelman (1987), Plasticity in the organization of adult cerebral cortical maps: A computer simulation based on neuronal group selection. *Journal of Neuroscience, 7,* 4209–4223.

W. Penfield and E. Boldrey (1937), Somatic motor and sensory representation in the cerebral cortex of man as studied by electrical stimulation. *Brain, 60,* 389–443.

D.P. Phillips, S.S. Orman, A.D. Musicant, G.F. Wilson, and C.-M. Huang (1985), Neurons in the cat's primary auditory cortex distinguished by their responses to tones and wide spectrum noise. *Hearing Research, 18,* 73–86.

W. Pitts and W.S. McCulloch (1947), How we know universals: The perception of auditory and visual forms. *Bulletin of Mathematical Biophysics, 9,* 127–147.

J.W. Prothero and J.W. Sundsten (1984), Folding of the cerebral cortex in mammals: A scaling model. *Brain, Behavior and Evolution, 24,* 152–167.

N. Rochester, J.H. Holland, L.H. Haibt, and W.L. Duda (1956), Tests on a cell assembly theory of the action of the brain, using a large digital computer. *IRE Transactions on Information Theory, IT-2,* 80–93.

K.D. Roeder (1962), The behavior of free flying moths in the presence of artificial ultrasonic pulses. *Animal Behavior, 10,* 300–304.

M.H. Schieber and L.S. Hibbard (1994), How somatotopic is the motor cortex hand area? *Science, 261,* 489–492.

A.B. Schwartz, R.E. Kettner, and A.P. Georgopoulos (1988), Primate motor cortex and free arm movements to visual targets in three-dimensional space. I. Relations between single cell discharge and direction of movement. *Journal of Neuroscience, 8,* 2913–2927.

E.L. Schwartz (1984), Anatomical and physiological correlates of visual computation from striate to infero-temporal cortex. *IEEE Transactions on Systems, Man, and Cybernetics, SMC-14,* 257–271.

C.J. Shatz (1992), Dividing up the neocortex. *Science, 258,* 237–238.

J.A. Simmons (1989), A view of the world through the bat's ear: The formation of acoustic images in echolocation. *Cognition, 33,* 155–199.

C.A. Skarda and W.J. Freeman (1987), How brains make chaos in order to make sense of the world. *Behavioral and Brain Sciences, 10,* 161–195.

M.I. Skolnik (1980), *Introduction to Radar Systems,* 2nd ed. New York: McGraw-Hill.

N. Suga (1985), The extent to which biosonar information is represented in the bat auditory cortex. In Gerald M. Edelman, W. Einar Gall, and W. Maxwell Cowan (Eds.), *Dynamic Aspects of Neocortical Function.* New York: Wiley Interscience.

J. Szentagothai (1978), Specificity versus (quasi-) randomness in cortical connectivity. In M.A.B. Brazier and H. Petsche (Eds.), *Architectonics of Cerebral Cortex.* New York: Raven Press.

K. Tanaka, H.-A. Saito, Y. Fukada, and M. Moriya (1991), Coding visual images of objects in the inferotemporal cortex of the macaque monkey. *Journal of Neurophysiology, 66,* 170–189.

R.J. Tusa, L.A. Palmer, and A.C. Rosenquist (1978), The retinotopic organization of area 17 (striate cortex) in the cat. *Journal of Comparative Neurology, 177,* 213–235.

A. Tversky (1977), Features of similarity. *Psychological Review, 84,* 327–352.

D.C. Van Essen, C.H. Anderson, and D.J. Felleman (1992), Information processing in the primate visual system: An integrated systems perspective. *Science, 255,* 419–423.

F.W. Weymouth (1958), Visual sensory units and the minimal angle of resolution. *American Journal of Ophthalmology, 46,* 102.

D. Whitteridge and P.M. Daniel (1961), The representation of the visual field on the calcarine cortex. In R. Jung and H. Kornhuber (Eds.), *The Visual System: Neurophysiology and Psychophysics.* Berlin: Springer.

M.P. Young and S. Yamane (1992), Sparse population coding of faces in the inferotemporal cortex. *Science, 256,* 1327–1331.

Applications of Simple Associators: Concept Formation and Object Motion

It is not possible to step twice into the same river.
Heraclitus (Greek)

Eppur si muove!
Galileo (Italian)

It ain't nothin' *till I call it.*
Bill Klem (Umpire)

This chapter suggests solutions to two problems using a neural network. In the first, a model for simple concept formation is developed. In a special case, the learning of patterns of random dots, a representation can be suggested, and the network simulated and compared with human performance. In the second, a network model developed by Sereno is described that computes the motion of extended moving objects. Many important practical applications require a system to integrate information from many ambiguous elementary receptors to arrive at a consensus interpretation.

How many associations can be stored accurately by a neural network such as the linear associator? Does a bigger brain automatically become a better brain because it can store more things? An important cultural distinction exists between the way an engineer or physicist views capacity and the way a cognitive scientist does. Most discussions of capacity we have seen were engineering oriented. The idea is to put things in and hope to retrieve them. If what is retrieved is not the same as what was put in, it is an error. Capacity measures how much can be put in before serious mistakes start being made. For simple neural networks this kind of capacity usually is some fraction of

---

This chapter covers two main topics. Many of the details on the concept-related material are taken from a paper by Knapp and Anderson (1984). The material on object motion is described at greater length in Margaret E. Sereno's doctoral thesis (1989) and in a book arising from the thesis (1993). The treatment of the model here is an expansion of that given in Anderson, Viscuso, Rossen, and Sereno (1990).

the number of elements, around 10% to 15% seems to be common. This figure usually causes a little handwringing about the low capacity of the system, but, of course, we have lots of elements ...

A cognitive scientist is often not impressed with the many computations of capacity that involve keeping patterns separate from one another. This is rarely the task of a cognitive system. Computers do it better. So do books, records, and all the traditional physical adjuncts to human memory. A neural network is a computational memory. The computation is the most interesting part. The memory part may not be very accurate.

The task of the human cognitive apparatus is rarely to memorize facts, but to respond appropriately to a complex environment. Humans often tend to do this by simplification. Given a vast number of facts and situations, they will form best examples, rules, and concepts. Humans systematically distort and simplify reality so it is understandable. This may involve supressing distinctions between similar items. One table is mahogony, big, and old, and has a number of scratches. Another table is plastic, is medium size, has a few dents, and is new. When speaking the sentence, "I left the book on the table," is it necessary to describe the table in detail? Usually it is not, and if it is, the description can be expanded.

So the interesting capacity question becomes more like, how many things can we *usefully* tell apart? We want to ignore irrelevant detail. We want things that belong together to be processed together and even to have the same internal representation. Given potentially very many inputs, we want to find how many really *different* signals there are among them. The things we actually see may be distortions of one of a small number of signals. In this kind of system, we may actually *like* distortions and errors. Bad examples of an object might mutate and make contact with a uniform internal representation of the object. It may be a step on the path of wisdom to observe that the natives are speaking a strange form of a language you already know.

Concepts

Psychologists have proposed the name *concept* to cover some of those instances in which many different examples are seen as variants of the same thing. The word concept itself is tricky, and has been used in many ways by many groups. Here, we will specifically discuss simple concepts in psychology.

One reason to use a concept-based computation is simplicity. Learning can be done once for relations between the concepts, thereby avoiding learning many special cases. Concept learning can be a sophisticated and useful form of generalization. We will be concerned here only with the simplest form of categorization and concept learning, but the way in which these elementary units are combined and used is central to cognition.

Models of Categorization

One model of categorization is particularly popular with computer scientists and some philosophers. It is usually called the classical model, or sometimes Aristotelian concepts. The idea behind it is simple, elegant, and wrong. Suppose categories are defined by a list of properties that have to be present or absent. A "boy" is "human," "animate," "male," "not old," and so on. If a new object is presented to the system, it is represented by a list of properties. These properties are compared with the various category lists, and a match signals that the new object is an example of that category, that is, of the concept "boy" or "bird" or "airplane."

Such an approach to categorization is easily mechanized. Unfortunately, natural categories can rarely be described this way; they are usually hazy around the edges and have too many special cases to capture in such a rigid format. Most birds fly, but, of course, penguins do not. Birds are small or medimum size, except ostriches, California condors, and moas. Penguins swim and don't fly. Bats fly, are warm blooded, but are not birds. It is possible to give unambiguous differences between birds and bats, porpoises and tunafish, because we know a great deal about DNA and cell biology, although these unambiguous descriptors are rarely used. However, even more difficult determinations, for example between Democrats and Republicans, or small trees and large bushes, or boulders and pebbles, are common.

This anecdote will illustrate the practical difficulties of the classic approach. A conference on psychological representation was held in Providence, Rhode Island. A highlight was an after-lunch discussion, quite heated and lasting the better part of an hour, among several eminent academics, including a computer scientist and a famous physicist, about how to categorize a three-legged dog. Is a three-legged dog any more or less a dog than the normal variety? (Besides having fewer legs, of course.) This discussion was especially interesting because, in fact, a three-legged dog lived on the east side of the city at that time. Even though it walked somewhat strangely, no one, even college professors, had any trouble telling what it was.

Since the strongest form of the Aristotelian concept is inadequate, the requirement for exact matching must be relaxed. This approach gives rise to various similarity-based models for categorization and concepts. One form or another is nearly universally accepted by psychologists. However, similarity is less straightforward than it seems at first. The end of chapter 10 pointed out some of its less obvious properties.

A well-known book on the theory of concepts is *Categories and Concepts* by Smith and Medin (1981). (A more recent review is Smith, 1989.) According to Smith and Medin, models of categorization fall broadly into two families, *probabilistic* and *exemplar*, depending on how categories are assumed to be represented in memory.

Probabilistic models assume that the representation of a category consists of a description of the category members. However, they are flexibile, so that properties of the description are not necessarily true of all category members. Category membership then becomes a continuous rather than a two-valued function: most birds fly, but not all of them.

Prototype Models

We will treat in detail in this chapter one class of probabilistic models that represents a concept by a *prototype* that best represents the category instances. A considerable body of psychological evidence supports the use of prototypes, or best examples, as a model for many kinds of categorization (Rosch, 1978; Mervis and Rosch, 1981). McClelland and Rumelhart (1985) discussed a concept-forming model similar to the one we will address.

One of the early sources of support for prototype theory had to do with the naming of colors. A great deal is known about the neurophysiology of color, and all members of the human species with normal color vision are roughly equivalent in how their retinas respond to the wavelength of light. Therefore it was surprising to early anthropologists when they found that different cultures named segments of the spectrum differently. At first this observation led to claims that color names (and, by radical extension, all concepts) were arbitrary and strictly a function of societal convention. On further investigation, the story turned out to be more reasonable. Human societies have a small number of primary color names, that is, red, blue, and green. When members of different societies are asked to point to a standardized color sample that corresponds to the best example of red, they point to roughly the same color. When they are asked to determine, for example, the boundaries between red and orange, the classification of boundary colors can vary considerably (Rosch, 1973).

This behavior has a nice connection to the physiology of color. The human eye has three different kinds of retinal cones, all with different wavelength absorption charateristics. However, most color specialists now hold that the signals from the three cones are combined early in the visual system to give rise to three color channels: a channel signaling black-white, one red-green, and one blue-yellow. It is as if a color channel acted like a volt-meter, with a positive deflection giving rise to the sensation green, say, and a negative deflection to the sensation red. This means that a particular color at a point in space can be described as a combination of the output from the three channels. Meters (and the channels) pass through zero, so at some points in the spectrum the contribution from, say, the blue-yellow channel is zero. Experimentally, the prototype green is close to the point at which the red-green channel is signaling green and the blue-yellow channel has zero value, and so on.

Color names seem to have a direct connection to neurophysiology. More complex concepts do not allow this direct connection, and the prototypes must presumably be learned by experience. For example, whether or not an animal is a dog or a bird depends on a series of biological properties. But when humans classify an animal, they often seem to do so by having a

Typicality scores (Rosch, 1975)

Fruit		Vegetables		Birds	
orange	1.07	pea	1.07	robin	1.02
apple	1.08	carrot	1.15	sparrow	1.18
banana	1.15	spinach	1.22	bluejay	1.29
peach	1.18	broccoli	1.28	canary	1.42
strawberry	1.61	cauliflower	1.62	dove	1.43
blackberry	2.05	cucumber	2.05	finch	1.66
lemon	2.16	beets	2.08	seagull	1.77
watermelon	2.39	turnip	2.37	pigeon	1.81
lime	2.45	eggplant	2.38	hawk	1.99
papaya	2.58	onions	2.52	parrot	2.07
mango	2.88	potato	2.89	owl	2.96
cranberry	3.22	peppers	3.21	pelican	2.98
date	3.35	parsley	3.32	vulture	3.06
persimmon	3.63	mushroom	3.56	duck	3.24
coconut	4.50	avocado	3.62	chicken	4.02
avocado	5.37	sauerkraut	4.18	turkey	4.09
pumpkin	5.39	pumpkin	4.74	ostrich	4.12
olive	6.21	garlic	5.07	penguin	4.53
squash	6.55	rice	5.59	bat	6.15

(Numbers are subjective judgments of how good
an example is of a category.)

Figure 11.1 Goodness of example scores selected from tables in Rosch (1975). Other examples were not included here. Subjects were asked to judge how good a member of a category an example was. They were instructed that a 1 was a very good example of the category and a 7 meant that the member fit into that category very poorly.

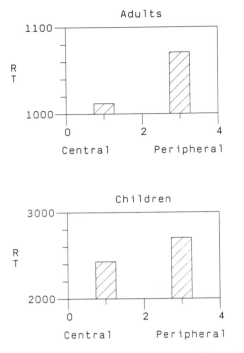

Figure 11.2   Reaction time for adults and children to verify an assertion involving typicality; for example, an apple is a fruit. Data are plotted from Rosch (1973).

notion of what constitutes a "best" dog or bird, and use distance from that prototype as a way to categorize the new animal. For example, the best bird for most Americans is probably something like a sparrow or a robin; that is, a small bird that is found on or near lawns. The best dog seems to be a medium-size mongrel. A chihuahua is perceived as a "bad" dog, that is, not a very good example of dogs, just as an ostrich or a penguin is perceived as a "bad" bird and not typical of birds. Biologically, of course, each is a perfectly fine example of its grouping and no better or worse than other dogs or birds. Much more than freqeuency of presentation is involved. A bird commonly encountered in urban America is a chicken, but a chicken is considered an atypical bird. Chickens, of course, are often encountered in pieces.

Support for the prototype notion comes from several lines of evidence. First, when humans are asked to judge typicality they can do it quite easily. For example, the most typical fruits are apples and oranges and the most typical vegetables are peas and carrots (figure 11.1) A less typical fruit would be a mango or a coconut and a less typical vegetable would be garlic or a pumpkin (Rosch, 1975).

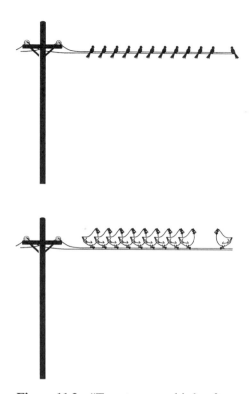

Figure 11.3 "Twenty or so birds often perch on the telephone wires outside my window and twitter in the morning." From Rosch (1978, p. 39).

Second, when asked to verify an assertion like "An apple is a fruit," subjects are faster to respond than when given the assertion "A mango is a fruit." The effect of typicality is stronger in children than adults, but is present in both (figure 11.2).

Third, and probably the most significant, Rosch, among others, pointed out that the prototypes tend to be the default meanings used when understanding sentences. To use one of her examples, when we interpret a sentence such as, "Twenty or so birds often perch on the telephone wires outside my window and twitter in the morning," we form a mental image containing prototypical birds such as sparrows and not, for example, chickens (Rosch, 1978, p. 39) (figure 11.3). This is a deep observation, and suggests a major practical use of concepts. They become a way of economically communicating a large amount of information, and then manipulating that information in simple and convenient ways. However, communicating in this fashion means that a lot of sometimes critical specific information is discarded, leading to ambiguity, controversy, and gainful employment for lawyers.

Exemplar Models

An alternative model for concept formation superficially seems to be quite different from prototype models. Smith and Medin called this class *exemplar* models, and they make up the other main family of categorization models. (For obscure reasons, examples of a concept are often called "exemplars" in the psychological literature. Here, we will usually call them examples.) According to this view, no single description of a category exists. Rather, a collection of separate descriptions of some or all category members represents the category. Stimuli are categorized according to the number of stored examples they retrieve. A recent neural network approach to concept learning using a modified exemplar model is Kruschke's ALCOVE model (1992).

As an aside, both these approaches are used in engineering pattern recognition. Exemplar models are close to what is called *nearest neighbor* models, discussed in chapter 13. In these, many examples are stored, and a new item is given the category of the nearest, or the several nearest, stored items, that is, an exemplar model. Prototype models are close to a common variant of nearest neighbor models. From a practical point of view, if a great many examples have to be stored, it is cumbersome to keep track of them and compute with them. It is far more efficient from almost every point of view to replace dense clusters of examples with a strategically located single example that categorizes—that is, is the nearest neighbor to—a large region of state space. Choosing the appropriate location for this special stored prototype pattern depends on the details of the application, but the average of a category is not a bad choice in many cases.

Figure 11.4 is a schematic of these two related approaches. Practically, it is hard to tell their predictions apart. We will see that a neural network approach can act like a prototype model when it stores many similar examples, or like an exemplar model when it stores a few widely separated examples.

The models that we proposed in earlier chapters can function as categorizers by making an additional assumption consistent with most of what we have said. We discussed representation of information in more detail in chapter 10, but we can make one major assumption explicit. *Let us make the fundamental representation assumption that the activity patterns representing similar stimuli are themselves similar, that is, their state vectors are correlated.* This means the inner product between the two similar patterns is large, and the more similar they are, the larger the inner product is. It is this assump-

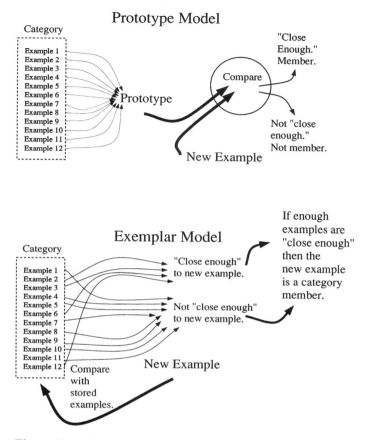

Figure 11.4   Schematic version of exemplar and prototype models.

tion that makes nearest neighbor models work and that makes statistical pattern recognition, in general, work.

Consider an associative network that has made the single association $(\mathbf{f} \rightarrow \mathbf{g})$. Let us restrict our attention to the magnitude of the output vector that results with various input patterns. To review the results obtained in chapters 6 and 7, for the linear associator, when $\mathbf{f}$ occurs at the input, an associated $\mathbf{g}$ occurs at the output.

output pattern $= \mathbf{g}[\mathbf{f}, \mathbf{f}']$.

If $\mathbf{f}$ and $\mathbf{f}'$ are uncorrelated, their inner product $[\mathbf{f}, \mathbf{f}']$ is small. If $\mathbf{f}$ is similar to $\mathbf{f}'$, the inner product will be large. If the network has learned several associations, it can categorize novel input patterns according to their similarity to the patterns already encountered. This is a limited form of categorization based strictly on similarity of representation.

Multiple-Member Categories

We will now apply this network to the more realistic situation in which a category contains many similar items. Here, an entire set of correlated activity patterns (representing the category members) becomes associated with the same response, for example, the category name. It is convenient to discuss such a set of vectors with respect to their mean. Specifically consider a set of correlated vectors $\mathbf{f}_1 \ldots \mathbf{f}_n$ with mean $\mathbf{p}$ ("$\mathbf{p}$" for prototype, of course). Each individual vector in the set can be written as the sum of the mean vector and an additional noise vector, $\mathbf{d}_i$, representing the deviation from the mean, that is,

$$\mathbf{f}_i = \mathbf{p} + \mathbf{d}_i.$$

When these n input patterns occur and are all associated with the same output, $\mathbf{g}$, the final connectivity matrix will be

$$A = \sum_i \mathbf{g}\mathbf{f}_i^T$$

$$= \mathbf{g} \sum_i (\mathbf{p}^T + \mathbf{d}_i^T)$$

$$= n\mathbf{g}\mathbf{p}^T + \mathbf{g} \sum_i \mathbf{d}_i^T.$$

The term containing the sum of the noise vectors ($\sum \mathbf{d}_i$) is particularly important. Suppose that this term is relatively small, as would happen if the network learned many randomly chosen members of the category. In that case, the connectivity matrix reduces to

$$A = n\mathbf{g}\mathbf{p}^T.$$

The network behaves as if it had repeatedly learned only one pattern, $\mathbf{p}$, the mean of the correlated set of vectors it was actually exposed to, that is, like a prototype model, because the most powerful response will be to the pattern $\mathbf{p}$, which may not have been seen.

However, if the sum of the $\mathbf{d}_i$ terms is not small, as might happen if the system only sees a few widely differing patterns, the response of the network will depend on the similarities between the novel input and each of the learned patterns; that is, the network behaves more like an exemplar model. If the perturbing noise vectors are of roughly constant size, the number of patterns in the correlated set of inputs will be the primary factor determining whether the distributed memory model behaves more like a prototype model or an exemplar model. One or two examples will be tend to be learned individually; 1000 or 2000 will be represented by their average.

Continuum

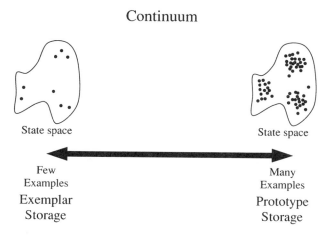

Figure 11.5 Exemplar and prototype models are extremes on a continuum of intermediate models.

Figure 11.5 suggests the intuition between this shift in properties from what appears to be an exemplar model to a prototype model. The seemingly different approaches are actually based on common intuitions about similarity, and can be realized by the same mechanisms operating in different parameter regions.

If more than one category appears in the experiment, the same kinds of analysis we applied previously can be used, suggesting that several categories can be formed. If the inputs are reasonably well separated, that is, the inner product between different inputs is small, the distortion of the associated output will be small.

The Prototype Effect

This result suggests that *in the act of storage* the examples merge together and generate a new pattern, the prototype, that may never have existed. Few users of a computer memory would like their computers to store spontaneously a new memory different from those the system actually learned. This behavior would seem to destroy memory rather than preserve it. However, a body of empirical evidence suggests that humans act exactly this way. Effects of this type are sometimes referred to as *prototype effects*. They occur as follows.

Suppose we teach a subject many examples. Suppose the average of the examples is a pattern—the prototype. Experiments are constructed so that the prototype can never appear as an example. After learning, subjects are tested with the patterns they learned, the unseen prototype pattern, and new patterns that follow the rules used to generate examples but that were not learned. A very common result is that subjects respond fastest to the prototype, categorize it most accurately, and, most striking, are more sure they saw the prototype than the patterns they actually did see. Is this an error? Or is it a useful adaptive property of a biological memory?

Simulations

We will show the prototype effect first in simulation, and then using human data from an experiment. Our character program, AMCHARDEMO, given in the last chapter, is ideally suited to do a few simple simulations along these lines. Suppose we have three categories, say, a string of digits, a string of lower-case letters in alphabetical order, and a string of capital letters in the order in which they appear on a typewriter keyboard, that form three prototypes. We can create distortions of a prototype by changing characters at random.

We generated five examples of each prototype by replacing half of the characters in each string by random characters chosen from the character list by the random number generator. Table 11.1 gives the input patterns that were learned and the appropriate associated responses. The input

Table 11.1   Input and output character strings

```
vC{8efg.i@kpmnLpqrsO?CVcy ⎫
abcPujfP#jklino:qkl-IO7xy  ⎪
T}Bd/mg5xjMomo)pqr":uDoqV  ⎬ → Alphabet. First 25 char.
lbcdefQhijklsyYpqrst\v!ry  ⎪
aacdefBjij\Ot9oFq5sUQgwby  ⎭

%7%op67\901B34767\T^?2e45  ⎫
V[3G/6k8Z0X2%406|8>01\3/5  ⎪
A2j4q6789A=23458X8q012t45  ⎬ → Twenty-five 25 digits.
rA3^5n08TX1934*6r*93<534<  ⎪
Y73456y8b0%^3.767)90123@@  ⎭

Qe3xTYUIOvASDFGeJKLZXCqB5  ⎫
1WeRTOUWO{ASDqGHJLLtxCV\N  ⎪
ZW-RTX.IOPASD9]1%KLZU1VB_  ⎬ → Capital letters, keyboard
QWE1TpQI?PASDF\D\KLZXOHBN  ⎪
QQE]|YUIr1iSDFG)J\L/rS::N  ⎭
```

Prototypes:
```
abcdefghijklmnopqrstuvwxy  → Alphabet. First 25 char.
1234567890123456789012345  → Twenty-five 25 digits.
QWERTYUIOPASDFGHJKLZXCVBN  → Capital letters, keyboard
```

Half of each input string is replaced by random characters.

strings are quite corrupted. There are five examples for each of the three outputs. These 15 examples are learned using the simple linear associator programs described in chapter 7. The resulting system correctly classifies all the old items. It also classifies the prototype correctly. Note that the length of the output vector, a measure of goodness of category membership we used before, is significantly longer for the prototype than for any old example.

We also constructed a new set of examples that were generated by the same rules as the old set, except that the system had not learned the new items. The response of the network to these new patterns was tested. This test with new examples and the prototype is one direct test of *generalization*, that is, the ability to respond appropriately to items that are related to old items but are not identical to them. One of the major practical virtues of neural networks lies in their ability to generalize. Unfortunately, it is not always possible to define exactly what is meant by correct generalization, because an appropriate inference from past experience might, and, in the real world, sometimes does, lead to disaster. In this simple test, generalization has a clear meaning: responding appropriately to a new distortion of a prototype.

The responses to the new examples were all correct. In table 11.2 we give the lengths of the output vectors of the network in response to old examples, new examples, and the prototypes. Note the order of output vector lengths is prototype, old examples, and new examples, demonstrating a prototype effect. Detailed information about the old examples is still present, in that they give significantly longer output vectors than new examples.

It is worth noting how effective the network is at this generalization task. This is the same system that had trouble accurately recalling a few character strings in chapter 7.

Table 11.2   Length of output vectors

Category	Output Average Length, Old Examples	Output Average Length, New Examples	Output Length of Prototype
1	30.9	25.3	40.9
2	28.3	20.4	37.0
3	33.4	23.9	41.4
Average	30.9	23.2	39.8

Ratio of old examples to new examples 1.33.
Ratio of prototype to old examples 1.29.
Ratio of prototype to new examples 1.72.

Random Dot Pattern Stimuli

It is possible to test our ideas about concept formation in some simple cases. These experiments are described in detail in a paper by Knapp and Anderson (1984) in which the artificial categories first conceived by Posner and Keele (1968, 1970; Posner, 1969) were used. The stimuli consist of random arrangements of dots, and they have been used as materials in many experiments. At Brown University, a version of the experiments we will describe is performed in an undergraduate cognitive psychology laboratory with consistent success.

Typically, several patterns are produced by randomly distributing nine dots within a display area. The original patterns are called *prototypes*. A family of *examples* of each prototype can then be generated by moving dots random distances in random directions. By controlling the distance moved, category examples can be either grossly distorted versions or only slight variations of the prototype. Each prototype and its progeny form an artificial category.

In the experiments a prototype pattern was created by randomly placing nine dots within a 300 by 300-unit grid centered in a 512 by 512-pixel display area. Examples of a prototype were constructed by displacing the dots short distances. For each dot, the direction of displacement was chosen at random, that is, uniformly distributed on the range zero to 360 degrees. The distance moved was determined by a probability distribution specified by the experimenters. Figure 11.6 shows a prototype and five examples at increasing levels of distortion. Each example is the result of moving the prototype dots distances drawn from a normal distribution. The average displacement in display screen units is given above each example. Because the prototypes were generated toward the middle of the display screen, the dots had ample room to move.

Representation of the Dot Patterns

Subjects viewed pairs of sequentially presented dot patterns. The first member of each pair was a prototype and the second was an example derived from that prototype. (Obviously, the words prototype and example are arbitrary here.) In the distorted patterns, dots were moved 15, 30, 45, 60, or 90 pixels in random directions. Sometimes there was no distortion at all, so the two dot patterns were identical. Subjects were told they would be shown a sequential series of dot pattern pairs. They were instructed to rate each pair on an eight-point similarity scale (1-highly similar, 8-very different) by pressing a key on a keyboard.

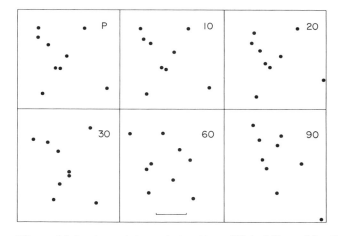

Figure 11.6 A prototype dot pattern (P) is followed by five examples at various levels of distortion. Dots were generated on a 512 by 512 array and presented to subjects on a CRT screen. The number refers to the average number of locations moved on the array. A distance of 100 locations is indicated. From Knapp and Anderson (1984). Reprinted by permission.

The first member of each pair was presented for five seconds, followed by a one-second interval, folowed by the second pattern, which remained visible until the subject responded. Another trial occurred two seconds after the response was made. Figure 11.7 shows the results of the similarity judgments plotted with the theoretical simulations, described next. This figure shows that the perceived similarity between an example and its prototype decreases as the average dot displacements used to create the example increase.

Based on our assumptions about similarity, the perceived similarity between two stimuli should be related to the inner product $[\mathbf{f}, \mathbf{f}']$, where $\mathbf{f}$ and $\mathbf{f}'$ are the activity patterns representing the two different stimuli. To compute similarity we must know how the dot patterns are represented in the network. We must give a rule for constructing a state vector from a presented dot pattern. Once we have state vectors representing the dot patterns, we should be able to predict the experimental similarity curve.

One possible data representation is suggested by neuroscience. We know that a topographic map of visual space is present on the surface of the cerebral cortex (see chapter 10). Let us consider the activity generated due to a single dot in visual space. We assume that this dot in space will map onto a "bump" of activity on a hypothetical surface composed of many elementary neurons that represents neural coding. Since real neurons have receptive fields of varying width, even in a single cortical region, we assume a fall-off of activity from the central location corresponding to the exact topographic

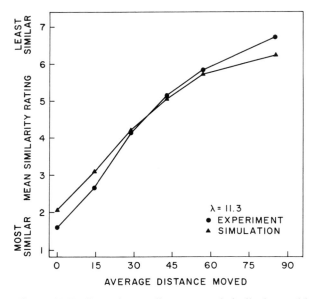

Figure 11.7 Experimentally measured similarity and best fitting curve for theoretical similarity. A response of 1 is most similar and 8 is least similar. From Knapp and Anderson (1984). Reprinted by permission.

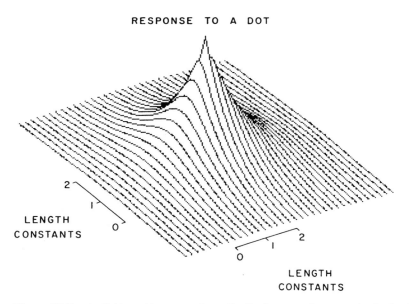

Figure 11.8 Activity pattern on a hypothetical cortex due to a single dot in the real world. There is an exponential decay with distance from a central point. Height corresponds to activity. The length constant of the exponential is shown. From Knapp and Anderson (1984). Reprinted by permission.

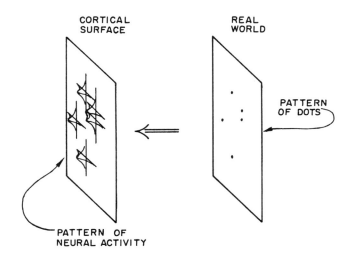

Figure 11.9 A group of dots in the real world maps into a group of bumps of activity on a surface of units. This surface generates the state vector that is used in the simulations. Given any dot pattern, a state vector can now be constructed if the activity pattern due to a single dot is known.

location of the dot. Some cells have large receptive fields and might respond to a dot even if the center of their field is far away from the dot location; other receptive fields are much smaller and must be centered on the dot location. This representation is consistent with the idea of neural *point images* (McIlwain, 1975), that is, the set of neurons that sees a point in space. The simplest and most effective representation of a point in visual space was the exponential decay of activation (figure 11.8). That is, the activity at a point, $a(r)$, is given by

$$a(r) = e^{-r/\lambda},$$

where λ is the length constant of the exponential and r is the distance from the center of the distribution to the point whose activity is to be computed. A pattern of several dots gives rise to several bumps of activity on the topographic map. Given any dot pattern, a state vector can be constructed (figure 11.9).

Interactions Between Activity Patterns

We have to compute inner products between two dot patterns. We can start by computing the inner product between the activity patterns of any two single dots. If our activity patterns are composed of 9 dots (as ours were), the basic linearity of the network lets us then compute interactions between all

81 pairs of dots, followed by the addition of the resulting inner products. This will give the overall result. It is not necessary to match up dots from the two stimuli; any activity patterns that interact will contribute to the inner product.

If we use the exponential activity distribution shown in figure 11.8, we can obtain a closed form solution for the inner product if we assume a continuous distribution of activity. This would correspond to the limit of a two-dimensional array containing so many units that the activity of each unit can be accurately approximated by continuous functions on a plane. In this limit, the network is essentially infinite-dimensional. This lets us approximate the inner product by an integral which we can do in closed form.

Suppose we have two activity patterns due to single dots separated by a distance d. We assume the distribution is continuous. For convenience let the length constant equal 1, and consider two exponential distributions, $a(x, y)$ and $b(x, y)$, separated by a distance, d, along the x axis. We want to evaluate the integral, $I(d)$, which will only be a function of displacment, d:

$$a(x, y) = e^{-\sqrt{(x+d/2)^2+y^2}}$$

$$b(x, y) = e^{-\sqrt{(x-d/2)^2+y^2}}$$

$$I(d) = \iint_{-\infty}^{\infty} a(x, y)b(x, y)\, dx\, dy.$$

Let us normalize the function so that $I(d) = 1$ when $d = 0$. This integral can be computed exactly and is given by

$$I(d) = (1/2)d^2 K_2(d),$$

where $K_2(d)$ is a modified Bessel function of order 2 (Abramowitz and Stegun, 1964). For connoisseurs of integration, this integral can be done by observing that the loci of constant product are ellipses. The equations are converted to elliptical coordinates (Korn and Korn, 1968) and then integrated with the tables in Gradshteyn and Rhyzik (1965). Doing such integrations provides much innocent amusement.

Computer Simulations of the Similarity Experiments

In this simulation we have 81 pairs of dots (9 dots taken 2 at a time, 1 from a pattern and the other from its distortion) among which we must compute inner products. In the simulation we computed all distances between dots. Our assumption about similarity says magnitude of the inner product between activity patterns is directly related to similarity. The only free parameter in the simulation is the length constant. Representations of the patterns were constructed that were statistically identical to those presented to subjects. A dot pattern and a distortion were generated, and the inner products between the representations of the two patterns were computed. Then the experimental and theoretical values for similarity were compared. Several measures of goodness of fit were used in the computations. First, the best-fitting straight line was computed between predicted similarity and experimental similarity. The length constant that minimized the mean square distance between this line and the experimental data was found. Second, the correlation between predicted and experimental values was computed.

The results of the measures were close to each other. The maxima of the relation between length constant and the measures of goodness of fit was quite broad. There were no critical aspects of the simulation. Several functions representing falloff of activity due to a single dot were investigated at various times: Gaussians, exponentials, laterally inhibited functions, and others. The best fits were obtained with simple exponentials, and this function was used in the simulations and in the figures. Exact shapes of the falloff were not very critical.

Figure 11.7 plots the experimental data along with the best-fitting theoretical curve. Note what are probably systematic deviations at the ends of the two curves, but overall, the shapes are similar.

Number of Examples

Next, let us look at a memory task using this representation. The linear associator develops a response to a category by associating an entire set of correlated activity patterns with the same response. The behavior of the network is a function of the number of activity patterns it learns. If the network learns many different category members, it behaves if it has seen the mean pattern, **p**, alone. If it learns only a few members, the network acts as if single examples were stored.

In a test of this aspect of the network model, subjects learned three categories containing 1, 6, and 24 members, respectively. The learning stimuli were distorted examples of one of three prototypes. The experiment had a *learning* phase, in which subjects classified category examples with feedback, and a *testing* phase, in which they classified old, new, and prototype patterns without receiving any feedback. The dot displacements were drawn from a normal distribution with mean 24 pixels and standard deviation 8 pixels. This distance was chosen in pilot studies because it seemed to maximize the experimental prototype enhancement. Because of the way the example patterns were generated, an example was exceedingly unlikely to duplicate its prototype.

The learning stimuli consisted of one example of the first prototype displayed 24 times, six examples of the second displayed four times each, and 24 examples of the third shown only once each, for a total of 72 learning trials, with each category being represented 24 times during learning. In the testing phase, the single learned example from the first category was presented eight times, four of the old examples from the second category were presented twice each, and eight of the old examples from the third catagory were each presented once. In addition, eight newly constructed examples of each prototype were presented once each and each prototype was shown eight times. The subjects had encountered neither the prototypes nor the new examples during learning. Finally, nine unrelated random control patterns were presented once each.

Procedure

The subjects were told that they would be shown a series of dot patterns, and that their task would be to determine which patterns were to be grouped together under the same response. In the testing phase, subjects were instructed to respond as quickly as possible, based on what they had learned in the first part of the experiment, and were told that no feedback would be given.

Figure 11.10 shows the percentage of correct classifications for each test trial combination of stimulus type (old, new, prototype) and number of

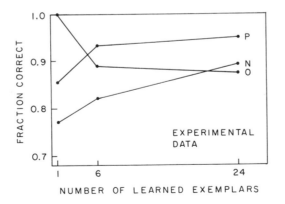

Figure 11.10 Percentage of correct test trial classifications for memory experiment. P, O, and N stand for prototype patterns, old patterns, and new patterns, respectively. Subjects always had the same number of trials, but saw different numbers of individual examples. From Knapp and Anderson (1984). Reprinted by permission.

learned examples (1, 6, 24) collapsed across subject groups. A ceiling effect is apparent for the single old example in the one-instance category: all the subjects classified this pattern correctly all the time. A prototype effect was apparent for both 24- and 6-example categories.

The network's behavior is understandable if we begin by considering the summed representation of a single dot in the nine dot patterns (figures 11.11 and 11.12). Figure 11.11 shows activity patterns due to four example dots equally spaced from each other and from the location of a prototype dot. As the separation between the example dots increases, the peaks of activity in the sum separate. At first, the distribution of activity in the sum as the dots

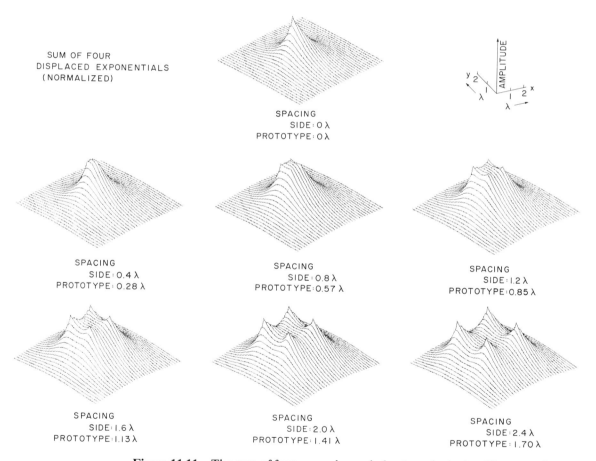

Figure 11.11 The sum of four examples each due to a single dot. The examples are located at the corners of a square; the center of the square is the location of the prototype in this example. As the dot spacing increases, there is less relative enhancement of the prototype and more representation of single examples. The plots in the figure equate the maximum activity at each displacment so the shapes of the curves can be compared.

MEMORY FORMATION

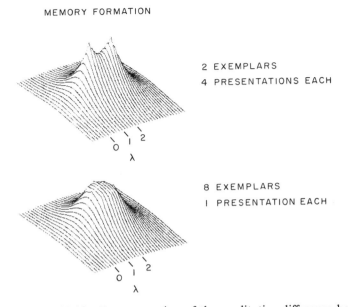

2 EXEMPLARS
4 PRESENTATIONS EACH

8 EXEMPLARS
1 PRESENTATION EACH

Figure 11.12 Demonstration of the qualitative difference between a memory sum constructed from two different examples and one constructed from eight different examples. Examples in both cases had the same average separation from the prototype location. A very lumpy memory is produced when only a few examples are stored. The representation at the prototype location is the same in both instances. The relative representation of old and new examples changes markedly between the two cases. From Knapp and Anderson (1984). Reprinted by permission.

separate simply seems to broaden the peak located at the prototype location. Then bumps due to the individual examples appear. Even with dots well separated, substantial representation still is present at the prototype location. It is clear that some representation of variability (the width of the activity pattern) is present in the network as well as representation of central tendency. In figure 11.11 the figure has the maximum values in the pattern drawn at the same height, so relative curve shapes can be compared.

Figure 11.12 shows the formation of a sum from two examples (above) or from eight examples (below). In the diagrams, the example dots were chosen to be equally spaced about a central location corresponding to the dot in the prototype pattern. In both sums, the activity at the prototype location is equal and substantial. In the two-example sum, the activity is greatest at the location of the individual examples, but in the eight-example sum it is largest at the prototype location. The smaller number of examples produces a "lumpy" sum dominated by learned examples, whereas the larger number yields a smoother sum with the prototype enhanced.

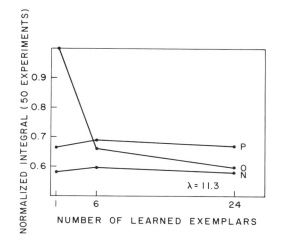

Figure 11.13 Simulation of the memory experiment (figure 11.10) using the best fitting length constant as determined in the similarity experiment. The y axis gives the inner product between a test input and the memory. P, O, and N stand for prototype patterns, old patterns, and new patterns, respectively. From Knapp and Anderson (1984). Reprinted by permission.

Simulation

Prototype dot patterns were generated with parameters identical to those used in the experiments. Examples were formed according to the rules followed in the experiments. Sums of examples were constructed and a memory was formed. If a particular experiment had three prototypes, three sums were constructed. An input was classified as to which sum it was most similar to by computing the inner product with each sum in turn and choosing the largest. The sum with greatest similarity was the classification of the input. (This approximation to the behavior of the full matrix memory model was required because of computer limitations at that time.)

In the experiments modeled in the simulations, we used groups containing 1, 6, and 24 examples of particular prototypes. Of course, a prototype cannot be formed when only a single example is presented. The model will generate similarities for new examples and the prototype, however, and we can predict the responses in this special case with the same model we used for numerous examples.

Figure 11.13 shows the results when the length constant for best correlation in the similarity experiments was used in the prototype-simulation program. Although it is not possible to make a direct mapping of inner product into percentage of correct responses without numerous additional assumptions, there should be a monotonic relation between similarity and

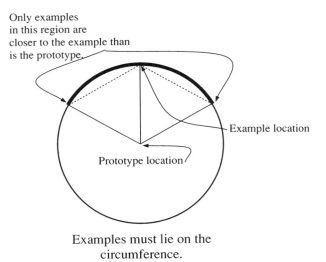

Only examples
in this region are
closer to the example than
is the prototype.

Example location

Prototype location

Examples must lie on the
circumference.

Figure 11.14   Even if only a single example is learned, the prototype pattern will still show better response than new examples because of the geometry used to construct the dot patterns.

percentage of correct classifications. The results in figure 11.13 show considerable similarity to those in figure 11.10. In particular, a crossing over of the responses to old examples is shown for the 6-example case and the coming together of the responses to old and new examples in the 24-example case. There is a strong response to the single old example in the one-example case, but the responses to the prototype and new examples for this seemingly different case fall right in line with the data.

The prototype in the single-example case is enhanced relative to new examples. This is a direct prediction of the geometry that generates examples (figure 11.14). Let us assume that the set of examples is generated from points distributed uniformly on the circle, that is, the dots move a radius, r, in a random direction. The prototype is a distance, r, from any example actually seen. All new examples lying within r of the old example will give a stronger output than the prototype. These points are on the arc contained within the two equilateral triangles having one corner at the old example. These two triangles cover an arc of 120 degrees. Therefore new examples are twice as likely to lie in the 240-degree arc that is more than a radius r from the old example than is the prototype, and the prototype will be enhanced relative to new examples.

Sometimes prototypes are not enhanced. We can predict when this occurs. We are interested in the relative sizes of the inner product between the prototype and an old example. Let assume that several examples are

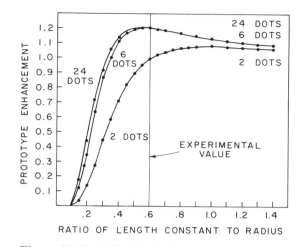

Figure 11.15 This simulation assumed varying numbers of dots, from 2 to 24, arranged on the unit circle. The prototype was located at the center of the circle. The graph gives the ratio of the system response to a dot at the prototype location and the system response to a dot at the location of an old exemplar. The y axis gives the radius of the circle in length constants. The average displacement of patterns used in the memory experiments is indicated by a mark on the figure. This displacement is used based on extensive informal experiments because it produced the maximum prototype effect. From Knapp and Anderson (1984). Reprinted by permission.

located equally spaced on the unit circle, with the prototype at the center of the circle. The number of examples in the simulations varied from two to a large number. We varied the length constant and computed inner products between an input pattern located at the prototype location and at the location of one of the dots on the circle. Figure 11.15 presents the ratio of the memory inner product at the two locations.

It can be seen that when the distance to the examples is very small in length constants, the values at prototype and example locations are almost identical, because everything adds without significant falloff. This case corresponds to small average dot movement.

When the distance to the examples is large in length constants, activity from one example decays to almost zero before it encounters activity from another example. There is essentially no representation at the prototype location. This case corresponds to storage only of single examples with no prototype formation and to experiments with large average movement of dots.

A region of optimal prototype enhancement exists that reaches a peak value around 20% greater than the value at the location of any individual example. Extensive pilot studies determined the experimental distortion that

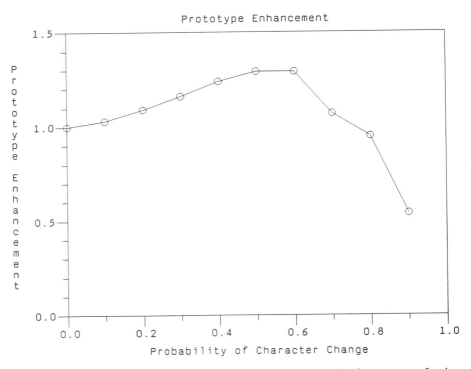

Figure 11.16 Simulation using the character-based system. As the amount of noise, that is, the probability of substituting a random character in the prototype string, increases, the prototype effect passes through a maximum. This effect is similar to what is seen with dot patterns (figure 11.15).

maximizes the prototype effect. These studies were done before anyone was aware of theoretical reasons for expecting there to be a maximum in the prototype effect.

Character-Based Simulations

We can close this brief discussion of the prototype effect by returning to our character-based simulations. Most of the qualitative effects noted for dot patterns are also seen for these, for example, the maximum in the prototype effect shown in figure 11.16. Suppose we change the probability of changing a character in the strings used in table 11.1. If the probability is zero, the examples and the prototype will be identical, and one will not be enhanced relative to the other. Just as in the analysis of the dot patterns, the prototype effect for character strings passes through a maximum as the strings get noisier and noisier. The prototype effect is the ratio of the response (length

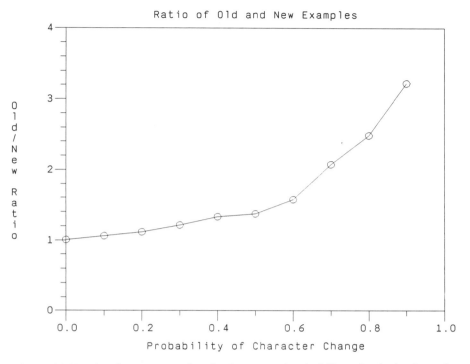

Figure 11.17 As the amount of noise increases (probability of substitution of a random character), the network responds more strongly to old examples and less strongly to new ones. The network is becoming biased toward learning specific examples, and less generalization is occurring.

of output vector) of the network to the prototype to old examples. When we have no character substitutions, that is, no distortion, this ratio is 1.0, since the input vectors and the prototype are the same.

When distortion becomes so large as to produce examples that are completely new strings, there is no prototype effect, since the prototype will be learning a new random string. At an intermediate value of distortion, as shown in figure 11.16, the prototype effect is maximum, since cooperative averaging occurs to form the prototype, and the response to the old examples is reduced because they differ significantly from each other. In this simulation the maximum enhancement is about 30% when a probability of about 0.5 exists that a character will be substituted in the prototype string. This is comparable in magnitude to what was found for the dot patterns.

As the examples become less and less like the prototype string, as more and more random characters are substituted, the network effectively learns the old examples as individuals. Therefore the response to old examples should be stronger than to new examples, as the probability of a character substitution increases. Figure 11.17 plots the ratio of the lengths of learned vectors (old examples) to new examples. When the chance of changing a character in a string is 0.9, new examples (average output length 5.7) are nearly indistinguishable from completely random strings (average output length 5.0). This shift reflects the underlying shift from an averaging prototype model at low distortions to an exemplar model at high distortions, as predicted.

Moving Objects

In this part of the chapter we will discuss a model for determining the speed and direction of motion of a large moving object. This model was developed by Sereno (1989, 1993) and provides an elegant example of how to construct a very large neural network to do a computation. It was originally developed as a model for human performance and has been used to explain human psychophysical data on motion perception. It has also been used as the basis for interpreting neurophysiological data and for suggesting physiological experiments. More details can be found in Sereno (1993). It should be possible to adapt some of these ideas to other kinds of sensory processing. This model may serve as a particularly understandable example of how a neural network might reliably and quickly compute a complex perceptual property.

The Aperture Problem

Single units early in the visual system have small receptive fields, that is, they see only small parts of the visual field. Somehow the perceptual apparatus has to put together information from many receptors to give rise to the perception of the large-scale structure of the visual world. Imagine looking at the world through a straw to get an idea of the problem involved. Exactly the same problem occurs in other sensory systems and in many artificial systems in which local information from a number of modalities must be integrated.

The visual system could compute object movement in a number of ways. Anyone who has taken a calculus course could suggest one: take two snapshots of a scene in quick succession and look for changes between them. The distance an object moved divided by the time between snapshots would be a measure of object velocity. However, this idea is hard to implement in practice. To determine the distance moved, it is necessary to identify corresponding points in the two images. Solving the *correspondence problem* turns out to be remarkably difficult for real images.

It seems easier to build receptors that detect local motion. When neuroscientists first looked at responses of single units in the retina of animals, and by extension our own, they discovered many cells that were far more sensitive to *changes* in light intensity than to a constant light level. We saw in chapter 4 that the *Limulus* responds most strongly to spatial boundaries. The same is true in the time domain; cells in both the *Limulus* and the vertebrate retina respond most vigorously to changes in light intensity with time. Our eyes are so concerned with intensity changes, and not intensity, that if an image is *stabilized* on the retina so it does not shift with respect to the receptors, the unchanging image will subjectively disappear after a few seconds.

Neurons in the retina of many animals respond to the speed and direction of *local* motion. We have a number of ways to compute local motion. One technique involves asymmetrical delayed lateral inhibition (Barlow and Levick, 1965). Figure 11.18 shows a simple neural circuit that will provide motion selectivity in the direction shown by the arrow. In one direction, a small moving object will excite the receptor. By the time it excites the in-

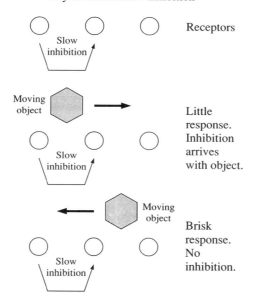

Figure 11.18 One of many simple neural circuits producing direction selectivity for local motion. It uses asymmetrical inhibition. This simple model is similar to ones proposed by Barlow and Levick (1965) and Hassenstein and Reichardt (1956).

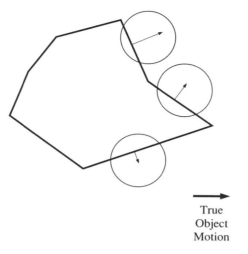

True
Object
Motion

Figure 11.19 The *aperture problem*. A two-dimensional pattern moves rightward. Motion is assumed to be in a plane perpendicular to the viewer. At each contour, local motion detectors can measure only the perpendicular component of motion.

hibiting neuron, the receptor has already responded vigorously. In the other direction, the moving object first excites the inhibiting cell. By the time the object gets to the receptor, the receptor is inhibited and does not respond. This produces units that respond to speed of motion as well as direction, because a rapidly moving object might reach the receptor before the inhibition arrives, or a slowly moving object might excite the receptor after the peak of inhibition has occurred.

This technique works when estimating direction and speed of motion for objects that a receptor sees in its entirety. However, a large object will be seen by many receptors. Consider a moving edge. Suppose we view it as a local motion detector would see it through a small sperture (figure 11.19). The edge is represented as a featureless line. Without local texture to go on, a local motion-sensitive unit will respond only to the *normal* component of motion; that is, it will give the strongest response when the edge is moving perpendicular to the cell's best orientation, in the direction of the arrows in figure 11.19. If the line is moved in different directions while its orientation is held constant, it appears to move in a direction perpendicular to its orientation with the speed given by the component of motion in the perpendicular direction. Subjectively, we as observers respond to motion perpendicular to the edge; single units in our retina behave the same way.

A problem now arises. Consider the complex moving object in figure 11.19 that extends over a number of receptors. We have drawn this object as a polygon, where each edge is larger than the receptive field of a single

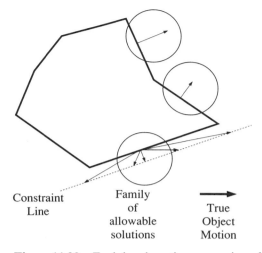

Constraint
Line

Family
of
allowable
solutions

True
Object
Motion

Figure 11.20   Each local motion sensor is *ambiguous* with respect to object motion. A family of possible pattern motions consistent with the observed motion is represented by vectors whose tips lie along a *constraint line*.

receptor. The set of receptors responding to local movement will be responding to movement of the edges, but they will each respond best to a *different* direction of movement. They are signaling movement, but not the actual movement of the object. The task of detecting the true motion of an oriented line segment in a complex extended object is known as the *aperture* problem.

The aperture problem affects other sensory systems. Large objects contacting the skin, and in the auditory system, correlated movements of bands of frequencies, common in speech signals, have similar difficulties.

We will consider only rigid movement in a plane in front of the observer, perpendicular to the line of sight. In figure 11.20, a two-dimensional pattern consisting of oriented contours moves rightward with speed and direction indicated by the arrow located at the corner of the pattern. At each contour, local movement detectors, represented by circular apertures, measure the perpendicular component of motion. A *family* of possible pattern motions consistent with the perpendicular motion is represented by vectors whose tips lie along what is called the *constraint line*. The length of each vector gives its speed, and its angle gives its direction. Therefore, a given local motion signal is consistent with *many* different object motions. The object could be moving in the direction of the normal. Or it could be moving at a higher velocity in a direction away from the normal (Adelson and Movshon, 1982; Albright, 1984; Hildreth, 1984; Movshon et al., 1985).

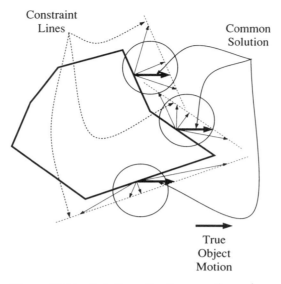

Constraint
Lines

Common
Solution

True
Object
Motion

Figure 11.21   Solution of pattern motion using constraint lines. Even though each local motion sensor is ambiguous, one solution is common to every family (Adelson and Movshon, 1982).

A simple formal solution to the aperture problem is shown in figure 11.21 (Fennema and Thompson, 1979; Adelson and Movshon, 1982). This solution is also consistent with what humans perceive when their motion perceptions are studied psychophysically. A given local motion signal is consistent with many different object motions, that is, the sensor output is highly ambiguous. The problem for the network is one of disambiguation. At each contour, the constraint lines indicate the family of global pattern velocities consistent with the local motion. Suppose we look at the families of pattern velocities produced by several edges of the same object. *Only one vector is common to all the families.* This common vector correctly describes the perceived motion of the entire pattern.

A great deal is now known about the physiology of the parts of the nervous system that seem to be doing the computation. See Van Essen (1985), Maunsell and Newsome (1987), and Van Essen, Anderson, and Felleman (1992) for relevant anatomy and physiology.

Neurons in primary visual cortex (cortical area V1) respond to the orientation of a line segment. Many of them are also selective for movement perpendicular to that edge, that is, they have both a best speed and a best direction. Units in V1 respond to the component of motion perpendicular to the preferred orientation of the units, that is, they are subject to the aperture problem. Cortical area MT is upstream of V1, and extensive evidence indicates that it is deeply involved in motion processing. Area MT receives a

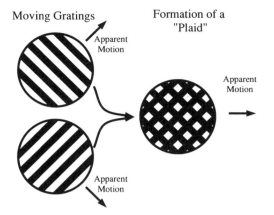

Figure 11.22 Formation of a plaid from moving gratings. The perceived direction of motion is shown by the arrows.

projection from V1 (see figures 10.8 and 10.9). Many cells in area MT are selective for the speed and direction of an object, while having little selectivity for spatial structure. Area MT has larger excitatory receptive fields than V1, suggesting converging inputs from V1 units.

Physiological evidence seems to suggest that MT units see the way we do, in that they have solved the aperture problem, whereas V1 units are subject to the aperture problem. (Speculations about why and how our conscious perceptions seem more like MT than another area of cortex, V1, a few centimeters away, will be firmly discouraged.) We can infer this result from data obtained by some elegant physiological experiments (Movshon et al., 1985).

Consider the two moving gratings in figure 11.22. When a grating moves in the direction of the arrow, we have the sensation of movement in that direction. However, when gratings are superimposed, a new pattern is formed, called a *plaid*. The plaid gives the impression of a uniform object moving to the right. It makes no subjective sense that this pattern is the superposition of two component movements.

We can record from single units responsive to motion in areas MT and V1 and see how they respond to these stimuli. When we show the animal a simple moving grating, the cells show a maximum response in the direction of motion. However, when we show the animal a plaid, the responses of cells in the two regions differ. Cells in V1 show not one but *two* maxima in their response. In the V1 cell whose responses are shown in figure 11.23, the maximum responses are 90 degrees apart, corresponding to the angular separation of the two components of the plaid. The interpretation of this result is that the cell is responding to the component motions of the plaid.

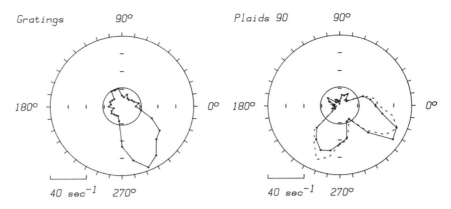

Figure 11.23   Directional selectivity of a special complex cell recorded in area 17 (V1) of a cat. The spatial frequency of the stimulus grating was 1.2 c/deg, and the drift rate was 4 Hz. On the left is the neuron's tuning for the direction of motion of single gratings and on the right is the neuron's response to moving 90-degree plaids. The dashed curve on the right shows the expected response of a component direction selective neuron. The outer circles in each plot show the expected response of a component direction-selective neuron. The inner circles in each plot show the neuron's maintained discharge level. The distance from the origin indicates the amount of unit activity in spikes per second (see scale), and the direction is the direction of motion of the stimulus. From Movshon, Adelson, Gizzi, and Newsome (1985). Reprinted by permission.

This *does not* correspond to our subjective perception. The MT cells shown in figure 11.24 respond to gratings the same way V1 cells do. However, their response to the plaid displays a single maximum, unlike the V1 cells, which is like our subjective impression of the plaid. Somehow MT has solved the aperture problem and gives the appropriate response to large moving objects.

There are not many synapses between V1 and MT. This suggests that the object motion computation may not be very difficult. Sereno noted that the constraint line model can be realized by a simple neural network. The structure of that neural network can be inferred from the nature of the task; learning is not actually necessary. However, the network can develop a similar solution by seeing examples; the transformation seems to be an easy one to learn with the proper network structure. The learned and inferred solutions are essentially identical. Therefore, there is no mystery in the way the computation is performed. We will describe the inferred solution first, and then show how it can be learned.

Units in both V1 and MT are responsive to both speed and direction. Figure 11.25 shows experimental data taken from MT units to the direction of motion. Figure 11.26 shows experimental data from MT units for the

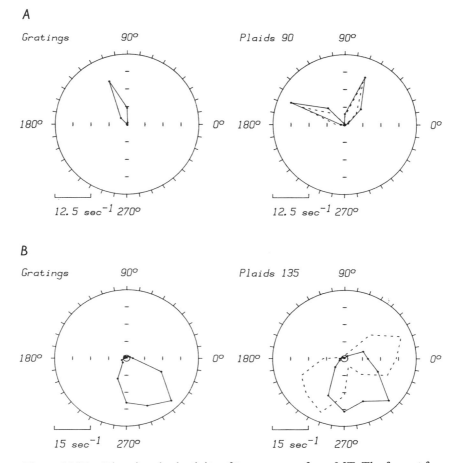

Figure 11.24 Directional selectivity of two neurons from MT. The format for each figure is the same as in figure 11.23. A, Spatial frequency 3.6 c/deg, drift rate 4 Hz. B, Spatial frequency 2.7 c/deg, drift rate 4 Hz. The distance from the origin indicates amount of unit activity in spikes per second (see scale), and the direction is the direction of motion of the stimulus. From Movshon, Adelson, Gizzi, and Newsome (1985). Reprinted by permission.

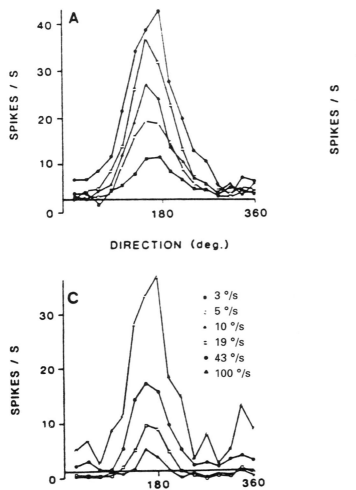
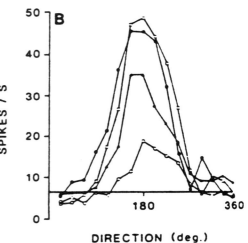

Figure 11.25 Experimental response of units in MT to direction of motion. Direction tuning curves for three representative MT neurons were tested with a moving slit at different speeds. Position of the *x* axis indicated level of spontaneous activity. The labeling conventions given in plot C apply to all three cells. For purposes of comparison the set of curves for each cell has been shifted so that the cell's best response corresponds to a stimulus moving at 180 degrees relative to an upward movement. From Rodman and Albright (1987, figure 3). Copyright 1987, with kind permission from Pergamon Press Ltd.

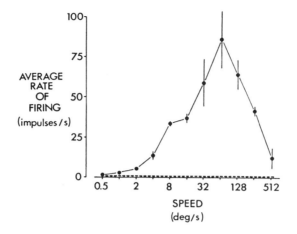

Figure 11.26 Typical experimental response of units in MT to speed of motion. Responses of a representative unit in MT to stimuli moving in its preferred direction at different speeds. The speed axis is logarithmic. Bars indicate the standard error of the mean for five repetitions of each speed. A dashed line indicates the background rate of firing. This unit, like most in MT, had a sharp peak in its response curve. The receptive field was 15 degrees across and each stimulus traversed 20 degrees. Response of MT units to different speeds of motion. From Maunsell and Van Essen (1983). Reprinted by permission.

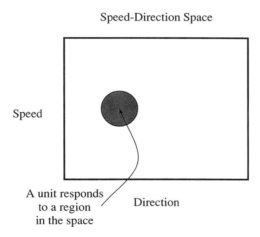

Figure 11.27 Units respond to two aspects of the stimulus: speed of motion and direction of motion. Therefore a unit responds best to a region in two-dimensional speed-direction space.

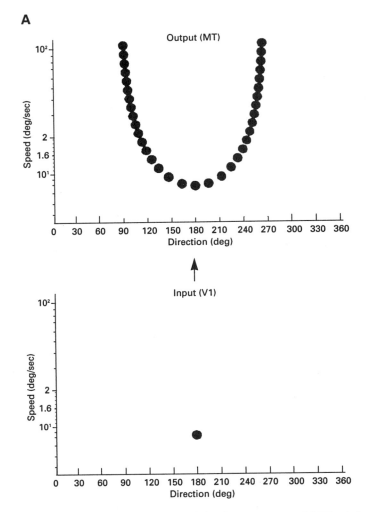

Figure 11.28 Inferred connectivity between a model V1 region and a model MT region. A possible neurophysiological implementation of the aperture problem. Each coordinate in the lower box represents an input layer component motion neuron with speed and direction selectivity defined by its value of *x* (direction) and *y* (speed) coordinates. Each coordinate in the upper box represents an output layer pattern motion neuron with particular speed and direction selectivity. A, Projective field of a single component motion neuron. This neuron has a direction preference of 180 degrees and a speed preference of 8 deg/sec, and projects to units belonging to a family of possible pattern velocities consistent with its velocity. B, Receptive field of a single pattern motion neuron. This neuron has a preferred direction and speed of 180 degrees and 64 deg/sec, respectively. It receives inputs from units with component velocities that may be contained in a pattern moving with the velocity of the pattern motion unit. From Sereno (1993). Reprinted by permission.

B

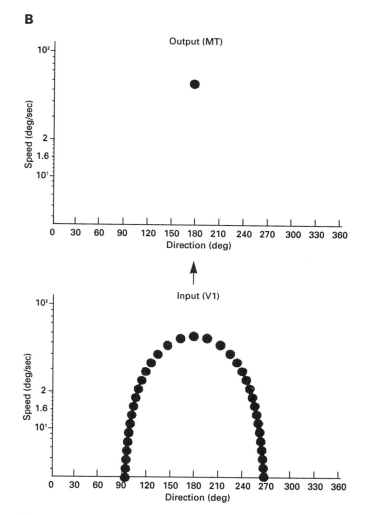

Figure 11.28 (continued)

speed of motion. The V1 units are similar but less broadly tuned (Rodman and Albright, 1987; Maunsell and Van Essen, 1983). The response of a single unit could be considered to be a response to a stimulus that falls in a *region* of a two-dimensional speed-direction space (figure 11.27).

When an input unit responds, because of the ambiguity in its response, many MT units are consistent with its activity. To be consistent with the constraint line computation, an input unit in V1 should project to a large number of MT units (figure 11.28A). The horseshoe shape is given by a cosine function, since the unit is responding to the projection of a motion vector onto the vector normal to the edge that is what the input unit actually "sees." This connectivity pattern could be called the *projective field* of

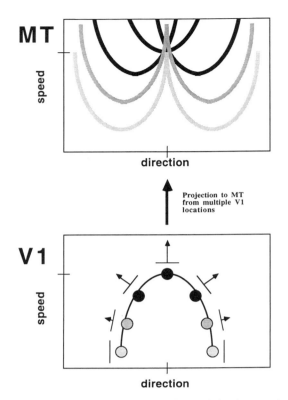

Figure 11.29   Basic mechanism of the integrative solution found by the network model. Schematic diagram depicts a neural implementation of a constraint line solution to the aperture problem. The velocity of a translating textured pattern is represented by the tick marks on the speed and direction axes of the V1 and MT maps. Such a pattern activates a set of units in the V1 map. Each unit is selective for a particular component motion, represented by the vector adjacent to each circular unit. Each V1 unit in turn projects to a set of units in the MT map represented by a U-shaped arc of the same shade of gray. The intersection of the U-shaped projective fields in the MT map represents the most active MT unit, which has a velocity preference identical to the pattern's velocity. From Sereno (1993). Reprinted by permission.

the V1 unit. We can also determine the *receptive field* of the MT unit by an inverse argument, that many different local signals from V1 are consistent with a given MT speed and direction. The receptive field of an MT unit is shown in figure 11.28B.

When V1 units represent an object with many unit activities, each ambiguous, the "horseshoes" will overlap at one point in MT, which is the true speed and direction, the common interpretation (figure 11.29). This inferred connectivity pattern is the inferred solution to the problem. Obviously, the more input activity and the more texture on the object, the more cells cooperate and the better the integration of speed and direction to form a consensus interpretation.

One can make a couple of additional comments about the individual units. We saw that both V1 and MT units are quite broadly tuned. The unit responses, and, by extension, the data representation, are also nonlinear and nonmonotonic, referred to the input parameters. For example, a unit will start to respond as speed increases, respond maximally, and then cease responding. This pattern does not present any particular conceptual problems for speed and direction; the term "best" speed and direction suggests it. However, our intuitions can lead us into serious difficulties in other sensory systems, for example, the auditory cortex units we discussed in chapter 10 (see figure 10.23).

The Motion Network

It is now possible to make a network model for object motion. The network Sereno proposed had two layers, an input layer of V1 neurons responding to local motion and an output layer of MT neurons responding to object motion. Because the units are quite broadly tuned, a particular velocity at a particular point in space will excite many units and is represented as a distributed pattern of activity across the units. To signal local motion at a point in visual space, we must have a group of cells, each tuned to part of the entire space of speed-direction parameters. Then the local motion signal can be determined from a population response, with some units in a group active, and with others silent. There are many points in visual space, and the object motion determination will require the cooperation of a number of groups of input units, each group doing the local computation at a specific point.

In figure 11.30, a number of such groups of units form the input layer. The individual units are the circles within each cluster. The input layer, modeled after area V1 neurons, has component motion units selective for velocities perpendicular to their preferred orientation. Each group of input units has

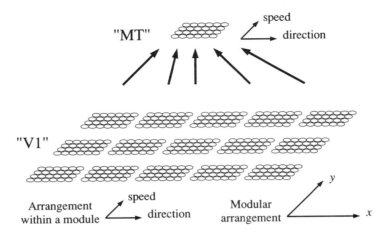

Figure 11.30 Network model of the aperture problem. There are two layers of units. The input layer has component motion units with different preferred directions and speeds. A number of different spatial locations are represented each with a full complement of receptor speeds and directions. The output layer, modeled after MT neurons, has pattern motion units selective for different directions and speeds of pattern motion. Redrawn from Sereno (1989).

units tuned to a complete spectrum of directions and speeds. Different clusters respond to different regions of the visual field, that is, a cluster has x and y coordinates. The output layer, modeled after area MT neurons, has pattern motion units selective for different directions and speeds. The spatial resolution of the output layer is poor because it is integrating information from a number of different input locations. The output representation is distributed. The consensus output speed and direction computed in the simulation are computed from the weighted output of the model MT units. The overall response of each unit is the multiplication of its individual responses to direction and speed. The units are broadly tuned so that a single unit responds to a wide range of speeds, and a single speed is represented by several active units. The output representation is important in this network. Trying to train the network with a grandmother cell output representation does not work. (It has been attempted.) A distributed output representation is essential. Figure 11.31 shows the response patterns of the artificial units used in the simulations.

We should note that the idea that each point in space is looked at by a large group of analyzers occurs in neurophysiology, where it is sometimes referred to as a *hypercolumn*. The implication is that every point is fully analyzed by several classes of selective units. This idea also suggests that there should be many more cells in V1 than there are retinal inputs. The human retina has perhaps 1 million fibers and V1 has perhaps 150 million

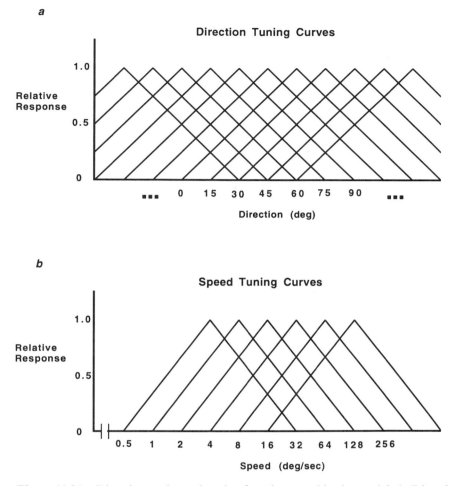

Figure 11.31 Direction and speed tuning functions used in the model. A, Direction tuning curves with one-half response bandwidths of 60 degrees and peaks at 15-degree intervals. B, Speed tuning curves range from 4 to 128 deg/sec on a log scale. From Sereno (1993). Reprinted by permission.

units, although it is hard to draw precise implications from such numbers because some cortical units may serve other functions.

Simulation Results

Sereno used the linear associator as described in chapters 6 and 7, combined with simple gradient descent error correction, to see if it was possible to learn a good solution to the aperture problem by adjusting the set of weights between the V1 and the MT units. We already have an idea of the connectivity of the ideal solution. The network is determined in that we know what the desired output should be in response to a given input. The network was presented with a large number of input patterns. For each pattern, activity in the input layer was propagated linearly to the output layer:

$$\mathbf{g}'_i = \mathbf{A}\mathbf{f}_i,$$

where $\mathbf{g}'_i$ is the vector of output layer neuron activities resulting when a pattern $\mathbf{f}_i$ is input to the system, and $\mathbf{A}$ is a matrix of connection strengths between input and output layers. Actual output ($\mathbf{g}'$) is compared with correct output ($\mathbf{g}$) for that pattern, and all weights in the system are then adjusted to reduce error between actual and correct output using the vector form of the Widrow-Hoff learning rule, as described in chapter 9.

Because these simulations involved so many units, they were initially done on a supercomputer at the late lamented John von Neumann Supercomputer Center in Princeton, New Jersey. A representative simulation used units having 24 direction tuning curves, spanning 360 degrees with peaks at 15-degree intervals, and half-response widths of 60 degrees, and with 6 speed tuning curves, spanning 128 degrees/second with peaks at 1-octave intervals and half-response widths of 3 octaves. Since each unit was sensitive to a different combination of preferred speed and direction, a total of 144 units was used for each spatial location. The model used 16 sets of V1 units, that is, 16 locations, and 1 set of MT units. The system was trained on 360 patterns and tested on the same patterns plus 360 new ones. The input patterns were textures composed of 16 line segments, 1 line segment per location in the input network (figure 11.32).

Sereno found that this was an easy transformation to learn. After only 15 passes through the training set, performance reached a plateau. The mean cosine of the angle between the correct output vector and the actual output vector that the system produced was 0.98; the mean cosine of new, non-associated vectors was 0.93. This represents good performance, and demonstrates that the system generalizes well to stimuli it has never seen.

These values correspond to an error in estimating speed of about 10% and of direction of about 15%, roughly consistent with human performance. One network prediction is that the more input groups used to estimate object motion, the more accurate should be the estimation. A series of simulations systematically changing the number of groups from 1 to 32 shows this effect clearly (figure 11.33). This point is worth making, because the network becomes more accurate as it grows larger. Because of simple parallelism, no increase in processing time for a large network is necessary if parallel hardware can be used. The largest simulation in this series used nearly 4000 units to provide an estimate of object motion in a single region of space.

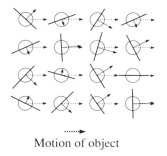

Motion of object

Figure 11.32   Example pattern used to train the system in the simulations. In this case 16 hypercolumns are looking at lines on the moving object. The motion of the pattern is rightward, indicated by the dotted arrows. The component motion detected by each aperture, that is, the normal component of the actual object motion to the line, is represented by the length and orientation of the solid arrows. Redrawn from Sereno (1989).

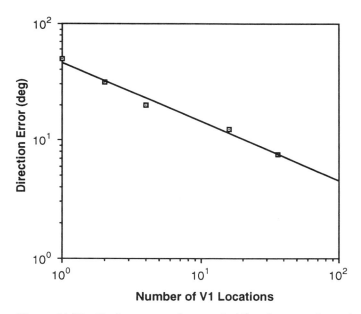

Figure 11.33   Performance of network (direction error) on the test set of vectors as a function of the number of input locations. Direction error for new patterns, plotted against number of input locations using log/log coordinates. A best-fitting line was drawn through the data points and extrapolated to 100 input locations. From Sereno (1993). Reprinted by permission.

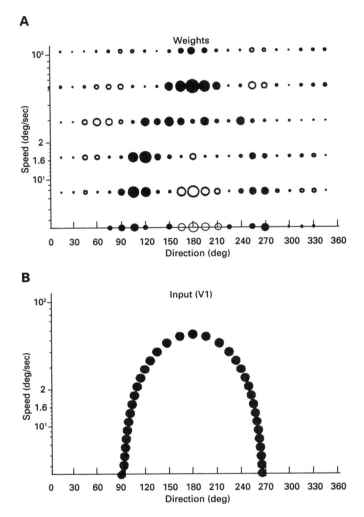

Figure 11.34   This set of connection strengths developed by the network should be compared with the inferred pattern of connections in figure 11.28. Receptive field of a single pattern motion unit. A, Actual connection strengths between one output-layer unit (with a preferred direction of 180 degrees and speed of 64 deg/sec) and a complete set of 144 V1 units at one location. Each circle represents the weight between a V1 unit and MT unit: the V1 unit's direction and speed is determined by its *x* and *y* coordinates, the strength of connection by the size of the circle, and the kind of connection (positive or negative) by whether the circle is filled or open, respectively. The positive weights have an inverted V-shaped appearance. B, Predicted connection strengths. From Sereno (1992). Reprinted by permission.

Obviously we can simultaneously perceive object motion in several regions of the visual field. Therefore this network forms only a single module of a complete motion system. Presumably it is duplicated many times. This computation may indicate why it is so important to use large networks, and perhaps suggests one reason the brain is so large. Here it is obvious that bigger is better. Increased network size does not give rise to a combinatorial explosion of difficulty in the computation, but to increased accuracy.

Sereno examined the resulting network weights to determine whether the solution developed by the system to solve the aperture problem resembled the inferred solution. Figure 11.34 shows the weights between a complete set of 144 V1 units at one location and a single MT unit. Each circle represents the weight between a V1 unit and MT unit: the V1 unit's direction and speed is determined by its x and y coordinates, the strength of connection by the size of the circle, and the type of connection (positive or negative) by whether the circle is filled or open, respectively. If the system found the guessed solution, the MT unit's receptive field should be similar to the one illustrated in figure 11.28. The connection patterns are very similar. The MT unit has strong positive weights to V1 units, with component motions corresponding to consistent inputs.

It is possible to use this model to make predictions about a number of experimental effects in both physiology and psychophysics with encouraging results. For example, the simulation suggests that MT units should have a wider range and higher preferred speeds than V1 units, a finding consistent with neurophysiological data. It suggests that object motion estimation results from integration of local motion responses over an area instead of along connected contours. There are also some strong predictions about abilities of both the network and humans to estimate velocities of movement. We will not develop this aspect of the model further, since we were interested in the network as an example of how an important perceptual computation using a very large network might be organized. Application of the model to human performance is described in detail in Sereno's 1993 book.

Conclusions

In this chapter we applied a simple neural network, the linear a sociator, to two difficult problems. In both applications, the data representation was more important than details of the learning rules. In both, the solution makes good use of the ability of neural networks to integrate information from a number of input units. Although determination of motion is superficially quite unlike concept learning, in fact the critical mechanisms are identical: adding up a bunch of noisy and/or ambiguous inputs can give the correct answer.

The key requirement in both models is the proper form for the data representation. Once that is determined, most reasonable network learning rules would be able to come up with an adequate set of connection strengths. In the motion network, Sereno first used the Widrow-Hoff rule in the network described here. However, her brother (M.I. Sereno, 1989) showed the network would also work using straightforward outer product Hebbian learning, although the connection patterns developed by the model's MT units did not reflect the guessed solution quite so neatly. The model has also been applied to detection of rotation, performed in the area beyond MT (Sereno and Sereno, 1991).

In these, our first real examples of network computation, we tried to show that some basic principles for representing information in simple neural networks are useful. First, topographic maps with localized activities to code continuous magnitudes are effective in practice. They allow natural generalization and estimation. Unfortunately, these data representations require very large numbers of units. They also are highly nonlinear *relative to the input stimulus dimensions*. It may make more sense to have a highly nonlinear representation combined with a linear or simple nonlinear network than to represent stimulus magnitudes directly, and to place all the significant nonlinear operations in the network. Second, simple neural networks can make effective use of very large amounts of learned information. The computations can be truly consensual and cooperative. This result should bode well for the functioning of extremely large simple nets.

More complex multilayer nets need not have this desirable scaling property. If the representation of the data or the problem can be made good enough to avoid the use of complex networks, very large nets can be intelligently designed. They can be particularly effective for neural network signal processing. Both our examples demonstrate a particular approach, although it is not the kind of signal processing commonly discussed in textbooks.

References

M. Abramowitz and I.A. Stegun, Eds., (1964), *Handbook of Mathematical Functions. National Bureau of Standards Applied Mathematics Series, No.* 55. Washington, DC: U.S. Government Printing Office.

E.H. Adelson and J.A. Movshon (1982), Phenomenal coherence of moving visual patterns. *Nature, 300,* 523–525.

T.D. Albright (1984), Direction and orientation selective neurons in visual area MT of the macaque. *Journal of Neurophysiology, 52,* 1106–1130.

J.A. Anderson, M.L. Rossen, S.R. Viscuso, and M.E. Sereno (1990), Experiments with representation in neural networks: Object motion, speech, and arithmetic. In H. Haken and M. Stadler (Eds.), *Synergetics of Cognition,* Berlin: Springer-Verlag.

H.B. Barlow and R.W. Levick (1965), The mechanism of directional selectivity in the rabbit's retina. *Journal of Physiology, 173,* 477–504.

C.L. Fennema and W.B. Thompson (1979), Velocity determination in scenes containing multiple moving objects *Computer Graphics and Image Processing,* 9, 301–315.

I.S. Gradshteyn and I.M. Ryzhik (1965), *Table of Integrals, Series, and Products 4th ed.* New York: Academic Press.

E.C. Hildreth (1984), *The Measurement of Visual Motion.* Cambridge. MIT Press.

A.G. Knapp and J.A. Anderson (1984), Theory of categorization based on distributed memory storage. *Journal of Experimental Psychology: Learning, Memory, and Cognition, 10,* 616–637.

G.A. Korn and T.M. Korn (1968), *Mathematical Handbook for Scientists and Engineers, 2nd ed.* New York: McGraw-Hill.

J.K. Kruschke (1992), ALCOVE: An exemplar based connectionist model of category learning. *Psychological Review.* 99, 22–44.

J.L. McClelland and D.E. Rumelhart (1985), Distributed memory and the representation of general and specific memory. *Journal of Experimental Psychology: General, 114,* 159–188.

J.T. McIlwain (1975), Visual receptive fields and their images in the superior colliculus of the cat. *Journal of Neurophysiology, 38,* 219–230.

J.H.R. Maunsell and W.T. Newsome (1987), Visual processing in monkey extrastriate cortex. *Annual Review of Neuroscience, 10,* 363–401.

J.H.R. Maunsell and D.C. Van Essen (1983), Functional properties of neurons in the middle temporal visual area of monkeys. I. Selectivity for stimulus direction, velocity, and orientation. *Journal of Neurophysiology, 49,* 1127–1147.

C.B. Mervis and E. Rosch (1981), Categorization of natural objects. *Annual Review of Psychology, 32,* 89–115.

J.A. Movshon, E.H. Adelson, M.S. Gizzi, and W.T. Newsome (1985), The analysis of moving visual patterns. In C. Chagas, R. Gattas, and C.G. Gross (Eds.), *Pattern Recognition Mechanisms.* Rome: Vatican Press.

M.I. Posner (1969), Abstraction and the process of recognition. In J.T. Spence and G.H. Bower (Eds.), *Advances in Learning and Motivation*, Vol. 3. New York: Academic Press.

M.I. Posner and S.W. Keele (1968), On the genesis of abstract ideas. *Journal of Experimental Psychology, 77*, 353–363.

M.I. Posner and S.W Keele (1970), Retention of abstract ideas. *Journal of Experimental Psychology, 83*, 304–308.

H.R. Rodman and T.D. Albright (1987), Coding of visual stimulus velocity in area MT of the macaque. *Vision Research, 27*, 2035–2048.

E. Rosch (1973), On the internal structure of perceptual and semantic categories. In T.E. Moore (Ed.), *Cognitive Development and the Acquisition of Language*, New York: Academic Press.

E. Rosch (1975), Cognitive representations of semantic categories. *Journal of Experimental Psychology: General, 104*, 192–233.

E. Rosch (1978), Principles of categorization. In E. Rosch and B.B. Lloyd (Eds.), *Cognition and Categorization*. Hillsdale, NJ: Erlbaum.

M.E. Sereno (1989), A neural network model of visual motion processing. Doctoral thesis, Brown University, Providence, RI.

M.E. Sereno (1993), *Neural Computation of Pattern Motion*. Cambridge: MIT Press.

M.I. Sereno (1989), Learning the solution to the aperture problem for pattern motion with a Hebb rule. In D.S. Touretsky (Ed.), *Advances in Neural Information Processing Systems 1*. San Mateo, CA: Morgan Kauffman.

M.I. Sereno and M.E. Sereno (1991), Learning to see rotation and dilation with a Hebb rule. In R.P. Lippmann, J. Moody, and D.S. Touretsky (Eds.), *Advances in Neural Information Processing 3*. San Mateo, CA: Morgan Kauffman.

E.E. Smith (1989), Concepts and induction. In M.I. Posner (Ed.), *Foundations of Cognitive Science*, Cambridge, MA: MIT Press.

E.E. Smith and D.L. Medin (1981), *Categories and Concepts*. Cambridge: Harvard University Press.

D.C. Van Essen (1985), Functional organization of primary visual cortex. In A. Peters and E.G. Jones (Eds.), *Cerebral Cortex*, Vol. 3. *Visual Cortex*. New York: Plenum Press.

D.C. Van Essen, C.H. Anderson, and D.J. Felleman (1992), Information processing in the primate visual system: An integrated systems perspective. *Science, 255*, 419–423.

Energy and Neural Networks: Hopfield Networks and Boltzmann Machines

Down in the valley
Valley so low ...
Trad. song

The bear went over the mountain,
The bear went over the mountain,
The bear went over the mountain,
To see what he could see ...
Trad. song

Hopfield networks are autoassociative nonlinear networks with dynamics that can be interpreted as minimizing a system energy *function. A simple* asynchronous updating *rule can be shown to decrease system energy monotonically, leading the system state to a local energy minima. A simple Hebbian learning rule, among others, can place the local minima at the location of stored patterns. When too many patterns are stored, the network randomizes and can no longer serve as a memory. An extension of the Hopfield network, the* Boltzmann machine *has dynamics capable of finding the true global energy minimum by a process akin to* simulated annealing. *Probabilistic neurons sometimes can act to increase system energy, and the chances of doing so are dependent on a parameter usually called* temperature. *A simulation of both a Hopfield network and a Boltzmann machine is presented.*

In 1982 John Hopfield published a short, clearly written paper that had a major impact on the field. It presented a sophisticated, coherent, theoretical picture of the way a neural network might work.

Neural networks are complex and often contain nonlinear components. Their behavior is difficult to analyze. Hopfield applied some ideas to them from an important and developing area of mathematics called *nonlinear dynamical system theory*. He was able to show that sometimes a nonlinear neural network evolving in time could be analyzed qualitatively, by showing that as it evolved, it was minimizing a particular function related to what would be called energy in a physical system. From extreme complexity, a kind of simplicity emerged. Others had made one or another of these points before, but Hopfield brought them together in a visible and prestigious forum.

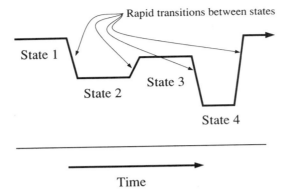

Figure 12.1   Perhaps mental life proceeds by way of movement from stable state to stable state, with quick transitions between them. Nonlinear dynamical systems can display qualitative behavior like this since they can develop stable *attractors* in state space where the system state remains for a long period.

One of the interesting aspects of the Hopfield paper is the order in which he presents his ideas. One approach to understanding a network is the one we have followed to some degree in this book. We suggest a learning rule, or a network architecture, and show what happens. The focus is on input-output relations. A common design goal is to construct a network that gives the most accurate output for a given input. Hopfield also started by discussing the desired goal of a network computation, but he emphasized the temporal aspect of network behavior. We can view the activity pattern, the state vector, in a network as changing in time. Suppose the dynamics of the network are such that the state vector spends most of its time at or near a few locally stable states. Suppose we can control where these locally stable points are by adjusting connection strengths in the network (figure 12.1).

This idea has an agreeable feel. Certainly, subjectively we believe that meaningful mental states somehow are solid, coherent, and long lasting. Something like this behavior has been suggested by a number of modelers. Psychological measurement suggests that most significant mental operations take 100 msec or more, suggesting a time course involving network operations with significant temporal integration and relatively long-term stability.

Nonlinear dynamical systems can show this kind of behavior. If the system started at other points in state space, the dynamics of the system cause the state to flow into the stable points, called *attractors* in nonlinear system theory. The system state will move away from other points, sometimes called *repellors*.

Perhaps we could arrange it so the stable points would give the results of computations. Hopfield provided a couple of examples of how such a system

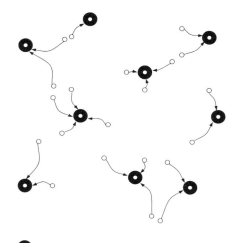

● = Legal Computer Word (Attractor)
○ = Word with Errors

Figure 12.2 An attractor network such as a Hopfield network could be used to correct errors. Suppose a network is constructed so only widely spaced legal bit patterns are attractors. If a few bits are in error in a computer word, the erroneous system state will move to the nearest legal bit pattern, correcting the error.

might perform a practical function. One application in particular is worth mentioning: error-correcting codes. Suppose data in a computer are constrained to allow only certain patterns of ones and zeros. An example of this is the parity bit in some personal computers. Memory is organized in eight-bit bytes. Sometimes, an additional *parity* bit is present to be set to 1 if the number of 1s in the byte is odd (odd parity) or even (even parity). If the parity bit does not agree with the number of ones in the byte, there must have been an error. The parity bit can detect single-bit errors, but, of course, fails if two bits are in error.

One can generalize this idea. Suppose we have a data word or group of words. Computations based on abstract algebra can be used to attach numerous bits to the data that not only detect errors, but also tell which bit is in error and thereby correct it. Therefore only some patterns of ones and zeros are allowable. For parity, only half the possible patterns exist; for error correction a very much smaller number is allowable.

A Hopfield network could be used to perform error correction, although in fact, it is not very good at it. (Algebraic techniques are faster and more reliable.) If the allowable bit patterns are placed at attractors, a pattern with errors will be converted to the nearest allowable word by the dynamics of the system, correcting the error. If there are too many bits in error, the word may go to the wrong allowable pattern, true for any error-correcting system (figure 12.2).

Attractors can also be used to reconstruct missing information, since the stable point may complete missing parts of an intially incomplete or noisy state vector. What is often called *associative memory* in the Hopfield network literature is based on this reconstruction property. From the point of view of a cognitive scientist, the definition of associative memory in this literature and the resulting view of human memory is a little naive (see chapter 16). Because of the importance of the stable points—attractors—Hopfield networks and a number of other networks that display similar behavior are sometimes called *attractor networks*. Amit (1989) provides the most detailed description of some of the qualitative properties of attractor networks.

Analysis of the Hopfield Net

To maintain consistent notation, in this book we denote the connection matrix as **A** and the state vector as **f**. In Hopfield's original paper, and in many papers since in this tradition, the connection matrix is denoted as **T** and the activity pattern as **V**.

In 1982 Hopfield assumed that the basic elements of the network were threshold logic units, essentially McCulloch-Pitts neurons. The units that summed synaptic inputs compared the sum with a threshold, θ, and responded 1 if the sum was at or above the threshold, and zero otherwise. If $f[i]$ is the activity of a single unit and $A[j,i]$ is the connection between the ith unit and the jth unit, the equation for the state of a single unit was

$$f[i] \rightarrow 1 \quad \text{if} \sum_j A[j,i]f[j] \geq \theta$$

$$f[i] \rightarrow 0 \quad \text{if} \sum_j A[j,i]f[j] < \theta.$$

In Hopfield's work the threshold was almost always zero. The choice of two state neurons, with states zero and 1, caused unnecessary algebraic complexity since these states are not symmetrical about zero, although for those who like logic they have a pleasing resemblence to the logical states true and false. Life and algebra become much simpler if the allowable states are made $+1$ and -1 (figure 12.3). Therefore, the unit property we use is

$$f[i] \rightarrow 1 \quad \text{if} \sum_j A[j,i]f[j] \geq 0$$

$$f[i] \rightarrow -1 \quad \text{if} \sum_j A[j,i]f[j] < 0.$$

Computing Unit for a Hopfield Net

Stage 1   Stage 2

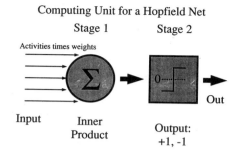

Input   Inner   Output:
             Product   +1, -1

Figure 12.3 The elementary computing unit used in the Hopfield net is a small variation on the generic neural network computing unit. Allowable output states are $+1$ and -1. In the original Hopfield papers McCulloch-Pitts neurons were used with possible outputs zero and 1. This assumption complicates algebra.

Hopfield Network

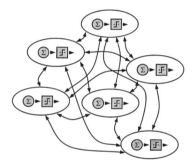

Figure 12.4 The connectivity of a Hopfield network is autoassociative. The network can be fully connected, that is, any unit can connect to any other unit, with the important restriction that a unit cannot connect to itself.

The overall architecture of a Hopfield network is simple. A group of units projects to itself, as it does in the autoassociators we saw before. An important restriction on connectivity is that a single unit does not connect to itself; that is, the connection matrix has zeros down the main diagonal (figure 12.4).

Energy Minimization

Hopfield suggested an important technique for analyzing neural networks by viewing them as minimizing an energy function. Consider a quantity called E, given by

$$E = -\frac{1}{2}\sum_i \sum_j A[i,j] f[i] f[j].$$

Each term in this sum is composed of a three-part product, the state of the ith unit, the state of the jth unit, and the strength of the connection between them. Suppose this system were formed from a group of interacting magnets instead of a neural network. Each magnet interacts with all the other magnets with some strength of interaction. If the above expression were to describe such a physical system, the function E would then be related to the *energy* of the system. Of course, a neural network, where the terms are state vectors and the interactions are connection strengths, does not have an energy in the sense that the physical system does. However, it turns out that we can do the analysis anyway, and our intuitions for the physical system carry over to the artificial one. The term energy in neural networks is common, although unfortunate. In nonlinear dynamical system theory, such a special function based on the system state is usually called a *Lyaponov* function.

An alternative interpretation of the energy function views it as a measure of *constraint satisfaction*. If the states of units are identified as the truth or falsehood of propositions, a low energy state would correspond to the state of maximum agreement between pairs of coupled assertions. States would be in agreement if, say, they were both true (or false) and the interaction strength between them was positive, suggesting that they should be in agreement. Agreement would lead to a positive triple product, and therefore a negative contribution to the energy function. Similarly, if the two states disagreed and the interaction strength was positive, energy would increase because the states should be in agreement, and they were not.

Dynamics

An unchanging energy function in a static system is of little interest. We must next propose a way that the network evolves in time, its *dynamics*, and see what happens to the energy function. For the discussion here we will consider the special case of a symmetric matrix, that is, one where

$$A[i,j] = A[j,i].$$

Virtually all the formal results for Hopfield networks have assumed a symmetric matrix, although symmetric connections are extremely unbiological; however, it is a convenient assumption.

Asynchronous Updating

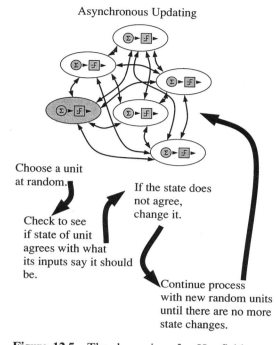

Choose a unit at random.

Check to see if state of unit agrees with what its inputs say it should be.

If the state does not agree, change it.

Continue process with new random units until there are no more state changes.

Figure 12.5 The dynamics of a Hopfield net are controlled by a process called *asynchronous updating*. A unit is chosen at random and its current state is made to agree with the rule governing the neuron state. If the unit state does not agree with its input, it is changed so that it does. The process is repeated.

States are changed by a process called *asynchronous undating*. Only a single unit changes state at a time. A unit is chosen randomly. The inner product between the input pattern and the connection strengths is computed. The rules for changing the state of a unit have already been given. The inner product is checked to see whether or not the value is greater than or equal to zero, when, according to the rule for state changes, the unit goes to the $+1$ state, or less than zero, when the unit goes to the -1 state. If the unit is not in agreement with what it should be, given by the rule for changing state, its state is changed so it is in agreement (figure 12.5).

Suppose a unit, say $f[k]$, changes state. This means only the terms in the energy expression involving $f[k]$ will change. There are two classes of terms, those involving the inputs to unit k and those involving outputs from unit k. Energy is the sum of the products of pairs of units times their connection strengths. Consider a unit, k, connecting to another unit, j, with strength $A[j,k]$. Unit j connects with k, with strength $A[k,j]$ (figure 12.6). The total energy contribution from these two interactions is then

$$E_{\text{interaction}} = -\tfrac{1}{2}f[j]f[k]A[k,j] - \tfrac{1}{2}f[k]f[j]A[j,k].$$

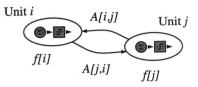

Figure 12.6 Interactions between two computing units in a Hopfield net.

The two partial contributions are equal because the two matrix terms are equal by the assumption of symmetry, so that

$$E_{\text{interaction}} = -f[j]f[k]A[j,k].$$

To get the total energy contribution due to unit k, $E[k]$, we have to add up all the contributions from the other units, j, so we have,

$$E[k] = -\sum_{j \neq k} f[k]f[j]A[j,k].$$

Since $f[k]$ is the same for all the terms in the sum, we can take it outside the summation, so,

$$E[k] = -f[k]\left(\sum_{j \neq k} A[k,j]f[j]\right).$$

The summation term in brackets is the term for sum of synaptic inputs to unit k; that is, this is the term we use to decide whether or not to change unit state.

Suppose we are going to look at unit k during our updating process. If the unit does not change its state, obviously no change in energy occurs. But suppose the synaptic inputs to a unit determine that it should switch state because the sum, the inner product between input activities and weights, is now above or below threshold. What happens to the energy then? Let us compute the energy, $\Delta E[k]$, due to a change in state of unit k, $\Delta f[k]$, where

$$\Delta E = E[k](\text{new}) - E[k](\text{old})$$

and

$$\Delta f[k] = f[k](\text{new}) - f[k](\text{old}).$$

Then,

$$\Delta E[k] = -\Delta f[k]\left(\sum_{j \neq k} A[k,j]f[j]\right).$$

Again, the term in brackets is the inner product that determines the state of the unit.

Suppose $f[k]$ is $+1$ and goes to -1. Then $\Delta f[k]$ is negative. But if the neuron changed state, the sum of synaptic terms must also be negative, otherwise it would not have changed state. The product of the two negative terms is positive, and there is an overall *decrease* in energy since $\Delta E[k]$ is negative.

Suppose $f[k]$ is -1 and goes to $+1$. Then $\Delta f[k]$ is positive. But for the neuron to change in the positive direction, the sum of synaptic interactions must be positive and their product is positive. Therefore an overall *decrease* in energy also occurs for this case.

Changes in state during asynchronous updating always decrease the energy of the system, and the dynamics of the system always act to reduce the energy. Since there is a lower bound to the energy because all the terms in the sum are finite, the energy function must have at least one minimum. In general there are many minima. As the system evolves with time, its energy decreases to one of the minimum values and stays there, since no further changes in energy are possible. The points at which the system stops changing are *local energy minima*.

This simple analysis gives a powerful description of the dynamics of the system in terms of a global energy function defined for the network. The use of energy functions as an analytical tool for characterizing complex networks is common.

Ising Models

Hopfield made the portentous comment. "This case is isomorphic with an Ising model," thereby allowing a deluge of physical theory (and physicists) to enter neural network modeling. This flood of new participants transformed the field. In 1974 Little and Shaw made a similar identification of neural network dynamics with the Ising model, but for whatever reason, their idea was not widely picked up at the time.

Theoretical physicists are an unusual lot, acting like gunslingers in the old West, anxious to prove themselves against a really good problem. And there aren't that many really good problems that might be solvable. As soon as Hopfield pointed out the connection between a new and important problem (network models of brain function) and an old and well-studied problem (the Ising model), many physicists rode into town, so to speak, with the intention of shooting the problem full of holes and then, the brain understood, riding off into the sunset looking for a newer, tougher problem. (Who was that masked physicist?) Unfortunately, the problem of brain function turned out to be more difficult than expected, and it is still unsolved, although a number of interesting results about Hopfield nets were proved. At present, many of the traveling theoreticians have traveled on.

Learning

We have merely shown that minima exist. We want to arrange it so that the energy minima are in the right places. For a memory, ideally they would be the memories that the system was storing. We have a number of ways to construct **A**, the connection matrix. The learning rules already discussed can accomplish this. In Hopfield's original paper, the outer product Hebb rule was used.

Suppose we are only storing one memory, say, **f**. We assume the initial weights of the network are zero. We want to arrange to have an energy minimum at **f**. Suppose we suggest that the change in the matrix strength, $\Delta A[i,j]$, when a particular vector, **f**, is stored is

$$\Delta A[i,j] = f[i]f[j].$$

It is now clear why Hopfield assumed that a unit had no connections to itself, that is, $\Delta A[i,i] = 0$. This term would always be positive since $f[i]f[i]$ is always positive, and if learning continues, the diagonal term could increase without bound. This is a problem for any autoassociative system with simple Hebbian learning, but is easy to avoid.

We can show the outer product learning rule generates a stable state at **f** with only one learned state. Suppose the network is in the state, **f**, the learned state, and we want to show that **f** is an energy minimum. Consider a unit, i. The inner product between the weights connecting to unit i and the pattern shown by the other units is

$$(\text{inner product at } i) = \sum_{i \neq j} A[j,i]f[j]$$

$$= \sum_{i \neq j} f[j]f[i]f[j]$$

$$= f[i] \sum_{i \neq j} f[j]^2.$$

But this expression will always have the same sign as $f[i]$, since the sum of squares is always positive. If it always has the same sign as $f[i]$, it obviously agrees with what its inputs are saying. Since this is true for every unit, this result shows that **f** is an energy minimum.

As in the linear associator, if more items are stored, recall accuracy degrades. If the added stored patterns are orthogonal, exactly the same argument we used for the linear associator shows that the energy minima stay at the stored patterns. Exact pattern recall implies the signs are the same, which implies an energy minimum. However, if other patterns are stored, *crosstalk* is present between the nonorthogonal patterns, and the locations of the attractors wander away from the desired locations. As more nonorthogonal patterns are stored, the more likely errors become, as we found for our other networks. The thresholding behavior of the Hopfield neuron, which allows only two output values, provides a degree of noise resistance not shown in linear systems. This is because small errors, as long as they do not change the sign of the inner product for a unit, introduce no error at all. This property suggests that error behavior for a Hopfield net may be very good when small numbers of items are stored, and then degrades rapidly as the noise at more and more units reaches threshold. This observation turns out to be correct, but something much more interesting happens in a Hopfield network as the number of random patterns stored increases, something that has no parallel in linear systems.

Capacity and Phase Transitions

A number of computer simulations and some analysis led Hopfield and others to conclude that the number of random binary vectors that could be stored by a network was between 10% and 15% of the dimensionality. Chapter 1 of Hertz, Krogh, and Palmer (1991) briefly shows that capacity is proportional to the number of units if some errors can be accepted. If no errors are tolerable in the location of the minima, the capacity is proportional to the ratio of the number of units over the log of the number of units.

We showed that linear associator degraded slowly, that is, the more patterns stored, the less well it worked. The decrease in accuracy was quite gradual. This property of graceful degradation is one of the purported strengths of neural networks, although it can occur to a lesser or greater degree in any particular network.

Hopfield networks show an unexpected abrupt change in behavior as more and more patterns are stored. This calculation was first performed by Amit, Gutfreund, and Sompolinsky (1985) and is discussed in considerable detail by Hertz, Krogh, and Palmer (1991). We will not perform the derivation here, since it is involved, but the final result is worth mentioning.

If we assume the network only has to store random binary vectors, with on the average as many $+1$ as -1 terms, an abrupt discontinuity is seen in the accuracy of recall of the network. If more patterns are stored than 13.8% of the number of units, the system *randomizes* its energy minima. That is,

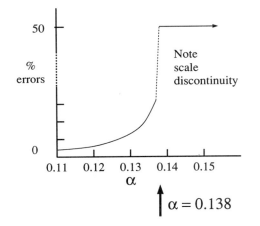

Figure 12.7   If too many patterns are stored in a Hopfield network, it undergoes a *phase transition*. This graph indicates the percentage of bits in error as the number of patterns increases. From a very small number, the error percentage suddenly jumps to 50% when critical numbers of random vectors are stored. An error of 50% means that the attractors are completely uncorrelated with the stored patterns. Redrawn from Amit (1989).

above this critical value, *the retrieved vector has no relation to the stored vector.*

Figure 12.7 shows how the number of bits in error increases with number of patterns stored. Error rates are extremely low when the number of stored patterns is less than 10% of the number of units, and then the rise in errors is spectacular. In one simulation, under 1% of bits were in error when a number of random patterns equal to 13% of the number of units was stored. The simulation found 50% of bits in error when a number of patterns equal to 13.8% of the number of units was stored. A 50% error rate indicates that the output is statistically a new random vector. This abrupt transition is similar to what is called a *phase transition* in physics, where a fundamental change in system properties occurs, for example, going from a liquid to a gas.

With lower numbers of stored vectors, there are significant numbers of *spurious attractors*, that is, energy minima that do not correspond to stored patterns. The interpretation of spurious attractors depends on who is doing the interpreting. They might be random errors. Or, they might be interpreted as combinations of stored items that may provide the answers to questions that no one has yet thought to ask!

Dynamical systems can show more complex kinds of behavior than simply reaching a fixed energy minimum. Sometimes closed orbits called limit cycles occur, where the state of the system constantly changes, but approaches

Continuous Hopfield Unit

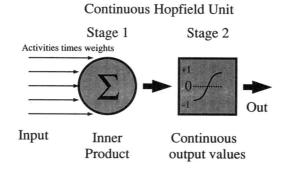

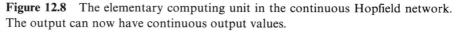

Figure 12.8 The elementary computing unit in the continuous Hopfield network. The output can now have continuous output values.

a fixed trajectory. Such states cannot be reached with the simple network considered here, but more complex nonlinear networks that incorporate mechanisms such as adaptation can develop rich oscillatory behavior. Studying the kinds of qualitative behavior that such nonlinear systems can show is a fascinating and important part of modern theoretical biology. It is fair to say that this branch of modern mathematics may revolutionize the way that we look at the behavior of complex systems in neurobiology.

Continuous Networks

One class of extensions to the Hopfield net is worth mentioning briefly: energy-minimizing networks using units with continuous activities. These were discussed by Hopfield in 1984, when he presented them as an extension of the model he described in 1982. The **BSB** network discussed in chapter 15 is another example.

Hopfield proposed a more realistic neuron model and showed that the results he reported in 1982 still held. The neuron contained an internal variable, u, the linear sum of its excitation and inhibition. This sum took the form of the sum of the product of presynaptic activity weighted by the appropriate connection strength, that is, the familiar inner product box in our generic neuron. The internal variable was converted into an output activity by a sigmoidal nonlinearity. The input-output relationship was given by $g(u)$ (figure 12.8).

This sigmoidal nonlinear function, $g(u)$, was assumed to be monotone increasing as the sum of inputs increased. However, the slope of u was very low for large values of the sum of inputs, so large increases in the sum made a small effect on output activity. The slope of the sigmoid was low for small values of the sum as well. In place of the sigmoid, $g(u)$, Hopfield found it convenient to use the inverse function (the output-input function), which he called $g^{-1}(f[i])$, where $f[i]$ is the unit output.

Hopfield used the same technique as in his first paper: he set up an energy function and showed that the evolution of the system in time, given the properties of the neurons, decreased energy. The equation was similar to the one that appeared in the discrete system, with the addition of a term that tended to keep the state vector out of the limits of the state space. In the simple integrator model of the neuron similar to the one we discussed in chapter 2, voltage changes require energy to charge the membrane capacitor through the membrane resistance. The amount of energy required to reach a particular voltage is given by the integral of the inverse function weighted by the inverse of the membrane resistance. This term adds significant energy at extreme values of activity and tends to keep the attractors well inside the limits of the sigmoid. (This behavior is different than the BSB model, discussed in chapter 15, which has no such charging term and prefers to be at the limits.)

The overall expression for the energy function was given by

$$E = -\frac{1}{2}\sum_{j \neq i}\sum A[i,j]f[i]f[j] + \sum_i \frac{1}{R}\int_0^{f[i]} g_i^{-1}(f[i])\,df[i].$$

The first term above is identical to the discrete network we just analyzed. The second term is the charging term. There is a connection between the stable points of the discrete model proposed in 1982 and the continuous one. The slope of the sigmoid corresponds to the gain of the system. If the gain is high, the stable points of the continuous system correspond to the stable points of the discrete system, and are near the limits of the sigmoid in the state space. Lower gains cause the energy minima to move inward from the limits but still to stay near them. As gain becomes small enough, the minima start to disappear. Figure 12.9 shows the attractor structure of a small continuous Hopfield net.

As has been pointed out on many occasions, neural net hardware acts more like *analog* than *digital* circuits. The analog-digital distinction has important implications for designers of very-large-scale integrated (VLSI) circuitry. We know a great deal about digital VLSI design. It is far more difficult to build analog VLSI. Carver Mead (1989) discusses some of the problems arising when building analog neural network VLSI through a series of impressive examples specifically designed to model biological sensory systems. Most of the commercial neural network chips designed as of 1994 are still digital because the technology is so reliable, fast, and well understood. Examples

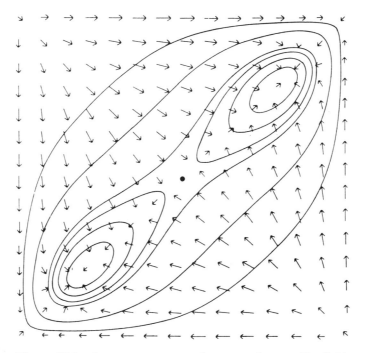

Figure 12.9 Attractor structure for a continuous Hopfield network. An energy contour map for a two-neuron two-stable state system. The ordinate and abscissa are the outputs of the two neurons. Stable states are located near the lower left and upper right corners, and unstable extrema at the other two corners. The arrows show the motion of the state. This motion is not in general perpendicular to the energy contours. From Hopfield (1984). Reprinted by permission.

of current digital neural network VLSI are the work of Hammerstrom at Adaptive Solutions, Inc., (Beaverton, OR) and the Bell Laboratories group (Graf, Jackel, and Hubbard, 1988). INTEL (Beaverton, OR) currently has a commercially available neural network chip (ETANN) and recently announced a chip realizing a version of the Nestor algorithm described in chapter 13. Other VLSI projects from companies such as Bellcore, Siemens, Hitachi, and Fujitsu are in various states of development, and events in this area are moving very rapidly.

Optimization Using Neural Networks

The fact that a network acts as if it is minimizing an energy function suggests a strategy for solving some classes of problems. Suppose we arrange it so that desired solutions occupy low energy points in state space. Then we design a network that seeks out these low energy points. This idea has found application in a number of areas.

An early example of solving a difficult problem by network energy minimization was by Hopfield and Tank (1985). The problem involved was the famous traveling salesman problem. Suppose a salesman must visit a number of cities, visiting each city, but each city only once on the tour. The problem is to find the minimum distance trip. A great many important optimization questions can be shown to be closely related to the traveling salesman problem, and it has led to a large literature in computer science and mathematics. Finding the true shortest tour seems to require inspecting all possible legal tours. This gives rise to a combinatorial problem since the potential number of different tours gets very large very rapidly as the number of cities on the tour increases. A number of algorithms are known that can find reasonably good solutions quickly. To test the abilities of neural networks to do optimization, Hopfield and Tank set up a network where, essentially, a short tour would have a smaller energy than a long tour and the dynamics of the network would therefore tend to find short tours. Their results suggested that neural networks can perform optimization to some degree. The network that they proposed did not find the best solution, but pretty good ones. Unfortunately, the exact technique they used did not work very well in practice. It depended critically on parameter values (Wilson and Pawley, 1998; Peterson, 1990). More generally, simulated annealing methods seem to produce better results in practice than the simplest energy-minimizing networks, as we will see next.

Boltzmann Machines: Extensions of Hopfield Networks

The Hopfield net strictly decreased energy and found a local minimum of an energy function. Something more is necessary if the true best solution to a problem is to be found. It is possible to combine a Hopfield network with a technique called *simulated annealing* to produce a system that is much better at finding global minima.

Hinton and Sejnowski (1984) and Ackley, Hinton, and Sejnowsi (1985) suggested an extension of the Hopfield network. Geman and Geman (1984), in a landmark paper, discussed image processing with an almost identical approach (although presented in terms of probability maximization rather than energy minimization) and proved some important theorems providing a firm mathematical footing for the techniques. Simulated annealing was popularized in 1983 by Kirkpatrick, Gelatt, and Vecchi, although the essen-

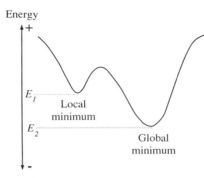

Energy

E_1 .... Local
minimum

E_2 .... Global
minimum

Figure 12.10   A simple energy landscape, with two minima, one with energy E_1 and one with energy E_2. If only these two minima exist, then E_2 is the global minimum and E_1 is the local minimum.

tials of the idea were proposed in the early 1950s (Metropolis, Rosenbluth, Rosenbluth, Teller, and Teller, 1953).

When we talked about energy functions and the Hopfield network, we pointed out that a neural network did not have "real" energy. However, some other useful ideas from physical systems can be applied to computation. Consider the energy landscape in figure 12.10. In a Hopfield network, energy monotonically decreased as the dynamics operated. In an energy landscape with many local minima, this meant that with a random starting point, we are likely to end up in a local minimum. The standard intuitive image is of a rock rolling downhill. Unfortunately, the rock may end up on the floor of a mountain valley, well above the global minimum.

Finding Minima

The general problem of finding minima (or maxima) is of major importance, as we saw in chapter 9 in relation to gradient descent learning algorithms. Hopfield networks only go downhill. What we need is a way to allow temporary *increases* in the energy of the system so we can move over a mountain ridge and find a better energy minimum on the other side (figure 12.11).

To see how this can be accomplished, we must return to the physics of the energy landscape. Although it seems superficially reasonable to claim that energy always decreases, this assumption leads to some misleading predictions. Consider, for example, a gas molecule in the atmosphere. The gas molecule has weight. Its lowest energy position is on the ground. Yet we know that even on the top of the World Trade Center we can still breathe; the atmosphere is not lying immobile in the streets of Manhattan. Physical

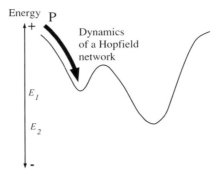

Figure 12.11   If we start at point P, a Hopfield net will find only the local minimum. However, if we climb over the ridge, the network can find the global minimum. This would require temporarily *increasing* the system energy.

systems have *temperature* associated with them. Temperature has the effect of giving elements in the state vector a certain average *energy* so they do not always occupy the lowest energy state. The elements in a system spend all their time in the lowest accessible energy state only at zero temperature.

If the temperature of the system is high, the state vector has considerable energy and can move out of the local minimum. Qualitatively (figure 12.12), if the average energy of the state vector is large enough, the state vector can move freely between the local minimum and the global minimum; it does not see the barrier. Even when the temperature is above zero, however, the system energy still has an effect on the state vector since it determines how likely the state vector is to be in various energy states. Suppose we have two states in our system, state 1 with energy E_1 and state 2 with energy E_2 (figure 12.13). If the system can move freely between states, the ratio of the probability of finding the state in state 1 (p(state 1)) as opposed to state 2 (p(state 2)) is given by the *Gibbs distribution* of physics; that is,

$$\frac{p \text{ (state 1)}}{p \text{ (state 2)}} = \frac{e^{-E_1/T}}{e^{-E_2/T}}$$

$$= e^{-(E_1 - E_2)/T}$$

$$= e^{-\Delta E/T}.$$

We define ΔE so that

$$\Delta E = E_1 - E_2.$$

If $E_1 = E_2$, that is, $\Delta E = 0$, the system is equally likely to be in state 1 as in state 2. If $\Delta E/T$ is very positive, the system will most often be found in state 2, since the exponential will be very negative. If $\Delta E/T$ is very negative,

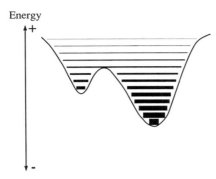

Figure 12.12 A high system temperature means the system state will spend some time in both minima, but proportionally *more time* in the lower energy state.

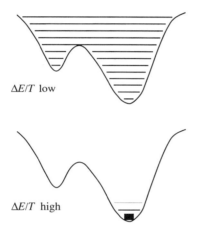

Figure 12.13 If $\Delta E/T$ is small, the low energy state will be only slightly more probable than the high energy state. If $\Delta E/T$ is very large, the network will spend most of its time in the lowest energy state.

the exponential will be very positive and the system will be much more likely to be in state 1. If the temperature becomes very high with respect to ΔE, the probabilities of both states will become nearly equal since the exponent is nearly zero. If the temperature is very low compared with ΔE, the low energy state is chosen most often.

The Gibbs distribution can also be used in nonphysical constraint satisfaction networks. The most popular way of incorporating it into a system is to use the energy difference to generate decision probabilities. That is, we compute the energy in one configuration and the energy in another configuration, for example, with one state changed in the asynchronous updating procedure for a Hopfield network. We then choose one configuration over the other with probability of choice based on the Gibbs distribution. If the

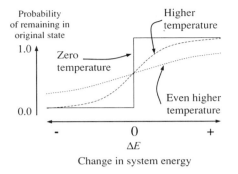

Figure 12.14   Probability of acceptance of a state according to the system temperature. At zero temperature, the Boltzmann machine always accepts the lower energy
state and acts like a Hopfield network.

energies are equal, the probabilities are equal, and so on. A small energy
difference between the states relative to temperature means that we are
nearly equally likely to choose either state; a big energy difference means we
will tend to accept the lower energy more often. Obviously, temperature is
not physical in these abstract systems.

Figure 12.14 shows the probabilities of acceptance of a particular state
with a given energy difference, based on the Gibbs distribution, as a function
of temperature, T. As the temperature drops, the lower energy state is more
and more likely to be chosen, until the step function at $T = 0$ realizes a
Hopfield unit, where the lower energy state is always chosen.

This change in behavior with system temperature suggests a way of finding the global minimum (Kirkpatrick, Gellatt, and Vecchi, 1983). Suppose
we start the system off at a high temperature so the state vector can move
throughout the energy landscape because its energy is so high; that is, $\Delta E/T$
is very small. Then we gradually reduce the temperature. As the temperature
gets lower, the system state will be more and more likely to be in the lower
energy states. At the end of this process, the final frozen state at very low
temperature will be the global energy minimum.

The rate at which a metal is cooled is called *annealing*. The process of
reducing the temperature of an abstract system is therefore called *simulated
annealing*. The *annealing schedule* has a major effect on the strength and
hardness of a metal. A difficult problem for the abstract system is to determine the best *annealing schedule*, that is, how slowly the temperature should
be reduced to let the system freeze out into the global minimum. If the
annealing schedule is too fast, the global minimum may not be found; if it is
too slow, huge quantities of computer time can be wasted.

Although simulated annealing often seemed to work in practice, it was
not unil 1984 that Geman and Geman proved that annealing schedules exist

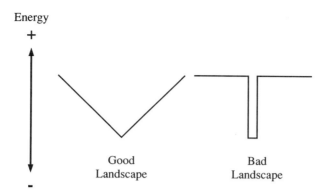

Figure 12.15   Good and bad energy landscapes. It is a lot harder to find the minimum in the bad landscape if the annealing schedule does not fully search the state space.

that find the global minimum in probability. Unfortunately, the schedule having this property is exceedingly slow. Geman and Geman showed that if the temperature at the kth state, $T(k)$, is kept above the value

$$T(k) > \frac{c}{\log(1 + k)},$$

the system will converge in probability to the minimum energy distribution. The parameter c is a large constant. In practice, a good deal of experimentation is involved, trading off goodness of solution against computer time. In many problems, the practical difference between a very good solution and the absolute best solution is extremely small. It is often better to run the simulation a number of times and choose the best answer than to run it only a few times with a very long annealing schedule. Sometimes other distributions can give faster settling in practice than the Gibbs distribution (Szu, 1986).

A schedule sometimes used in practical simulated annealing is an exponential schedule. In one example, the temperature on the ith state is given by

$$T(i) = T_o e^{-i/n},$$

where T_o is an initial temperature and n is a scaling constant.

In practice, the shape of the energy surface has a major influence on how short an annealing schedule is required. Figure 12.15 shows a good energy surface, where finding a minimum is not difficult and a very bad energy surface like a putting green with the hole in the middle. As an aside, a properly designed code designed to conceal the contents of a message might have an energy surface like the bad one with the meaningful message having the lowest energy. There should be no "nearness" or "similarity" relations in a good code, unlike the simple substitution code used for cryptic puzzles in newspapers. The way simple substution codes (i.e., 'a' → 'r', 'y' → 'g', etc.) are solved is to try substitutions for frequent letters, until the text starts to look more and more like a meaningful message and additional letters can be guessed. In an effective code, the message should be completely meaningless until the correct key is used.

The important criterion determining the annealing schedule is the time necessary to search the state space adequately. The system must spend enough time to sample all regions so it can obtain estimates of the relative energies in different regions.

An Example of Dynamics

Let us combine a Hopfield neural network with simulated annealing. Before, during asynchronous updating, we always either chose the state with lower system energy or did nothing. Now, we will look at the two states of the unit and decide which one to change based on the Gibbs distribution. This means that figure 12.14 can be interpreted as the probability that the unit will change state, if its state change produces different system energies. Thus, sometimes the unit will go to a state that disagrees with its inputs, and the system state will go to a higher energy.

We will set up the system with the appropriate local interactions, the annealing schedule and the energy function. We will start with an initial state that can be random or have some of the initial values clamped so that they cannot change. Clamping values effectively changes the energy landscape. It is one way to generate cognitively interesting behavior, because the clamped units dictate the lowest energy completion that can be designed to perform the computation required of the system.

These systems are easy to work with, although computation intensive, and many of their properties can be demonstrated with small systems. Let us consider a simple Boltzmann machine program called BOLTZ using only nearest neighbor interactions. (The listing of the complete BOLTZ program is in appendix K.)

Suppose we have a 15 by 15 rectangular array of units arranged in toroidal geometry; that is, the two sides are connected and the top and bottom are connected, so column 1 is the neighbor of column 15 (figure 12.16).

The units have only two states, $+1$ and -1. A unit is connected to its four nearest neighbors (up, down, right, left) by weights, w. The weights in the

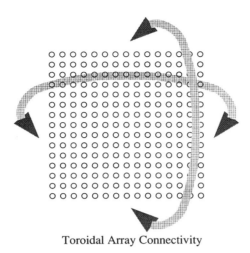

Toroidal Array Connectivity

Figure 12.16   Array of units in the Boltzmann machine example. This 15 by 15 array of units is connected as a toroid, that is, row 1 is next to rows 2 and 15, and column 1 is next to columns 1 and 15.

program are called w(up), w(down), w(left), and w(right). Every unit is connected to all four of its neighbors. The interaction energy of unit $[i,j]$ is defined as the product of the state of unit $[i,j]$, which we will call $f[i,j]$ times the synaptic weight to the neighboring unit, times the state of the neighboring unit, based on the simplest Hopfield energy function.

$$E[i,j] = -1/2(w(\text{up}) * f[i,j] * f[i,j-1] +$$

$$w(\text{down}) * f[i,j] * f[i,j+1] +$$

$$w(\text{right}) * f[i-1,j] * f[i,j] +$$

$$w(\text{left}) * f[i+1,j] * f[i,j]).$$

To make the system simple, let all the weights w equal 1. Figure 12.17 shows this configuration. Then, the energy of $E[i,j]$ is given by

$$E(i,j) = -1/2(f[i,j] * f[i,j-1] + f[i,j] * f[i,j-1]$$

$$+ f[i-1,j] * f[i,j] + f[i+1,j] * f[i,j]).$$

By inspection, we can see that pairs of units have lowest energy when the neighboring units are in the same state, either $+1$ or -1. That is also true for the entire system. The all $+1$ and all -1 configurations have the same energy, so there are two global minima (figure 12.18).

Suppose we start off with a system with roughly half the units in the $+1$ state and half in the -1 state. We change the state of a unit at random and

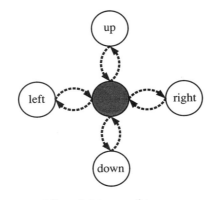

All weights equal to one.

Figure 12.17 Connections of units in the example problem. A unit is connected only to its four nearest neighbors. All the weights have the value 1.

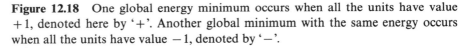

Figure 12.18 One global energy minimum occurs when all the units have value +1, denoted here by '+'. Another global minimum with the same energy occurs when all the units have value −1, denoted by '−'.

compute the new system energy. It is easy to decide whether or not we raised the energy. Five units, the one we changed and its four neighbors, have their energies changed. The change in energy due to interaction of the right neighbor with the one we changed is given by their product. The interaction of the one we changed with its right neighbor is the same. So the total change in energy is twice as great as that obtained by looking at the energy of only the one we changed.

A Hopfield net analysis of the system might go like this: choose a unit at random, change its state, and see if system energy goes up or down. If it goes down, change the state to the new value, otherwise leave the state the same. By this rule, no change in state will occur if system energy is unchanged. (This is a slight modification of the original Hopfield rule, necessitated by

```
- - - - -        - - - - -
- + + + -        - - + + -
- + + + -        - + + + -
- + + + -        - + + + -
- - - - -        - - - - -

   (a)              (b)
```

Figure 12.19   Two different configurations with the same system energy.

```
- - - - -        - - - - -        - - - - -
- - + + -        - - + - -        - - - - -
- + + + -        - + + + -        - + + + -
- + + + -        - + + + -        - + + + -
- - - - -        - - - - -        - - - - -

   (b)              (c)              (d)
```

Figure 12.20   If the network can explore different configurations with the same energy it will eventually find configuration (c), which leads to configuration (d) with significantly lower system energy.

the simplicity of the network in the simulation.) This simple system has many configurations of units where single changes leave its energy unchanged. Consider the two configurations shown in figure 12.19. They have the same energy, as can be seen by considering the changed unit. The upper left corner in either case is surrounded by two $+1$ and two -1, and therefore both (a) and (b) have the same energy. Either of these configurations would not change by single state changes, yet both are well above the ground state of all $+1$ or -1.

Let us now see how a Boltzmann machine might work. Suppose start with configuration (b) in figure 12.20. We now allow a unit to change state, even with no change in system energy. Then we have a 50% chance that configuration (c) would arise if we change the upper right corner. But configuration (c) is unstable, because if the upper $+1$ is flipped, configuration (d) will have lower energy than (a), (b), or (c), all of which are the same. Unless we are able to change the system state without changing the energy or by increasing the energy in other configurations, we could never have found the lower energy configuration, (d). Obviously also, we have to count on several events occurring in proper sequence to find the lower energy state, which may be unlikely and will take time. If the temperature is high, such moves are common, because it is definitely probable that the system will come to occupy a higher energy state than it had before, with probability given by the Gibbs distribution. Some states even in our simple system are very stable local minima in that any state change actually increases the system energy. Because of the toroidal geometry, any state that has a complete band, up and down or left and right, is such a local minimum. Figure 12.21 shows such a

```
- - - - - - - - - - -        - - - - - - - - - - -
- - - - - - - - - - -        - - - - - - - - - - -
+ + + + + + + + + + +        + + + - + + + + + + +
+ + + + + + + + + + +        + + + + + + + + + + +
+ + + + + + + + + + +        + + + + + + + + + + +
- - - - - - - - - - -        - - - - - - - - -. - - -

         (a)                          (b)
```

Figure 12.21   A band leads to deep local minimum in toroidal geometry. The only way the true global minimum could be found would be by passing through temporary configurations such as (b) that considerably increase system energy.

very deep local minimum, confirmed in simulation. (See figures 12.26 and 12.27.)

Configuration (a) in figure 12.21 is lower energy than configuration (b). Once the configuration (b) has arisen, corners serve as points of attack where the row can be nibbled away. Configuration (a) is a deep local energy minimum. All configurations that have corners (boxes, rectangles, etc.) are eventually nibbled away and destroyed by the dominant sign.

Demonstrations

Suppose we start with the initial random configuration of plus and minus values shown in figure 12.22. Three edited demonstration runs of the Boltzmann program (appendix K) are shown in later figures. The parameters and starting configurations are identical except for the initial temperature.

In figure 12.23 we set the temperature at zero and have what is essentially a Hopfield net. The final system state energy (-258 units) is far from the lowest energy state (-450 units). The simulation found a shallow local minimum.

When the starting temperature becomes finite, that is, forming a Boltzmann machine, sometimes the energy increases. Figure 12.24 shows that system energy actually increases as the system evolves. Such behavior is impossible in a Hopfield network. In figure 12.25 a global minimum, all -1, is found after about 17,000 asynchronous state changes. The region of $+1$s is nibbled away and shrinks to nothing. This behavior is rather entertaining to watch and might make a nice screen saver for a desk-top computer.

In figures 12.26 and 12.27 a slightly lower starting temperature finishes with the system state ending in a local minumum. Once a band of $+1$s is established running across the whole field at about 10,000 iterations, the system is trapped into a local minimum because the temperature is too low

```
Iteration  :     0

+ - - - - + + - + - + + - + -
+ + + - + + + + + - + + + + -
+ + + - + + + + + + - - - - +
- + + - - - + + + - - - + + +
- + - - - - + + + - - + + + -
+ + + - - - + + - + - + + - +
+ - + + + - - + - - - - - + +
+ - + + + + + - - - - - + + -
+ - - - - - + + + + - + - - -
- + - + - - + + + - + + + - +
- - + - - - + - + + - - + + -
- + - + + + + - - - + + - - -
- - + - + - + - + + + - - - +
- + - - - + - + - - + + - + -
- - + + + - + + - + - - - - +

Temperature   :      0
Current energy:      6
Ground state  :   -450
```

Figure 12.22   This random initial configuration is used for all the simulations in the
next few figures.

```
Iteration  :  5000

+ - - - + + + + + + + + + + +
+ + + - + + + + + + + + + + +
+ + + - + + + + + + - - + + +
+ + + - - - + + + - - - + + +
+ + + - - - + + + - - + + + +
+ + + - - - + + - - - + + + +
+ + + + + + + + - - - - + + +
- - + + + + + - - - - + + + -
- - - - - - + + + - - + + - -
- - - - - - + + + - - + + - -
- - - - - - + + + + - - - - -
- - - - + + + + + + + - - - -
- - - - + + - - + + + - - - -
- - + + + + - - - + + - - - -
- - + + + + + - - + + - - + +

Temperature   :      0
Current energy:   -258
Ground state  :   -450
```

Figure 12.23   When a Boltzmann machine at zero temperature is applied to the
random configuration, it finds this local minimum. This Hopfield network differs
from the one we discussed only in that the state of a unit does not change if there is
no change in energy.

```
Iteration  :   500
+ + - + - - - + - - + + + + -
+ - - - - + - + + + + + + + +
- + + - + + - - + + + - + - -
- + - - - + - - - - + + + - -
- - - - - - - - + + - - - + +
+ + + + + + - + - - - - - - +
+ + + - + + - - + - - - - + -
+ + + - + - - - - + - - + + +
- + + + + - + - - - + - - - +
- - - + + - + + - + + - - + -
- + + - - - + + - - - + - - -
- - - - + + + - + - - - - - -
+ + - + - + + - + + + + + + +
- + + + - + + - + - - + - + +
- + + - - - - - + + + - + - +
```

```
Temperature   :       25
Current energy:      -42
Ground state  :     -450
```

```
Iteration  :   800
+ - - - + - - + - + + + + - -
- - - - + - + - + - + - - - -
+ - + - + + - + + + + - + - +
- + - + + + + + - + + - - -
- - - - - - + + + + + - + + +
- + - - + + - - - - - + - - +
+ + + + + + - - - - - + - + -
+ + + - + - + - - + - - + + -
- + - - + - + + + - + + + + +
- - - - + - + - + + + + + + -
+ + - - + - + - - + - - + + -
- + + - - + - + + - - + - - -
- + + + - + - - + - + + + + -
- - + + - - + - - - + + - + +
- + + - + - + - + + - + - +
```

```
Temperature   :       23
Current energy:        6
Ground state  :     -450
```

Figure 12.24 When the network starts with a high initial temperature, system energy can increase. Here the network has increased its system energy between asynchronous updates 500 and 800.

```
Iteration  : 10000
- - - - - - - - - - - - - - -
+ + + + + - - - - - - - + + +
+ + + + + - - - - - - - + + +
+ + + + + - - - - - - - + + +
+ + + + + - - - - - - - + + +
+ - - + + - - - - - - - + + +
- - - + - - - - - - - + + + +
- - - - - - - - - - - + + + -
- - - - - - - - - - - + + + -
- - - - - - - - - - - + + + -
- - - - - - - - - - - + + + -
- - - - - - - - - - - - - - -
- - - - - - - - - - - - - - -
- - - - - - - - - - - - - - -

Temperature   :      1
Current energy:    -362
Ground state  :    -450


Iteration  : 15000
- - - - - - - - - - - - - - -
- - - - - - - - - - - - - - -
+ + + - - - - - - - - - - + +
+ + + - - - - - - - - - - - +
+ + + - - - - - - - - - - - +
+ + + - - - - - - - - - - - +
+ + + - - - - - - - - - - - +
+ + + - - - - - - - - - - - -
- - - - - - - - - - - - - - -
- - - - - - - - - - - - - - -
- - - - - - - - - - - - - - -
- - - - - - - - - - - - - - -
- - - - - - - - - - - - - - -
- - - - - - - - - - - - - - -

Temperature   :      0
Current energy:    -410
Ground state  :    -450
```

Figure 12.25 The Boltzmann machine leads to the correct solution. The "+" are isolated in groups and nibbled away.

to allow the increase in energy required to reach the global minimum. The next 15,000 iterations clean up the edges, and at about 25,000 iterations, the system finds a stable local minumum (figure 12.27). This indicates that the annealing schedule is too fast. However, in the demonstration program, the true global minimum is found about two times out of three for different random starting configurations with $+1$ and -1 equally likely.

A little experimentation with the program suggests one of the main problems with the Boltzmann machine: it is very slow. However, this architecture always arouses a great deal of interest because it really does find a global minimum, unlike gradient descent. Simulated annealing has taken its place

```
Iteration  : 10000
 - - - + + + + + + + + + + + -
 - - - + + + + + + + + + - + -
 - - - - + + + + + + - - - - -
 - - - - + + + + - - - - - - -
 - - - - - + + + - - - - - - -
 - - - - - + + + - - - - - - -
 - - - - + + + + - - - - - - -
 - - + + + + + + - - - - - - -
 - - + + + + + - - - - - - - -
 - - + + + + + - - - - - - - -
 - - + + + + + - - - - - - - -
 - + + + + + + - - - - - - - -
 - - + + + + + + + - - - - - -
 - - - + + + + + + + + - - - -
 - - - + + + + + + + + + + + -
```

```
Temperature    :       1
Current energy:     -338
Ground state   :     -450
```

```
Iteration  : 15000
 - - - + + + + + + + + - - - -
 - - - + + + + + + + + + - - -
 - - - + + + + + + + + + - - -
 - - - + + + + + + + + + - - -
 - - - + + + + + + + + + - - -
 - - - + + + + + + + + - - - -
 - - - + + + + + + + - - - - -
 - - - + + + + + + + - - - - -
 - - - + + + + + + + - - - - -
 - - - + + + + + + + - - - - -
 - - - + + + + + + + - - - - -
 - - - + + + + + + + - - - - -
 - - - + + + + + + + - - - - -
 - - - + + + + + + + - - - - -
 - - - + + + + + + + - - - - -
```

```
Temperature    :       0
Current energy:     -378
Ground state   :     -450
```

Figure 12.26 A band of '+' form in a field of '−', leading to the deep local minimum of figure 12.27.

as a useful and general way of finding good solutions to some difficult problems. As a general comment, however, the more that is known about the details of a problem, the less useful such general statistical techniques are. There is always a trade-off between computation time and problem-specific knowledge. This holds strongly for neural networks as well.

A learning rule is associated with the Boltzmann machine described originally. However, learning rules arise from a very different approach to learning than we have seen up to this point. The aim of network learning in the Boltzmann machine is to reproduce the energy structure of the environment. In practice it is exceedingly slow because a number of probabilities must be

```
Iteration   : 25000

- - - + + + + + + + + - - - -
- - - + + + + + + + + - - - -
- - - + + + + + + + + - - - -
- - - + + + + + + + + - - - -
- - - + + + + + + + + - - - -
- - - + + + + + + + + - - - -
- - - + + + + + + + + - - - -
- - - + + + + + + + + - - - -
- - - + + + + + + + + - - - -
- - - + + + + + + + + - - - -
- - - + + + + + + + + - - - -
- - - + + + + + + + + - - - -
- - - + + + + + + + + - - - -
- - - + + + + + + + + - - - -
- - - + + + + + + + + - - - -


Temperature    :        0
Current energy:      -390
Ground state   :      -450
```

Figure 12.27   The stable local minimum. All such bands have the same energy. The frequent appearence of this local minimum suggests that the annealing schedule in the simulation is too fast.

estimated. It is also extremely unpsychological. However, it was the first learning rule that was capable of solving the credit assignment problem for multilayer networks, a major accomplishment. NETtalk (chapter 9) was originally done using Boltzmann machine learning. Further details can be found in Ackley, Hinton, and Sejnowski (1985).

Recently, some network modelers incorporated mechanisms that work somewhat like the Boltzmann machine in more traditional models. For example, McClelland (1990) incorporated stochastic elements in a deterministic model to obtain some of the desirable features we described in this chapter.

References

D.H. Ackley, G.E. Hinton, and T.J. Sejnowski (1985), A learning algorithm for Boltzmann machines. *Cognitive Science*, 9, 147–169.

D.J. Amit (1989), *Modelling Brain Function*. Cambridge: Cambridge University Press.

D. Amit, H. Gutfreund, and H. Sompolinsky (1985), Storing infinite numbers of patterns in a spin glass model of neural networks. *Physical Review Letters*, 55, 1530–1533.

S. Geman and D. Geman (1984), Stochastic relaxation, Gibbs distributions, and the Bayesian restoration of images. *IEEE Proceedings on Artificial and Machine Intelligence, 6,* 721–741.

H.P. Graf, L.D. Jackel, and W. Hubbard (1988), VLSI implementation of a neural network model. *IEEE Computer, 21,* 41–50.

J. Hertz, A. Krogh, and R.G. Palmer (1991), *Introduction to the Theory of Neural Computation.* Reading, MA: Addison-Wesley.

G.E. Hinton and T.J. Sejnowski (1984), Optimal pattern inference. *IEEE Conference on Computers in Vision and Pattern Recognition, 1984.* New York: IEEE.

J.J. Hopfield (1982), Neural networks and physical systems with emergent collective computational abilities. *Proceedings of the National Academy of Sciences, 79,* 2554–2558.

J.J. Hopfield (1984), Neurons with graded response have collective computational properties like those of two state neurons. *Proceedings of the National Academy of Sciences, 81,* 3088–3092.

J.J. Hopfield and D.W. Tank (1985), "Neural" computation of decisions in optimization problems. *Biological Cybernetics, 52,* 141–152.

S. Kirkpatrick, C.D. Gelatt, Jr., and M.P. Vecchi (1983), Optimization by simulated annealing. *Science, 220,* 671–680.

W. Little and G. Shaw (1974), The existence of persistent states in the brain. *Mathematical Biosciences, 19,* 101–120.

J.L. McClelland (1990), Stochastic interactive processes and the effect of context on perception. *Cognitive Psychology, 22,* 1–41.

C. Mead (1989), *Analog VLSI and Neural Systems.* Reading, MA: Addison-Wesley.

N. Metropolis, A.W. Rosenbluth, M.N. Rosenbluth, A.H. Teller, and E. Teller (1953), Equation of state calculations by fast computing machines. *Journal of Chemical Physics, 21,* 1087–1092.

C. Peterson (1990), Parallel distributed approaches to combinatorial optimization: Benchmark studies on the travelling salesman problem. *Neural Computation, 2,* 261–269.

H. Szu (1986), Fast simulated annealing. In J.S. Denker (Ed.), *Neural Networks for Computing.* New York: American Institute of Physics.

G.V. Wilson and G.S. Pawley (1988), On the stability of the travelling salesman problem algorithm of Hopfield and Tank. *Biological Cybernetics, 58,* 63–70.

Nearest Neighbor Classifiers

Example is the school of mankind, and they will learn at no other.
Edmund Burke

Caesar had his Brutus; Charles the First his Cromwell; and George the Third–may profit by their example. *If* this *be treason, make the most of it.*
Patrick Henry

This chapter discusses networks based on learning examples where the examples are kept distinct to some degree in the resulting system. These networks capture the intuition that patterns that are near classified examples are likely to share the properties of the previously encountered patterns. It is easy to build a simple nearest neighbor classifier *with a neural network because the inner product taken between the weights and the input activity pattern is closely related to geometrical distance. We also discuss the* sparse distributed memory *of Kanerva, which is based on the distributed storage of specific examples. We briefly mention the similarity of neural networks to the mathematical topics of* interpolation *and* approximation. *Finally, we look at the important class of cognitive models, now sometimes called* case based reasoning, *where a few very complex events are stored in memory and the problem becomes what part of the event to use for comparison and generalization.*

Both neural networks and humans learn by experience. However, knowing how to use experience to guide future performance is difficult, and there is no obvious right way to do it. A system that only memorized facts would be of limited value. The ability to give correct answers to questions that have not yet been asked is called *generalization*. Good generalization is probably the most valuable property in an adaptive learning system. Unfortunately, it is exceptionally slippery. The two extreme types of learning models both make strong, and sometimes contrary, claims about the appropriate way to generalize.

For the first kind of generalization, suppose we have a system that learns a very large number of correctly classified examples; a neural network that recognized handprinted digits, for instance. New inputs are likely to be close to something the system has already learned and classified.

The second form of generalization is encountered less frequently in the neural network literature but is far more important for human behavior. Suppose the learning system has encounted a small number of complex events. These events are rich in detail and we wish to use them as guides to future action. Which parts of the events are relevant to the future? What was Patrick Henry trying to say in the quotation at the beginning of the chapter? Why did the cry of "Treason" come from one of those listening to his words?

In this chapter we will discuss systems, sometimes neural networks, sometimes closely related to them, that learn by storing specific examples. We will be concerned primarily with the first type of generalization in which many examples are learned; they are not very complex, and the task is always clear. We will mention the second type at the end of the chapter. In that type of generalization the problem is not learning examples, but the far more difficult task of knowing what part of experience can properly be applied to a new problem. Some of these issues arose in our discussion of concepts in the first part of chapter 11.

Pattern Classifiers

At present, most applications, both scientific and engineering, of neural networks are as pattern classifiers. This has been true of research in the field since the beginning. It was clearly the model used in perceptron research and has been carried consistently through to the present. It now colors the way most neural network research is performed; however, it was not the model used by the earliest famous neural network, that of McCulloch and Pitts, where the primary goal was computation with logic functions.

Pattern recognition and pattern classification are active fields in statistics and engineering. They are of great practical importance. Duda and Hart (1973) is a classic reference. A book by Pao (1989) makes the connection between neural networks and pattern recognition explicit, as does much of the work of Kohonen (1984). Batchelor (1974) is particularly good for the models we discuss in this chapter. Many neural network algorithms have already been analyzed in the statistical and pattern recognition literature, although sometimes under different names.

Adaptive pattern classification is a natural way to use a neural network. We have an input that represents data of some kind, and we have an output category that could be the name of a character, for example, or one of the ten digits. By presenting the network examples of the category, the weights in the network change automatically to allow the network to classify cor-

Real-World Categories

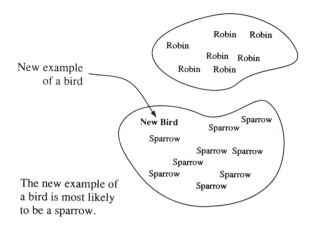

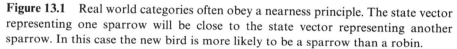

Figure 13.1   Real world categories often obey a nearness principle. The state vector representing one sparrow will be close to the state vector representing another sparrow. In this case the new bird is more likely to be a sparrow than a robin.

rectly new examples of the category; that is, the network can *generalize* by seeing a limited set of examples.

Practical pattern classification usually makes a critical assumption about the statistical structure of the world. Suppose we describe an input pattern as a point in state space, that is, as a set of input element activities, and we know or are told the classification of this point. It is often assumed, consciously or unconsciously, that nearby patterns are likely to have the same classification. Many of our experiences as humans are with a world that acts this way; similar things tend to have the same name and similar properties. Such an assumption about similarity suggests a rule that *points close in state space are likely to have the same classification.* We made this assumption in one of our rules for neural network data representation at the end of chapter 10. The underlying model of pattern classification that this represents has *regions* of some size in state space, each associated with a particular classification, as we mentioned in chapter 8 (see figure 8.1; figure 13.1).

Cases in which this assumption does not hold are likely to be hard for neural networks to learn. For example, as we saw in chapter 9, a notoriously hard problem is Exclusive-OR (X-OR), or, even worse, parity, a generalization of X-OR. Parity is classification based on whether the allowable number of 1s in a binary vector is even or odd (figure 13.2). Change of a single component in an allowable state vector *always* changes parity. Therefore, adjacent binary vectors are *never* classified the same. The infatuation of

Parity in Three Dimensions

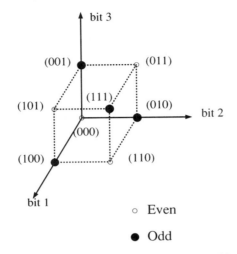

Figure 13.2 Parity, that is, whether a binary word has an even or an odd number of bits, violates the nearness principle. The nearest neighbor to a point *always* has the opposite classification. This figure diagrams the three-dimensional (three-bit) parity problem.

many researchers with networks capable of solving X-OR and parity is peculiar, since the world rarely is organized that way.

Nearest Neighbor techniques are based on an assumption about the world: nearby points are more likely to be given the same classification than distant ones. Suppose we have a set of points in state space that have been given classifications. Suppose a new pattern is to be classified. The *distance* from the new pattern to all the old ones is computed. The new pattern is given the classification of the previously classified point that is closest to it, that is, its *nearest neighbor* (figure 13.3).

A Nearest Neighbor Calculation Using Model Neurons

One version of a nearest neighbor classifier can be realized easily using the generic connectionist neuron (Nilsson, 1965). Neural network model neurons, the generic computing units discussed in chapter 2, can compute comparative distances between examples because of the close relation between the inner product of the weights and the input pattern and Euclidean distance.

Let us assume the potential input patterns are normalized, that is, of constant length. Consider a classified pattern, a point, **p**, and a new pattern,

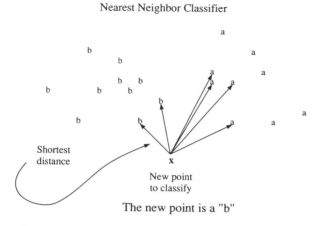

The new point is a "b"

Figure 13.3 A simple nearest neighbor classifier. The distance from a new point to classified old ones is computed. The new point is given the classification of the nearest old point.

$\mathbf{x}$, to be classified, both points in state space. The square of the distance between $\mathbf{p}$ and $\mathbf{x}$ is

$$[(\mathbf{p} - \mathbf{x}), (\mathbf{p} - \mathbf{x})] = [\mathbf{p}, \mathbf{p}] - 2[\mathbf{x}, \mathbf{p}] + [\mathbf{x}, \mathbf{x}].$$

This quantity is always positive. Since $[\mathbf{x}, \mathbf{x}]$ and $[\mathbf{p}, \mathbf{p}]$ are the same for all points, by the assumption of normalization, if we find the point $\mathbf{p}$ with the **maximum** value of $[\mathbf{x}, \mathbf{p}]$, this will be the point with *minimum* distance between $\mathbf{x}$ and $\mathbf{p}$.

Suppose we have a pattern, $\mathbf{p}_i$. Let us construct a unit whose weights are $\mathbf{p}_i$, that is, the weights are given the values shown by the activity pattern in state space. Then if the input pattern is $\mathbf{x}$, the output activity of the first stage of the unit is $[\mathbf{x}, \mathbf{p}_i]$. Suppose we construct many such units, one for every pattern the system has already seen, and whose classification is known. If a pattern to be classified is given as input, the unit *with maximum output* will have the shortest distance between the new and the old patterns given by the weights of that unit. The classification attached to this unit is then the classification of the new pattern (figure 13.4). Notice the similarity of this architecture to a multilayer neural network. The only nonstandard part of the network is providing a mechanism to pick out the single most active hidden layer unit and give as the output the classification of the pattern stored in the weights of this unit. Such a network could be built using the winner-take-all networks we discussed in chapter 4.

If we want to use more complex nonlinear units this technique will still work if the nonlinearity in the second stage is monotonic and the same for all units. The inner product is what provides the selectivity for the system.

Nearest Neighbor Network

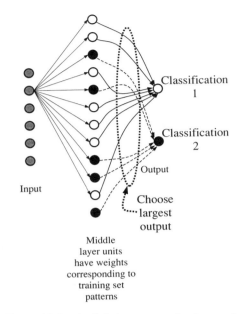

Figure 13.4 A slightly nonstandard neural network can realize a nearest neighbor classifier. The weights at the middle layer units reproduce the values of a classified input pattern. (The length of all the input patterns is the same.) If a new pattern is presented to the network, the classification of the most active unit will be the nearest neighbor classification.

Nearest neighbor classifiers are very good in practice. The best possible classifier in a probabilistic system is called the *Bayes classifier*. A Bayes classifier assumes that *everything* is known about the distribution of the possible inputs, that is, how probable it is that a particular point belongs to any given classification. Cover and Hart (1965) proved that, in the limit of very many stored patterns, the nearest neighbor technique will never do poorer than a factor of two worse than the Bayes classifier. Because of its simplicity and effectiveness, a nearest neighbor technique is almost always one of the methods used for benchmark comparisons of different pattern classification algorithms. It is sometimes disturbing to see how little improvement a complex pattern-recognition algorithm, including neural nets, will give over a nearest neighbor classifier.

Nearest Neighbor Variants

A commonly used variant of a nearest neighbor classifier is called the k-*nearest neighbor* classifier. Instead of looking only at the closest point, the

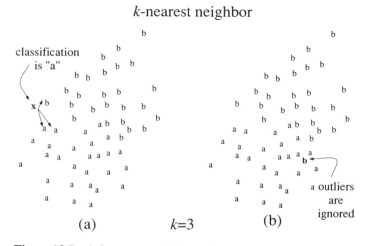

Figure 13.5   A *k*-nearest neighbor classifier. In this example, *k* = 3. (a) A new pattern is given the classification of the *majority* of the three nearest neighbors. (b) The *k*-nearest neighbor is more conservative than strict nearest neighbor. A pattern with one classification in the middle of a number of patterns with another classification will be ignored. This kind of averaging discriminates against noise and outliers.

k-nearest neighbor classifier looks at the *k* nearest classified patterns. The majority classification of the *k* patterns is given to the new pattern. The *k*-nearest neighbor acts somewhat like an averager, and reduces the effect of a single outlying point (figure 13.5).

One practical problem with nearest neighbor algorithms is that the more the system learns, the more computation it must perform to compute distances between new and old patterns. We can reduce the amount of computation dramatically for many practical situations by storing only a few classified patterns, but placing them strategically. Consider the common situation in which all the points in a sizable region of state space have the same classification. One reason this is common in practice is because it reflects the assumption that led to nearest neighbor techniques in the first place (figure 13.6).

We may have very many nearby points with the same classification. We can *approximate* a region in which all the points have the same classification by storing a *single* point somewhere in the middle of the region. This point is sometimes called a *prototype*. We need store only enough prototype points to approximate the regions accurately. How well this technique works depends on the shapes of the classification regions.

Prototype techniques are often used under other names. In some speech-recognition systems, patterns representing different speech sounds are stored in a code book. The initial classification of the sound takes the input

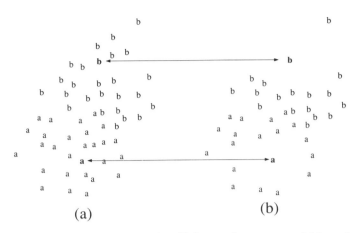

Figure 13.6   To improve the efficiency of a nearest neighbor classifier, not all examples need be stored. (a) The b classifications form a compact region. (b) A large number of b patterns can be replaced by a single b pattern near the center of the region. This particularly good b pattern is sometimes referred to as a *prototype*.

pattern and measures its distance from the classified patterns in the code book, the nearest identifying the sound.

The Nestor Algorithm

In 1982 Reilly, Cooper, and Elbaum described an effective version of a nearest neighbor algorithm in a form realized using a neural network. It is similar to the system we just described. In the mid-1970s Cooper and Elbaum founded Nestor, Inc., which is one of the better-known companies developing commercial applications for neural networks. The classification technique they developed was the basis for early Nestor products, including what became the *NestorWriter*, a system that recognized characters handprinted on a special graphical input device. After a little training with the printing of a particular user, *NestorWriter* was quite accurate.

As we saw, a network unit can be used to do the nearest neighbor distance computation. The Nestor algorithm used this idea, and stored learned patterns as the weights of the units. To make a practical system, however, a number of important details had to be worked out. The Nestor algorithm was realized by a three-layer network (figure 13.7).

The output layer was not taught, but implemented a simple, preprogrammed algorithm that ensured that the network functioned as a classifier. In correct operation one and only one element in the output layer was active, which corresponded to the classification. That is, the network used a grandmother cell output representation.

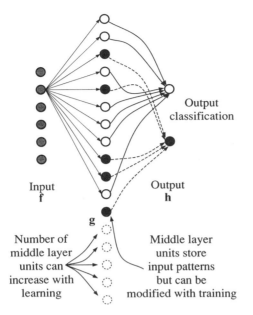

Input
f

h

classification

g

Number of
middle layer
units can
increase with
learning

Middle layer
units store
input patterns
but can be
modified with training

Figure 13.7   A neural network algorithm proposed by Reilly, Cooper, and Elbaum (1982) implements a prototype-based nearest neighbor categorizer. It has an input set of units, **f**, a middle layer, **g**, and an output layer, **h**. The number of middle layer units can increase with time. The output layer represents the categories. The middle layer units are given input weights corresponding to a useful subset of examples seen.

Units in the middle layer were referred to as *prototype* units because their weights corresponded to particular learned patterns whose classification was known. The output, **h**, layer was connected to the middle layer by connections arranged so as to ensure that only one output unit was active at a time, and that several middle, **g**, layer units were connected to a single output layer unit; that is, several different prototype-layer units could be given the same output classification. The selectivity and information-processing power of the system came from the connections between the input and the prototype layers. One unusual and important assumption was made. The number of units in the middle layer, where the classified examples were stored in the weights, was not fixed but could increase with time. Most networks assume the number of units to be fixed in the initital design, probably based on analogy with the nervous system, where it is difficult or impossible to create new neurons. Of course, in an artificial network, any convenient assumption can be made.

In operation, an input pattern was presented to the first layer of units. The input patterns were all normalized, that is, of length 1. The units in the middle layer computed the inner product between the input pattern and the weights. The result of the inner product was compared with a threshold, and if the inner product exceeded threshold, the output of the unit was the inner product, with a possible additive scaling constant; otherwise it was zero.

When the network started learning, the prototype, **g**, layer might contain no units at all, or preset starting weights might correspond to good initial values for a set of units to perform a particular function, for example, to classify letters or digits. Given an input pattern, every middle layer unit computed, potentially in parallel, the inner product between its weights and the input pattern. With a particular input pattern, many middle layer units could show output activity. Suppose that the unit that responded best has weights corresponding to an A. Then the input pattern was classified as an A. If the classification was correct the weights were not changed. If the pattern was not an A, something was done.

First, suppose that no unit responded to the pattern with activity above threshold. Then a new prototype unit was added to the middle layer. The weights of the new unit were set equal to a constant times the unrecognized input pattern. This learning rule is a disguised version of the Hebb rule because the output of the newly formed unit should be positive when the incorrectly classified pattern was presented. The newly formed prototype unit was given the appropriate classification connection to the output layer. If the misclassified input pattern recurred, it caused the newly formed proto-type unit to respond, and the connection to the output layer gave the correct classification.

Second, suppose the input pattern was recognized, but incorrectly. The incorrectly responding unit had its weights *reduced* by being multiplied by some constant, λ, less than 1. Future presentation of the input pattern would give less output from the unit that made the error.

This classifier has a very simple geometrical interpretation. The normalized input vectors lie on the unit hypersphere. Each middle layer (prototype) unit lies in a region of the hypersphere where it gives the largest response of all the middle layer prototype units. Therefore, all new input patterns falling into this region will be given the classification of that prototype unit. The size of the region is determined by how large the weights of the unit are. If the weights are multiplied by a constant, λ, less than, 1, the region will shrink. A misclassification shrinks the size of the region until the offending point is no longer within the region. When a new prototype unit is formed it gives rise to a new region centered on the pattern it just learned (figure 13.8).

The strength of this model, as well as other nearest neighbor techniques, is that by making enough prototype units in the middle layer, it can approx-

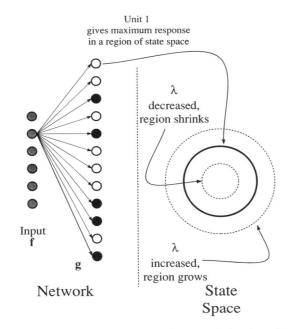

Unit 1
gives maximum response
in a region of state space

λ
decreased,
region shrinks

λ
increased,
region grows

Input
f

g

Network

State
Space

Figure 13.8 Each middle layer unit is given its classification to a region of the surface of the unit sphere. If the weights of the middle layer unit are multiplied by a constant, λ, less than 1, the region shrinks. If λ is greater than 1, the region expands.

imate decision regions in state space of any shape. As the system gains more and more experience, and more and more prototype units, it approximates complicated decision regions more and more closely. Figure 13.9 shows that the network can do a good job approximating the shapes of two complex decision regions.

A weakness of this system and nearest neighbor techniques in general arises from the same source. Strangely shaped decision regions may require many learned examples before patterns can be classified correctly, especially near the boundaries of the regions. Even worse are noisy systems in which boundaries can overlap, that is, in which the same point may be first given one classification and then another. Unfortunately, this occurs frequently in practice. The complete Nestor system looks for these special cases and handles them with some explicit rules. Overall the system is robust and simple.

The LVQ Algorithms

In the Nestor algorithm, the number of middle layer units increased as more patterns were learned and the size of the region associated with each unit changed with experience. Kohonen, Barna, and Chrisley (1987) suggested

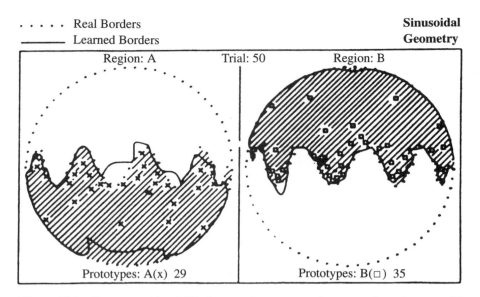

Figure 13.9 for Real Borders / Learned Borders, Sinusoidal Geometry, Region: A, Trial: 50, Region: B, Prototypes: A(x) 29, Prototypes: B(□) 35

Figure 13.9 Because each middle layer unit corresponds to a region of space with a given classification, complex decision regions can be formed. In this case, the initial data were taken from two regions separated by a sinusoidal boundary. These categories are not linearly separable. The lefthand figure shows the middle layer patterns (circles) corresponding to weights from units classified as from one region; the right-hand region shows middle layer patterns (squares) from the other. The shape of the decision region the network found after 50 patterns were learned is shown. The decision regions found by the network is starting to approximate the shape of the actual distribution giving rise to the data. Many more units are required to define the boundary than the large uniform regions near the centers of the categories. The more examples the network sees, the better the definition of the boundary regions. From Reilley, Cooper, and Elbaum (1982). Reprinted by permission.

another nearest neighbor algorithm in which the number of examples stayed constant but the locations of the examples in state space changed with learning. They called their algorithm *LVQ*, for *learning vector quantizer*. Many problems work with data that have a large amount of noise, so that the distributions of possible examples can actually overlap. That is, a particular input pattern could belong to either one category or to another. When this occurs, what is usually done is to try to estimate in which category a particular example is most likely to be. That is, if the probability that a pattern is a member of category A is 99% and of category B is 1%, it is almost certainly a member of category A. To apply this criterion, it is necessary to know the underlying probability distribution of the categories. If we knew everything about the category probabilities, we could make the best decisions. Unfortunately, we almost never have this knowledge, and we must estimate probabilities based on events we observe. A number of tech-

LVQ uses a fixed number of stored patterns.

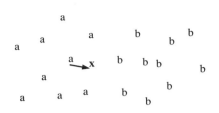

If **x** is classified correctly as an a,
the nearest pattern is moved toward x.

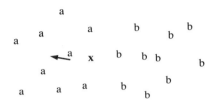

If **x** is classified incorrectly as an a,
the nearest pattern is moved away from x.

Figure 13.10 The LVQ algorithm described by Kohonen, Barna, and Chrisley (1988). There is a fixed number of classified points for a nearest neighbor classifier. During training, if a new pattern is classified the same as an old pattern, the weights of the nearest old pattern will be shifted slightly *toward* the new example. If the nearest neighbor is a different classification, the nearest neighbor is moved *away* from the new example.

niques in statistical pattern recognition are based on estimation of probability distributions, for example, Parzen windows.

The LVQ algorithm starts with a fixed number of processing units, each of which stores a classified pattern. Initialization requires a reasonable set of starting categorizations. The strategy is to move classified points around; that is, in a neural network implementation, to change the weights of the unit responding to the pattern.

During the training of LVQ, suppose an example with a known classification is to be learned. If the nearest neighbor to the example is classified the same way as the example, the stored pattern will shift slightly *toward* the example. If the nearest neighbor has the wrong classification, the stored pattern is modified and shifted *away* from the example it classified incorrectly. The patterns stored in the changed units minimize the number of incorrect categorizations (because they moved away from them) and maximize the number of correct categorizations (because they moved toward them). Units develop stored patterns that are good representatives of their categories by being moved toward other examples of the category, and by being moved away from regions with other classications (figure 13.10).

In the benchmark study, the LVQ algorithm did significantly better overall than back-propagation, the most popular neural network classification algorithm, for two difficult noisy test cases. In addition, LVQ learning uses only a fraction of computer time used by backpropagation or the Boltzmann machine.

Sparse Distributed Memories

Kanerva wrote a lucid, short book in 1987 describing a memory model that is related to nearest neighbor models, although not quite like them. The ideas are so original as to be unclassifiable, but it is appropriate to address them in this chapter. The model is an example of what has been called a weightless network because it works without weights as we traditionally define them.

Let us consider how to store information in a computer memory. We have some data in the form of a long binary word, and we want to put them somewhere so we can retrieve them later. We must now ask why we want to retrieve the data. A computer scientist would respond, to get back exactly what we put in, because we could potentially use any part of those data, and they have to be exact. So we will put the data in a convenient memory location, one of a large number of potential locations, make a note of where we put them in a list somewhere, and when we need the data, find them and use them. This approach places a heavy burden on the index but it makes no particular assumptions about the contents of the memory or what the data might be needed for.

In many neural network models it is assumed that both the input and output are represented by distributed activity patterns. To represent information, the element activities generate the basic system entities, the state vectors. Because these state vectors represent the data, it is often implicitly assumed that small differences in the state vectors *usually* do not make major, discontinuous changes in the meaning that the vectors carry. This is the assumption behind nearest neighbor classifiers.

The importance of nearness suggests a number of useful memory operations. One that is central to the operation of the Kanerva memory is the *best match*: given a new state vector, which are the stored data that most closely match the new data? The most straightforward way to answer this ques-

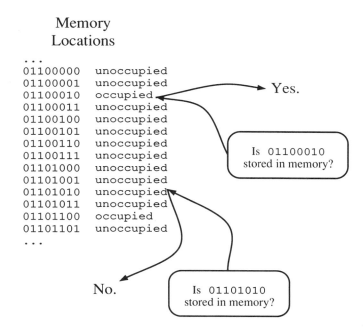

Memory
Locations

Figure 13.11   In the sparse distributed memory (Kanerva, 1987, 1993), the contents of an item to be stored are identical to its location in memory. Data are assumed to be binary vectors of high dimensionality. To see if an item is stored, all that is required is to see if that memory location is occupied. This is an unusual way to view a computer memory, but well suited to finding the *best match* between a new item and a stored item. In a traditional approach, every potential memory location must be tested to see if it contains the best match. However, in the simplest form of a sparse distributed memory, the nearest occupied location to the new item will contain the best match.

tion simply is to check all the stored information for the best match. This can involve a very large number of operations.

A much different way of solving best match uses a single step. To use one of Kanerva's examples, suppose we have 100 bits in the data we wish to store or retrieve. That means 2^{100} different possible binary vectors could be stored. Suppose we assume an enormous memory of 2^{100} memory locations. Then we can use the data to determine their own storage location. That is, we use the contents of the state vector as their position in memory (figure 13.11).

Checking for an exact match with stored data now becomes trivial: simply see if that memory location is occupied. Best match becomes a matter of looking at nearby locations: there are 100 one-bit differences, 9900 two-bit differences, and so on. If we assume relatively few memories compared with the size of the state space, then, as Kanerva says (p. 51). "... *if we construct the memory as we store the words of the data set*, we need only one location ... for each word of the data set." In Kanerva's terminology the memory is *sparse*, that is, most memory locations are unoccupied.

This insight now gives rise to a different way of looking at memory. Suppose that instead of 2^{100} locations, the hardware designer gives us a set of only 2^{20} 100-bit location designators, assumed to be randomly distributed across all the 2^{100} possible words. These prewired locations Kanerva called *hard locations*. Suppose we want to store a data word. One way to proceed would be to find the nearest hard location to the value of the word and put the word there. If the distribution of locations is uniform, a location close to the contents of the word should be available. A simple version of the best match retrieval operation is the same as before. When a word is to be checked for best match, the nearest hard location that contains something is most likely to contain the best match. Suppose we looked at all locations some distance around the word to be checked for the best match. The probability that the answer is correct can be computed. Simply looking at some radius around the word is not very accurate.

To make the memory work better, the storage procedure can be modified. Instead of storing a single example of the word, we store several examples,

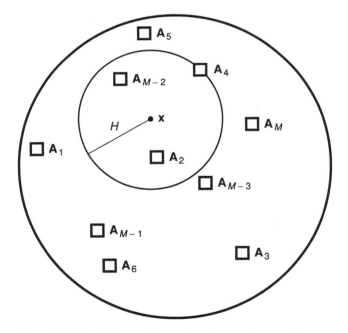

Figure 13.12 When an input data item, **x**, is stored, numerous copies are stored in all the hard locations within a certain distance of **x**. Location **x** itself may or may not exist. From Kanerva (1993). © 1993 by Pentti Kanerva. Reproduced by permission.

that is, we put copies of the input data into all the hard locations within some radius of the word to be stored (figure 13.12). Kanerva called this technique *distributed storage*. However, an immediate problem arises: we are likely to store information from several data words on top of each other, creating interference among different words. Kanerva suggested several ways to cope with this problem, all requiring an additional assumption of some kind about the information stored in the hard locations. The most common assumption was that each hard location did not store a single word, but formed instead the *sum* of all the words it stored during training.

Assume that an input word is an exact match to a stored word. Suppose that the radius that is to be checked for matches contains 1000 hard locations. In the act of storing the word, a copy of the input word was put into each of those 1000 locations. Other unrelated words from other stored data also will be stored at some of these locations. The most common one, however, will be the stored word because the radius of storage and the radius of

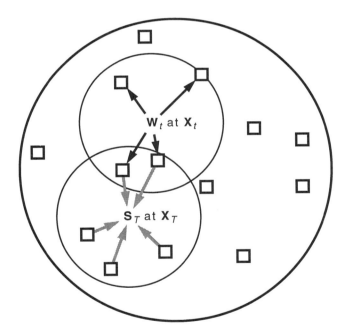

Figure 13.13 During storage, the word $\mathbf{w}_t$ was stored at many locations near the location $\mathbf{x}_t$. Suppose a new location $\mathbf{x}_T$ is used for retrieval that is close to $\mathbf{x}_t$. The retrieval process causes the words stored at the hard locations within a certain distance of the word $\mathbf{x}_T$ to become activated and to add their contents together, forming the sum $\mathbf{s}_T$. Many of these units will have have the data $\mathbf{x}_t$ stored in them because the regions overlap. By a process very much like averaging, a good approximation of the stored data can be retrieved from the sum. From Kanerva (1993). © 1993 by Pentti Kanerva. Reproduced by permission.

retrieval are coincident. Therefore, all the 1000 activated hard locations will contain a copy of the stored word. If we sum up all the data from the 1000 activated hard locations, we will have 1000 copies of the stored data, but fewer copies of other data that had a different radius of storage (figure 13.13). The stored pattern will dominate the sum simply because of the statistics of the items used to form the sum. The statistics of this summation process are similar to the summed vector memory described in chapter 7.

If a word to be checked and the stored word are not identical but close, the number of examples of the stored word within the radius of retrieval will still be large because the two circles—storage and retrieval—will be nearer

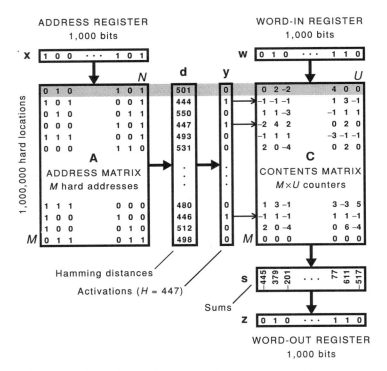

Figure 13.14 A detailed example of the structure of the memory words in a hard location in a sparse distributed memory. Each element in the word contains *integer* information, not binary. If a new data item is stored, a 1 at an element location adds 1 to the value of the element, a zero subtracts 1 from the value stored at that element. The righthand set of numbers lists a (subset!) of the million memory locations (hard locations) in the memory. The Hamming distance between the hard location and the input word is given in column d. If the Hamming distance is less than 447, column y is set to 1; that is, the hard location is within a certain (Hamming) radius of the input word. The active locations (with y = 1, marked with small arrows) are then added together to form a sum (s), which is converted into binary in the word-out register. From Kanerva (1993). © 1993 by Pentti Kanerva. Reproduced by permission.

to coincidence than for other, more distant words. Figure 13.14 presents a detailed example of such a memory.

Kanerva also proposed and analyzed a simple iterative technique for constructing the location of the best match, as long as the input and the best match were not too far apart. Suppose the average data word, generated by the summing process from the input word, is used as the input word for another retrieval step. If the target word and the input word are not too far apart, the resulting average data word will move closer to the target word. Repeated iterations will eventually converge to near the location of the best match. This technique can also be used to average out the noise from noisy data as in figure 13.15.

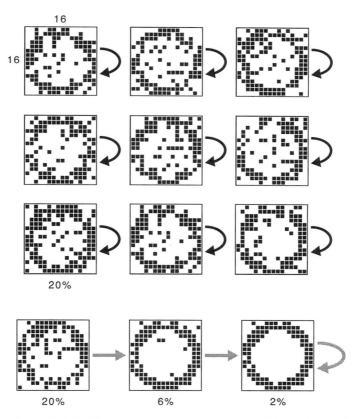

Figure 13.15 The averaging operation in recall can be used to reconstruct the prototype of a noisy set of stored patterns. These images are 256-bit patterns displayed as a two-dimensional pattern so accuracy of retrieval can be evaluated subjectively. Nine noisy words (20% noise) are stored and the tenth is used as a retrieval cue. By taking the output after one retrieval operation and using it as the input to another retrieval operation, the error rate can be used. A modification of this idea can be used to form an associative memory. From Kanerva (1993). © 1993 by Pentti Kanerva. Reproduced by permission.

The notion of distributed storage also allows for storage of other kinds of data in the stored word, in addition to the part that is used to determine the memory location. Large blocks of data can be retrieved based on similarity to the part of the data used to construct the memory location.

It is worth emphasizing the simplicity and reliability of a sparse, distributed storage system, as long as the requirement for completely perfect accuracy in retrieval is discarded. Kanerva showed that retrieval based on such a system was adequate for many purposes, and the amount of error potentially introduced in storage could be computed.

The lack of connections between processing units leads to an important practical point. In many neural networks the number of connections scales as the square of the number of elements. For those interested in building network hardware, this rapid increase in connections causes serious difficulties. The simplicity of the sparse distributed memory makes it exceptionally easy to build. As a bonus, Kanerva shows that the storage capacity is a linear function of the number of hard locations, and therefore, it will scale well to large systems.

Approximation and Interpolation

Poggio and Girosi (1989, 1990; Girosi and Poggio, 1990) popularized the idea that a multilayer neural network could be usefully viewed as an interpolating and approximating machine. They proposed a simple architecture to realize this behavior.

We have emphasized the associative nature of neural networks. Another way to formulate the problem is to consider a network as a mathematical function that transforms an input state vector to an output state vector. The training set provides a number of known examples of the function. Instead of having weights trained by an algorithm such as back-propagation, units in the hidden layer of a multilayer network derive their output values from functions based on the presented examples. We saw this idea used in nearest neighbor networks where middle layer units reproduced in their weights the pattern of previously classified examples.

Approximation

A well-known problem in mathematics is asking how well a function can be approximated as a sum of other functions. Fourier analysis is the best-known example of approximating a function as the sum of other functions. Another is principal components analysis, which does close to what a feedforward neural network does. Consider the two-layer linear associator. This device tranforms an input pattern, **f**, into an output pattern, **g**, by way of the connection matrix and the simple equation

g(f) = Af.

That is, **g** is a function of the input pattern, **f**. Suppose the true function we want to approximate is **G(f)**. How close can **G(f)** and **g(f)** get?

As we discussed in chapters 6 and 7, for the linear associator, a single output unit response is given by the inner product

$$g[i] = \sum_{i=1}^{n} a[j,i]f[j],$$

that is, the function, **g(f)**, is computed by inner products between sets of weights and input activities. Unfortunately, the desired function, **G(f)**, may be very poorly approximated by an inner product. As we saw in chapter 7, if we want to compute logical functions like X-OR or parity for **G(f)**, they cannot be approximated well, no matter how the matrix **A** is constructed.

Other feedforward networks we have seen can also be put in this form. For example, back-propagation can be regarded simply as a set of nested sums of sigmoidal functions. Going back to our discussion in chapter 9, we see that the output of a three-layer backpropagation network with layers **f**, **g**, and **h** is simply, for the kth unit in **h**, $h[k]$,

$$h[k] = f\left(\sum_{j} w_{\mathbf{gh}}[k,j] f\left(\sum_{i} w_{\mathbf{fg}}[j,i] \right) \right),$$

where $f(.)$ is the sigmoidal function. This expression can be extended to any number of layers by summing and applying the sigmoid function.

Consider the more general three-layer network shown in figure 13.16. The generic neural network neuron computes the inner product between an input pattern and the weights, followed by a nonlinear function, to generate the output of the unit. This assumption was made initially because it corresponded to what we thought was physiologically plausible. We can always make a new assumption about what these units are doing. Suppose we assume instead that units in the hidden layer, which we will call **g**, are computing some other function between stored information and the input pattern, **f**. We can call this function **g(f)**. For any input pattern, **f**, the output of the hidden layer will be the set of functions **g(f)**. The output layer units in the **h** layer weight these functions by a set of coefficients, $w_{\mathbf{gh}}$, and transform them with possibly another function, **h(g)**, to generate the output. There are many possibilities for weights and functions.

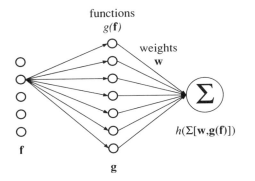

Figure 13.16 An adaptive feedforward neural network can be viewed as an approximating device. An input pattern, **f**, gives rise to an output pattern **h(f)**. Ideally, the output pattern would exactly equal some desired output pattern for all possible values of **f**. Because the individual output elements in this network act as summers of many input contributions, the output functions they can approximate are those that can be constructed from a function of a sum of individual functions.

Interpolation

The supervised neural network learning strategy depends on having known examples of an input pattern together with the associated correct output pattern. Consider, as an analogy, mathematical tables, which are like a training set for a neural network in that they provide a set of known input-output values. Often we want to know values of the function that lie between the values in the table. This means we must use the known values of the function and *interpolate* between them to arrive at an estimated value for the function at the unlearned input point. We can do this in many ways.

We could use linear interpolation, and simply say that the output for an intermediate input must lie on the line between two known points. For example, if the unlearned input value is half way between two input values, the output value is halfway between the two known output values (figure 13.17a). This is called a piecewise linear interpolation. The resulting function is not very smooth since the different line segments can have different slopes. We could require that the output function must be smooth by requiring that the functions drawn between known points must have matching derivatives at the points (figure 13.17b).

There are other ways to do interpolation. For example, if we require that both first and second derivatives of an interpolating polynomial function must be equal at the known points, it is referred to as a *cubic spline*. Any book on numerical analysis will describe many useful ways to perform interpolation. Both interpolation and approximation are central parts of

Interpolation

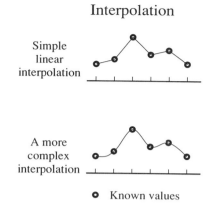

Simple
linear
interpolation

A more
complex
interpolation

○ Known values

Figure 13.17 Interpolating from known values. Suppose we know the values of a function for a few points, and we want to estimate the value for an intermediate point. The classic example of this is a table of mathematical functions. (Top) Simple linear interpolation from the two nearest values. (Bottom) A more complex interpolation might require derivatives to be matched as well as values, forming a smoother interpolating curve.

numerical analysis, and the techniques can become complex, although the intuitions behind them are usually simple. See chapter 1 of Carnahan, Luther, and Wilkes, 1969, or chapter 3 of Press, Flannery, Teukolsky, and Vetterling, 1986, for examples and programs.

We know that networks approximate an output function as the weighted sum of elementary functions of the input. We know the values of output for the members of a training set. Ordinarily, we simply choose a number of hidden units in a network based on experience or a good guess, and turn loose the learning algorithms. Suppose, however, we are more systematic. Suppose we use the known values to determine the functions computed by the middle (hidden) layer of a neural network. This approach is similar to the one we suggested earlier for a nearest neighbor classifier, and the one used in the Nestor algorithm (Reilly, Cooper, and Elbaum, 1982) where each hidden unit had its weights determined by a single classified example.

The simplest version of this approach says that each hidden layer unit computes a function determined by a single known input pattern. In this case we must have as many hidden units as we have classified examples. In extensions of the idea, we can apply the same ideas we used for nearest neighbor models. For example, we can choose to learn only best examples, or we can move the example patterns around based on network performance. These variations are discussed in detail in Poggio and Girosi (1989, 1990) and Haykin (1994). Let us consider only the simplest version.

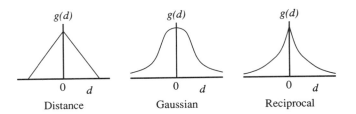

d is the distance between a learned pattern
and a new pattern.

Figure 13.18   Some choices for radial basis functions: a Gaussian, a linear function, and the reciprocal of a power of the distance. These functions are only a function of the distance between a learned pattern (a member of the training set) and a new pattern.

Suppose we have a known set of patterns, $\{\mathbf{p}_i\}$. Each middle layer unit computes a function, $\mathbf{g}$, that is a function of a known pattern, $\mathbf{p}_i$, and the input pattern, $\mathbf{f}$. There are many choices for the function $\mathbf{g}$; however, a particularly valuable set is referred to as *radial basis functions*. They depend only on the *distance* between the known pattern, $\mathbf{p}_i$, and the new pattern, $\mathbf{f}$, that is the square root of the inner product, $[(\mathbf{p}_i - \mathbf{f}), (\mathbf{p}_i - \mathbf{f})]$. Let us call this distance d, that is,

$$d = ([(\mathbf{p}_i - \mathbf{f}), (\mathbf{p}_i - \mathbf{f})])^{1/2}.$$

Figure 13.18 shows some common choices for functions, $g(d)$: the Gaussian, and various linear and polynomial functions. These assumptions have determined the computation performed by the hidden layer: there is no learning beyond inserting the training set into the functions in the hidden layer. The overall network learns by adjusting the weights between the hidden and output layers. Extensive theoretical justification exists for this technique, which gives excellent results in many practical problems. Radial basis functions are called that because the hidden layer functions serve as a *basis* set for the function to be approximated, and the functions display *radial* symmetry, being only a function of distance between learned and input patterns.

The close relation between this approach and interpolation can be seen in the trivial network diagrammed in figure 13.19. Here we have a single input unit, a three-unit middle layer, and a single output unit. We assume three known input data points, f_1, f_2, and f_3, with known output values, h_1, h_2, and h_3. The input points are equally spaced d units apart. We have assumed

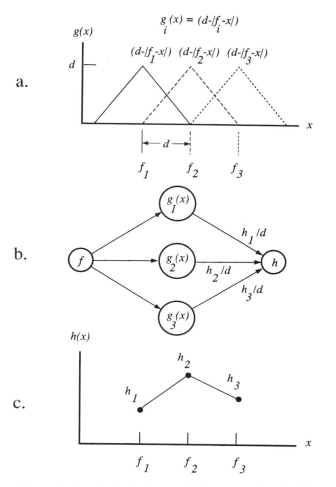

Figure 13.19   A simple interpolation network going from an input scalar value, f, on the x axis, to an output scalar value, h. We assume there are three known input-output pairs, spaced a distance, d, apart. Correspondingly, there are three units in the middle layer. The function being computed by the middle layer units, $g(x)$, is graphed in part a. The network itself is shown in part b. The weights connecting the middle layer to the output layer are shown next to the connections. This network does linear interpolation, as shown in part c.

a middle layer function $g(x)$, which is

$$g_i(x) = d - |f_i - x| \qquad \text{for } |f_i - x| < d$$

$$= 0 \qquad \text{for } |f_i - x| \geq d.$$

This function is graphed in figure 13.19a for the three input data points. It has a maximum value of d when the input point, x, coincides with a particular input data point, f_i, and is zero outside a distance, d, from the particular input data point. The function, g_i, is local, being zero outside a region of width $2d$ centered on a data point. We assume three data points, so there are three hidden unit functions, each centered on one of the data points.

If f_1 is input, the output value of the first unit in the hidden layer is d and the other two units have output zero. To obtain an output value of h_1, the weight connecting unit 1 in the hidden layer and the output should be h_1/d. The other connection strengths are found the same way. Suppose an input value of x appears halfway between f_1 and f_2, that is, it has value

$$x = \frac{f_1 + f_2}{2}.$$

It is easy to see the value of the two hidden layer functions are both $d/2$. Then the value of the output, h, will be given by

$$h = \frac{h_1 + h_2}{2}.$$

This network is doing simple linear interpolation between the equally spaced data points. Other functions that could be used include the Gaussians, reciprocals, as shown in figure 13.18, and many others.

Poggio and Girosi (1989) pointed out that the useful properties of the Gaussian lead to what they call science fiction neurobiology, in which the behavior of Gaussians is held to bear some resemblance to receptive field codings found in the visual system. Our simple example could also be modeled using radial Gaussian functions for the hidden unit. The resulting interpolation would not be linear, but might look a little wavy, depending on the values used for the Gaussian.

Even this simple example shows that important system parameters are related to the spacing of data points, and the shape of the functions and their behavior at near and far distances. With many points and different functions, complex kinds of interpolation can be modeled. Some obvious problems exist with networks that learn particular examples. If the training set includes noisy points, it is easy to end up with a network that accurately fits the fine details of the specific samples of learned noisy data; that is, the network has overlearned. We discussed this problem in connection with backpropagation (Geman, Bienenstock, and Doursat, 1992). A carefully chosen test set is of considerable importance. It is also possible to change adaptively the location of the centers of the basis functions instead of simply letting them be learned examples. The adaptive mapping algorithms described in the next chapter are very effective at generating a middle layer representation—a good set of examples—that captures the essentials of the data set and ignores noise.

Example-Based Reasoning in Cognition

The approximation-interpolation approach emphasizes the strong connections neural networks have with two well-developed disciplines in traditional applied mathematics. It can be dangerous to let someone else define your problems for you. If cognition or perception involves nothing more than getting input-output relations correct, the analysis in the last section shows a good way to do it. However, for networks more complex than simple feedforward networks, and for computations of a different type than function approximation, the analysis is less useful and may even be misleading. Many cognitive scientists are uncomfortable with the function approximation approach to learning, because it seems to offer little place for the flexibility of human performance. Although it captures some of the flavor of human learning, models using it take a long time to learn and seem unduly rigid. Much human learning is fast, and seems to make essential use of complex higher-level structures such as concepts and rules. Something important is being missed.

In our discussion in chapter 11 we made the distinction between exemplar-based and prototype-based concept models. The most common versions of exemplar-based concept models both look and act like nearest neighbor models, and some variants of them look like k-nearest neighbor; the prototype models resemble efficient nearest neighbor models, where a few particularly good examples organize a large region of state space. We can easily have many tens of thousands of examples of correctly classified letters or digits. But usually only a few examples are available for the complex events that humans encounter in their daily lives. Complex events by their nature contain a great many attributes and relations. When two of them are compared, it is difficult to know what parts are being used for the comparison.

Consider, for example, the quotation from Patrick Henry at the beginning of this chapter. Where the dash is seen, the Speaker of the House of Burgesses was noted to have shouted, "Treason!" In a few seconds he had extracted the relevant facts from a large set of historical information he had about Caesar, Brutus, Cromwell, Charles I, and George III, found the common relationship, and drawn a conclusion he disliked enough to provoke his outburst in a legislative chamber.

Case-Based Reasoning

An artificial intelligence technique based on small numbers of examples is called *case-based reasoning*. (See Riesbeck and Schank, 1989; and Stanfill and Waltz, 1987, for memory-based reasoning.)

One common way of making "intelligent" systems is to use what are called *expert systems*. Expert systems consist of a set of IF-THEN rules that are applied to input information and to previous results. For example, suppose we have a television that does not light up when the power switch is turned on. A useful rule for troubleshooting the set might be IF (power switch is on) AND (nothing happens) THEN (check to see if power cord is plugged in). Large assemblies of hundreds of rules form useful systems for reasoning about limited domains, for example, fault finding or configuring compatible groups of computer parts. In practice, however, expert systems work poorly in less constrained domains, and also when very many rules are used. Part of the trouble is psychological. Human experts have great difficulty formulating rules to describe what they are doing. It has been pointed out that novices use explicit rules, whereas experts operate with a combination of experience and intuition, neither of which they find easy to describe.

Rules and prototypes are abstractions often based on a great many examples. However, major events that contain significant lessons may not recur often and not in a clean form that allows abstraction. To use another example, in response to the question, "What do you think of our increased assistance in El Salvador?" many callers to a radio talk show gave answers along the lines of, "What happened in Vietnam should have taught the American people a lesson" (Reisbeck and Schank, 1989, pp. 5–6). The mode of thinking here is clear and understandable. However, it is hard to describe clearly what is being done and to model the sophistication of the analogy being drawn.

Memory for specific complex events plays a large role in human mental life. As Resibeck and Schank commented, "Human experts are not systems of rules, they are libraries of experiences" (1989, p. 15). The question then becomes how a case-based reasoner can become flexible enough to apply what may be a small number of specific old experiences to new problems: "A case based reasoner solves new problems by adapting solutions that were used to solve old problems" (Reisbeck and Schank, 1989, p. 25). The problem is suggesting good ways that memorized instances, rich in irrelevant detail, can be used to determine appropriate future behavior. For example, the Vietnam example clearly did not draw its lessons from the fact that both countries had hot climates. In later chapters we will suggest a few ways that simple memory-based associators can be used to do more flexible computation than just providing correct input-output relationships. General techniques for using memory-based reasoning in a flexible way to provide reliable generalization are an important unsolved research problem.

Summary

Although powerful neural network algorithms, such as back-propagation and the Boltzmann machine, have real virtues, for many practical classification problems simple models using nearest neighbor techniques make good engineering sense. They are reliable, fast, and understandable, and can be realized in parallel with the same kinds of elementary computing units used by more complex neural networks. They also have connections with human reasoning, as suggested by concept-forming models and case-based reasoning. It is tempting to say that rules and averages can be used when many examples fall into clean groupings. The fewer examples that are seen, and the more complex they are, the less we can rely on statistics and the more on other ways of drawing appropriate lessons.

References

B.G. Batchelor (1974), *Practical Approach to Pattern Classification*. London: Plenum.

B. Carnahan, H.A. Luther, and J.O. Wilkes (1969), *Applied Numerical Methods*. New York: Wiley.

T.M. Cover and P.E. Hart (1965), Nearest neighbor pattern classification. *IEEE Transactions on Information Theory, IT-13*, 21–27.

R.O. Duda and P.E. Hart (1973), *Pattern Classification and Scene Analysis*. New York: Wiley.

S. Geman, E. Bienenstock, and R. Doursat (1992), Neural networks and the bias/variance dilemma. *Neural Computation, 4*, 1–58.

F. Girosi and T. Poggio (1990), Networks and the best approximation property. *Biological Cybernetics, 63,* 169–176.

S. Haykin (1994), *Neural Networks.* New York: Macmillan.

P. Kanerva (1987), *Sparse Distributed Memory.* Cambridge: MIT Press.

P. Kanerva (1993), Sparse distributed memory and related models. In H.M. Hassoun (Ed.), *Associative Neural Memories: Theory and Implementation.* New York: Oxford University Press.

T. Kohonen (1984), *Self-Organization and Associative Memory.* Berlin: Springer.

T. Kohonen, G. Barna, and R. Chrisley (1987), Statistical pattern recognition with neural networks: Benchmarking studies. *Proceedings of the IEEE International Conference On Neural Networks, San Diego, 1988.* San Diego, CA: IEEE.

N. Nilsson (1965), *Learning Machines.* New York: McGraw-Hill.

Y.-H. Pao (1989), *Adaptive Pattern Recognition and Neural Networks.* Reading, MA: Addison-Wesley.

T. Poggio and F. Girosi (1989), *A Theory of Networks for Approximation and Learning.* Cambridge: MIT AI Laboratory.

T. Poggio and F. Girosi (1990), A theory of networks for learning. *Science, 247,* 978–982.

W.H. Press, B.P. Flannery, S.A. Teukolsky, and W.T. Vetterling (1986), *Numerical Recipes: The Art of Scientific Computing.* Cambridge: Cambridge University Press.

D.L. Reilly, L.N. Cooper, and C. Elbaum (1982), A neural model for category learning. *Biological Cybernetics, 45,* 35–41.

C.K. Riesbeck and R.C. Schank (1989), *Inside Case Based Reasoning.* Hillsdale, NJ: Erlbaum.

C. Stanfill and D.L. Waltz (1987), *The Memory-Based Reasoning Paradigm.* Cambridge, MA: Thinking Machines Corporation.

14 *Adaptive Maps*

Data representations using topographic maps are common in the nervous system. Neural networks can organize themselves to form maps in several ways. The most frequent way uses learning rules designed so that neighborhoods will learn to be similar in their responses. Map-forming algorithms are robust, fast, and effective, and have a number of applications, either alone or as part of more complex systems. The biological data, however, suggest that adaptive algorithms are only part of a more complex picture.

Real brains, as well as real computers, have a severe connectivity problem. Elementary parts have to be connected to other elementary parts to do the computation. The problem is especially severe for very-large-scale integrated (VLSI) chips, which are essentially two-dimensional. When circuit topology requires that wires must cross over each other, insulating layers must be provided, costs increase substantially, and yields of good chips decrease. One of the most powerful theoretical attractions of optical computation is the ability to ignore many connectivity problems because light waves can pass through each other without interacting.

Although the brain is three dimensional, connectivity is just as much of a problem for it. Axons of nerve cells take up a substantial amount of the mass of the mammalian brain. The white matter is composed of axon bundles, and perhaps 75% of cortical mass is "cabling" in the white matter. Although many of the neural networks we have seen assume full connectivity, that is, with every unit connecting to every unit, real neurons have severely limited numbers of inputs. The fractional connectivity of a real neuron is very low, at most perhaps a few thousand connections from tens of billions of neurons.

One of the first rules of wiring up any device is to try to arrange it so that things that have to connect to each other are close together. As we have mentioned, maps of various kinds are ubiquitous in cortex (see chapter 10), and one reason for mapping relationships is almost surely to ease connectivity problems. Some of the first theoretical studies of cortical self-organization looked for the formation of appropriate topographic relationships. Von der Malsburg (1973) modeled the development of visual cortex and remarked on the similarities between the arrangement of orientation selective units that he saw in simulation with the arrangement of orientation selective neurons recorded from cat primary visual cortex by physiologists. The key observation in both simulation and experiment was that neighboring cortical cells had similar, although not identical, orientation responses, and the orientation preference of cells changed smoothly across cortex. Nass and Cooper (1975) showed similar simulation results from a slightly different theoretical perspective. Figure 10.21 (Blasdell, 1992) shows a picture of the spatial arrangement of unit orientation preferences in visual cortex produced with modern imaging techniques.

There are many large-scale maps of a stimulus parameter on the cortical surface. In one example, which we discussed in chapter 10, the visual field maps onto visual cortex by a regular transformation. There is a huge over-representation of the fovea, the area of high-resolution vision, and under-representation of peripheral vision. But topographical relations are preserved on average, both in primary visual cortex and in some higher visual areas. Other sensory systems seem to use maps as well, for example, audition in terms of frequency, and the somatosensory and motor systems in terms of body surface.

It is not hard to get topographic self-organization in networks. Two discussions of this problem analyzed simple map-forming systems in computer simulation and with formal analysis (Amari, 1980; Willshaw and von der Malsburg, 1976). The basic result of both is that if the inputs have a correlational structure, the output layer can pick it up in the form of the spatial arrangement of units, given an output layer that shows simple Hebb learning combined with short-range excitation and long-range inhibition. This combination tends to make nearby cells learn to respond similarly, because of excitation, and to make distant cells respond differently, because of inhibition. A topographical map is often the resulting solution.

Kohonen Maps

Kohonen (1982) suggested a simple algorithm for map formation along these lines, and described the maps' potential information-processing abilities. Adaptive maps that reflect the statistical and similarity structure of the input patterns using the techniques we will describe are often referred to as Kohonen maps, although this is a little unfair to extensive earlier work.

One good aspect of map-forming algorithms is that they are amazingly robust. Almost any variant that respects the basic ideas will work. Some may work better than others, but mapping algorithms, like associative networks using Hebb synapses, "want to work." Another good feature is rapid learning; the outlines of the map are developed quickly.

The Basic Algorithms

Let us sketch the basic ideas of an adaptive map-forming algorithm. The development follows that in Kohonen's 1982 paper. In a topographically organized system, nearby units must respond similarly, at least to some parameters. The essential effect of the learning rule is to force the system to modify its connections so that this occurs; that is, neighborhoods of similarly responding units develop with as few discontinuities as possible. Intuitively, this means smooth changes in the properties of the units across the spatial arrangement of units.

First, full connectivity is generally assumed with respect to the input, so all input information is connected to all units. This assumption reduces the physiological realism of the models.

Second, it must be possible for neighborhoods to be recognized. Kohonen proposed that lateral connections could accomplish this. He assumed that cells are surrounded by short-range lateral excitation and long-range lateral inhibition. When a cell is activated, it tends to produce what he called a "bubble" of excitation around its location where cells are maximally excited, surrounded by an area where cells are turned off. His suggestion was that the bubble formed the biological basis for the neighborhood, for example, high excitation somehow enabled the modification rule to operate.

Third, connections in the neighborhood are modified so that all the units in the neighborhood start to respond more alike. Many learning rules can accomplish this; the one that is chosen is something of a matter of convenience and taste.

In operation, we start with an array of typical network neurons (figure 14.1). Each cell has small random connections to a set of input lines. The network is fully connected so that all cells see every input. An input pattern appears on the input lines. Units in the array of units respond to the parameter of interest. One unit, by chance, responds *best* to that input. The neighborhood around the unit then becomes active, perhaps by the bubble of lateral excitation. All the units in the neighborhood that is formed have their synaptic weights changed, so they respond more strongly to that input than they did before the change (figure 14.2). During learning, the shaded cells in

Kohonen Map Architecture

Array of Units

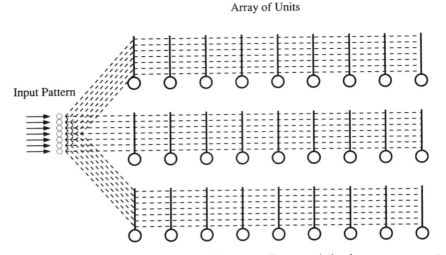

Input Pattern

Figure 14.1   Basic adaptive map architecture. Every unit in the array connects to an input pattern.

Kohonen Map Architecture

Array of Units   Neighborhood

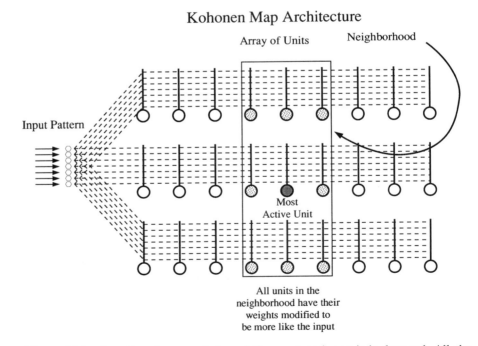

Input Pattern

Most
Active Unit

All units in the
neighborhood have their
weights modified to
be more like the input

Figure 14.2   A pattern is presented, and the most active unit is detected. All the units in a neighborhood around the most active unit are modified so that they will respond more strongly to that pattern in the future.

Excitatory Region

```
Initial state: +.  Final state: *
Inh Length Const: 6.00 Max Inh: 0.50 Exc Strength: 0.20 Ex Radius:  2
Neuron  L  Firing Rate: Spikes/Second |        |        |        |          L
        0........10........20........30........40........50........60........70
    12  *        +
    13  *        +
    14  *        +
    15  *        +
    16  *        +
    17  *                 +
    18                             +                                          *
    19                                     +                                  *
    20                                             +                          *
    21                                     +                                  *
    22                             +                                          *
    23  *                 +
    24  *        +
    25  *        +
    26  *        +
    27  *        +
    28  *        +
    29  *        +
    30  *        +
```

Produced by lateral interactions:
- **excitatory center region**
- **inhibitory surround region**

Final state has units fully on or fully off.

Figure 14.3 The lateral inhibition program discussed in chapter 4 can be easily modified to include local lateral excitation. In this case, a small region of lateral excitation is surrounded by a larger region of lateral inhibition. In response to a peak of activity, the final activity goes to either the fully on or fully off state. Kohonen (1982) referred to this as a bubble.

the neighborhood have their connections modified so that they tend to respond more vigorously to future patterns like the input pattern than they did before. One can modify connections in many ways to give this property. One way would be to add the input pattern to all the weights in the neighborhood and then renormalize the weights.

All that is strictly necessary to form the map is the formation of neighborhoods where learning ensures that the cells in the neighborhoods become more similar. The formation of a localized bubble of activity, presumably enabling learning in the region, can easily be demonstrated using a modification of the lateral inhibition program developed in chapter 4. Suppose we assume a large region of lateral inhibition surrounding a small center region where there is lateral excitation. A localized region of input activity rapidly leads to the formation of the bubble, an area of maximum excitation surrounded by a region of zero activity. Figure 14.3 shows the formation of such a region from an input given by a peak pattern used in our earlier discussion of winner-take-all networks. The effect is robust, and such an output pattern—center region fully on, surround region fully off—occurs for many parameter values.

These simple assumptions are adequate to do the topographical ordering. Neighborhoods become similar in their response properties, and global order follows from the the smooth gradation of one neighborhood into the next. It is possible to prove rigorously in a very simple one-dimensional system that self-organization proceeds properly. (See Kohonen, 1984, for more details.) The adaptive mapping algorithm is well suited to computer experimentation. In most cases formal analysis of the algorithm would be beside the point even if it was possible.

MAPS: An Adaptive Mapping Program

A simple program can be created based on Kohonen's original paper. Let us start by defining a two-dimensional square array of model neurons that we will call MAPS. We will define a Pascal TYPE Neuron, which contains a RECORD of the location of the unit, its activity, and its connections to the input pattern lines. We will start with two-dimensional input patterns, with only two connections between the input and the neuron. This gives an input that matches the dimensionality of the array. Many of the most interesting results for adaptive maps occur when the array dimension and the pattern dimensions do not match. The program is given in appendix L.

During initialization, all the connections are set to small random values. When a pattern is presented to all the model neurons, it provokes the most activity in one of them. It might seem more in the spirit of neural networks to look for the largest unit response. However, what could also be done is to look for the *closest* response, that is, the unit whose connections most closely match the input pattern. In our program, finding this unit is a two-step process. First, we compute the distance between the connections of every unit in the array and the input pattern. Second, we find the *smallest* value of distance, which gives the unit with the closest match to the input.

Once the closest unit is found, we use it as the center of a neighborhood. Initially the neighborhoods are very large, as large as 80% of the entire array, but shrink as the simulation goes on, until only the six closest array members are modified after many iterations. Using a very large initial neighborhood seems to be essential for proper formation of the maps. The modification rule multiplies the existing weights by 1 minus a constant between zero and 1, and adds the constant times the input pattern to the connection. This ensures that the weights in the neighborhood will become more like the input. Effectively, this rule moves the vector corresponding to the connection strengths of cells in the neighborhood toward the input vector. The procedure is repeated many times.

The rest of the program is concerned with human factors. Several initial probability distributions for the input pattern are provided for experimentation—a bar, a triangle, a square, and a circle. In the display, the program marks the unit that responds best when a particular pattern is presented. The display can present best responses either to points spaced throughout the original distribution, or points that were never presented to the algorithm.

Formation of the Map

The adaptive map picks up the distribution of probabilities of the input patterns. For example, consider input patterns drawn randomly with uniform distribution from the unit square. Units want to be close in their responses to their neighbors. One way to ensure this would be if the best response of the units was also arranged topographically in a square array, reflecting the input pattern distribution.

The initial input distribution is shown in figure 14.4. Each pair of letters denotes a point in the square, $[(-0.5, 0.5), (-0.5, 0.5)]$. *Capital letters denote columns and lower-case letters denote rows.* As many points are plotted as there are units in the array, so a perfect match between the unit best response and points in the distribution would be obvious.

Let us run the program for 1000 iterations, which is one time constant for the neighborhood decay function. Figure 14.5 shows the result. We have only plotted every third column for clarity. Even with only 1000 iterations,

Map Formation
Simple Uniform Distribution

```
                    Distribution of Input Points
    Alpha:    0.1 Nbhd Decay:  1000. Seed:  12345 Distribution: Square.
          Aa Ba Ca Da Ea Fa Ga Ha Ia Ja Ka La Ma Na Oa Pa Qa Ra Sa
          Ab Bb Cb Db Eb Fb Gb Hb Ib Jb Kb Lb Mb Nb Ob Pb Qb Rb Sb
          Ac Bc Cc Dc Ec Fc Gc Hc Ic Jc Kc Lc Mc Nc Oc Pc Qc Rc Sc
          Ad Bd Cd Dd Ed Fd Gd Hd Id Jd Kd Ld Md Nd Od Pd Qd Rd Sd
          Ae Be Ce De Ee Fe Ge He Ie Je Ke Le Me Ne Oe Pe Qe Re Se
          Af Bf Cf Df Ef Ff Gf Hf If Jf Kf Lf Mf Nf Of Pf Qf Rf Sf
          Ag Bg Cg Dg Eg Fg Gg Hg Ig Jg Kg Lg Mg Ng Og Pg Qg Rg Sg
          Ah Bh Ch Dh Eh Fh Gh Hh Ih Jh Kh Lh Mh Nh Oh Ph Qh Rh Sh
          Ai Bi Ci Di Ei Fi Gi Hi Ii Ji Ki Li Mi Ni Oi Pi Qi Ri Si
          Aj Bj Cj Dj Ej Fj Gj Hj Ij Jj Kj Lj Mj Nj Oj Pj Qj Rj Sj
          Ak Bk Ck Dk Ek Fk Gk Hk Ik Jk Kk Lk Mk Nk Ok Pk Qk Rk Sk
          Al Bl Cl Dl El Fl Gl Hl Il Jl Kl Ll Ml Nl Ol Pl Ql Rl Sl
          Am Bm Cm Dm Em Fm Gm Hm Im Jm Km Lm Mm Nm Om Pm Qm Rm Sm
          An Bn Cn Dn En Fn Gn Hn In Jn Kn Ln Mn Nn On Pn Qn Rn Sn
          Ao Bo Co Do Eo Fo Go Ho Io Jo Ko Lo Mo No Oo Po Qo Ro So
          Ap Bp Cp Dp Ep Fp Gp Hp Ip Jp Kp Lp Mp Np Op Pp Qp Rp Sp
          Aq Bq Cq Dq Eq Fq Gq Hq Iq Jq Kq Lq Mq Nq Oq Pq Qq Rq Sq
          Ar Br Cr Dr Er Fr Gr Hr Ir Jr Kr Lr Mr Nr Or Pr Qr Rr Sr
          As Bs Cs Ds Es Fs Gs Hs Is Js Ks Ls Ms Ns Os Ps Qs Rs Ss

    Iteration:        0. B)est rsp D)ist I)t until M)iss Q)uit X)cute >
```

Initial Distribution

Figure 14.4 A demonstration of the adaptive mapping program developed in the text. Possible input patterns are drawn from a simple uniform distribution. The distribution is continuous, but the same number of points is plotted here as the number of units in the array. In the unit array, capital letters mark columns, and lower case, rows.

the unit best responses are clearly arranged in a pattern like the uniform square distribution of the input pattern. On the righthand side of figure 14.5 units respond best to column A, largely in order, although the column takes a bend between points Ar and As. The other columns are curved somewhat, but are clearly picking up local neighborhood relations. Overall, the unit best responses fall into a skewed square pattern.

As learning continues, figure 14.6 shows the pattern of best responses after 10,000 iterations and figure 14.7 after 100,000 iterations. The pattern of best responses has become steadily more square-like, and has regularized, but has not changed fundamentally.

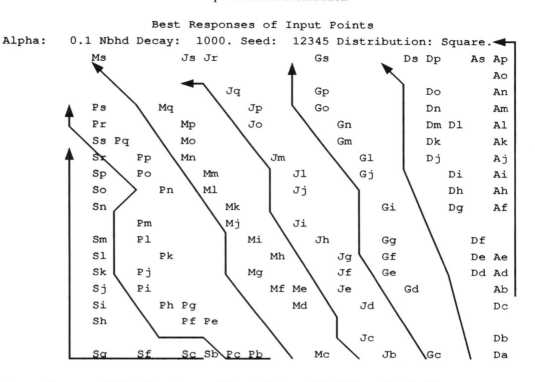

Map Formation
Simple Uniform Distribution

```
                 Best Responses of Input Points
Alpha:    0.1 Nbhd Decay:   1000. Seed:   12345 Distribution: Square.◄
```

```
Iteration:    1000. B)est rsp D)ist I)t until M)iss Q)uit X)cute >
```

1,000 Iterations

Figure 14.5 After 1000 iterations of samples taken from the uniform distribution in figure 14.4, the best response of units in the array of units is as shown. Note that nearby units respond to similar points in the distribution, indicating a map is being formed. Only every third column is shown for clarity.

Other Distributions

Suppose if, instead of having a continuous distribution of input patterns, we choose an input distribution that has holes in it. Consider a set of input patterns drawn from a distribution like that shown in figure 14.8. Only points in the bars can appear as input patterns. There is a large region of missing input points. After 1000 iterations, the network has developed best responses that look like a skewed square (figure 14.9), as before, but now we see distinct signs of a gap between columns N and F. If we plot units responding best to the *missing* patterns, we can see that the units in the gap respond best to these patterns, which of course, were never presented, but

Map Formation
Simple Uniform Distribution

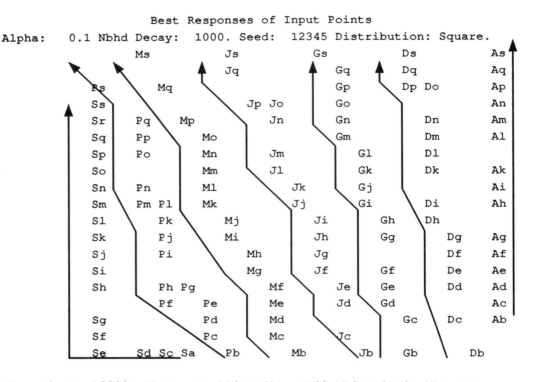

10,000 Iterations

Figure 14.6   The uniform distribution after 10,000 iterations.

the gap is somewhat smaller that it should be, based on the geometry of the input distribution. That is, these units respond best to patterns they never actually saw. After 100,000 iterations (figure 14.10) the gap appears further compressed. Units generalize across the gap in that unit best responses are developed to missing patterns. There has been a degree of spatial interpolation across a missing portion of the input distribution.

Map Pathologies

Map pathologies often occur, usually involving twists or islands. For example, the simulation presented in figure 14.11 had the same input statistics as the first one, that is, uniform on the square. But after only 1000 iterations the square has become twisted, so points from column A run up and points

Map Formation
Simple Uniform Distribution

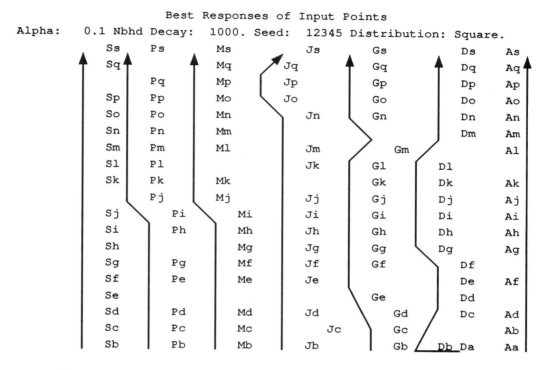

Best Responses of Input Points
Alpha: 0.1 Nbhd Decay: 1000. Seed: 12345 Distribution: Square.

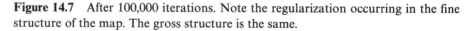

Iteration: 100000. B)est rsp D)ist I)t until M)iss Q)uit X)cute >

100,000 Iterations

Figure 14.7 After 100,000 iterations. Note the regularization occurring in the fine structure of the map. The gross structure is the same.

from column S run down. After 100,000 iterations (figure 14.12) column G has developed a cusp, apparently near where the twist occurs. This twist is difficult to remove, and is still present after 1 million iterations. Such pathologies are common, and Kohonen (1982) provided examples of several others. Local learning cannot be counted on to eliminate completely global errors in organization; initial mistakes can be impossible to overcome.

High-Dimensional Maps

Although it is easy to understand a mapping relation that takes a two-dimensional space into a two-dimensional space, the most interesting applications of adaptive maps are those in which a high-dimensional space is

Map Formation
Bar Distribution

```
                Distribution of Input Points
Alpha:    0.1 Nbhd Decay:   1000. Seed:   12345 Distribution: Bars.
          Aa Ba Ca Da Ea Fa                    Na Oa Pa Qa Ra Sa
          Ab Bb Cb Db Eb Fb                    Nb Ob Pb Qb Rb Sb
          Ac Bc Cc Dc Ec Fc                    Nc Oc Pc Qc Rc Sc
          Ad Bd Cd Dd Ed Fd                    Nd Od Pd Qd Rd Sd
          Ae Be Ce De Ee Fe                    Ne Oe Pe Qe Re Se
          Af Bf Cf Df Ef Ff                    Nf Of Pf Qf Rf Sf
          Ag Bg Cg Dg Eg Fg                    Ng Og Pg Qg Rg Sg
          Ah Bh Ch Dh Eh Fh                    Nh Oh Ph Qh Rh Sh
          Ai Bi Ci Di Ei Fi                    Ni Oi Pi Qi Ri Si
          Aj Bj Cj Dj Ej Fj                    Nj Oj Pj Qj Rj Sj
          Ak Bk Ck Dk Ek Fk                    Nk Ok Pk Qk Rk Sk
          Al Bl Cl Dl El Fl                    Nl Ol Pl Ql Rl Sl
          Am Bm Cm Dm Em Fm                    Nm Om Pm Qm Rm Sm
          An Bn Cn Dn En Fn                    Nn On Pn Qn Rn Sn
          Ao Bo Co Do Eo Fo                    No Oo Po Qo Ro So
          Ap Bp Cp Dp Ep Fp                    Np Op Pp Qp Rp Sp
          Aq Bq Cq Dq Eq Fq                    Nq Oq Pq Qq Rq Sq
          Ar Br Cr Dr Er Fr                    Nr Or Pr Qr Rr Sr
          As Bs Cs Ds Es Fs                    Ns Os Ps Qs Rs Ss

Iteration:          0. B)est rsp D)ist I)t until M)iss Q)uit X)cute >
```

Initial Distribution

Figure 14.8 A nonuniform input distribution. Only the regions marked (two bars) are possible input patterns.

Figure 14.9 After 1000 iterations, the best responses of the units reflect the input distribution to some extent. If inputs from the "forbidden region" are presented, they form a compressed map, although smaller than its actual proportion of the input distribution. Some cells do not have best responses to any presented patterns, but respond best to patterns that they can never have seen.

```
              Best Responses of Input Points
Alpha:    0.1 Nbhd Decay:   1000. Seed:  12345 Distribution: Bars.
          Sg Sf Se Re    Sd Sb Qc    Qb Pb Ob        Fb Eb Db
             Rg Rf          Qd       Pc Oc Nc
    Si Sh          Qe                             Ec Dc
    Sj    Qh    Qg Qf    Pe    Pd Oe Nd    Fd          Cc
    Rj    Qi    Ph    Pg Og Of Nf          Ee De Dd Cd
    Sk    Qj  Pi          Ng          Ff             Ce
    Rk                Oh             Ef Df          Be
    Sl    Qk Pk Pj    Oi    Ni       Fh    Eg       Cf
       Ql          Oj                Eh Dg          Bf
    Sn       Pl    Ok    Nk          Fi    Dh       Cg
       Qm Pm    Ol             Fk Ej       Di Ch    Bh
    Rn Qn          Om Nm       Fl El Ek    Dj    Ci    Bi
       Pn    On                         Dk    Ck Cj Bj
    So Qo    Po    Nn       Fm       Em    Dl    Cl    Bk
       Qp       Oo          Fn       En Dn Dm Cm
    Sq    Pp    Op Np          Eo    Do    Co Cn    Bm
    Rq Qq Pq                Fo
          Oq Nq          Fp Ep    Dp          Bo
    Ss Qs Ps    Os Ns       Fs Es Dq Cr Cq Cp Bp Ap    Bn

Iteration:    1000. B)est rsp D)ist I)t until M)iss Q)uit X)cute >
```

```
              Best Responses of MISSING Input Points
Alpha:    0.1 Nbhd Decay:   1000. Seed:  12345 Distribution: Bars.
                                          Mb Ja Ga
                                          Lc Hc
                                          Kc Gd
                                    Md Mc Ld
                              Mf Me Ke Ie He
                              Mg Kg If Hf Gf
                           Mh Lg Jg Ig Gg
                           Li Ki Ih Gh
                        Mj Ji Ii Hi
                        Lj Jj    Gi
                     Mk Kl Ik Hk
                  Mm Jm Il Gl
                  Mn    Im
               Mo Kn Jn Hn Gn
               Lo Jo In
               Mp Io Ho
               Kp Ip Hp
               Lr Jq Hq
            Ms Ls Js Is Gr

Iteration:    1000. B)est rsp D)ist I)t until M)iss Q)uit X)cute >
```

1,000 Iterations

Bar Distribution

```
                    Best Responses of Input Points
Alpha:    0.1 Nbhd Decay:  1000. Seed:  12345 Distribution: Bars.
          Sd Sc Sb Sa Qa Qb Pb Ob Oa Na    Fb Fa Eb Db Ca    Ba
          Se Rd                   Nb    Fc    Ec Dc Cb Bb Ab
          Sf    Re    Qd Qc    Pc Oc    Nc    Fd Ed    Dd Cc    Bc
                Rf    Qe    Pd    Od Nd    Fe Ee       Cd Bd Ad
          Sg    Rg    Qf Pf Pe Oe    Ne    Ff       De    Ce Be Ae
          Sh    Rh Qg    Pg    Of    Nf       Ef Df    Cf
          Si Ri    Qh          Og Ng       Fg    Eg Dg       Bf Af
                Qi Pi Ph    Oh Nh    Fi Fh Eh    Dh    Cg Bg Ag
          Sj Rj    Qj Pj    Oi    Ni    Fj    Ei    Di    Ch Bh Ah
          Sk Rk Qk    Pk    Oj    Nj    Fk    Ej    Dj    Ci Bi Ai
          Sl Rl Ql       Ok Nk       Fl    Ek    Dk    Cj Bj Aj
          Sm Rm Qm    Pl    Ol Nl       El Dl       Ck Bk Ak
          Sn Rn Qn    Pm Om    Nm    Fm    Em    Dm    Cl
             Ro       Pn On    Nn    Fn    En    Dn    Cm    Bl Al
          So    Qo    Po Oo    No    Fo    Eo Do       Cn    Bm Am
          Sp Rp Qp    Pp Op    Np    Fp Ep    Dp    Co       Bn
          Sq Rq Qq    Pq Oq Nq    Fq Eq    Dq Cq Cp Bp Bo    An
             Rr       Nr       Er    Cr    Bq    Ap Ao
          Ss Rs Qs Ps    Os Ns    Fr Fs Es Ds    Cs Bs Ar Aq

Iteration: 100000. B)est rsp D)ist I)t until M)iss Q)uit X)cute >
```

```
                    Best Responses of MISSING Input Points
Alpha:    0.1 Nbhd Decay:  1000. Seed:  12345 Distribution: Bars.
                                     Ma Kb Gb
                                     Mb Jd Gc
                                     Mc Id
                                     Me He
                                     Lf Gf
                                     Kg
                                  Mh Ih Gg
                                  Li Gi
                               Mi Kj Gj
                                  Jk
                               Ml Il
                               Lm Gm
                            Mm Km
                            Mn Lo Gn
                            Mo Kp Go
                            Mp Jp
                            Mq Hr
                            Mr Hs
                         Ms Ks

Iteration: 100000. B)est rsp D)ist I)t until M)iss Q)uit X)cute >
```

100,000 Iterations

Twisted Map Formation
Simple Uniform Distribution

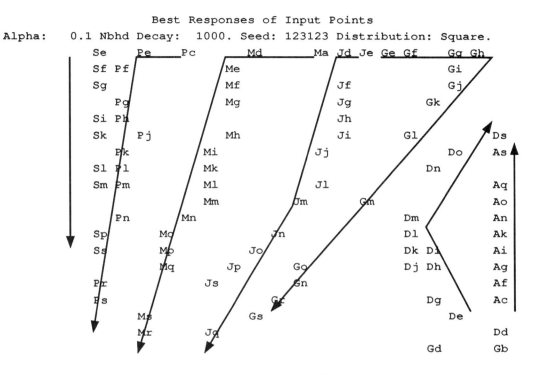

Best Responses of Input Points
Alpha: 0.1 Nbhd Decay: 1000. Seed: 123123 Distribution: Square.

Iteration: 1000. B)est rsp D)ist I)t until M)iss Q)uit X)cute >

1,000 Iterations

Figure 14.11   A map pathology. In response to a simple uniform distribution of inputs, the map has become twisted, so rows are moving upward on the right and downward on the left.

◄   **Figure 14.10**   After 100,000 iterations, the number and extent of cells responding well to the missing region of inputs has decreased.

Twisted Map Formation
Simple Uniform Distribution

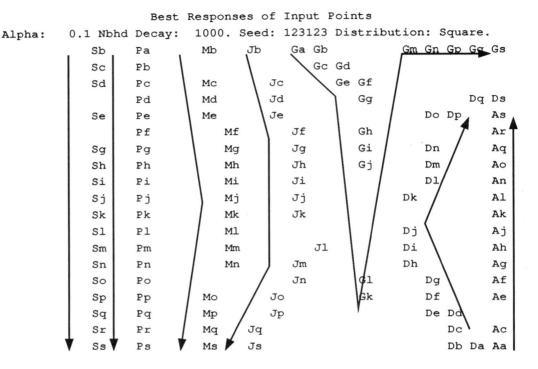

Best Responses of Input Points
Alpha: 0.1 Nbhd Decay: 1000. Seed: 123123 Distribution: Square.

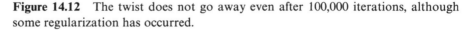
Iteration: 100000. B)est rsp D)ist I)t until M)iss Q)uit X)cute >

100,000 Iterations

Figure 14.12 The twist does not go away even after 100,000 iterations, although some regularization has occurred.

mapped onto a much lower-dimensional space. A number of statistical techniques, for example, principal component analysis and cluster analysis, are known optimum ways of doing this in some cases. It is not clear at present what relation adaptive mapping algorithms have to these more traditional techniques.

Kohonen's "phonetic typewriter," a simple isolated word speech-recognition system, is a good example of an adaptive map reducing a high-dimensionality space to a two-dimensional array of units. Recognition of speech by machines, like other perceptual tasks that humans do well, seems easy but is very hard. Neural net speech recognition systems in general were reviewed by Lippman (1989).

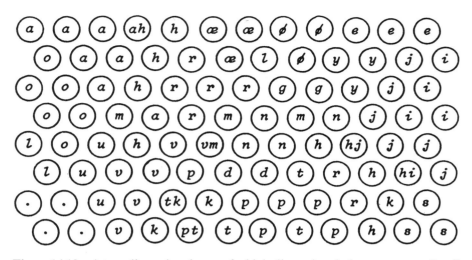

Figure 14.13 A two-dimensional map of a high-dimensional phoneme space. Small regions of units respond best to one or another phoneme. A spoken word corresponds to a trajectory on the map. By identifying phonemes with regions on the map, a phonetic typewriter can be produced. From Kohonen (1987). Reprinted by permission.

Kohonen used a two-dimensional array of about 100 units to analyze a much higher-dimensional input vector representing the speech signal. He used a straightforward spectral representation of the input waveform, derived from a fast Fourier transform, of a segment of the speech waveform a few milliseconds long. Fifteen numbers characterize the spectrum during that interval. This representation is sometimes modified slightly by adding a component representing amplitude, or by incorporating two spectra taken from different times, 30 msec apart.

It is not clear what the proper features of the speech signal are. This subject has been a matter for debate among linguists for decades. A phoneme is roughly defined as an elementary sound in speech, say, /t/ or /o/. One reason speech recognition is hard is that a great deal of variation exists among different examples of the same phoneme from the same speaker, and much more among the same phoneme spoken by different speakers. In addition, a phoneme is strongly influenced by its neighbors (*coarticulation*), so what is called the "same" phoneme from the same speaker in different contexts varies considerably in the physical signal. For example, much of the information identifying the phoneme /k/ in a consonant-vowel pair such as /ka/ is actually carried by the signal corresponding to the vowel. The physical signals perceived as /k/ can be quite different in different vowel contexts, for example, /ka/ versus /ki/. Much effort has been devoted to identifying invariant features that might unambiguously characterize individual speech sounds, with little success.

Despite these complexities, the adaptive algorithm produced good mapping relationships. Even though phonemes were never explicitly taught to the system, different examples of the same phoneme tended to occur near each other on the map. The process used to form the map seemed to make contiguous regions on the array of units correspond to the same phoneme. Once the map displayed this behavior, the speech problem became finding which region on the map corresponded to which phoneme. A word containing several phonemes is represented by a trajectory on the map. Figure 14.13 shows the map resulting from one learning run. If its regions are connected to appropriate letters on a typewriter, we can construct a *phonetic typewriter*. Because the correspondence between phonemes and written letters is so regular in Finnish, spelling produced this way is often correct. This simple technique would not work well in English, which has notoriously irregular spelling. (See the discussion of NETtalk in chapter 9.)

High-Dimensional Simulation

Our simulations using Kohonen networks used input patterns taken from a low-dimensional continuous distribution. In general, in such a system input patterns will never repeat and the map will organize itself statistically. This is different from the way a traditional supervised or even unsupervised neural network learns. It has a discrete training set of vectors, and uses learning to build the right input-output relations. Perhaps learning distinct state vectors, instead of a distribution, will form a different kind of mapping system.

Schyns (1992) looked at Kohonen maps that learned discrete patterns coded as high-dimensional state vectors. Such systems are organized in a simple way, with a unit giving best response to one of the learned vectors—a prototype vector—surrounded by units that respond less well to the prototype. Units half-way between two prototype units responded best to patterns half-way between the prototypes. This result bears an interesting similarity to the simple neural network prototype-forming systems we analyzed in chapter 11.

We can reproduce Schyn's experiment easily, making a small modification of our mapping program to do a high-dimensional simulation. We will use our character coding technique in a 384-dimensional (48-character) space. The main modification simply involves changing the definition of the Pascal RECORD Neuron in the program to have 384 connections instead of 2.

We will use a set of test vectors that were generated for a set of simulations on the learning of elementary arithmetic (see chapter 17 for details). Strong psychological evidence suggests that human number representation contains a symbolic part, say the number name, which is somewhat arbitrary, and an analog part. The analog part explains the many experiments suggesting that, to humans, numbers seem to have a large sensory component such as weight or light intensity.

The pattern files for the 10 digits were represented as strings of characters, as in chapter 7. To test the effect of correlation of input patterns on the formation of maps, 48 character strings were converted into 384-dimensional state vectors.

The pattern materials chosen were based on material we already had used for our simulations of arithmetic learning and are shown in figure 14.14. These vectors representing "number" were two-part vectors representing the digits from 1 through 0 (0 was considered to be 10 for the simulation). The first part of the vector coding was a symbolic string. The first four characters coded the number name, followed by a random-character string. The second part was a topographical bar code representing number magnitude. As the magnitude of the number increased, the area of activity in the state vector moved to the right. The analog part of the representation assumed that 3 is bigger than 2, just as a bright light is more intense than a dim one.

We would expect different number symbols to be uncorrelated. However, the bar codes should produce high correlations between numbers close to each other. In this coding, the underline, '_', corresponds to eight zero-vector element values. These test strings contain many zeros.

We made three sets of patterns to be learned, with three groups of ten 48-character strings. One group, the number patterns, had both the random part of the string and the bar-coded part. The second group, the symbol patterns, had only the random strings. The third group, the bar patterns, had just the bar codes.

A pattern emerged during learning that was similar to that observed by Schyns. Figure 14.15 shows the 19 by 19-unit map after it learned the set of number patterns. The 10 input patterns corresponded to the 10 codings in figure 14.14. Figure 14.15 presents the array resulting from the use of the number patterns, and for each unit gives the number of the input vector that was closest to the weights of the unit. Note that best responses to a particular input group together. The asterisk marks the unit that gives the overall best response to a particular vector. These prototype units are surrounded by units that respond more weakly to the same vector. It is clear that nearby numbers are nearby on the map, that is, 1 is next to 2, which is next to 3. The

```
                          Stimulus Files


Complete number representation.
Contains both random and bar coded parts.
file: numbers.rno

One1  #$mL%<'EW'  =====_____
Two2  ^H^5L&8UWI  ___=====_____
3Thr  g,P'Ty\Ava  _____=====_____
For4  |[unw.`X@L   _____=====_____
5Fve  jYjdj/tQ|3  _____=====_____
Six6  "P?%|]<a8S   _____=====_____
7Svn  lHz#X=i^0V  _____=====_____
Eig8  0\{HLct*Gv  _____=====_____
Nin9  8u"</(zV?g  _____=====_____
10Tn  BfC`!VM$&(  _____=====_____


Contains random string ("symbolic") part of representation.
file: symbols.rno

One1  #$mL%<'EW'  _____
Two2  ^H^5L&8UWI  _____
3Thr  g,P'Ty\Ava  _____
For4  |[unw.`X@L   _____
5Fve  jYjdj/tQ|3  _____
Six6  "P?%|]<a8S   _____
7Svn  lHz#X=i^0V  _____
Eig8  0\{HLct*Gv  _____
Nin9  8u"</(zV?g  _____
10Tn  BfC`!VM$&(  _____


Contains bar coded part of number representation.
file: bars.rno
```

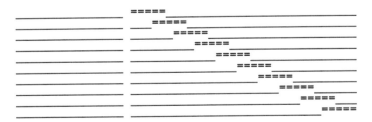

Figure 14.14 Input patterns used in a high-dimensional mapping experiment. These patterns are like those used in the arithmetic simulations described in chapter 17. The representation of number is a hybrid, partly symbolic and partly analog. The symbolic codes are uncorrelated; the analog code is highly correlated.

Regions for Numbers

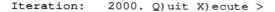

```
            Best Responses    (* is maximum response)
Iteration:  2000    Data File: numbers.rno   Seed: 123123
        10 10 10 10  9   9   9   9  9*  9   9   9  8   8   8  7   7   7  7*
        10 10 10 10 10   9   9   9   9   9   9   8   8   8   8  7   7   7
        10 10 10 10 10  10   9   9   9   9   8   8   8   8   8  7   7   7
        10 10 10 10 10 10 10   9   9   9  8   8   8* 8   8  8  7   7
        10 10 10 10 10 10 10 10   9   9  8   8   8   8   8   8  4   4
        10 10 10 10 10 10 10 10 10   5  8   8   8   8   8   8  4   4   4
        10 10 10 10 10 10 10 10   5   5   5   5  8   8   8  4   4   4
        10 10 10 10 10 10 10   5   5   5   5   5   5   5  4   4   4  4*
        10 10 10 10 10 10 10   5   5   5  5*  5   5   5   5  4   4   4   4
        10   1   1   1   1   1   1   1   5   5   5   5   5   5   5  4   4   4   4
         1   1   1   1   1   1   1   1   5   5   5   5   5   5   5   5  4   4   4
         1   1   1  1*  1   1   1   1   1   5   5   5   5   5  6   6   6   6   6
         1   1   1   1   1   1   1   1   1   5   5   5   5  6   6   6   6   6
         1   1   1   1   1   1   1   1   1   5   5   5  6   6   6   6   6   6
         3   3   1   1   1   1   1   1   2   2   2  6   6   6   6   6   6   6
         3   3   3   3   1   1   1   2   2   2   2   2  6   6   6  6*  6   6   6
         3   3   3   3   3   3   2   2   2   2   2   2   2  6   6   6   6   6
         3*  3   3   3   3   3   2   2   2   2   2   2   2  6   6   6   6   6
         3   3   3   3   3   3   2   2   2  2*  2   2   2   2  6   6   6   6   6

Iteration:   2000. Q)uit X)ecute >
```

Three Groupings

Figure 14.15 Maps formed with the 10 discrete number patterns contain regions corresponding to the numbers. Units that respond best are marked with an asterisk. Note that nearby numbers tend to be in the same parts of the map, suggesting the map responds to the higher-level structure in the patterns. This typical simulation had three groupings in which successive numbers were adjacent to each other.

number patterns have three contiguous groupings. Numbers in each group are connected by lines in the figures so the groups can be seen clearly.

If the input vectors were completely uncorrelated, we would expect the clumps to display no particular structure, that is, the clump for 7 might be right next to the clump for 1 and the clump for 6 might be on the other side of the array. The map for the input file using essentially random strings shows this behavior, since the representations are uncorrelated (figure 14.16). Lines are drawn between adjacent numbers for a representative simulation. There are seven separate groups.

The map for the input file containing the bar codes is highly correlated, as a glance at figure 14.14 will show. Adjacent numbers form adjacent regions,

Regions for Symbols

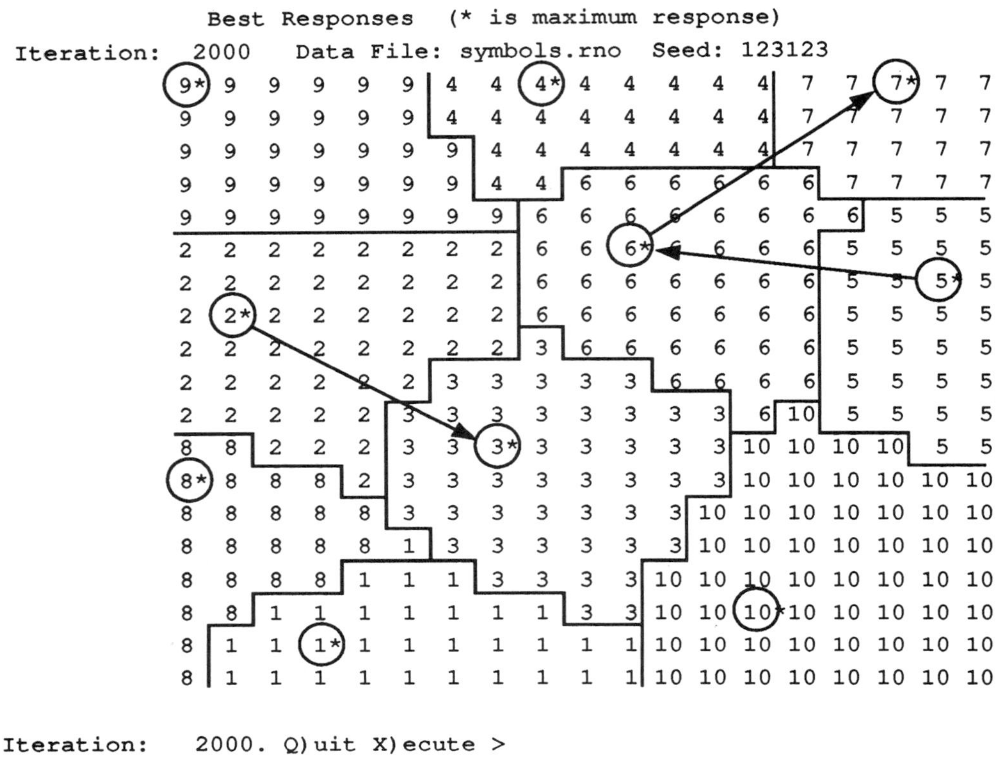

```
        Best  Responses   (* is maximum response)
Iteration:  2000    Data File: symbols.rno   Seed: 123123
          9*  9   9   9   9   9 | 4   4  4*  4   4   4   4   4 | 7   7  7*  7   7   7
          9   9   9   9   9   9 | 4   4   4   4   4   4   4   4 | 7   7   7   7   7   7
          9   9   9   9   9   9   9 | 4   4   4   4   4   4   4 | 7   7   7   7   7
          9   9   9   9   9   9   9 | 4   4   6   6   6   6   6   6 | 7   7   7   7
          9   9   9   9   9   9   9   9 | 6   6   6   6   6   6   6 | 6 | 5   5   5
          2   2   2   2   2   2   2   2 | 6   6  6*  6   6   6   6 | 5   5   5   5
          2   2   2   2   2   2   2   2 | 6   6   6   6   6   6   6 | 5   5  5*  5
          2  2*  2   2   2   2   2   2 | 6   6   6   6   6   6   6 | 5   5   5   5
          2   2   2   2   2   2   2 | 3   6   6   6   6   6   6 | 5   5   5   5
          2   2   2   2   2   2 | 3   3   3   3   3 | 6   6   6   6 | 5   5   5   5
          2   2   2   2   2   3   3   3   3   3   3   3 | 6 | 10  5   5   5   5
          8   8   2   2   2 | 3   3  3*  3   3   3   3   3 | 10  10  10  10 | 5   5
         8*  8   8   8   2 | 3   3   3   3   3   3   3   3 | 10  10  10  10  10  10
          8   8   8   8   8 | 3   3   3   3   3   3   3 | 10  10  10  10  10  10
          8   8   8   8   8   1 | 3   3   3   3   3 | 10  10  10  10  10  10  10
          8   8   8   8   1   1   1 | 3   3   3   3 | 10  10  10  10  10  10  10
          8   8 | 1   1   1   1   1   1   1 | 3   3 | 10  10 10* 10  10  10  10  10
          8 | 1   1  1*  1   1   1   1   1   1 | 10  10  10  10  10  10  10  10
          8 | 1   1   1   1   1   1   1   1   1 | 10  10  10  10  10  10  10  10
```

Iteration: 2000. Q)uit X)ecute >

Seven Groupings

Figure 14.16 The map formed from the uncorrelated symbol patterns shows no sign of higher level structure.

and there are only two discrete groups. Clearly, the map is picking up the correlational structure of this high-dimensional input pattern set and using it to structure the map. The map for the input file for numbers containing both parts of the representation (figure 14.15) is intermediate in organization between the maps for the subparts, the highly correlated bars (figure 14.17), and the uncorrelated symbols (figure 14.16). The transformation between the high-dimensional input space and the two-dimensional map is geometrically quite complex, although simple in practice: it is learning best examples— prototypes—and generalizing and interpolating among them as best it can. We can see from this simulation that the map is responding to both the local and global structure in the input patterns. Schyns used this result as the basis for some interesting cognitive speculations.

Regions for Bars

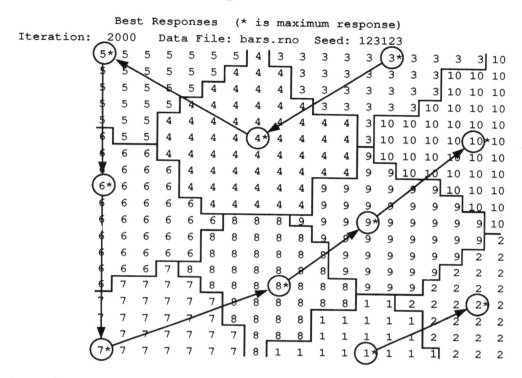

Best Responses   (* is maximum response)
Iteration: 2000   Data File: bars.rno   Seed: 123123

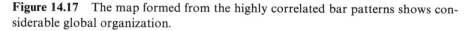

Iteration: 2000. Q)uit X)ecute >

Two Groupings

Figure 14.17 The map formed from the highly correlated bar patterns shows considerable global organization.

Our high-dimensional simulation program can demonstrate another important aspect of simple mapping algorithms. Mapping algorithms can be exquisitely sensitive to the earliest learned items. The first few items can have a disproportionate effect in structuring the final form of the map. This also means that Kohonen maps respond strongly to prewired initial structure in the map. We will not demonstrate this here, but it is easy to show using modifications of the simulation programs.

To show sensitivity to initial learning, we will use the 10 bar patterns discussed before. We looked at the maps produced from 4000 trial learning sequences. Suppose we generate sequences that differ only in the very first item. We generated 10 such sequences that differed only in that the first presented pattern was set to be 1, 2, 3, and so on. The 10 different sequences gave rise to 9 completely different final maps. (Two sequences produced

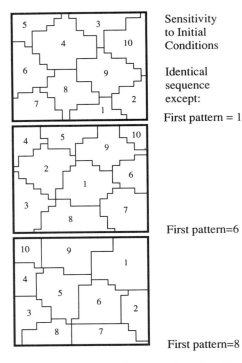

Sensitivity
to Initial
Conditions

Identical
sequence
except:

First pattern = 1

First pattern=6

First pattern=8

Figure 14.18 Adaptive mapping algorithms are exquisitely sensitive to the first few items they learn. These three maps were formed from identical 4000-pattern sequences that differed only in one pattern: the first pattern in one case was the vector coding 1, in the second a 6, and in the third an 8. The resulting maps look very different.

identical final maps.) Figure 14.18 shows three of the maps; inspection of the entire set suggested no obvious commonalities among maps from almost identical sequences.

Schyns also provided an interesting high-dimensional example of map formation. For this simulation, he used 64 by 64-pixel images of a teapot, presented uniformly for all angles between two extreme rotations, 180 degrees apart. Image rotations were in a plane between the extreme orientations. The system had only seven units. The initial neighborhood size was very large at the beginning of training and very small at the end. Figure 14.19 shows the images formed by the connection strengths to the seven units, from the random connections at the start of training to the final set of connection strengths after 1000 learning trials. At the end of learning, a map organized by image rotation was formed, with the two extreme orientations represented by the connection strengths at the two ends of the map. The connections of the other units represented roughly equally spaced rotations between the extreme values. We could view this set of seven units as a

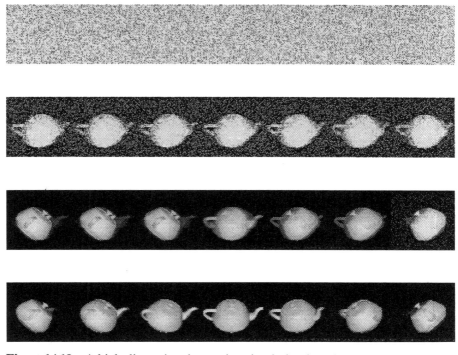

Figure 14.19 A high-dimensional mapping simulation based on rotated images of a teapot. The system had seven units. The images learned were 64 by 64-pixel arrays derived from rotations. The two extreme images were 180 degrees apart, and the example rotations were uniformly distributed between the extremes. The four rows of images (top to bottom) correspond to the connection strengths of the seven units during different phases of learning. The top row shows the initial random connectivity; the bottom row after about 1000 images were used to form the map. The bottom row contains seven images of the teapot showing roughly equal rotations between the extreme orientations. Intermediate image orientations can be estimated accurately by the amount of activation of the units representing nearby orientations. Reproduced courtesy of Dr. Philippe Schyns.

distributed representation for orientation. Intermediate values of image orientation can be estimated accurately by the amount of activity of the tuned units. The sensitivity to initial conditions seen before does not occur here because of the small number of units and the highly constrained input set.

Biological Caveat

The learning process used to form maps depends on the formation of regions where cells are similar in responses; that is, an input tends to excite a region of cells so that neighborhoods can develop. It is true that everywhere one looks in the brain there are maps of sensory surfaces, as we discussed in chapter 10. However, the biological picture is far more complex than that suggested by simple adaptive models.

First, biological mapping relations are initially sketched in during the early development of the nervous system. It is not certain how this is done. (See Udin and Fawcett, 1988 for a review of the neurobiology that contains references for some of the following comments.) Several major theories have been put forth, each with some experimental support. Probably several mechanisms are involved. One suggestion is that chemical signals—that is, addresses—on the target structure are actively searched for by the growing connections. (Alan Turing, of computing machine fame, was one of the first to analyze this idea.) Some evidence exists for such chemicals for a few developmental systems.

Another suggestion is that axons grow out in temporal sequence and innervate the target structure in order; or axons could physically maintain neighborhood relationships as they grow. Structures could even self-organize from zero, based strictly on network learning rules and system interactions, as discussed here. This is extremely unlikely to be the only mechanism involved, despite the high expectations of neural modelers. Given the sensitivity of the mapping algorithms to initial conditions, it would be hard to explain similarities in maps between brains if they organized de novo. Biologically, it would make no sense to trust something as important as the fundamental structure of the sensory computation to a random process.

Once a map is formed, however, some biological systems show its plasticity in response to various environmental events. Some of this plasticity might be explainable by the kinds of learning rules and self-organization studied in neural networks. A particularly striking example is the connections between the retina and the optic tectum in both goldfish and frog. A topographic map of the visual world is on the surface of the tectum. If half of the tectum is destroyed after regeneration, the *entire* visual field will be mapped onto the remaining *half*. Somehow all the topographic map relations are maintained, even though the actual connections from cell to cell have changed markedly. Such effects are seen in mammals, but only in very early development. Sometimes, however, results do not conform to such a nice picture, and plasticity does not occur. For example, the chick shows no such ability to expand and contract the visual field representation.

Sometimes plasticity occurs in one structure in a sensory pathway and not in other structures in the same system. In the frog, a tectal map is produced by superimposing maps from the *ipsilateral* (same side) eye and the *contralateral* (opposite side) eye. If a *Rana pipiens* frog has one eye rotated surgically, the two maps develop out of register. Yet in *Xenopus* (a toad), the same manipulation leads to maps in register. Fawcett and Udin (1988) gave many more examples of the complexities of map formation.

In normal cats, except for a narrow vertical strip that includes the fovea, one half of the visual field is represented in one hemisphere, and the other half in the other hemisphere. Some Siamese cats have severe miswiring of the visual system, apparently caused by errors in connection from the eyes to the lateral geniculate body. The miswiring is not corrected, suggesting that chemical or other cues are used in the initial connections. Some axons do not cross from the left side to the right side of the brain properly. In one class of miswired Siamese cats, a point of the visual field *and its mirror image on the other side of the vertical strip* are mapped onto the same point in cortex, causing severe conflict between out-of-register images (Kass and Guillery, 1973; Hubel and Wiesel, 1971; Schatz, 1987). Substantial behavioral visual defects are produced by this miswiring that do not seem to be correctable by plastic mechanisms. Yet other visual projections in the same aberrant animals seem to be connected correctly.

Another class of predictions is made by the adaptive mapping algorithms if we apply them to biology. The algorithms strongly suggest that nearby cells would be correlated in their discharges. Superficially, this seems consistent with the biological maps. In visual cortex, nearby cells are sensitive to stimuli located in nearby regions in the visual field. However, cells in cortex respond to many other stimulus parameters as well. For example, they respond best to stimuli moving with a certain velocity, or to bars with a preferred orientation, or to input from just one eye or to input from both eyes, and to several other independent parameters. Often many maps are superimposed; for example, the visual field map is present as well as a map of orientation preference. In primary visual cortex, certain cells, the so-called color blob cells (Livingstone and Hubel, 1984), partake of the spatial map but have no orientation selectivity and are responsive to color. These cells are found in groups (blobs) interspersed amid direction-selective cells. As we mentioned, many neural mechanisms such as lateral inhibition operate to sharpen cell responses and make them *less* like their neighbors. In the relatively small number of cases in which this question has been specifically studied experimentally, nearby cortical cells are usually not highly correlated in their discharges when the animal is awake and alert.

Does it make theoretical sense to have nearby cells correlated in their discharges? It does not make sense from the point of view of information on theoretical efficiency to use highly correlated units. If two units are highly correlated in their responses to input patterns, they convey less information than two uncorrelated units would. Suppose two cells are perfectly correlated over the entire set of possible input patterns: Then one cell is redundant and can be removed with no loss of information-carrying capacity. If the cells are uncorrelated, both are necessary because the behavior of one cannot be predicted from the other. As we have mentioned, neurons have a high biological cost, and considerable evolutionary pressure is exerted to use them efficiently.

It has been conjectured that one of the functions of lateral inhibition and numerous other physiological "sharpening" mechanisms is to make cells behave less like each other, to become less correlated, thereby making more efficient use of the cells and transmission pathways. Barlow (1972, 1985), in particular was a strong advocate of this point of view. Although the idea of mapping based on similarity is important, many other processes are going on in the nervous system as well. However, from a practical point of view, the adaptive mapping algorithms form a very useful and powerful set of techniques for network applications. They are especially well suited to pre-processing of complex inputs, such as the speech signals used by Kohonen.

References

S.-I. Amari (1980), Topographic organization of nerve fields. *Bulletin of Mathematical Biology*, *42*, 339–364.

H. Barlow (1972), Single neurons and sensation: A neuron doctrine for perceptual psychology? *Perception*, *1*, 371–394.

H. Barlow (1985), The twelfth Bartlett memorial lecture: The role of single neurons in the psychology of perception. *Quarterly Journal of Experimental Psychology*, *37A*, 121–145.

G.B. Blasdel (1992), Orientation selectivity, preference, and continuity in monkey striate cortex. *Journal of Neuroscience*, *12*, 3139–3161.

D.H. Hubel and T.N. Wiesel (1971), Aberrant visual projections in Siamese cats. *Journal of Physiology (London)*, *218*, 33–62.

J.H. Kaas and R.W. Guillery (1973), The transfer of abnormal field representations from the dorsal lateral geniculate nucleus to the visual cortex in Siamese cats. *Brain Research*, *59*, 61–95.

T. Kohonen (1982), Self-organized formation of topologically correct feature maps. *Biological Cybernetics*, *43*, 59–69.

T. Kohonen (1984), *Self-Organization and Associative Memory*. Berlin: Springer.

T. Kohonen (1988), The "neural" phonetic typewriter. *Computer*, *21*, 11–22.

R. Lippman (1989), Review of neural networks for speech recognition. *Neural Computation*, *1*, 1–38.

M.S. Livingstone and D.H. Hubel (1984), Anatomy and physiology of a color system in primate visual cortex. *Journal of Neuroscience, 4,* 309–356.

C. von der Malsburg (1973), Self-organization of orientation sensitive cells in the striata cortex. *Kybernetik, 14,* 85–100.

M.M. Nass and L.N. Cooper (1975), A theory for the development of feature detecting cells in visual cortex. *Biological Cybernetics, 19,* 1–18.

C.J. Schatz (1987), Visual system, Siamese cats. In G. Adelman (Ed.), *Encyclopedia of Neuroscience, II.* Boston: Birkhauser.

P. Schyns (1992), A modular neural network model of concept acquisition. *Cognitive Science, 15,* 461–508.

S.B. Udin and J.W. Fawcett (1988), Formation of topographic maps. *Annual Review of Neuroscience, 11,* 289–327.

D.J. Willshaw and C. von der Malsburg (1976), How patterned neural connections can be set up by self organization. *Proceedings of the Royal Society of London, B, 194,* 431–445.

15 *The BSB Model: A Simple Nonlinear Autoassociative Neural Network*

It is, sir, as I have said, a simple network, and yet there are those who love it.
Modified from Daniel Webster (1818)

This chapter discusses the brain-state-in-a-box (BSB) model, a simple nonlinear, autoassociative, energy-minimizing neural network. It is relatively weak in terms of its information-processing power, but it captures some of the flavor of network computation in its dynamics, associativity, and parallelism. It has been used primarily to model effects and mechanisms seen in psychology and cognitive science. Matrix feedback models have strong similarities to the power method of eigenvector computation, and some numerical experiments with the power method are described. The basic BSB equations use a simple associator with a limiting nonlinearity in the model neurons. Stable attractors are usually corners of a box of limits in hyperspace. Some simulations are presented to illustrate basin structure. The time it takes the system state to reach an attractor is computed for one simple case. The use of BSB as a clustering algorithm is illustrated with an application to radar analysis.

The last three chapters in this book suggest one direction the development of neural network models may take. They are also somewhat more technical than the preceding chapters. Up to this point, we have been concerned primarily with presenting and justifying some of the key ideas behind neural networks. Now we offer a more advanced and detailed discussion of an approach that includes descriptions of a good deal of research done at Brown University. Our bias is that the most significant line of development for neural networks over the next decade will be in the development of models that use extremely large sets of neurons, devote more effort to data representation and problem-specific detail and less to clever learning algorithms, and deliberately set out to develop practical network software based on ideas taken from neuroscience and cognitive science.

A modified version of this chapter appeared as a chapter in a book edited by Mohammed Hassoun, 1994, *Associative Memory*, published by Oxford University Press.

If we are to take neural networks seriously as models of human computation, it is necessary to have some idea of their computational powers and limitations. The brain-state-in-a-box (BSB) model was proposed in 1977 (Anderson, Silverstein, Ritz, and Jones) as a simple, nonlinear network that was rich in behavior but easy to simulate and to analyze formally. One reason it is easy to work with is that it is two networks combined: a simple associative network coupled to nonlinear dynamics. It is weak in that it does not have the intrinsic power to form correct input-output relationships shown by more powerful networks using complex learning rules such as backpropagation.

Since it is weak we cannot count on the BSB network to learn difficult discriminations. It is not a magic network that can theoretically solve any problem. We can make a virtue of this weakness and say that since there are so few places to tweak the network, we must worry instead about externals: the data representation, what we want to compute, and what we want the answer to look like. We also must focus on network size. A small weak network can solve very little, but a large weak one can be powerful if it scales well. It is hard not to be impressed by the huge size of biological networks, especially in structures such as mammalian cerebral cortex.

A useful network to perform cognitive functions must be at least adequate for a very wide range of tasks, although not necessarily outstanding at any of them. Versatility is a major virtue in a biological system. Perhaps BSB captures enough of the essentials of mental computation in its parallelism, associativity, nonlinearity, and temporal dynamics to be useful as a low-order approximation to several human mental operations. In any case, focusing on models for human cognitive computation would seem to be more important than designing yet another network capable of solving X-OR or parity.

In one sentence, a **BSB** network is an autoassociative, energy-minimizing, nonlinear dynamic system. This chapter is a description of the theory behind it and an attempt to describe informally some of its mathematical properties. We provide some psychological simulations, one with a practical cast, at the end as examples.

It is worth remembering that state vectors in a neural network are real in that they correspond in some way to unit discharges. They are also real in that they contain the "meaning" in the system and the results of the network computation. Therefore we have to be careful when we discuss operations on the state vectors in terms of abstract mathematical operations: these operations are describing information processing by a real entity.

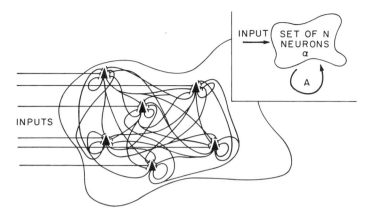

Figure 15.1   A group of neurons feeds back on itself. The basis of autoassociative feedback in the BSB model. From Anderson, Silverstein, Ritz, and Jones (1977). Reprinted by permission. © American Psychological Association.

Feedback Models

The BSB is a feedback model. Its connectivity is shown in figure 15.1. The nervous system is full of recurrent systems of various kinds, at all levels. For example, the recurrent collateral system of cortical pyramids provides widely distributed local excitatory feedback over a few millimeters. At a higher level, cortical areas have strong downward as well as upward projections. As Van Essen (1985) put it in a review article on visual cortex, "A particularly significant feature of cortical connectivity is a strong, perhaps universal tendency for pathways to occur as reciprocal pairs.... [T]here are as yet no convincing counterexamples to what appears to be a basic organizational principle in macaque visual cortex" (p. 284).

Little evidence suggests that cortical information processing is done purely by simple feedforward networks. Actual anatomy implies the substrate for powerful feedback loops at many levels, and that computation may be done by many richly interconnected functional modules settling on a consensus answer. A recent paper captures in a bewildering picture (reprinted in figure 10.9) the degree of modular interconnectivity and reciprocal connectivity now believed to exist in the visual system (Van Essen, Anderson, and Felleman, 1992). The BSB is a simple example of a formal model that tries to capture some of this flavor.

Autoassociative Reconstruction

Consider the autoassociative system, which is a variant of the standard associative structure. We showed in chapter 6 that we can use autoassociative feedback to reconstruct the missing part of a state vector.

The easiest way to analyze an autoassociative feedback system is in terms of the eigenvectors and eigenvalues of the connection matrix, **A**. Let us assume we have several items stored in the connection matrix **A**. Let us now assume that **f** is an *eigenvector* of **A**, with eigenvalue λ, that is,

$$\mathbf{Af} = \lambda\mathbf{f}.$$

Suppose a set of units with autoassociative feedback is receiving an input, say, **f**, and also feedback through the connection matrix, **Af**, and adds them. Let us assume the feedback pathway operates with a time delay, say, one time step (figure 15.2). Then, if we assume that the input is constant and always present,

at $t = 0$, the activity on the set of elements is **f**,

at $t = 1$, activity $= \mathbf{f} + \lambda\mathbf{f} = (1 + \lambda)\mathbf{f}$,

at $t = 2$, activity $= (1 + \lambda)\mathbf{f} + \lambda(1 + \lambda)\mathbf{f} = (1 + \lambda)^2\mathbf{f}$,

at $t = 3$, activity $= (1 + \lambda)^2\mathbf{f} + \lambda(1 + \lambda)^2\mathbf{f} = (1 + \lambda)^3\mathbf{f}\ldots$

It can be seen that as long as λ is positive, activity keeps on growing geometrically.

The enhancement is proportional to the magnitude of the eigenvalue, which becomes a kind of strength of self-association, effectively a gain parameter. Here we have a simple example of a system in which a large and complex pattern of unit activities, in this case an eigenvector, behaves in a simple way under the control of a single scalar variable, the eigenvalue. For someone familar with the classic mathematical psychology tradition, it is hard to avoid the suggestion that the eigenvalues of the connection matrix in this system have similarities to the concept of trace strength as used in many formal psychological models. Trace strength is usually conceived of as a scalar value that is related to the retrievability and overall importance of an item stored in memory.

Recent interest in nonlinear dynamic systems in physics and mathematics has led to productive attempts to predict and work with global system properties, rather than focusing on the behavior of the small parts of the system. As one of many examples, a "synergetics" group in Germany has

BSB Feedback Network

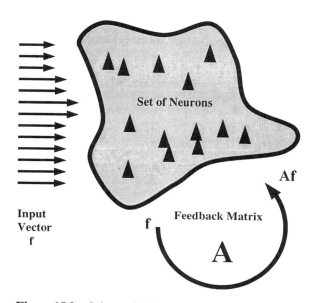

Figure 15.2 Schematic diagram of the vector feedback used in the BSB model. For a single iteration, the previous state vector, **f**, is fed back through the connection matrix, **A**, and adds to the previous state of the system.

followed this approach for years. Recently, they turned their attention to complex biological and psychological phenomena (Haken and Stadler, 1990). In one analogy, a laser is best understood by considering the behavior of the entire system: "In a laser there may be 10^{16} degrees of freedom of the atoms and 1 degree of freedom of the field mode. . . . [A]t the lasing threshold, the whole system is governed by a single degree of freedom" (Haken, 1985, p. 5).

Similarly, in the feedback systems we will describe, the behavior of a large, nonlinear, and complex system can be understood to some degree by use of well-defined, large-scale entities such as its eigenvectors. Therefore this approach suggests a simple way of linking the behavior of many low-level units with overall system behavior. The most desirable and useful psychological model would be one in which the richness of the world could be represented by extremely complex patterns of unit activation, but its overall system behavior would be governed by a small number of global parameters.

If an input is composed of the sum of many eigenvectors, the eigenvector with largest-magnitude eigenvalue will receive the greatest relative enhancement. This is the basis of the *power method* of eigenvector computation, which can be used to pick out the eigenvector with the largest-magnitude eigenvalue of a matrix, **A**, by repeatedly passing a random starting vector through the matrix. After a number of iterations, the eigenvector with largest-magnitude eigenvalue will dominate.

An Experiment with the Power Method

Let us do a numerical experiment to see if we can get some feeling for how matrix feedback behaves. More general discussions of the power method can be found in many books on numerical analysis (Householder, 1964; Wilkinson, 1965; Hornbeck, 1975; Young and Gregory, 1973). The unmodified power method is most often used for computing a few eigenvectors, those with the largest-magnitude eigenvalues. Variants such as the inverse power method are capable of computing eigenvectors with eigenvalues near any value. The more widely spaced the eigenvalues, the better and faster the system works.

The idea is simple. We have a matrix, **A**. Let us assume the eigenvectors are orthogonal. Let us start with a random vector, **v**(0), which we will repeatedly pass through **A**. At the ith iteration,

$$\mathbf{v}(i + 1) = \mathbf{A}\mathbf{v}(i).$$

But **v**(0) can be represented as a weighted sum of eigenvectors, $\mathbf{e}_i$,

$$\mathbf{v}(0) = \sum a_i \mathbf{e}_i,$$

where a_i are weighting coefficients. After the first iteration, $v(1)$ is given by

$$\mathbf{v}(1) = \sum \lambda_i a_i \mathbf{e}_i.$$

Consider the ratio of coefficients of two terms, i and j. After a single iteration, the ratio will be given by

$$a_i \lambda_i / a_j \lambda_j,$$

and after n iterations, the ratio will be

$$(a_i/a_j)(\lambda_i/\lambda_j)^n.$$

If $|\lambda|_i > |\lambda|_j$ the magnitude of this ratio will grow without bounds.

The eigenvector with the eigenvalue with largest magnitude is called *dominant*. After a large number of iterations, the vector, **v**(n), will be almost entirely in the direction of the dominant eigenvector. The only time this would not happen would be in the improbable event that the initial random vector happened to have no component in the direction of the dominant eigenvector.

By successive iterations we can find the direction of the dominant eigenvector to whatever precision we want. We can also estimate the eigenvalue. Suppose we have waited long enough so that the vector, $\mathbf{v}(n)$, is in the direction of the eigenvector, $\mathbf{e}_i$. Then,

$$\mathbf{v}(n + 1) = \mathbf{A}\mathbf{v}(n) = \lambda\mathbf{v}(n).$$

By looking at the ratio of lengths of $\mathbf{v}(n + 1)$ and $\mathbf{v}(n)$ we can estimate the eigenvalue, λ_i.

We can easily demonstrate this process. First, we have to know how to construct a matrix $\mathbf{A}$ with known eigenvectors and eigenvalues. An outer product matrix has this property. Consider a set of orthogonal normalized vectors, $\{\mathbf{e}_i\}$, with eigenvalues, $\{\lambda_i\}$. Then we can construct a matrix $\mathbf{A}$ with desired eigenvectors and eigenvalues by

$$\mathbf{A} = \sum \lambda_i \mathbf{e}_i \mathbf{e}_i^T.$$

It is easy to check that $\mathbf{e}_i$ is an eigenvector and the λ_i is its eigenvalue.

For the first simulation, the vectors used are in table 15.1. There the '+' correspond to +1 and the '−' to −1. These vectors are examples of Walsh functions, a set of orthogonal functions often used in image processing. A matrix with known eigenvectors and eigenvalues was constructed from these 32-dimensional vectors by normalizing each vector and forming the autoassociative outer product of the vector with itself. The resulting outer product matrix for each vector was multiplied by the eigenvalue, and the six outer product matrices were summed to form the final matrix.

We chose a random starting vector. For the first simulation, the state vector was renormalized after each iteration. After 100 passes through the matrix, table 15.2 gives the cosine between the six eigenvectors and the

Table 15.1   Vectors used to construct a matrix with known eigenvectors. Vectors are Walsh functions

Eigenvalue	State Vectors
1.1	+ + + + + + + + + + + + + + + + − − − − − − − − − − − − − − − −
1.0	+ + + + + + + + − − − − − − − − + + + + + + + + − − − − − − − −
0.9	+ + + + − − − − + + + + − − − − + + + + − − − − + + + + − − − −
0.8	+ + − − + + − − + + − − + + − − + + − − + + − − + + − − + + − −
0.7	+ − − + + − − + + − − + + − − + + − − + + − − + + − − + + − − +
0.6	+ − + − + − + − + − + − + − + − + − + − + − + − + − + − + − + −

The '+'s correspond to vector elements of +1 and the '−'s to −1.

Table 15.2 State vector after 100 iterations starting from random vector

Estimated Eigenvalue	Cosines Between True Eigenvector and Final State Vector After 100 Iterations					
	1	2	3	4	5	6
1.0999998	0.9999999	0.0004442	0.0000000	0.0000000	0.0000000	0.0000000

The estimate of the eigenvalue is the ratio of lengths of input vector and vector after an iteration through the matrix.

resulting vector. The final vector is pointing in the direction of e_1, the eigenvector with largest eigenvalue. The ratio of lengths between the input and output at the hundredth iteration is a very accurate (six places) estimate of the eigenvalue of the dominant eigenvector. There are no components from any eigenvectors except the two with the largest-magnitude eigenvalues, and the contribution from the eigenvector with the second-largest eigenvalue is very small, decreasing in magnitude with each iteration.

We need not stop with computing only the dominant eigenvector. We can modify our matrix construction method, which used the sum of outer products, in an obvious way, and a natural way for a neural network, to find the other eigenvectors. (For use of this technique in a cognitive simulation, see Kawamoto and Anderson, 1985.)

We have just estimated the eigenvector and eigenvalue by the power method. Suppose we form the outer product of the estimated normalized eigenvector, multiply it by the estimated eigenvalue, and *subtract* it from the matrix. (To a neural network modeler, the outer product is simple Hebbian learning, in this case with a negative sign.) If e' and λ' are the estimates of eigenvector and eigenvalue, we can modify $\mathbf{A}$, forming matrix $\mathbf{A}'$, so that,

$$\mathbf{A}' = \sum \lambda_i \mathbf{e}_i \mathbf{e}_i^T - \lambda' \mathbf{e}' \mathbf{e}'^T.$$

Assuming the e' and λ' are good estimates, the new matrix, $\mathbf{A}'$, will have reduced the eigenvalue of the dominant eigenvector, $\mathbf{e}_i$, say, to

$$(\lambda_i - \lambda') \approx 0.$$

By repeating this process for each eigenvector as it is computed by the power method, it is possible to compute the eigenvectors in order of the magnitude of their eigenvalue. The anti-Hebb adaptation is an extremely robust technique. The estimate of the dominant eigenvalue and eigenvector can actually be quite poor. Adaptation merely has to reduce the magnitude of the dominant eigenvalue enough so that it is no longer dominant.

Table 15.3 shows the results of this process applied to the matrix we constructed earlier. The estimates of eigenvectors and eigenvalues are accu-

Table 15.3 Power method demonstration: Computation of successive eigenvalues and eigenvectors

| Pass | Estimated Eigenvalue | Cosines Between True Eigenvector and Final State Vector After 100 Iterations | | | | | |
		1	2	3	4	5	6
1	1.0999998	0.9999999	0.0004442	0.0000000	0.0000000	0.0000000	0.0000000
2	1.0000001	−0.0004887	0.9999999	0.0000134	0.0000000	0.0000000	0.0000000
3	0.9000000	0.0000000	0.0000149	−1.0000000	0.0000557	0.0000000	0.0000000
4	0.8000001	0.0000000	0.0000000	−0.0000626	−0.9999999	0.0000052	0.0000000
5	0.7000000	0.0000000	0.0000000	−0.0000001	−0.0000060	−1.0000000	−0.0000002
6	0.6000001	0.0000000	0.0000000	0.0000000	0.0000000	0.0000002	−0.9999999

rate to five or six decimal points. These estimates, since they are statistical, will contain small errors. If the process is continued indefinitely, the estimates will become inaccurate enough to be useless.

The values in table 15.3 give the estimated eigenvalue and the cosine between the true eigenvector and the eigenvector computed by this process. These can be compared with the actual eigenvalues used to construct the matrix in table 15.1. One hundred iterations were performed for each estimation. Eigenvectors are determined only to a multiplicative constant. A cosine of −1 simply means that the eigenvector is pointing in the opposite direction to the reference direction; it is still a correct solution.

Shifting Eigenvalues

Because the ratio of eigenvalues determines how rapidly the computation converges, it is possible to speed up the computation and improve accuracy if we have some idea of the value of the eigenvalues. Sometimes this eigenvalue can be estimated from other information in the problem. For example, when we make the connection with neural networks, we often use a learning rule, for example, Widrow-Hoff, that wants to produce eigenvalues with maximum values near 1 and minimum values near zero.

The complete set of eigenvalues is called the *eigenvalue spectrum*. It is easy to translate the eigenvalue spectrum up or down by subtracting a multiple of the identity matrix, $\mathbf{I}$, from the matrix $\mathbf{A}$. For any vector, $\mathbf{Ix} = \mathbf{x}$. The identity matrix is all zeros except for 1s down the main diagonal. Suppose we form the matrix $\mathbf{A}'$ so that

$$\mathbf{A}' = \mathbf{A} + \gamma \mathbf{I}.$$

Consider an eigenvector of **A**, **e**, and its associated eigenvalue, λ. We know that $\mathbf{Ae} = \lambda\mathbf{e}$. But **e** is also an eigenvector of $\mathbf{A}'$, because

$$\mathbf{A}'\mathbf{e} = \mathbf{Ae} + \gamma\mathbf{Ie}$$

$$= \lambda\mathbf{e} + \gamma\mathbf{e}$$

$$= (\lambda + \gamma)\mathbf{e}.$$

The new eigenvalue of **e** is $(\lambda + \gamma)$. The entire eigenvalue spectrum has been translated upward by an amount, γ. Consider the ratio between two eigenvalues, a quantity that dominates the power method computation. The larger this ratio is, the faster is convergence. By using a proper γ this ratio can often be made larger.

As an aside, in numerical analysis it is often more convenient to compute the eigenvectors of the inverse matrix, which will be the same as for the uninverted matrix. The eigenvalues of the inverse matrix are the reciprocal of those of the uninverted matrix. By shifting the eigenvalue spectrum, the dominant eigenvector of the inverse matrix can be placed at any point by moving its uninverted eigenvalue closest to zero. Eigenvectors in any region of the eigenvalue spectrum can be computed using this method, which is fast and accurate (Wilkinson, 1965). It is difficult to see how to apply this trick to neural networks, however.

Although the pure power method simply picks out the dominant eigenvalue, positive and negative eigenvalues behave differently in the feedback model. A negative eigenvalue will cause the sign of the eigenvector component to change after each iteration. When we discussed neural network feedback models, we assumed that the previous state vector remained present. Suppose the connectivity matrix is **A**. If the current state vector is the eigenvector, $\mathbf{e}(t)$, the next iteration gives

$$\mathbf{x}(t + 1) = \mathbf{e}(t) + \mathbf{Ae}(t) = (\lambda + 1)\mathbf{e}(t).$$

This process has shifted the eigenvalue by $+1$, increasing the magnitude of positive eigenvalues and decreasing the magnitude of negative ones.

Learning

In the simplest version of the BSB model, the matrix is formed using a Hebbian rule, giving a matrix that is a sum of outer products. If a set, $\{f_i\}$, is to be learned, then

$$\mathbf{A} = \eta \sum_i \mathbf{f}_i\mathbf{f}_i.$$

This matrix (except for the constant) has the form of the covariance matrix of statistics, as we pointed out in chapter 9. The eigenvectors of this matrix are used to form the *principal components* of statistics. This matrix is symmetrical, and therefore has orthogonal eigenvectors and real eigenvalues. The eigenvalues are zero or positive. The principal components, ordered in terms of size of eigenvalue, contain the most variance; the first one contains the largest amount of the variance of the system, the second contains the next largest, and so on.

The eigenvectors that are relatively enhanced by matrix feedback are therefore the ones that are best to use, if it is desired to make discriminations among members of the input set. We once suggested that such a system when combined with simple Hebbian learning acted like analysis in terms of what are called *distinctive features* in psychology, criterial features that are particularly useful in making discriminations (Anderson et al., 1977). Many other neural networks turn out to have close connections to principal components, as was mentioned in chapter 9 (Cottrell, Munro, and Zipser 1988; Baldi and Hornik, 1989; Linsker, 1988; MacKay and Miller, 1990; Oja, 1982, 1983; Foldiak, 1989; Krogh and Hertz, 1990). The similarity between the rule for Hebbian synaptic modification and correlation seems to make a connection between principal components and neural networks natural and inevitable. In any case, matrix feedback combined with a variant of Hebbian learning can serve as valuable and statistically valid preprocessing.

At the present time, besides simple Hebbian learning, another useful associative learning algorithm to construct the connection matrix is the well-known least mean squares algorithm, discussed in chapter 9 (Widrow and Hoff, 1960). See Proulx and Begin (1990) and Begin (1991) for further comments about learning. In the context of autoassociative matrix feedback, that is, where the input and output are the same, if the Widrow-Hoff technique works perfectly, we will develop a system that has eigenvectors corresponding to the learned set and with eigenvalue 1, that is,

$$A\mathbf{f}_i = \mathbf{f}_i.$$

Perfect simple Widrow-Hoff learning ultimately aims to produce a degenerate system, with numerous eigenvectors having eigenvalue 1. Degeneracy means that any linear combination of eigenvectors with the same eigenvalue will also be an eigenvector.

As learning progresses, the eigenvalue spectrum will contract so that the differences between the largest eigenvalues will decrease. This makes vector feedback and the power method enhancement less efficient. However, because of the nonlinearities in BSB, this is not a major problem in practice. One can deal with it in a number of ways. One possibility, with plausible biological support, would be to assume the presence of adaptive processes that constantly shift the eigenvalues of a particular eigenvector up and down, depending on history. One could also use Widrow-Hoff learning where the error measure is not simple error, but is computed from a constant multiple of the desired vector, for example,

$$\Delta \mathbf{A} = \eta(c\mathbf{g} - \mathbf{g}')\mathbf{f}^T.$$

The target eigenvalue will then become the constant, c, instead of 1. In addition, a connection exists between principal components and error correction learning. Widrow and Stearns (1985) noted that the eigenvectors of the covariance matrix define the principal axes of the error surface. Corrections will be largest to the principal components with the largest eigenvalue because the error surface is steepest there.

Introducing BSB Nonlinearity

Since positive feedback may make the output vector grow without bounds, the simplest way to contain the vector is to put upper and lower limits on the allowable values of its elements. We have discussed the virtues of the sigmoid or squashing function used in the generic neuron model of neural networks. If a simple limiter is used, it has the effect of limiting the state vector to restricted region of space, hence the name brain-state-in-a-box (BSB) (Anderson et al., 1977; Anderson and Mozer, 1981; Golden, 1986; Hui and Zak, 1992).

We will describe the simplest BSB variant. The activity pattern in a set of neurons feeding back on itself is assumed to be composed of the input to the neurons, the feedback coming back through recurrent connections from past states of the neurons, and the persistence of activity in the state. That is, activity does not drop to zero immediately but decays with some kind of time constant. We assume time is quantized.

Suppose $\mathbf{x}(t + 1)$ is the state vector at the $t + 1$ time step. In general, we will assume the input is presented at $t = 0$, and that the system processes that input. The feedback is given by $\alpha \mathbf{A}\mathbf{x}(t)$, where α is a feedback constant to control strength of feedback. The decay of the previous state is given by some constant, γ, multiplying the previous state, that is, $\gamma \mathbf{x}(t)$. Sometimes it is desirable to keep the initial input, $\mathbf{f}(0)$, present, perhaps multiplied by a constant, δ, usually either 0 (not present) or 1 (present). So we have the next state, $\mathbf{x}(t + 1)$ given as

$$\mathbf{x}(t + 1) = \text{LIMIT}(\gamma \mathbf{x}(t) + \alpha \mathbf{A}\mathbf{x}(t) + \delta \mathbf{f}(0)).$$

The LIMIT operation clips values that exceed upper or lower limits. In most BSB simulations the limits are symmetrical, that is, the upper and lower limits are equal.

Constants are chosen so feedback is positive. This equation produces a simple dynamic system. Positive feedback causes the vector to lengthen, just as we saw for the power method in the linear case. However, in BSB the lengthening slows when the state vector components start to limit, and ceases when all components are at a limit. In the terminology of dynamical system theory, the final stable states are called *attractors*. The BSB dynamics minimize an energy (Lyaponov) function, so the stable states are points of minimum energy (Golden, 1986).

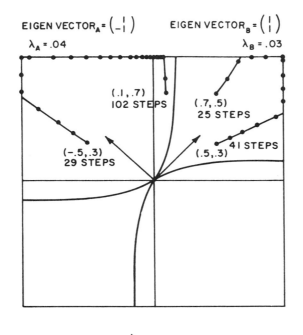

Figure 15.3 A two-dimensional BSB simulation taken from an early paper. The basin boundaries are curved lines. Different starting points take various numbers of iterations to reach stable attractor states in corners of the box of limits. The arrows indicated the direction of the eigenvectors. The larger basin is associated with an eigenvalue of 0.04 and the smaller with an eigenvalue of 0.03. From Anderson and Silverstein (1978). Reprinted by permission. © American Psychological Association.

We saw other networks that minimized an energy function when we talked about Hopfield networks in chapter 12. Some corners in BSB tend to be stable attractor states because feedback is always pushing them outward. All the points in a region end in a final stable attractor state. These points are referred to as the *basin of attraction* of the attractor. Figure 15.3 shows a simple example of a two-dimensional BSB model in action (Anderson and Silverstein, 1978). Other dynamic system models are described in the neural net literature, such as the models of Hopfield (1982, 1984), which we discussed in chapter 12, the ART models (Carpenter and Grossberg, 1987), and the BAM model of Kosko (1988). The qualitative behavior of these different neural networks is often similar, and they have a number of formal similarities (Cohen and Grossberg, 1983).

For the BSB network, it is easy to show that there are three classes of attractors in interesting cases that have more than one attractor. The origin can be an unstable attractor, that is, the state vector there does not change. Any shift, however, no matter how small, away from the origin causes the state vector to shift to a stable attractor. The boundaries between basins are also unstable attractors. Finally, there can be stable point attractors, which are generally corners of the hypercube of limits. Simple BSB models cannot normally generate the limit cycle attractors found in more complex dynamical systems (Greenberg, 1988; Hui and Zak, 1992). A careful stability analysis of a generalized BSB model that allows each unit to have its own bias and limits has been presented by Golden (1993).

Corner Stability

If all the eigenvalues are positive, and if we add the previous state to the current state, that is, if $\gamma = 1$, all corners are attractors. However, most applications of BSB incorporate decay of the previous state. If we are interested in BSB as a classification and categorizing algorithm, we want to have only a few stable attractors with large basins of attraction.

Consider the simple BSB model,

$$\mathbf{x}(k + 1) = \text{LIMIT}(\gamma \mathbf{x}(k) + \alpha \mathbf{A}\mathbf{x}(k)),$$

where $\mathbf{A}$ is the weight matrix. Suppose we teach this network one state vector, $\mathbf{c}$, which is a corner of the box of limits. We will assume the limits, and therefore, all elements of $\mathbf{c}$ are $+1$ or -1. Let us assume that $\mathbf{A}$ is formed by simple outer product learning of $\mathbf{c}$,

$$\mathbf{A} = \mathbf{c}\mathbf{c}^T.$$

Note that corner **c** is an eigenvector of **A**, since

$$\mathbf{Ac} = \mathbf{c}(\mathbf{c}^T\mathbf{c})$$

$$= n\mathbf{c},$$

where n is the dimensionality of the space, since the inner product $[\mathbf{c}, \mathbf{c}] = \mathbf{c}^T\mathbf{c} = n$. If the sum of the off-diagonal elements is less than the diagonal elements, the matrix is referred to as row dominant, and it is possible to show that then all corners are stable (Hui and Zak, 1992). The identity matrix, an extremely uninteresting matrix for information processing, is row dominant. (Uninteresting for most, except possibly for the very spiritually evolved; perhaps the identity matrix in a memory instructs us that one should accept the world exactly as it is.) The identity matrix can arise with Hebbian learning when large amounts of random noise are learned, providing a natural neural network senility (or enlightenment) mechanism, if one is required.

Attractor Basins and Stability

The basic mode of processing done by BSB is to turn an unlimited number of possible input vectors into a small number of stable attractors. The basins of attraction of the BSB model are, in general, well-behaved, and for most situations, only a few attractors are present. The qualitative behavior of the basin boundaries is intuitive. For a simple two-dimensional case (Anderson et al., 1977) the boundary has an exact solution: segments of a logarithmic spiral. Although we have no exact solutions for higher-dimensionality boundaries, we would expect similar simplicity.

Let us do some simulations to check boundary behavior. Suppose we start with 10 binary vectors, constructed with values $+1$ and -1 equally probable, and normalize them. We assume a high-dimensional space; between 250 and 1000 dimensions were used for the following simulations. In the first simulation, in a 1000-dimensional system, we constructed a connection matrix using outer product (Hebbian) learning from these 10 vectors, with a weighting constant ranging from 0.4 for the first vector to 0.22 for the tenth, and weighting constants spaced 0.02 apart. As we would expect, the initial vectors are approximately orthogonal, and the initial vectors are within a few degrees of eigenvectors with eigenvalues spaced approximately from 0.4 to 0.22. That is, the eigenvalues were close to the initial weighting constant, and the 10 eigenvectors with largest positive eigenvectors were approximately in the direction of the original 10 vectors.

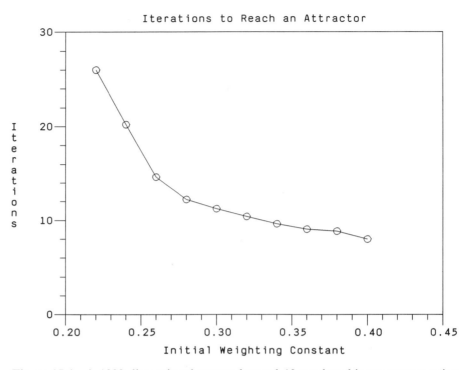

Figure 15.4 A 1000-dimensional system learned 10 random binary vectors using outer product learning. As expected, the initial vectors with larger weighting constants (approximately eigenvalues) reached an attractor more quickly than vectors with smaller weighting constants.

When tested, each input vector was associated with a different BSB attractor. The number of iterations required to reach an attractor was a function of the eigenvalue associated with that vector. Figure 15.4 shows the number of iterations required in one simulation to reach an attractor for various initial weighting constants (approximately eigenvalues).

The next simulation also used a 1000-dimensional system. Suppose we next consider the plane between two of the initial vectors. Suppose we construct a normalized starting vector for BSB that lies in this plane. The simulation used 100 starting positions at roughly 1-degree intervals. If the system is well behaved, we would expect that starting points close to one starting vector would end in the attractor associated with that vector, and when close to the second starting vector would end in the attractor associated with that vector.

Figures 15.5 and 15.6 show typical results for two pairs of vectors. The number of iterations required to reach the attractor increases slowly as the starting point moves away from the initial vector. Despite the different start-

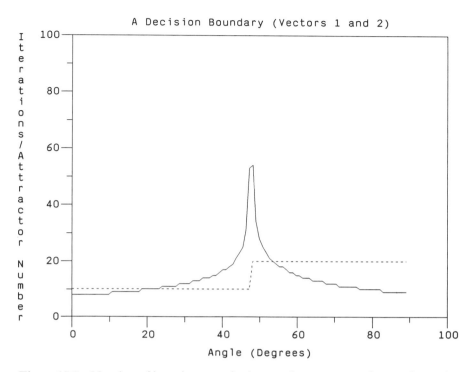

Figure 15.5 Number of iterations required to reach an attractor for starting points lying in the plane between two learned random vectors. A 1000-dimensional system learned 10 random binary vectors using outer product learning. Normalized test vectors were constructed falling in the plane between vector 1, the initial vector with the highest weighting constant (0.4, approximately the eigenvalue) and vector 2 with the second highest weighting constant (0.38). The dashed line shows the number of the attractor corresponding to the final state of the system. (The y-axis value divided by 10 gives the attractor number.) All starting points in the plane are attracted into the attractor associated with vector 1 or vector 2, although the number of iterations to reach that final state varies depending on starting point.

ing points, however, the final state of the system is always in either the first or second attractor. In one large region the attractor is reached in a roughly constant number of iterations, although the number of iterations required increases slowly and then more rapidly as the boundary is approached. As the final state shifts from one attractor basin to the other, very near the boundary, many iterations are required to reach the attractor.

Occasionally, the boundary has a slightly more complex structure (figure 15.7). The starting points used here were in the plane between two initial vectors with relatively small associated eigenvalues. The two initial vectors were numbers 5 and 9. Starting points in the plane between them gave rise to a small interpolated region associated with vector 3. In addition, a single

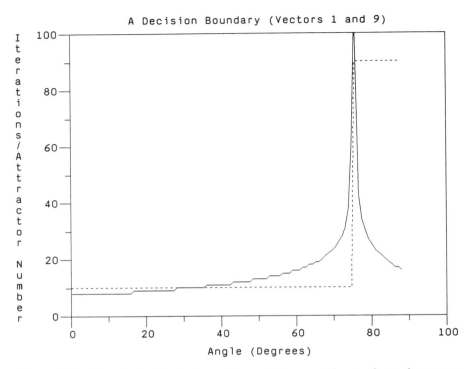

Figure 15.6   Same basic simulation as figure 15.5, except the starting points are on the plane between vector 1 with initial weighting constant 0.4, and vector 9 with weighting constant 0.24. The dashed line shows the number of the attractor corresponding to the final state of the system. (The y-axis value divided by 10 gives the attractor number.) All starting points are attracted into the attractor associated with vector 1 or vector 9. Note the boundary between the attractor basins has shifted away from vector 1, compared with figure 15.5.

point on the plane gave rise to a new attractor that was not associated with any of the learned vectors (shown as attractor 0). The BSB system has *spurious* attractors, although the word spurious is misleading. These attractors are what can give the system some ability to generalize in cognitively oriented simulations.

As we expected, larger eigenvalues have larger attractor basins. If we look at the boundary between two basins, we can see it shifts toward the initial vector with the smaller associated eigenvalue. This shift can be seen clearly in figures 15.5 and 15.6. Figure 15.8 shows the location of the boundary as the starting vector moves from the plane between vectors 1 and 2, where the initial weighting constants (close to the eigenvalues) are 0.4 and 0.38 to the location of the boundary on the plane between vectors 1 and 10, which have eigenvalues approximately 0.4 and 0.22. When the eigenvalues of the two vectors forming the plane are nearly the same, the boundary is nearly 45

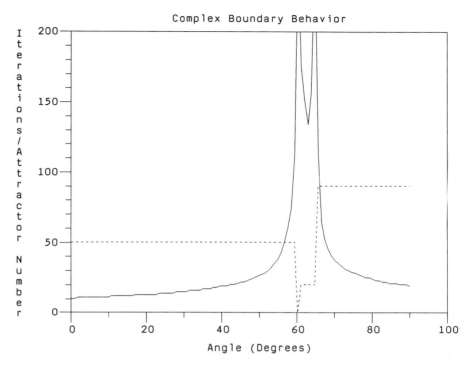

Figure 15.7 A more complex boundary structure. Same basic simulation as figure 15.5, except the starting points are on the plane between vector 5 with initial weighting constant 0.32, and vector 9 with weighting constant 0.24. The dashed line shows the number of the attractor corresponding to the final state of the system. (The y-axis value divided by 10 gives the attractor number.) A small region near the boundary contains points that are attracted into the final state associated with learned vector 3, and there is a "spurious" attractor, which is not associated with any of the learned vectors, around 60 degrees.

degrees. A 2:1 ratio of eigenvalues shifts the boundary to only 10 degrees away from the smaller eigenvalue.

Effects of Limits

Basin size is affected by the size of the box of limits. For example, the linear power method could be considered to realize a nonlinear BSB system in which the limits are far from the origin. In the power method, the direction of the dominant eigenvector becomes the attractor, in that every starting point eventually comes asymptotically close to it. Therefore we should expect that the basins associated with large eigenvalues should grow relative to small ones as the limits increase. This is what is seen in simulations.

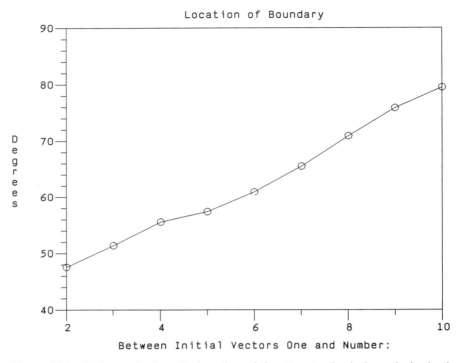

Figure 15.8 This graph gives the location of the attractor basin boundaries in the planes between vector 1, the initial vector with the largest weighting constant, and the other nine initial vectors. Notice the nearly linear shift of the angle of the basin boundary away from the vector 1 with the largest initial weighting constant, as the weighting constants decrease linearly from 0.4 to 0.22.

Figure 15.9 shows the results of a 250-dimensional simulation using 1000 different normalized random binary vectors as initial starting points. This system had five learned normalized binary vectors, each of which had an associated attractor. Almost all the random vectors finished in one of these five attractors; there was a small number of new attractors. (For this simulation, an attractor and its antiattractor, that is, the corner diagonally across the box, were considered the same attractor.) The limits were increased from 1 times an element magnitude to 35 times an element magnitude. In the first case, the starting point was actually at a corner; in the last, it was far away from a corner. It can be seen that as the box increases in size, the basin associated with the larger eigenvalues increases to some degree. The magnitude of this effect depends on the spacing of the eigenvalues. If the box is asymmetrical, so that the positive and negative limits differ, behavior is more complex.

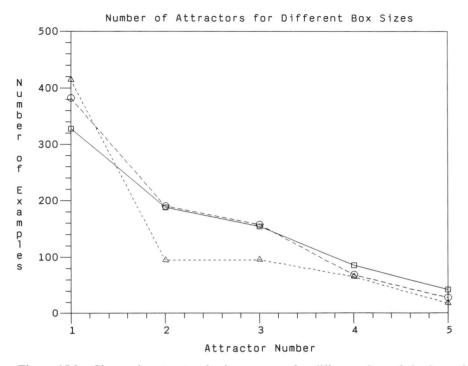

Figure 15.9 Change in attractor basin structure for different sizes of the box of limits. Using the outer product rule, this 250-dimensional system learned five binary random vectors. Attractors 1 through 5 are the attractors associated with the five learned vectors. Attractor 1 had the largest weighting constant during learning (approximately the associated eigenvalue) and attractor 5 had the smallest. Attractors were computed for 1000 normalized binary random starting vectors. (An attractor and its antiattractor were counted as the same attractor.) As the limits increase from 1 times the starting element size (solid line) to 10 times the starting element size (dashed line) to 50 times the starting element size (dotted line), the attractor basins changed in relative size. The attractor basins associated with the largest eigenvalue (number 1) grow at the expense of the basins associated with the smaller eigenvalues.

Decay Rate

The general BSB equation contains a term, γ, that multiplies the previous state vector; that is, it governs the decay of the state vector in the absence of feedback. (If the connection matrix was zero, the state vector would decay exponentially with a time constant of γ.) If γ is less than 1, we would expect that attractors involving the smallest eigenvalues would be most strongly affected, since the eigenvalue spectrum is shifted downward. Qualitatively, as the state vector decays more and more quickly, there are fewer and fewer

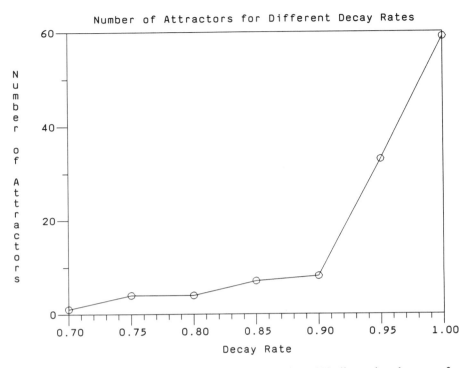

Figure 15.10 Number of different attractors seen in a 500-dimensional system for various rates of decay, γ. Using the outer product rule, the network learned 10 binary random vectors. Decreasing γ simplifies the attractor structure by eliminating attractors associated with eigenvalues near zero. (An attractor and its anti-attractor were counted as the same attractor.)

attractors; γ less than 1 simplifies the attractor structure. If the sum of the largest positive eigenvalue and the decay term is less than 1, there will be no stable corners at all, and the only stable attractor will be at zero; that is, all starting points will end at the origin. In general, if we have a system (as is often found in applications) with a small number of eigenvectors with large eigenvalues and a number of other eigenvectors with small eigenvalues, produced, perhaps, by random noise processes, reducing γ will have a disproportionate effect on the small eigenvalues and will tend to eliminate attractors associated with them. For example, if the sum of γ and a small eigenvalue is less than 1, the related eigenvector cannot support a stable attractor. This effect can be used as a technique to suppress noise.

Figure 15.10 shows the number of different attractors seen in a 500-dimensional system, where 10 learned starting vectors were given weighting constants from 0.4 to 0.22 when the connection matrix was constructed, as before. One thousand random normalized binary vectors were used as starting points. All the conditions had the same random number seed, so the same set of random starting vectors was used for each value of γ. The attractor corner and the anticorner were considered to be the same attractor. The simplest BSB model, with γ of 1.00, showed about 60 different attractors from 1000 random starting vectors. When γ was lowered to 0.70, only one stable attractor attracted almost all the initial vectors. (A few starting vectors did not reach an attractor at all in 400 iterations.) Small decreases in γ were effective at eliminating large numbers of attractors with small basins of attraction.

A Simulation

Figure 15.11 presents the output from a BSB simulation using a 200-unit network as an example of the system operating. In this case, the learned material was the state data base used for simulations in chapters 7 and 9. Reference to figure 9.6 indicates a small number of retrieval errors, even for a network using the Widrow-Hoff procedure to learn the 10-state data base. The BSB can do slightly better than this, because the entire state vector can cooperate effectively during the retrieval iterations. The BSB network retrievals were completely correct for a 15-state data base. After 150 random pattern presentations, the Widrow-Hoff learning procedure placed attractors at the desired locations that BSB dynamics allowed the system state vector to reach. In figure 15.11, the number after "Limited:" gives the number of units that have activities at their upper or lower limits. As the state vector approaches the attractor, this number increases to 200, where the state vector reaches a corner of the box of limits in state space. This network was fully connected, so that every one of the 200 units connected to every other one. The parameters in the simulation were not critical, here they were $\gamma = 0.9$, $\alpha = 0.5$, and $\delta = 1.0$. More details are given in appendix M.

Connectivity

Many neural networks assume the network is fully connected, so every unit in a group connects to every other unit in the same or a different group. This is generally assumed in Hopfield networks, projections between layers in back-propagation, the ART models, as well as many others. In fact, as we mentioned in other chapters, connectivity in the real nervous system is small compared with the number of cells involved. A "typical" central nervous system unit might receive 1000 inputs, give or take an order of magnitude.

```
                    BSB procedure demonstration
                      15-state data dase
                     Probe: state name

0.  _____Oklahoma     Limited:    0
1.  _____Oklahoma     Limited:   64
2.  _____Oklahoma     Limited:   64
3.  _____Oklahoma     Limited:   71
4.  _k_____ity_____Oklahoma          Limited:  105
5.  Ok_ah___ City_____Oklahoma           Limited:  145
6.  Oklahoma City_____Oklahoma           Limited:  159
7.  Oklahoma City___ Oklahoma            Limited:  180
8.  Oklahoma City     Oklahoma           Limited:  194
9.  Oklahoma City     Oklahoma           Limited:  200

0.  _____Louisiana    Limited:    0
1.  _____Louisiana    Limited:   72
2.  _____Louisiana    Limited:   72
3.  _____ge_____Louisiana           Limited:   76
4.  B__o_ _o_ge_____Louisiana           Limited:  114
5.  B__on Ro_ge _____Louisiana           Limited:  148
6.  B_ton Rouge    __ Louisiana          Limited:  178
7.  Baton Rouge     Louisiana            Limited:  195
8.  Baton Rouge     Louisiana            Limited:  200
```

Threshold: 0.5

Figure 15.11 Demonstration of BSB procedure. This simulation used the state data base used in previous chapters but with 15 states instead of 10. The connection matrix used the Widrow-Hoff learning rule with 150 random presentations. The BSB retrieval used partial information (the state name) to reconstruct the name of the capital. The retrieval threshold was set at 0.5. Letters are displayed only when all of their elements are above 0.5. The number after "Limited:" gives the number of units whose values have reached an upper or lower limit. As the system state vector enters a corner, this number will grow to 200, when the iterations are terminated. The simple error-correcting system produced retrieval errors for these particular state names in the 10 state simulation in figure 9.6. The BSB procedure got all the characters correct even when 15 states were learned; the attractor state is a learned pattern. Parameters were not critical. For this simulation they were $\gamma = 0.9$, $\alpha = 0.5$, and $\delta = 1.0$.

That is a very small number compared with the total number of neurons in a mammalian brain. Therefore, network models that depend for their analysis and operation on full connectivity—and many do—are in trouble if they are to be taken seriously as brain models.

Because BSB is a feedback model with point attractors, serving primarily as an associator and categorizer, it works well with partial connectivity. All the qualitative results described above, and the cognitive simulations described in the references to applications, hold over a wide range of connectivities. (The simulations presented in figures 15.6, 15.7, and 15.8 gave identical qualitative results for all connectivities tested between 30% and 100%.) Cognitive simulations done at Brown University routinely use 50%

random connectivity or less, partly to save computer time and partly in the interests of (a little) realism. Qualitative effects observed when connectivity is incomplete are increase in time to reach attractors, somewhat lower accuracy, reduced storage capacity, and, with the lowest connectivity, occasional elements that change so slowly that they do not reach limit values in reasonable numbers of iterations.

The qualitative behavior of the attractor structure is remarkably resistant to changes in connectivity. However, when connectivity is changed, some obvious gain changes have to be taken into account. Consider, for example, the weighted outer product we used to construct matrices with known or approximately known eigenvectors and eigenvalues. With simple Hebbian (outer product) learning of normalized input vectors, partial connectivity reduces the eigenvalues proportionally (subject to some statistical assumptions).

In the real nervous system, the limited allowable connectivity is under tight biological control and is of the utmost importance for function. One major task of modelers in the next decade—it is already starting to occur—is to incorporate biologically realistic connectivity patterns into network models. A large part of the biological computation is done by the initial wiring, and biological learning modifies and tunes up this rather small number of preexisting connections. The brain seems to operate in a region of parameters that has huge numbers of units and a small number of connections per unit, and where units are arranged in groups (nuclei and subregions), each group subserving one or a small number of simple computational steps.

Study of networks of this type has not been popular with other than biologically motivated theorists. Part of the reason for this is that the large number of units required makes them difficult to simulate with existing computer technology and difficult to realize for practical devices. However, if they can be made, it is reasonable to believe that such large modular networks may be reliable and robust.

Corner Shifting and Adaptation

Many problems require the learning of sequences of state vectors. Since BSB, like many simple dynamic system models, has stable point attractors, it must be modified somehow to produce temporal sequences of attractors. There are *many* ways to do this for attractor networks and many published variants in the literature. Amit (1989) gave a number of examples in his book. One straightforward way to change the state vector from one attractor to another is to manipulate the matrix so that attractor's local energy minimum is raised up and the state vector "spills" into a nearby minimum.

As we saw in our power method demonstration, we can easily manipulate eigenvalues of the connection matrix using outer product learning with a negative sign. If we assume that an adaptive process of an anti-Hebbian type exists, the longer the state vector stays in an attractor, the weaker the eigenvalue(s) associated with that attractor become. At some critical value the state vector shifts to a new attractor. The technique can be applied in several ways; the one we will discuss is from Kawamoto and Anderson (1985).

Necker Cube Reversals

The Necker cube is the best known example of what is called *multistable perception* (Necker, 1832). The two-dimensional drawing of figure 15.12 is ambiguous in that it is consistent with a three-dimensional perception that goes into the paper or comes out of the paper. If it is looked at steadily for a period, first one interpretation will be seen, then the other, then the first again, and so on. Ambiguity is a common perceptual problem. For example, many words are ambiguous as to meaning, pronunciation, or part of speech. Multistability can be demonstrated for many types of stimuli, both visual and nonvisual.

Kawamoto and Anderson (1985) listed some of the major experimental findings related to multistability in vision:

1. A stimulus has two or more possible configurations; that is, the stimulus is ambiguous.

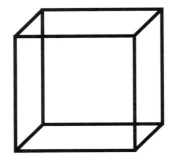

Necker Cube

Figure 15.12   The classic example of multistable perception: the Necker cube.

2. Once a particular configuration is perceived, it remains stable for a while before spontaneously changing to an alternative configuration.

3. The rate of fluctuation increases during the first 3 minutes of viewing before reaching an asymptote.

One natural model for such an effect is an attractor model with adaptation. Suppose we postulate adaptation in a BSB network in the form of Hebbian antilearning to let a state vector shift from attractor to attractor. Assume a set of initial conditions, $\mathbf{x}(0)$, the initial state vector, $\mathbf{A}(0)$, and the long-term values of the connection matrix, $\mathbf{A}$, that represent the stable component of the strength of the trace. We assume that the time constant for permanent modification of these values is slow relative to the dynamic modifications.

For simplicity, changes in matrix $\mathbf{A}$ were not allowed until the activity of the system reached a stable state. Once the activity of the network reaches an attractor, adaptation starts and continues as the activity remains in the corner. We assume that two eigenvectors correspond to the two perceptions of the cube, they point toward corners, and they are orthogonal.

If the system is in a corner, $\mathbf{f}$, change in eigenvalue magnitude is accomplished simply by Hebbian synaptic antilearning

$$\mathbf{A}(t + 1) = \mathbf{A}(t) - (1 - \sigma)\mathbf{f}\mathbf{f}^T,$$

where σ is a constant determining amount of adaptation. This implies that

$$\lambda(t + 1) = \sigma\lambda(t),$$

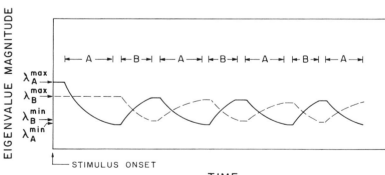

Figure 15.13 Behavior of the eigenvalues of the system during bistable perception. The system state will cycle between corners A and B and as the eigenvalues cycle up and down. Note that the first reversal takes the longest time. From Kawamoto and Anderson (1985). © Elsevier Science Publishers, B.V. Reprinted by permission.

where λ is the eigenvalue associated with eigenvalue **f**. If λ drops too much, the eigenvector will leave the corner. For a computer simulation to test this, simple stimuli were used that lay between the eigenstates, which were low-dimensionality Walsh functions.

With these relationships governing the decay and recovery of the eigenvalues, the values of λ_1 and λ_2 as a function of time are plotted in figure 15.13. The system state oscillated between the two eigenvectors. In conformity with the psychological result, the first reversal took a long time; the oscillations speeded up until a constant frequency of perceptual reversal was attained. Several other effects that correspond to the psychological data could also be demonstrated in this simple model, including hysteresis and adaptation.

Vector Adaptation

Perhaps the cleanest and most obvious way to shift states is Hebbian antilearning. To show just how robust a nonlinear system can be in practice, let us demonstrate an adapting technique that might be easy to implement, works poorly in a linear system, but is quite robust in the nonlinear network.

Hebbian antilearning requires detailed changes at synapses. Suppose we assume a much cruder form of adaptation: when a unit is active, after a while it becomes less responsive. This can be accomplished in a number of ways. The unit can become "tired," or any one of a number of habituation processes might occur. Or an active inhibitory process might occur, producing a negative afterimage of the cell's discharge. The exact details do not matter much.

Let us consider a system where the activity of the system for a period of time has been an attractor, a state vector, **c**. We will use the BSB model, but most attractor networks should show similar behavior. We will assume that adaptation occurs, so the attractor, **c**, times a constant, κ, is simply subtracted from the state vector. With $\delta = 0$ for simplicity, the BSB equation then becomes,

$$\mathbf{x}(t + 1) = \mathrm{LIMIT}(\gamma \mathbf{x}(t) + \alpha \mathbf{A}\mathbf{x}(t) - \kappa \mathbf{c}).$$

The qualitative result is that vector **c** is less likely to occur. Unfortunately, and what makes this attractively simple idea so hard to use in practice, the constant subtraction of $-\kappa \mathbf{c}$ tends to force the system strongly into the stable anticorner, $-\mathbf{c}$.

When this model is simulated in the linear system, it works very poorly. The simulation used the same 32-dimensional Walsh functions. Figure 15.14

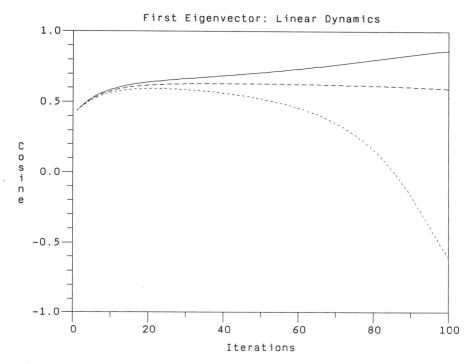

Figure 15.14   Vector adaptation with linear dynamics does not work well. The graph gives the cosine between the first (dominant) eigenvector and the system state vector after subtractive vector adaptation. Three values for adapting constant, κ, are plotted, corresponding to about a 5% range of values. Eventually every value of κ approaches a system state composed of one sign or the other of the dominant eigenvector. Near the transition point from positive to negative final system state, the approach to the dominant eigenvector may be slow.

shows the size of the components of the first eigenvector in the state vector for different values of κ in the linear system. The eigenvector with the second-largest eigenvalue grows a little, but the one with largest positive eigenvalue, or its antivector, rapidly dominates the system. Small differences in κ make large differences in the dynamics of the system; the three qualitatively different curves are from κ values that vary over a total range of only 5%.

When the BSB model is applied to the same system, the qualitative behavior becomes very different. The eigenvector with second-largest eigenvalue becomes a stable state over a wide range of κ. Nonlinearity keeps the system in a stable state corresponding to this eigenvector, and the size of the basin makes the model much less parameter sensitive. Figure 15.15 shows the dynamics of the first eigenvector component in the case where simple BSB is applied. Notice that this component drops to zero; the stable system state

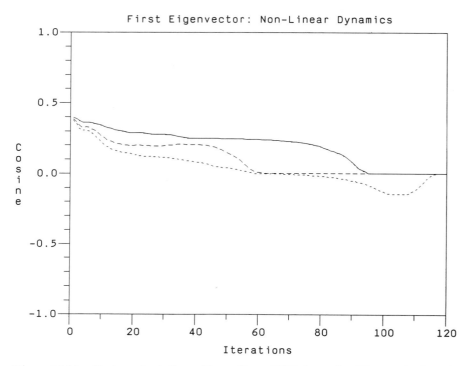

Figure 15.15 Vector adaptation with nonlinear BSB dynamics. The graph gives the cosine between the first (dominant) eigenvector and the system state vector after subtractive vector adaptation. Three values for the adapting constant, κ, are plotted, corresponding to about a 50% range of values. Note that the final system state vector for the three values is exactly the eigenvector with second largest eigenvalue. This is the desired result of adaptation. Time to reach the attractor varies with the fastest times in the middle of the range (dashes). The positive and negative extremes of the range (solid line and dots) take longer to reach the attractor.

vector at this point was entirely the second eigenvector, the one that we wanted to retrieve. The three curves represent a variation in κ of about 50% of the center of the range. Simple, but unstable, subtractive vector adaptation becomes a potentially usable technique that is rather insensitive to parameters.

Response Time

Most neural network research emphasizes the accuracy of the computation: whether or not the final system state is the correct answer to the question posed. The time required to reach that answer is sometimes of great importance. From the point of view of an experimental psychologist, the *response time* or *reaction time* to perform a task is a common experimental value and one that is easy to measure. It is also important to know the time required for a network computation to be completed for many practical applications.

Let us consider the time for a response to be generated. In feedforward neural models such as backpropagation, during *operation*, when an input state vector is presented at the input layer, the output state vector is generated in a few parallel steps. The function of various learning algorithms is to generate weights so the right input-output relations are produced, but the actual computation is not strongly time dependent without additional assumptions.

System models such as BSB have intrinsic time domain behavior due to their dynamics. Some early cognitive applications of neural networks, in particular the interactive activation model for letter perception (McClelland and Rumelhart, 1981; Rumelhart and McClelland, 1982), a complex dynamic system, showed time domain behavior that often displayed similarities to human reaction time data.

We will assume that response time—time for a computation to be completed—is related to dynamic system behavior. The key assumption is that *simple response time is related to the time required for an input vector to move to a point attractor*. Classic reaction time models that are most similar formally to this approach are also the ones that are the most successful in explaining experimental reaction time data: random walk and diffusion models.

Theory: Reaction Time Computation

Suppose that response time in experiments is related to the time required to reach an attractor in simulations. Computation of the time to reach an attractor can be a difficult calculation in the BSB model if the state vector starts to limit in one region of state space and not in others, because large, discontinuous alteration in rate of change of the state vector can occur. However, it is possible to compute easily the time required to reach an attractor for one interesting case.

For simplicity in analysis, we will assume that the connection matrix has only two eigenvectors and two attractors, which correspond to two possible responses. Because the dynamics of the BSB model are controlled by an equation with three constants (α, γ, δ), let us assume that the dynamics are represented by a single matrix, $\mathbf{B}$, with two eigenvectors and the associated eigenvalues, λ_1 and λ_2. If $\mathbf{f}$ is an eigenvector, and $\mathbf{A}$ is the connection matrix, and if λ_{1A} and λ_{2A} are the eigenvalues of $\mathbf{A}$, then the eigenvalues of $\mathbf{B}$ are

$$\lambda_1 = \alpha\lambda_{1A} + \gamma + \delta$$

$$\lambda_2 = \alpha\lambda_{2A} + \gamma + \delta.$$

The addition of a multiple of the identity matrix simply translates all the eigenvalues up or down together.

Suppose we start the system at some point, **P**, that lies along the direction of the eigenvector, at a distance, r_o, from the origin. We want to compute the time it takes to get from the point **P** into the attractor, which is assumed to be at a distance, d, from the origin. (The magnitude of d depends on the dimensionality and the limits of the box.) We assume that the eigenvector is exactly aligned with the attractor, so that all the element values limit simultaneously when the attractor is reached.

Qualitatively, the calculation below will *underestimate* the time required for starting points that are not multiples of the eigenvector to reach the attractor, because if some elements limit before others there are large discontinuous changes in the rate of approach to the attractor. However, the simulations presented earlier on basin boundaries suggest this effect is not large when there is a small number of learned vectors in a high-dimensional space. Overall, there will be more long times in the tails of the distribution than the calculation indicates, and an obvious mechanism exists for producing a few extremely long response times.

Suppose we are a distance, r, from the origin along one eigenvector. The next position of the point after one iteration will be given by λr, where λ is the eigenvalue of **B** associated with the eigenvector. The change in position, Δr, is given by

$$\Delta r = \lambda r - r = (\lambda - 1)r.$$

If each iteration takes time Δt, the velocity of the state vector is

$$\frac{\Delta r}{\Delta t} = (\lambda - 1)r,$$

or

$$\frac{\Delta r}{(\lambda - 1)r} = \Delta t.$$

When the distance to be traveled equals the sum of the Δrs, the state vector will have reached the attractor. The time will be given by integrating the above expression from the starting point, r_o, to the attractor, d, so

$$t = \int_{r_o}^{d} \frac{dr}{(\lambda - 1)r},$$

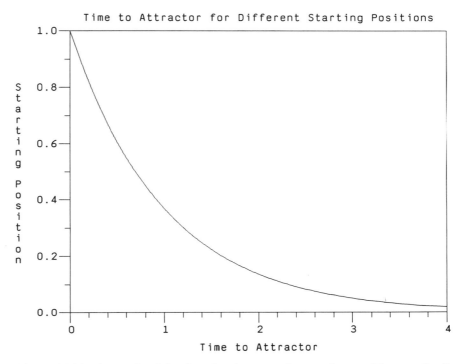

Figure 15.16 A graph of the function between the starting position on the line between the origin (0.0) and the corner attractor (1.0), and the time required to reach the attractor. From Anderson (1991). Reprinted by permission.

assuming we start at $t = 0$, then

$$t = \frac{1}{(\lambda - 1)}(ln(d) - ln(r_o)) = \frac{1}{(\lambda - 1)}ln\left(\frac{d}{r_o}\right).$$

A graph of this function is given in figure 15.16. Note that the function is the negative exponential function turned on its side.

Luce (1986) noted that many researchers suggested that reaction time data are experimentally well fit by a distribution called the ex-Gaussian, which is the convolution of an exponential distribution and a Gaussian. Such a distribution is skewed, with a long tail and a rapid rise. Several examples of the ex-Gaussian distribution are shown in figure 15.17.

Suppose we have a Gaussian distribution of starting points along the eigenvector. The values of different regions of the input distribution are multiplied by a negative exponential, and then spread out in time to generate the actual distribution of times along the horizontal axis. This function is not the convolution of a Gaussian and an exponential, but an exponential scaling of the Gaussian. The resulting time distribution looks identical,

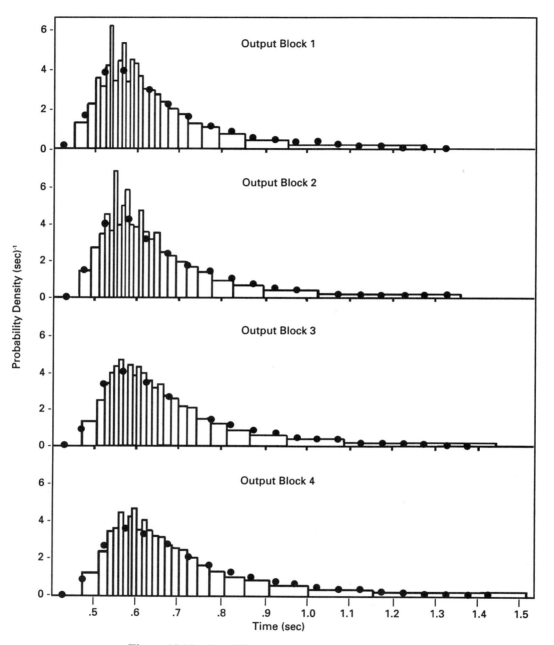

Figure 15.17 Ratcliff suggested that reaction time data is experimentally found to be well fitted by a distribution called the ex-Gaussian (Luce, 1986). Such a distribution is skewed, with a rapid rise and a long tail. The data contain both experimental reaction time distributions and the best-fitting distribution. From Ratcliff (1978). Reprinted by permission. © American Psychological Association.

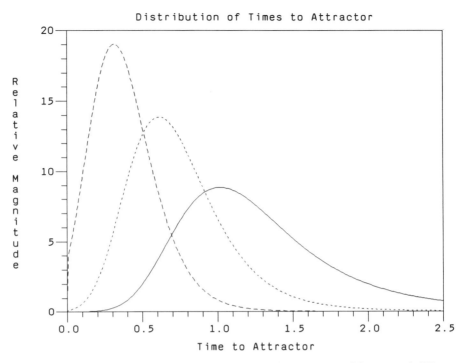

Figure 15.18   Theoretical reaction time distributions computed for several different mean starting points. The solid line corresponds to a mean starting position of 0.3, dots to 0.5, and dashes to 0.7. The input distribution is Gaussian with standard deviation of 0.15. Modified from Anderson (1991).

although moments will differ. Theoretical reaction time distributions computed for several different mean starting points and input distribution standard deviations are shown in figure 15.18.

Simulation

We can simulate this model as well. Let us consider a same-different task; that is, reporting whether or not two items presented to a subject are the same or different. Besides being a popular psychological experiment, this task is significant for neural networks for historical reasons. It realizes the truth table of X-OR, which is particularly difficult for neural networks to learn, as we discussed in chapter 8.

We assume the problem is solved by the data representation. Suppose we have a nonlinear mechanism that compares subject and object, say, with element by element matching. If the two items match, we have one coding; if they do not match, we have another coding. The additional component that violates the superposition rule for linear systems can be produced by nonlinearities, or by an extension to the vector coding used to represent the input data.

Data Representation

The bias in this book is that small neural nets are uninteresting. But looking at the outputs of large neural nets can be difficult to interpret. Strictly for convenience, as we discussed in chapter 7, we have often used in preliminary simulations a set of programs that construct state vectors from ASCII characters, so a character such as 's' is represented by eight vector elements. We can then concatenate characters. Although such a simulation is anything but efficient, it is easy to work with and captures some of the profligate way in which the nervous system uses the computing power of the computing elements. Such a coding was used for many of the cognitive simulations described in the references and in the next chapters.

We generated random-character strings of two characters to represent one or another value of a binary feature. For example, if '+' represents one value of a feature and '−' the other, we can easily construct a number of random pairs of strings.

In figure 15.19, strings of 12 characters were constructed. If the strings are identical, the first four characters are SAME; if the strings are different, the

```
         Identity   (12 Character Strings)

Response  First |  Second   | Matches
          string |  string   |

SAME+-++---++-++|+-++---++-++|++++++++++++------
SAME-------++++-|-------++++-|++++++++++++------
SAME++++-----++|+++++-----++|++++++++++++------

One different

DIFF+-++----+---|+-++--+-+---|++++++-+++++------

Two different

DIFF----++------|----+----+--|+++++-+++-++------

Eight different

DIFF---------++|+++---++++-+|---+++-----+------
```

Figure 15.19 Character strings used in the simulations. These strings have 12 elements to test for identity. When they are identical, the first four characters in the string are SAME, if they are different, the first four characters are DIFF. A '+' is used when the characters in that position in the first and second strings match, otherwise a '−' is used. The resulting state vector contained 384 elements, or 48 eight-element bytes, each representing a character. Modified from Anderson (1991).

first four characters are DIFF. The strings themselves are given in the next fields, separated (for our reading convenience) by the character '|'.

To indicate a match or mismatch between the strings, a '+' is used when the characters in that position in the first and second strings match, otherwise a '−' is used. This is a nonlinear function that allows the network to compute same-different. The computation is local, based on match or mismatch of pairs of characters. The characters at the right end of the string were not used in the simulations we will describe. Both the same and different strings were generated randomly.

The resulting system used 384 elements, or 48 eight-element bytes, each representing a character. All the simulations used 50% connectivity, that is, each element connected randomly to half of the other elements. The Widrow-Hoff error-correction procedure was used to form the connection matrix. The training set consisted of 100 examples of same and 120 examples of different in the 12 character comparisons. Each of the levels of difference (one element different, two elements different, and so on) was presented 10 times. During learning, 1000 learning trials were used, with a state vector presented randomly. During testing, the connection matrix did not learn. Two hundred fifty newly generated random stimuli were used at all levels of difference, as well as 250 new same state vectors. The testing procedure used input state vectors with the initial four characters replaced with zeros. Then BSB dynamics were applied, and the resulting strings were checked to see, first, if the correct four characters were reconstructed, and second, to note how long it took the system to reach the stable attractor where all elements were fully limited. The constants used in the simulation were $\alpha = 0.225$, $\lambda = 0.9$, and $\delta = 1$. Choice of constants was not critical.

The first thing we looked at from the simulations was the distribution of the number of iterations required to reach an attractor, defined here as number of iterations required to reach a corner of the hypercube of limits. Three typical patterns are shown in figure 15.20. Note the rapid rise and extended tail of this distribution, very like the distributions seen in human data for this and similar tasks, and also predicted by the theory. Same responses were fast. Although different response times where almost every element differed were faster than sames, the rest of different response times were slower.

It is possible to fit other aspects of the experimental data with this simple response time model. For example, both experimental data and theory find that the response time for a same response does not increase much as the number of characters to be compared increases, but the response times for different responses are strong functions of the number of differences and the overall number of characters. A canonical set of experimental data for

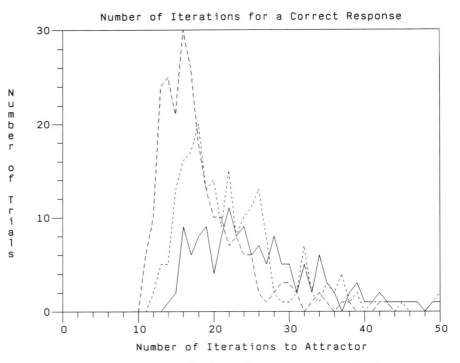

Figure 15.20 Three typical distributions of correct response time in a computer simulation. 250 random input different patterns were used and the number of iterations required to reach an attractor was recorded. The solid line has 4 of 12 characters different, the dotted line 6 of 12, and the dashed line 8 of 12. See figure 15.19 for the coding used. Modified from Anderson, 1991.

this task is from Bamber (1969; reprinted in Luce). More detailed discussion is found in Anderson (1991).

Connection with Theory

We should make one comment on the relation between the simulation, which had 32 zero elements out of 384 in the state vector, and this simple calculation. If we assume the nonzero part of the initial input is reasonably close to an eigenvector, and if δ is 1, that is, the input is constantly present, the 352 initially nonzero elements limit almost immediately, which is what is seen in the simulations. Consider the structure of the connection matrix in relation to the two pieces, initially zero and initially nonzero (NZ) regions of the input state vector. We wish to reconstruct the response in the zero block.

$$\begin{pmatrix} 0 \\ \hline NZ \end{pmatrix} \quad \text{and} \quad \begin{pmatrix} A|B \\ \hline C|D \end{pmatrix}.$$

In terms of its action on the state vector, the matrix has been partitioned into four parts. If the nonzero part is fully limited, it does not change. Therefore, block D, which connects the nonzero part of the state vector to itself, has no effect. Block B, which connects the reconstructing, initially zero part to the nonzero part, also has no effect. Part C, connecting the unchanging limited region to the reconstructing region, acts as a constant input to the reconstructing region. Part A acts as the feedback matrix driving the reconstruction. The dynamics of the system are being driven by the eigenvectors of this smaller block, which will follow the analysis given above, and where the constant input is provided by block D acting on the nonzero inputs. Many simulations have similar structure, so that this reaction time computation is actually reasonably general.

Clustering

A natural application of attractor networks is clustering because the basins of attraction divide up state space into regions, each associated with a particular attractor. The attractor basins in BSB are well behaved. If we assume that a set of data is generated from underlying discrete events or processes, the attractors might reflect this; that is, a basin might come to correspond to one particular class of events. Therefore it is theoretically possible to use an attractor network as a clustering algorithm, where an attractor can stand for a cluster of related items. This behavior is closely related to concept formation in psychology, as we discussed in chapter 10, since a single output state vector (the attractor) would represent a number of different examples of a concept.

It is possible to do both supervised and unsupervised clustering. The most common neural network architecture would use supervised learning, where a network was shown different preclassified examples and would learn correctly to classify members of the training set. However, attractor networks, since they form attractors no matter how they learn, also have the potential to do unsupervised clustering. Unsupervised learning, where the correct categories are not told to the network but must be developed by it, is, by nature, more difficult and more problematic than supervised learning. We are, after all, asking the network to divide up state space in an appropriate way, based solely on patterns presented to it. Part of the difficulty, of course, is that "appropriate" is a word that is highly task dependent.

A number of good traditional statistical clustering algorithms are available. Several neural networks have architecture that is well suited to unsupervised concept formation. A multilayer autoassociative attractor model was applied to clustering biomedical data with considerable success (Spitzer, Hassoun, Wang, and Bearden, 1990; Hasson, Wang, and Spitzer, 1992). The ART models (Carpenter and Grossberg, 1987, 1990; Carpenter, Grossberg, and Reynolds, 1991) are particularly well known. They use a combination of feedback, higher-level control, and nonlinearities to form regions in state space that correspond to concepts, based on the statistical structure of the input. The earlier ART models developed grandmother cells at an output layer. New patterns to be classified were judged either as new, if they were sufficiently far away from previously classified patterns, which led to formation of a new grandmother cell, or as new examples of old concepts if they were within a certain distance of previously classified patterns. The latest version of ART allows distributed output patterns to stand for categories.

It is possible to use BSB directly as a clustering network. An unmodified BSB network was used to cluster radar signals (Anderson, Gately, Penz, and Collins, 1990). The problem was an important practical one in the field of radar electronics.

The most common form of radar works by sending out a high-power microwave pulse. The pulse reflects from a target, and the return echo is detected by a sensitive receiver. The time it takes for the echo to return gives the distance of the target from the transmitter. Suppose we have a microwave receiver, and we listen to a region of the microwave spectrum. Near a busy airport, or in other areas where many radar transmitters are operating, the receiver detects a multitude of different radar pulses. The transmitters are all going simultaneously and their pulse trains overlap. The problem of knowing how many separate transmitters are actually present is called the *deinterleaving* problem in radar electronics (figure 15.21). The difficulty of this problem is not with lack of data or low signal to noise ratio. It is that there are *too many* data, and the network has to make sense of them. For several practical applications, it is important to be able to tell *quickly* how many emitters are present and what their properties are.

We could recast this problem quite easily as a psychological one. We are shown a great number of animals and we want to know how many different species are present. Or, we are given many different examples of dot patterns, and we want to know how many distinct classes of patterns there are, as we discussed for the Posner-Keele experiments in chapter 11.

Emitter Clustering

In this brief discussion we will consider only the deinterleaving network. (Transmitter identification with a neural network can be done using supervised learning of known examples of different transmitter types.) The net-

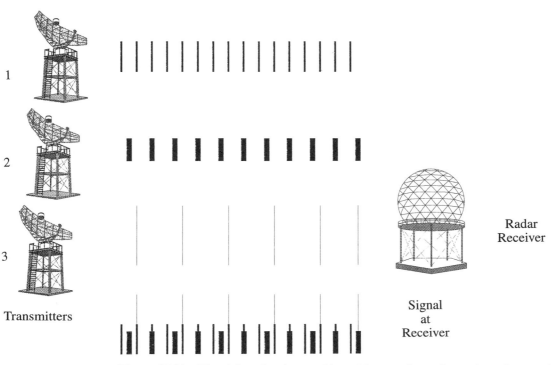

Deinterleaving Problem

Figure 15.21 The deinterleaving problem. Many radar pulse trains mix together and it is hard to tell which pulse belongs to which transmitter, and how many transmitters there are in a complex environment.

work is being asked to form a conception of the transmitter environment from the data themselves. A teacher does not exist.

In the simplest case, each transmitter emits pulses with constant properties; no noise is present. Then, determining how many transmitters are present is trivial: simply count the number of unique pulses. This would be the number of transmitters. Unfortunately, the data are moderately noisy because of receiver, environmental, and emitter variabilities. In a particular case of great interest, deliberate frequent change of one or another transmitter property may occur in an effort at concealment. Therefore, simple identity checks do not work. We want our network to form concepts of a transmitter so different pulses are seen as particular noisy examples coming from the same source.

The basic problem of a self-organizing clustering system has precedents in cognitive science. For example, William James (1892), in a passage well known to developmental psychologists, wrote:

... the numerous inpouring currents of the baby bring to his consciousness ... one big blooming buzzing Confusion. That Confusion is the baby's universe; and the universe of all of us is still to a great extent such a Confusion, potentially resolvable, and demanding to be resolved, but not yet actually resolved into parts. (p. 21)

We now know that the newborn baby is a highly competent organism with the outlines of adult perceptual preprocessing already in place. The baby is designed to hear human speech in the appropriate way and to see a world like ours: that is, a baby is tuned to the environment in which he or she will live. The same is true of the radar network, which must process radar pulses that have certain types of features with values that fall within certain parameter ranges. An effective feature analysis has been done for us by the receiver designer, and we do not have to organize a system from zero. This means that we can use a less general approach than we might have to in a less constrained problem. The result of both evolution and good engineering design is to build so much structure into the system that a difficult problem becomes tractable.

Neural Network Clustering Algorithms

Each pulse is different because of noise, but only a small number of transmitters are present relative to the number of pulses. We will take the input datum representing each pulse and form a state vector with it. As we discussed in chapter 12 for Hopfield networks, Hebbian learning in an energy-minimizing system tends to produce energy minima at the location of the learned pulse. The final low-energy attractor states of the dynamic system when BSB dynamics are applied tend to lie near or on stored information. Several hundred pulses are stored in a pulse buffer. We take a pulse at random and learn it, using the Widrow-Hoff error-correcting algorithm with a small learning constant. Since we have no teacher, the desired output is assumed to be the input pulse datum. *If the state vectors coding the pulses from a single transmitter are not too far apart in state space,* we will form an attractor that contains all the pulses from a single emitter, as well as new pulses from the same emitter (figure 15.22).

We can see why this is so from an informal argument. (Exact analysis is quite difficult when the emitters are close together or noise is large; then we must go to simulations.) Call the average emitter state vector of a particular emitter $\mathbf{p}$. Then, every observed pulse, $\mathbf{f}_k$, will be

$$\mathbf{f}_k = \mathbf{p} + \mathbf{d}_k,$$

where $\mathbf{d}_k$ is a distortion that is assumed to be different for every individual pulse, and different $\mathbf{d}_k$ are uncorrelated and are relatively small compared

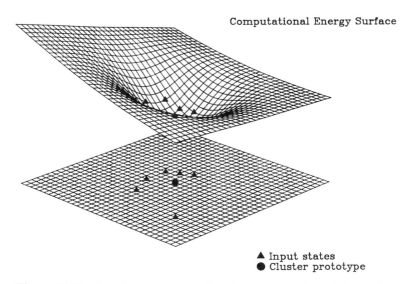

▲ Input states
● Cluster prototype

Figure 15.22 Landscape surface of system energy. Several learned examples may contribute to the formation of a single energy minimum that will correspond to a single emitter. This drawing is only for illustrative purposes and is not meant to represent the very high-dimensional simulations actually used. From Anderson et al. (1990). © 1990 IEEE. Reprinted by permission.

Early in Learning Process
Small Learning Constant, Many Examples

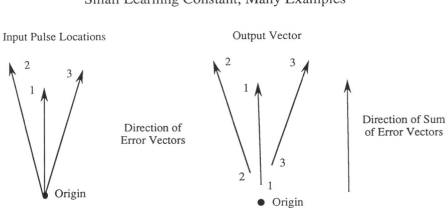

Figure 15.23 The Widrow-Hoff procedure learns the error vector. The error vectors early in learning with a small learning constant point toward examples, and the average of the error vectors points toward the category mean, that is, all the examples of a single emitter.

with **p**. With a small learning constant, and with the connection matrix **A** starting from zero, the length of the output vector, **Af**, also is small after only a few pulses are learned. This means that the error vector points outward, toward $\mathbf{f}_k$, that is, toward $\mathbf{p} + \mathbf{d}_k$. Early in the learning process for a particular cluster the error vectors (input minus output) all point toward the cluster of input pulses (figure 15.23).

Widrow-Hoff learning can be described as using a simple associator to learn the error vector. Since every $\mathbf{d}_k$ is different and uncorrelated, the error vectors from different pulses have the average direction of **p**. The matrix acts as if it is repeatedly learning **p**, the average of the vectors. At the start this system looks like simple Hebbian learning since the average error vector points toward the center of the cluster of pulses from a single transmitter.

After the matrix has learned so many pulses that the input and output vectors are of comparable length, the output of the matrix when $\mathbf{p} + \mathbf{d}_k$ is presented is near **p**. If error correction is successful, as we will assume, an eigenvector is near the cluster. Then, as in figure 15.24,

$$\mathbf{p} \approx \mathbf{Ap}.$$

Over a number of learned examples,

$$\text{total error} \approx \sum (\mathbf{p} + \mathbf{d}_k - \mathbf{A}(\mathbf{p} + \mathbf{d}_k)$$

$$\approx \sum (\mathbf{d}_k - \mathbf{Ad}_k).$$

Late in Learning Process
Small Learning Constant, Many Examples

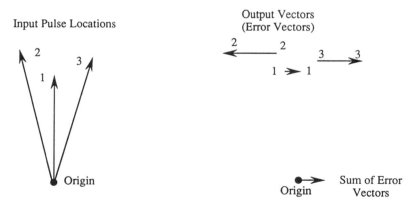

Input Pulse Locations

Output Vectors (Error Vectors)

Figure 15.24 Assume an eigenvector is close to a category mean, as will be the result after extensive error correcting, autoassociative learning. The error terms from many learned examples, with a small learning constant, will average to zero and the system attractor structure will not change markedly.

The maximum values of the eigenvalues of **A** are 1 or below, the **d**'s are uncorrelated, and this error term averages to zero.

As the system learns more and more random noise, the length of the error vector tends to grow longer and longer, as the eigenvalues of **A** related to the noise become large. For example, the covariance matrix of independent Gaussian noise added to each element is proportional to the identity matrix. That is, every vector is an eigenvector with the same eigenvalue, and this matrix is the matrix toward which **A** will evolve, if it continues to learn random noise indefinitely. When BSB dynamics are applied to matrices resulting from learning *very* large numbers of noisy pulses, the attractor basins become fragmented, so that the clusters break up. This is the intrinsic "senility" mechanism in this class of neural networks, as mentioned earlier. However, the period of stable cluster formation is very long, and it is easy to avoid cluster breakup in practice.

In BSB clustering the desired output is a particular stable state. Ideally, all pulses from one emitter are attracted to that final state. Therefore a simple identity check is now sufficient to check for clusters. This check is performed by resubmitting the original noisy pulses to the network that has learned them, and forming a list of the stable states that results. The list is then compared with itself to find which pulses came from the same emitter. The distinct states can be given names and conveniently characterized for further processing.

Stimulus Coding and Representation

We need a way to represent data from the microwave receiver so that this process can occur. The radar receiver gives us information about a number of continuous valued features: frequency, elevation, azimuth, pulse width, and signal strength. Our approach is to code continuous information as locations on a topographical map, a bar graph or a moving meter pointer. We represent each continuous parameter value by location of a block of activation on a linear set of elements. This bar coding of a continuous parameter is inspired by the frequent use of topographical maps of a parameter in the cerebral cortex, as we discussed in chapter 10. We used a similar map coding when we discussed psychological concept formation in chapter 11, and we will meet it again when we discuss Ohm's law in chapter 16 and a network to learn arithmetic in chapter 17. Although such a topographical coding uses (wastes?) many elements, it seems to be exceptionally effective in neural network practice and to capture many of the similarity relationships that networks can use effectively.

```
            Radar pulse fields: coding of input information
            Position of the bar of '=' codes an analog quanitity.


Azimuth   Elevation  Frequency    Pulse width  Pseudo-spectra
I<------->I<------->I<----------->I<---------->I<--------->I

...====....====.........====..............===..=.=.=.=.=...


In any field:  A move to the left decreases the quantity.
               A move to the right increases the quantity.
```

Figure 15.25 Input representation of analog input data uses bar codes. The state vector is partitioned into fields corresponding to azimuth, elevation, frequency, pulse width, and a field corresponding to additional information that might become available with advances in receiver technology. From Anderson et al. (1990). © 1990 IEEE. Reprinted by permission.

We represent the block/bar of activity value with a block of three or four '=' equals signs placed in a region of '.', periods. Single characters are coded by eight-bit ASCII bytes, as discussed elsewhere. Input vectors are purely binary. On recall, if the vector elements coding a character do not rise above a threshold size, the system is not sure of the output. Then that character is represented as the underline, '_'.

Neural networks can incorporate new information about the signal and make good use of it. This is one simple version of what is called the *data fusion* or *sensor fusion* problem. To code the various radar features, we simply concatenate the topographical vectors of individual feature into a single long state vector. Bars in different fields code the different quantities (figure 15.25). Below we will gradually add information to the same network to show the utility of this fusion methodology. The conjecture is that adding more information produces better clustering.

We describe a few representative simulations. The clustering network was robust, and with the bar-coded representation worked almost from the first try. The simulation used a BSB system with 480 units, 25% connected. It had a total of 10 simulated emitters with considerable added intrinisic noise. A pulse buffer of 510 different pulses was used for learning and, after learning, 100 new pulses, 10 from each emitter, tested the system. There were about 2000 total learning trials, that is, about 4 presentations per example. Parameter values were $\alpha = 0.5$, $\gamma = 0.9$, and $\delta = 0$. The limits for thresholding were $+2$ and -2. None of these parameters was critical.

Suppose we simply learn *frequency information*. Figure 15.26 shows the total number of attractors formed when 10 new examples of each of 10 emitters were passed through the BSB dynamics, using the matrix formed from learning the pulses in the pulse buffer. Only two attractors are formed.

```
                Clustering by frequency information only

Emitter                        Final output state
number

        Azimuth   Elevation  Frequency            Pulse width Pseudo-spectra
        I<------->I<------->I<--------------->I<------>I<--------->I

  1     ............................====...............................
  2     .............................====..............................
  3     ..........................====.................................
  4     ............................====...............................
  5     ..........................====.................................
  6     ..........................====.................................
  7     ............................====...............................
  8     ..........................====.................................
  9     ..........................====.................................
 10     ..........................====.................................
```

Figure 15.26 Final attractor states when only frequency information is learned. Ten different emitters are present, but only two different output states are found. From Anderson et al. (1990). © 1990 IEEE. Reprinted by permission.

```
        Clustering using azimuth, elevation, and frequency information

Emitter                        Final output state
number

        Azimuth   Elevation  Frequency            Pulse width Pseudo-spectra
        I<------->I<------->I<--------------->I<------>I<--------->I

  1     .====......====......====_...............................
  2     ....====.......====..........====......................
  3     .====......====......====_...............................
  4     ====_.....====...............====......................
  5     ._===......====......====_...............................
  6     ....====......====......====...........................
  7     ...====......====............==_=......................
  8     .__===_.....====_._.____=__..............................
  9     .__===_.....====.._.____=__..............................
 10     =__=..........====........=__=.........................
```

Figure 15.27 When azimuth, elevation, and frequency are provided for each data point, performance is better. However, two emitters are lumped together, and three others have very close final states. From Anderson et al. (1990). © 1990 IEEE. Reprinted by permission.

a)   . . . . . . = ...   Monochromatic pulse

b)   . . = . = . = . = . = . .   Subpulses with distinct frequencies.
                                          (or some kinds of FM or phase modulation)

c)   . . = = = = = = = = = . .   Continuous frequency sweep during the pulse
                                          (pulse compression)

Figure 15.28 Suppose we can assume that advances in receiver technology will allow us to incorporate a crude cartoon of the spectrum of an individual pulse into the coding of the state vector representing an example. The spectral information can be included in the state vector in only slightly processed form. From Anderson et al. (1990). © 1990 IEEE. Reprinted by permission.

All test inputs map into these attractors because of the close correlation of the 10 emitters in frequency.

One nice aspect of this system is that as we give it more information it forms better clusters. Suppose we now include information on *azimuth* and *elevation*. Clustering performance improves markedly (figure 15.27). We get nine attractors. The system still has some uncertainty, however, since few corners are fully saturated, as indicated by the underlines ('_') on the example of incorrect clumping as a result of insufficient information. Two other final states (8 and 9) are very close to each other in Hamming distance.

Just for the fun of it, suppose that future advances in receivers allow a quick estimation of the microstructure of each radar pulse. We used coding that is a crude graphical version of a Fourier analysis of an individual pulse, with the center frequency located at the middle of the field (figure 15.28). The spectral information can be included in the state vector in only slightly processed form: we have included almost a caricature of the actual spectrum.

We now combine *all* our information about pulse properties. None of the subsets of information could perfectly cluster the emitters. After learning, using all the information, we now have 10 well separated attractors, that is, the correct number of emitters relative to the data set (figure 15.29). The conclusion is that the additional information, even if it was noisy, could be used effectively.

The significance of this simulation is that it shows a simple network solving a much different problem than pattern classification, although its operation is based on good pattern classification. An unsupervised clustering algorithm is useful for many problems, especially those that look somewhat like the classic problem of psychological concept formation. In fact, this approach, used exactly as we described it here, solves the Posner-Keele concept-formation problem by developing attractors near the prototype patterns used to generate the examples. The reaction time patterns in simulations of this network are roughly similar to those experiments. We

```
                  Clustering with all information

Emitter                           Final output state
number

        Azimuth   Elevation  Frequency       Pulse width Pseudo-spectra
        I<------->I<------->I<--------------->I<------>I<--------->I

 1      .====.........====.....___=_.........===........======....
 2      ....====.........====..........====.....===....=.=.=.=.=...
 3      .====.........====.....====............===...........=.....
 4      ====.......====......====....===..........===...........
 5      ..._===_.....____====...._===_.......===.....__.=.=.=._...
 6      ....====......====....====...........===......=....=......
 7      ...====.........====......====..==_.....======....
 8      ...====....====........====..........===..=.=.=.=.=...
 9      ..._====__...._===_._.....__===_.........__==_......=......
10      ==_=...........===_.......==_=.........===.....======....
```

Figure 15.29 When all available information is used, 10 stable, well separated
attractors are formed. This shows that such a network computation can make good
use of additional information. From Anderson et al. (1990). © 1990 IEEE. Reprinted by permission.

should point out that the radar problem, as we described it, is also similar
to a problem in neurophysiological data analysis, where action potentials
from many neurons are recorded with a single electrode.

Historical note: A description of the genesis of the BSB model can be
found in *Neurocomputing* (Anderson and Rosenfeld, 1988).

References

D.J. Amit (1989), *Modelling Brain Function: The World of Attractor Neural Networks*. Cambridge: Cambridge University Press.

J.A. Anderson (1991), Why, having so many neurons, do we have so few thoughts. In
W.E. Hockley and S. Lewandowsky (Eds.), *Relating Theory to Data: Essays on
Human Memory*. Hillsdale, NJ: Erlbaum.

J.A. Anderson, M.T. Gateley, P.A. Penz, and D.R. Collins (1990), Radar signal categorization using a neural network. *IEEE Proceedings, 78*, 1646–1657.

J.A. Anderson and M.C. Mozer (1981), Categorization and selective neurons. In
G.E. Hinton and J.A. Anderson (Eds.), *Parallel Models of Associative Memory*.
Hillsdale, NJ: Erlbaum.

J.A. Anderson and E. Rosenfeld (1988), *Neurocomputing: Foundations of Research*.
Cambridge: MIT Press.

J.A. Anderson and J.W. Silverstein (1978), Reply to Grossberg. *Psychological Review, 85*, 597–603.

J.A. Anderson, J.W. Silverstein, S.A. Ritz and R.S. Jones (1977), Distinctive features, categorical perception, and probability learning: Some applications of a neural model. *Psychological Review, 84,* 413–451.

P. Baldi and E. Hornik (1989), Neural networks and principal component analysis: Learning from examples without local minima. *Neural Networks, 2,* 53–58.

D. Bamber (1969), Reaction times and error rates for "same"-"different" judgments of multidimensional stimuli. *Perception and Psychophysics, 6,* 169–174.

J. Begin (1991), Elaboration d'un nouveau modele de memoire associative applique au probleme de la categorisation. Doctoral thesis, Department of Psychology, University of Quebec, Montreal.

G.A. Carpenter and S. Grossberg (1987), ART 2: Self-organization of stable category recognition codes for analog input patterns. *Applied Optics, 26,* 4919–4930.

G.A. Carpenter and S. Grossberg (1990), ART 3: Hierarchical search using chemical transmitters in self-organizing pattern recognition architectures. *Neural Networks, 3,* 129–152.

G.A. Carpenter, S. Grossberg, and J.H. Reynolds (1991), ARTMAP: Supervised real-time learning and classification of stationary data by a self-organizing neural network. *Neural Networks, 4,* 565–588.

M.A. Cohen and S. Grossberg (1983), Absolute stability of global pattern formation and parallel memory storage by competitive neural networks. *IEEE Proceedings on Systems, Man, and Cybernetics, SMC-13,* 815–825.

G.W. Cottrell, P. Munro, and D. Zipser (1988), Image compression by back propagation: An example of extensional programming. In N.E. Sharkey (Ed.), *Advances in Cognitive Science,* Vol. 3. Norwood, NJ: Ablex.

P. Foldiak (1989), Adaptive network for optimal linear feature extraction. *Proceedings of the IJCNN, Washington, DC, June, 1989,* I-401–406, Piscataway, NJ: IEEE Service Center.

R.M. Golden (1986), The "brain-state-in-a-box" neural model is a gradient descent algorithm. *Journal of Mathematical Psychology, 30,* 73–80.

R.M. Golden (1993), Stability and optimization analyses of the generalized brain-state-in-a-box neural network. *Journal of Mathematical Psychology, 37,* 282–298.

H.J. Greenberg (1988), Equilibria of the brain-state-in-a-box (BSB) neural model. *Neural Networks, 1,* 323–324.

H. Haken (1985), Operational approaches to complex systems: An introduction. In H. Haken (Ed.), *Complex Systems—Operational Approaches in Neurobiology, Physics, and Computers.* Berlin: Springer.

H. Haken and M. Stadler (1990), *Synergetics of Cognition.* Berlin: Springer.

M.H. Hassoun, C. Wang, and A.R. Spitzer (in press), Robust decomposition of the electromyogram by a trainable dynamic neural network. Part I. Algorithm. Part II. Performance analysis. *IEEE Transactions on Biomedical Engineering.*

J.J. Hopfield (1982), Neural networks and physical systems with emergent collective computational abilities. *Proceedings of the National Academy of Sciences, 79,* 2554–2558.

J.J. Hopfield (1984), Neurons with graded response have collective computational properties like those of two-state neurons. *Proceedings of the National Academy of Sciences, 81,* 3088–3092.

R.W. Hornbeck (1975), *Numerical Methods.* New York: Quantum.

A.S. Householder (1964), *The Theory of Matrices in Numerical Analysis,* New York: Blaisdel. Reprinted in 1975 by Dover Publications, New York.

S. Hui and S.H. Zak (1992), Qualitative analysis of the BSB neural model. *IEEE Neural Network Transactions, 3,* 86–94.

W. James (1892/1984), *Psychology: Briefer Course.* Cambridge: Harvard University Press.

A.H. Kawamoto (1985), The (re)solution of semantic ambiguity. Doctoral thesis, Department of Psychology, Brown University, Providence, RI.

A.H. Kawamoto and J.A. Anderson (1985), A neural network model of multistable perception. *Acta Psychologica, 59,* 35–65.

B. Kosko (1988), Bidirectional associative memories. *IEEE Transactions on Systems, Man, and Cybernetics, 18,* 49–60.

A. Krogh and J.A. Hertz (1990), Hebbian learning of principal components. In R. Eckmiller, G. Hartman, and G. Hauske (Eds.), *Parallel Processing in Neural Systems and Computers.* Amsterdam: Elsevier.

R. Linsker (1988), Self-organization in a perceptual network. *Computer, 21,* 105–117.

R.D. Luce (1986), *Response Times.* New York: Oxford University Press.

D.J.C. MacKay and K.D. Miller (1990), Analysis of Linsker's application of Hebbian rules to linear networks. *Network, 1,* 257–297.

J.L. McClelland and D.E. Rumelhart (1981), An interactive activation model of context effects in letter perception. Part I. An account of basic findings. *Psychological Review, 88,* 375–407.

L.A. Necker (1832), Observations on some remarkable phenomena seen in Switzerland; and an optical phenomenon which occurs on viewing of a crystal or geometrical solid. *Philosophical Magazine, 1,* 329–343.

E. Oja (1982), A simplified neuron model as a principal component analyzer. *Journal of Mathematical Biology, 15,* 267–273.

E. Oja (1983), *Subspace Methods of Pattern Recognition.* Letchworth, UK: Research Studies Press.

R. Proulx and J. Begin (1990), A new learning algorithm for the BSB model. In M. Caudill (Ed.), *Proceedings of IJCNN-90 in Washington, DC,* Vol. I. Hillsdale, NJ: Erlbaum.

R. Ratcliff (1978), A theory of memory retrieval. *Psychological Review, 85,* 59–108.

D.E. Rumelhart and J.L. McClelland (1982), An interactive activation model of context effects in letter perception. Part 2. The contextual enhancement effect and some tests and extensions of the model. *Psychological Review, 89,* 60–94.

A.R. Spitzer, M.H. Hassoun, C. Wang, and F. Bearden (1990), Signal decomposition and diagnostic classification of the electromyogram using a novel neural network technique. Proceedings of the Fourteenth Annual Symposium on Computer Applications in Medical Care, IEEE Computer Society, Washington, DC, November 1990.

D.C. Van Essen (1985), Functional organization of primate visual cortex. In A. Peters and E.G. Jones (Eds.), *Cerebral Cortex*, Vol. 3. *Visual Cortex.* New York: Plenum.

D.C. Van Essen, C.H. Anderson, and D.J. Felleman (1992), Information processing in the primate visual system: An integrated systems perspective. *Science, 255,* 419–423.

B. Widrow and M. Hoff (1960), Adaptive switching circuits. *1960 IRE WESCON Convention Record.* New York: IRE.

B. Widrow and S.D. Stearns (1985), *Adaptive Signal Processing.* Englewood Cliffs, NJ: Prentice-Hall.

J.H. Wilkinson (1965), *The Algebraic Eigenvalue Problem.* Oxford: Oxford University Press.

D.M. Young and R.T. Gregory (1973), *A Survey of Numerical Mathematics.* Reading, MA: Addison-Wesley.

Any mental process must lead to error.
Huang Po (9th century)

Association is a natural operation for a neural network. This chapter discusses some of the history of association in psychology. Several applications of simple neural network associators are provided, including a small data base realized with a network, a simulation of the qualitative behavior of Ohm's law, a realization of a semantic network, and a discussion of disambiguation.

This chapter presents some examples of associative computations using neural networks. It does not include a pattern-recognition application, although all the examples use pattern selectivity as an essential step. In their excellent textbook, Hertz, Krogh, and Palmer (1991) say, "Associative memory is the 'fruit fly' or 'Bohr atom' problem of the field" (p. 11). They go on to give a formal definition of association:

Store a set of patterns ξ ... in such a way that when presented with a new pattern ζ_i, the network responds by producing whichever one of the stored patterns most closely resembles ζ_i. (p. 11)

This definition agrees with only a small part of what most cognitive scientists think of as association. It is a severely restricted definition of memory, and arises in the context of a Hopfield network in which a memory pattern is an energy minimum (see chapter 12). A more general definition of association includes the linking of patterns together. This can be done either by autoassociation, in which segments of a state vector are reciprocally linked, or in a feedforward pattern associator, in which an input pattern is linked to an output pattern, the architecture in the linear associator, and assumed for feedforward networks.

Association involves more than getting the input-output relationships right. It is powerful and complex computational tool. In this chapter we will present some simple examples of its use in computations, and a little of its history. Research on complex associative systems is at the forefront of research in cognitive science.

Pattern-recognition applications of neural networks form one simple kind of associative computation. An input state vector is associated with an output classification, with the output usually in the form of a grandmother cell representation. Let us be slightly more sophisticated, and consider what a more complex associative computer might look like. History, as it often is, is informative.

Aristotle On Memory

The major outlines of one way to use an associative computer were clearly expressed by Aristotle in 500 B.C. in an essay *On Memory* (*de Memoris et Reminiscenta*). Modern models of association usually look much like his. Aristotle made two important claims about memory structure: first, the elementary unit of memory is a *sense image*, and, second, associations and links between elementary memories serve as the basis for higher-level cognition. One English translation of this essay uses the words *memory* for the elementary unit and *recollection* for reasoning by associations between elementary units.

Aristotle placed emphasis on the *sensory* qualities of memory. He saw memory as belonging "to the perceptual part" and believed that it "involves an image in the soul, which is among other things, a sort of imprint in the body of a former sense image." Specifically, it does not consist of abstractions. Aristotle came up with a vivid visual image to express this: "For the change that occurs marks in a sort of imprint, as it were, of the sense image, as people do who seal things with signet rings."

Although Aristotle's image suggests that a raw sensory input is what is learned, we know now that the actual picture is much richer, even for situations where it seems as if only the raw image should be learned—visual images, for example. Visual memory is rarely an exact, veridical reproduction of the sense image but contains systematic distortions and omissions, as well as signs of participating in higher level abstract structures (Kosslyn, 1981).

As one example of the misleading nature of sensory memory, most of us have geographical maps available in our visual memories. Is Reno, Nevada, east or west of San Diego? Or, on our internal world map, is Paris north or south of Montreal? Interference exists between knowledge about the map and more abstract knowledge: Nevada is east of California, Reno is in

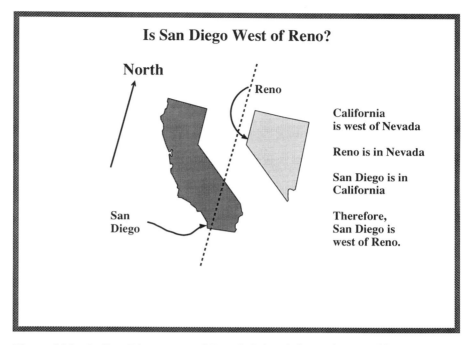

Is San Diego West of Reno?

California is west of Nevada

Reno is in Nevada

San Diego is in California

Therefore, San Diego is west of Reno.

Figure 16.1 Is San Diego west of Reno? Other information modifies the visual image.

Nevada, hence it is east of San Diego (figure 16.1). Montreal is northern because it is in Canada and very cold in winter, whereas Paris is warmer and in the heart of western Europe. In fact Reno is west of San Diego and Paris is north of Montreal (figure 16.2).

Visualization is a useful computational tool. Properly used, it saves storing a huge number of interrelated propositions. Like most good software packages, it comes with a set of elementary subroutines such as zoom, translate, scan, and the like that can be put together to construct small programs. But the spatial accuracy of our remembered images is much less than what exists in actuality, and it is contaminated with higher-level information. Also, there are many kinds of nonvisual remembered sense images. People with weak visual imagery may have powerful auditory or kinesthetic imagery and memories, as William James (1892/1984) observed in the nineteenth century.

Aristotle was aware of these qualifications and difficulties, as well as some others that are less obvious. For example, when we think of a triangle, do we think of the abstract geometrical definition that fits *all* triangles, or do we think of a specific, imaged triangle? Why *that* triangle? Suppose we need a triangle with a right angle for a proof and happen to think of a triangle with

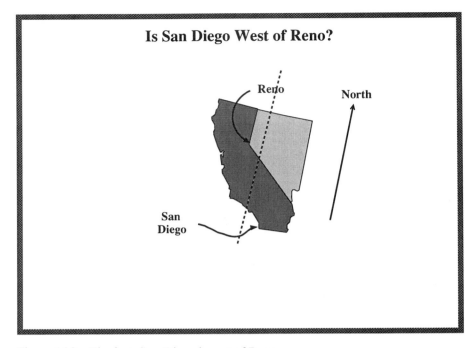

Figure 16.2   The fact: San Diego is east of Reno.

an obtuse angle (figure 16.3). Wouldn't this reduce the accuracy and generality of our geometrical reasoning? In understanding a sentence such as *the cat is near the mat*, is the cat to the left or right or in back of or in front of the mat? Or is the information purely abstract? Or is the cat in all locations at once? How can one have in image of one cat in all locations? and so on (figure 16.4).

The elementary units of memory are these sense images, but what can you do with them? Aristotle discussed at length how one "computes" with memorized sense images. The word *recollection* is used in the translation to denote this complex process: "Acts of recollection happen because one change is of a nature to occur after another." Clearly, Aristotle was proposing an (unexplained) linkage mechanism between memories. He suggested a number of ways that linkage could occur: by temporal succession or by "something similar, or opposite, or neighboring." The notion of the linkage of elementary memories is what is usually given the name *association* in the later psychological literature, and these passages are why Aristotle is often given credit for publishing the first work in the area (figure 16.5).

Aristotle saw recollection as a dynamic and flexible process: "[A] sort of reasoning. . . . [A] sort of search. And this kind of search is an attribute only

Triangles

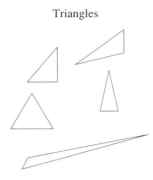

Figure 16.3   Many images of a triangle.

The cat is near the mat.

"CAT" "MAT"

Figure 16.4   A simple sentence such as, the cat is near the mat, can be visualized in several ways, or coded as purely linguistic information.

of those animals which also have the deliberating part." The modern scientific idea that comes closest to what he seemed to mean by recollection is now called a *semantic network* in which information is linked together to form complex structures.

A good deal of reasoning can be done using the semantic net to retrieve pertinent information by moving from node to node. The nodes in Aristotle's nets presumably corresponded to sense images, although modern semantic networks allow much more generality. One of the practical problems with semantic nets is branching, that is, what to do if there is more than one link leaving a node. Aristotle was aware of this problem: "[I]t is possible to move to more than one point from the same starting point." Sometimes the memory might move to the wrong branch, providing a mechanism for errors or failures of memory: "So a person is sometimes moved to one place and at other times differently." A general solution is not simple, even in modern

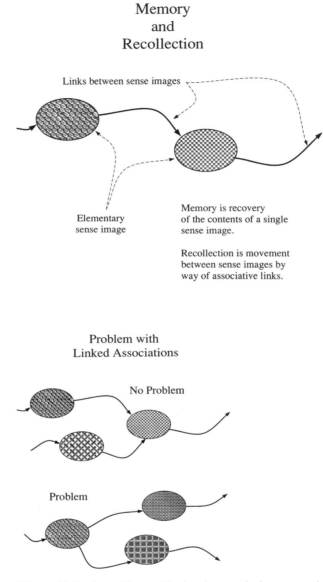

Memory
and
Recollection

Links between sense images

Elementary
sense image

Memory is recovery
of the contents of a single
sense image.

Recollection is movement
between sense images by
way of associative links.

Problem with
Linked Associations

No Problem

Problem

Figure 16.6   A problem with simple associative memories of many types comes up
when several possible associations arise from a single input.

Memory hierarchy

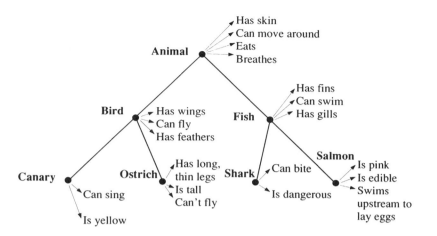

Figure 16.7   A memory hierarchy for factual information. Redrawn from Collins and Quillian (1969).

networks. It requires a nonlinear mechanism to select one or the other branch and to avoid the sum of output branches that a linear network would want to give (figure 16.6).

Because recollection is a creative process, it offers a computational mechanism for forming new ideas. Perhaps the most famous example of this process in the classical literature is Plato's dialog, *Meno*, in which Socrates led a slave boy to understand geometry by a series of small associative steps. Taken to the limit, properly directed recollection is capable of discovering new truths, using memorized sense images as the raw material.

In a modern example of a semantic network, information about birds, animals, and fish is stored in a hierarchy (figure 16.7). Specific information is stored at individual nodes, canaries are yellow, for example. General information about a class is stored at a higher level: birds fly. This network was viewed as both a device for storing information in a structured manner and as a way of answering questions, that is, performing a computation. To verify the assertion canaries are yellow, the information is looked for at the canary node. The answer to the assertion canaries have wings would not be found at the canary node but, by moving up one step, could be verified at the bird node, because birds have wings, canaries are birds, and therefore canaries have wings. Similarly, the assertion canaries have skin could be verified by moving up an additional link to the animal node (figures 16.8 and 16.9).

Canaries are yellow

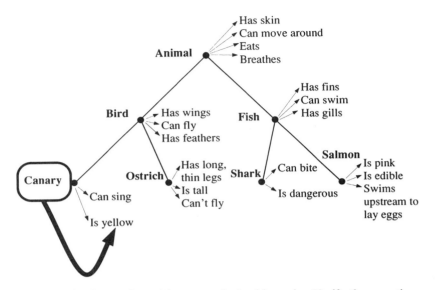

Figure 16.8   Reasoning with an associative hierarchy. Verify the assertion canaries are yellow.

Canaries have wings

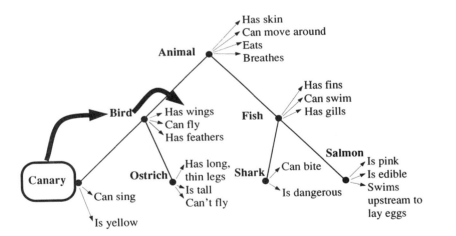

Figure 16.9   Reasoning with an associative hierarchy. Verify the assertion canaries have wings. It is necessary to move up one level.

It is tempting to apply this model of fact storage to human behavior. For example, one would expect that the length of time it takes to get information out of a system would be a function of the distance to the information. We have a little supporting evidence along those lines, but the actual picture is more complex. For example, unlikely counterfactual assertions (guppies have feathers) are rejected very quickly. One general conclusion from experiments is that human reasoning rarely involves specific application of logical rules, but is largely driven by analogy, plausibility, and memory (Collins and Michalski, 1989). The most significant claim of this approach is that it is possible to use systematically associative structures to perform reasoning by forming chains of associations, as well as more complex structures built from elementary associations.

Later Models of Association

A number of later theories of mental function used association as the basic element of mental computation. The most influential psychologists in this century were the behaviorists, in particular B.F. Skinner, the Harvard psychologist whose ideas about reinforcement learning dominated much of the theoretical discussion in psychology for several decades. We do not have space, time, or interest here for a history of the controversies surrounding Skinner; in my opinion, the relevant question is whether Skinner set back psychology 40 years or only 30 years.

To give a Psych 1 level caricature of the arguments, this school held that learning formed a link between a stimulus and a specific response. The link could be strengthened by positive reinforcement (to a first approximation, something useful or pleasant, or the cessation of something unpleasant) or weakened by negative reinforcement (absence of something pleasant, or something actively unpleasant) when the response followed the stimulus. This is the way animal trainers and politicians have worked for millenia, and when handled properly can be a very effective tool for shaping behavior in a particular direction. A number of careful experiments showed that animals followed accurate quantitative "laws of learning" in simple situations. The classic example is a rat learning to press a bar in a "Skinner box" to obtain a food pellet based on performance of some kind of behavior.

The problem is whether this simplistic view of association, a stimulus linked to a response, can be expanded to more complex situations. Many were critical of these ideas even as applied to rats, generally believing that if you ask a stupid question of a complex animal you are likely to get boring behavior. Recent work in animal behavior suggests that the mental abilities of rats and pigeons are marvelously subtle and complex when the right questions are asked. For example, pigeons seem to be able to develop concepts of specific people, or even of unlikely natural classes such as fish, and rats have amazing spatial memory.

The basic notion was that a stimulus (S) was coupled to a response (R), and all of psychology was based on the mechanistic laws that performed the coupling. Figure 16.10 has a brief quotation and a diagram that give a strong form of this idea. From the beginning, however, human behavior has seemed to humans to be far more complex than stimulus-response (S → R) association. In 1957 Skinner wrote a book attempting to explain language behavior by associative rules. In a famous review of that book, Noam Chomsky (1959) stated that S → R association *cannot* do some kinds of linguistic computations. His argument was lengthy, but one part stated that Skinner was proposing a well-defined computing machine with his model, and it was not powerful enough to do the computations we know language users perform. One of his specific examples had to do with the determination of grammaticality. Was it possible to construct a machine using S → R association that could determine whether a string of characters (or system states) was grammatical or ungrammatical? For example, simple S → R association *could not* determine the grammaticality of center-embedded strings generated by a simple rule, that is,

Grammatical   Ungrammatical

    abba   ab
   abbbba   abca
 accbbcca   bbbaaa
  cbaabc   bcca

… and so on   … and so on

Many human languages, including English, can use center embedding, but S → R models cannot compute this construction in general. Therefore Skinner's machine was inadequate.

Although this result was sometimes interpreted as causing fatal damage to models of cognition based on association, in fact it did no such thing. Skinner's S → R models had only as much computing power as the simplest heteroassociative neural networks, which no one had ever claimed were a general-purpose computer. More complex multilayer neural networks, and nonlinear networks with feedback are more powerful than necessary for this computation.

We may schematize our psychological problems as follows:

S...R
Given   ?(to be determined)

S...R
?(to be determined)   given

Your problem reaches its explanation always when:

S...R
has been determined   has been determined

Figure 16.10   A strong form of early S → R behaviorism. From Watson (1925).

This debate had striking echoes of Minsky and Papert and perceptrons. As we discussed in chapter 8, the initial stages of human perception show some of the computational limitations of perceptrons in that it is hard to determine the connectedness of complex figures at a glance. Similarly, the processing limitations of real language are in the direction suggested by simple associative models.

Taking center embedding as an example, it is true that English speakers can easily determine the meaning and grammaticality of two-layer embedded sentences such as

The rat the cat hunted hid.

But three-layer embedded sentences stop even sophisticated English speakers cold:

The rat the cat the dog chased hunted hid.

It is not possible to understand sentences of this structure if they are spoken. Again, as in the arguments over the perceptron, we as humans can clearly do better than the simplest machines, but we show weaknesses in the same areas that they do.

Associative Computation

Neural networks have an intrinsically associative architecture, in both the small, since unit activity influences other units, and in the large, since patterns of unit activity are linked to other patterns. Most modern work assumes that the entities linked and the links themselves can have internal structure. If all we want to do is reproduce a linked list of patterns, a neural network can do it. However, more flexible systems capable of complex reasoning can be produced with labeled links, for example a robin IS-A bird, an IS-A link, or Fred is the father of Herb, meaning that an associative link exists between Fred and Herb, and it carries the relationship Father-of. Complex and sophisticated computational models of cognition can be built from these pieces. Well known early examples of this approach in cognitive science are the HAM and ACT models (J.R. Anderson, 1976, 1983) and the LNR book (Norman and Rumelhart, 1975).

In the 1980s, many of those interested in semantic network models started working with neural networks. The term "connectionism" was often used to indicate the application of neural networks to high-level cognition. Recently, many attempts have been made to apply networks to reasoning, to complex concept structures, and, in particular, to language understanding. Influential early work (Hinton, 1981) discussed how to represent semantic network structures in neural networks. The second volume of the well-known PDP books (McClelland and Rumelhart, 1986) contained an entire section devoted to psychological applications of neural networks. Of particular note are chapter 14 on schemata (Rumelhart, Smolensky, McClelland, and Hinton, 1986) and chapter 19 on sentence processing (McClelland and Kawamoto, 1986).

A pitfall in applying neural networks to language is that networks are such effective statistical predictors. For example, it is possible to predict a great deal about a particular word in a sentence just by knowing a little about the words surrounding it. One can make remarkably accurate predictions of many aspects of English this way, but the language has an unusually rigid word order. It is doubtful if such a straightforward statistical approach would work for many languages that make extensive use of cases and that have a much more flexible word order.

Critics have argued that such a statistical approach avoids most of the really hard and interesting questions about representation of meaning and structure in language. The way children learn past tenses of verbs is often considered a good example of the application and misapplication of a rule. English has two basic classes of verbs, regular ones that add "ed" to the verb (walk → walked), and irregular ones that form the past differently (go → went). Formation of the regular past is more complex than the written form makes it seem. Depending on the last sound in the verb, "ed" can be pronounced /t/ by adding an unvoiced phoneme (walk → walked); /d/ by adding a voiced phoneme, (call → called); or /ed/ by adding a syllable (date → dated).

The irregular verbs are common in English. When children first start to learn the past tense they form it correctly for irregular verbs. As time goes on, however, they often start making errors in verbs they previously had spoken correctly, for example, go → goed. Eventually, they get it right again. This pattern usually is interpreted as the formation of a rule and its subsequent overgeneralization. With additional experience, the rule is applied correctly.

Rumelhart and McClelland (1986) suggested that a neural network could show this pattern without the need for an explicit rule-forming step. They argued that the first verbs learned were learned as special cases. The overgeneralizations were due to a nonrule-based mechanism such as the concept-forming models we discussed in chapter 11. Continued learning sorted things out.

To model a language phenomenon with a neural network requires proposing a representation for words. The representation chosen by Rumelhart and McClelland produced a network that worked. Simply stated, they used a context-based representation that started by coding triplets of sounds, for example, "test" might be coded as a spatial pattern of triplets '_te', 'tes', 'est', 'st_', where '_' marks end or beginning of the word. The representation actually used started with these triplets and ended with a more efficient representation based on them.

Perhaps because this model was such a direct attack on the existence of rules in language, a counterattack developed (Pinker and Prince, 1988; Lachter and Bever, 1988). Pinker and Prince (1988) finished their abstract with the sentence, "We conclude that connectionists' claims about the dispensibility of rules in explanations in the psychology of language must be rejected, and that, on the contrary, the linguistic and developmental facts provide good evidence for such rules" (p. 74). The vigor of the attack is perhaps due in part to the authors' belief that the connectionists had violated the "central dogma of modern cognitive science, namely that intelligence is the result of processing symbolic expressions" (p. 74). (Many other cognitive scientists think that the "central dogma" is actually more like a central question.)

The gist of one line of attack was that the results shown in the neural network were the result of structure built into the representation. As Lachter and Bever (1988) said, "Recently proposed connectionist models of acquired linguistic behavior have linguistic rule-based representations built. in" (p. 195). They argued that the network gave the right answers because the right answers formed part of the initial data representation. Therefore it was not surprising that the network found what its designers put there. Given the emphasis that this book has placed on data representation, however, perhaps we should not consider this comment a criticism but a statement of fact about the right way to build a neural network. The real problem is that the data representation for almost anything related to language is unknown, and is sure *not* to be as simple as assumed in the past tense model.

A more powerful critique of the computational power of neural networks is that of Fodor and Pylyshyn (1988), which addressed the area of flexibility. Their criticisms are discussed further in chapter 17. They claimed that current network architectures have fundamental restrictions on their power and cannot adequately address issues related to compositionality and creativity, questions such as how can we understand or produce a sentence we have never seen before? In chapter 17 we present as a simple test case of associative computation a network that learns elementary arithmetic, where we can make some supportable conjectures about the data representation. We then build and program a network to answer a variety of questions about unlearned classes of problems as one possible direction toward answering these justified and important concerns.

Simulations of Associative Computation

Questions of associative computation lie at the heart of much modern research on the applications of neural networks to cognition. There are no agreed-upon answers. A good presentation of recent work on connectionist models that deals with some of these difficult issues is Hinton (1991). Journals such as *Cognitive Science, Connection Science, Psychological Review,* and *Psychological Science* have all published numerous articles relating to network computation in cognition.

For the rest of this chapter, we will present some simpler and less controversial applications of association, regressing in time to the nineteenth century. It is hard not to be impressed with how sophisticated the associative models of the nineteenth century were compared with the twentieth, the heyday of simplistic S → R association. The chapter entitled "Association" by James (1892/1984), as we noted in chapter 7, contained a learning rule very much like a Hebbian rule and an architecture very much like a modern associative neural network. What is remarkable about this chapter is how *computational* James was in his later discussion of association. He was constantly concerned with how to use association to perform tasks that now would be called fact retrieval, disambiguation, or associative computation.

For example, James was well aware of the fact that what was being associated were complex assemblages of information, not a unitary S or R, and that by modulating the influence of parts of the assemblage differently, association could be directed one way or another. When he discussed association, he was concerned with its use as part of a flexible computational system. Computer simulations let us investigate these ideas in much more detail than was possible for him.

Information is sometimes represented as collections of small numbers of atomic facts relating pairs or small sets of items. James viewed mental entities as complex constructions of many more or less independent parts.

[T]he more other facts a fact is associated with in the mind, the better possession of it our memory retains. *Each of its associates becomes a hook to which it hangs, a means to fish it up by when sunk beneath the surface. Together, they form a network of attachments by which it is woven into the entire tissue of our thought. The "secret of a good memory" is thus the secret of forming diverse and multiple associations with every fact we care to retain.* (Original emphasis) (p. 257)

We will start by providing a simple demonstration of an autoassociative network capable of flexibly storing and retrieving information about drugs, diseases, and microorganisms (Anderson, 1986). We will use the slightly nonlinear BSB model as our neural network. As we did in chapter 7, we will code English words and sets of words as concatenations of the bytes representing their ASCII representations. In the outputs from the simulations the underline, '_', corresponds to all zeros or to an uninterpretable character whose amplitude is below an interpretation threshold. For our first demonstration, we will show how the network successively reconstructs information over time. Later we will merely provide information about the results.

As an example of how we might realize this, let us put information in the simple drug data base by representing information as large state vectors containing correlated information. We will assume a degree of localization of information in the state vector. We will structure our data representation so each state vector contains a number of atomic facts. We will teach a 200-dimensional network a series of connected facts about antibiotics and diseases. Figure 16.11 shows the strings of characters forming the data that we will store.

The data in the state vector are structured into fields; that is, contiguous blocks of characters correspond to different classes of information. Of the two field structures in the learned state vectors, one has to do with drugs and diseases and the other with side effects. The two share one field in common, that is, the name of the drug. Therefore, it is easy to find the side effects of a drug after its name has been retrieved (figure 16.12). Assumption of such a structured state vector is a critical representational assumption and determines to a large extent the kinds of computation that can be done with the network.

```
Staphaur+cocEndocaPenicil
Staphaur+cocMeningPenicil
Staphaur+cocPneumoPenicil
Streptop+cocScarFePenicil
Streptop+cocPneumoPenicil
Streptop+cocPharynPenicil
Neisseri-cocGonorhAmpicil
Neisseri-cocMeningPenicil
Coryneba+bacPneumoPenicil
Clostrid+bacGangrePenicil
Clostrid+bacTetanuPenicil
E.Coli  -bacUrTrInAmpicil
Enteroba-bacUrTrInCephalo
Proteus -bacUrTrInGentamy
Salmonel-bacTyphoiChloram
Yersinap-bacPlagueTetracy
TreponemspirSyphilPenicil
TreponemspirYaws  Penicil
CandidaafungLesionAmphote
CryptocofungMeningAmphote
HistoplafungPneumoAmphote
AspergilfungMeningAmphote
SiEfHypersensOralVPenicil
SiEfHypersensInjeGPenicil
SiEfHypersensInjeMPenicil
SiEfHypersensOralOPenicil
SiEfHypersensInje Cephalo
SiEfOtotoxic Inje Gentamy
SiEfAplasticAInje Chloram
SiEfKidneys++Inje Amphote
SiEfHypersensOral Ampicil
```

Figure 16.11 State vectors, interpreted as characters, storing factual information about drugs and diseases.

If the input state vector is broken into meaningful pieces, as in our simulations and as suggested in James, and the pieces can direct the resulting associative computation, perhaps we can give what is happening a new interpretation. A specific pattern in a region of state vector becomes one of a small number of discrete states allowed to be taken by the units in that region. The states in one region are linked associatively with states in other regions. When the network is settling into a final state, regions sum associative influences weighted by what are effectively strengths of associative coupling. A region looks much like a single neural network computing element, except patterns rather than simple scalar values are being combined. Perhaps connectionist models in psychology and cognitive science are successful because operations involving "cognitive" entities are akin to elementary neural network operations but at a much larger scale.

A *linking hypothesis* proposes how the properties of single neurons are related to the much more complex entities such as concepts that are considered primitives in psychology and cognitive science. A grandmother cell model assumes a linking hypothesis that identifies single units with concepts. Hebb (1949) suggested tightly coupled self-exciting groups of units

Drug data base
field structure

Drugs and diseases

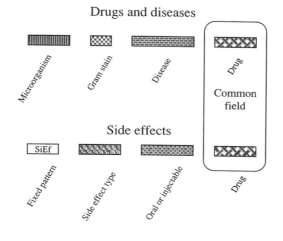

Partitioning of State Vector

Figure 16.12   The field structures used in the data representation of the information in figure 16.11.

(cell assemblies) made the connection between a neuron and a concept (see chapter 10). Perhaps these simple computer experiments on associative networks can be interpreted along similar lines. In some situations, complex patterns formed from many unit activities may behave quite simply, and can even follow rules of interaction similar to the elementary units themselves. Higher level structure emerges from the associative links between distinct patterns.

Pushed to its limit, this observation suggests mental operations might have what physicists sometimes call *self-similarity*; that is, the combining and weighting of patterns at high levels mirrors and is grounded in the combining and weighting of unit input activities at a low level. A more familar example of self-similarity would be a coastline where the bends and wiggles have roughly the same statistical structure whether you look at a continent or a small segment of coast. This is sometimes referred to as the "coastline of England" problem, because the length of the coastline depends on the scale of the ruler you use to measure it (Mandelbrot, 1982). Sutton (Sutton and Breiter, 1994) has made this point explicitly and it has considerable appeal.

Much of the traditional literature on association in psychology has successfully used models incorporating associative links between entities at a level far above the level of single neurons: the unitary $S \rightarrow R$ (stimulus-response) models are the extreme examples of this approach. Many of the network models we have discussed can display this behavior in the way they use activity patterns; special patterns (for example, eigenvectors), energy minima, and network dynamics make computations that are extremely involved at one level and become simple when observed at another level.

Returning to our simulation, the stored information is complex in that one bacterium causes many diseases, the same disease is caused by many organisms, and a single drug may be used to treat many diseases caused by many organisms. The information contained in the data base came from a classic medical school pharmacology text universally referred to as Goodman and Gilman's (Gilman, Rall, Nies, and Taylor, 1990). It is highly recommended for the general reader, and will quickly induce considerable unease about all kinds of pharmaceuticals, recreational and otherwise.

When the information in figure 16.11 was learned, the connection matrix was constructed using the vector form of the Widrow-Hoff rule. The network learned the state vectors easily, in a few hundred learning trials. The connection matrix was partially connected, with a random 50% of the weights set identically to zero. Once the data are learned, they form part of a flexible associative retrieval system. With a little initial information, we can reconstruct the associates of the given fact using any one of a number of associative techniques. We can then use *parts* of the reconstructed information to give us access to new sets of associations. In James' terminology, this can be done by "fading out," or ignoring, or setting to zero much or all of the information relating to the present mental object, and using the remaining information to generate new associations. James viewed this process as providing a range of possibilities.

If only part of the information was used to generate the association, James called it partial recall. To him, perhaps the most interesting case was one in which only a fraction of one thought was used to give rise to the next association. He called the extreme form of this process focalized recall. Modern terms such as "selective attention" capture part of this idea, but without the strong flavor of associative computation.

James' original words and figures are worth giving to describe how this flexible associative computation was performed. His original figures are collected in figure 16.13. Figure 16.14 suggests one way this system might be realized using our terminology.

The gradual passage from total to focalized, through what we have called ordinary partial recall may be symbolized by diagrams. Figure (58) is total, Figure (59) is partial and Figure (60) is focalized recall. A in each is the passing and B the coming thought. In "total recall" all parts of A are equally operative in bringing up B. In

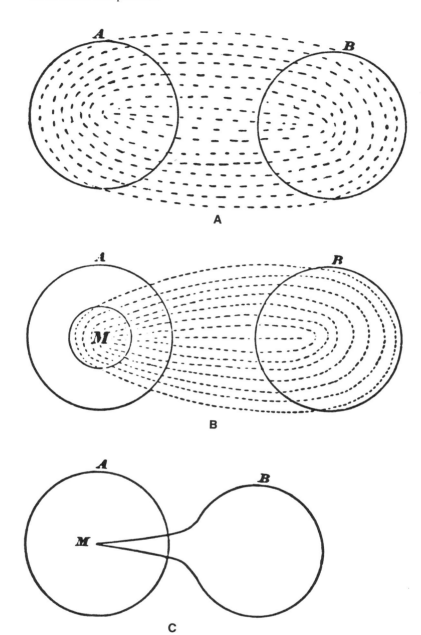

Figure 16.13   Diagrams of total, partial, and focalized association. From James (1892).

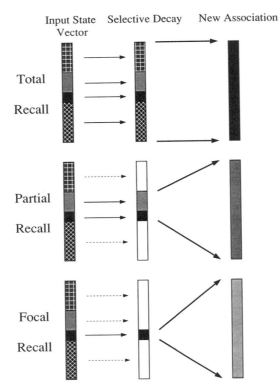

Figure 16.14 Reinterpretation of total, partial, and focalized association, represented as state vectors.

"partial recall," most parts of A are inert. The part M alone breaks out and awakens B. In ... "focalized recall" the part M is much smaller than in the previous case, and after awakening its new set of associates, instead of fading out itself, it continues persistently active among them.... Why a single portion of the passing thought should break out from its concert with the rest and act, as we say, on its own hook, why the other parts should become inert are mysteries which we can ascertain but not explain. (pp. 236–237)

Figures 16.15 and 16.16 show the simple retrieval of stored information. First, in figure 16.15, we probe the network with the string "FungMening" in the disease field, with the rest of the state vector set to zero. This might be interpreted as something like, "Tell me about fungal meningitis." The learned data include information about two kinds of fungal meningitis, that is, fungal meningitis is ambiguous, so the network chooses one kind, essentially arbitrarily. The same drug, amphotericin, treats all the fungal diseases the network knows about. A strong, unambiguous associative link exists between fungal infection and amphotericin. The string representing amphotericin is rapidly reconstructed in the drug field after about 20 iterations.

```
 1.      _____fungMening_____   Check:    80
...
11.      _____fungMening_m__ote     Check:    85
...
21.      _s_____fungMeningAm_hote     Check:   152
...
31.      _s_____i_fungMeningAmphote     Check:   173
32.      _s_____i_fungMeningAmphote     Check:   173
33.      _s_____i_fungMeningAmphote     Check:   173
34.      _s_____i_fungMeningAmphote     Check:   173
35.      _s_____i_fungMeningAmphote     Check:   173
36.      _s_____i_fungMeningAmphote     Check:   173
37.      As_____i_fungMeningAmphote     Check:   174
38.      As_____i_fungMeningAmphote     Check:   176
39.      As_____i_fungMeningAmphote     Check:   178
40.      As__p_i_fungMeningAmphote      Check:   179
41.      As__p_i_fungMeningAmphote      Check:   181
42.      As__p_i_fungMeningAmphote      Check:   182
43.      As__p_i_fungMeningAmphote      Check:   182
44.      As__pgi_fungMeningAmphote      Check:   185
45.      AspepgimfungMeningAmphote      Check:   185
46.      AspepgimfungMeningAmphote      Check:   186
47.      AspepgimfungMeningAmphote      Check:   186
48.      Aspepgi_fungMeningAmphote      Check:   188
49.      Aspe_gi_fungMeningAmphote      Check:   190
...
58.      Aspe_gi_fungMeningAmphote      Check:   198
59.      AspergilfungMeningAmphote      Check:   198
```

Figure 16.15 An example of associative reconstruction of information about fungal meningitis using the neural network data base. "Check" is the number of limited state vector elements.

```
 1.   SiEf_____Amphote   Check:    88
...
11.   SiEf_____Amphote   Check:   126
...
21.   SiEf__dney_++K__e_Amphote     Check:   160
22.   SiEf_idneys++K__e Amphote     Check:   164
23.   SiEf_idneys++K__e Amphote     Check:   169
24.   SiEf_idneys++K__e Amphote     Check:   174
25.   SiEf_idneys++K__e Amphote     Check:   175
26.   SiEf_idneys++K__e Amphote     Check:   178
27.   SiEfKidneys++K__e Amphote     Check:   183
```

Figure 16.16 Retrieval of information about the side effects of amphotericin.

Because the organism is ambiguous and could be either *Cryptococcus* or *Aspergillus*, the organism field is filled in about 60 iterations. In the simulation shown, the organism chosen was *Aspergillus*, but in other simulations with a different learning sequence, the equally correct *Cryptococcus* was chosen. We will see throughout this set of demonstrations, and as was discussed in chapter 15, that the number of iterations required to reach an attractor is a useful measure of certainty.

If we want more information about the drug, we can directly realize James's ideas about focalized recall. Amphotericin, the drug retrieved in the first simulation, can be checked for side effects by setting to zero all values other than the drug name and SiEf, which are maintained at their input values. This focalized recall of the side effects (figure 16.16) shows that amphotericin causes severe kidney damage, among others, but it may be the only way to treat some extremely dangerous fungal diseases.

The BSB constants involved were not critical. For this simulation they were $\gamma = 0.9$, $\alpha = 0.2$, and $\delta = 1.0$, as in the BSB equation in chapter 15. The upper and lower limits were $+1.3$ and -1.3. The number refers to the interation number, that is, how often the state vector has passed through the matrix. "Check" refers to to the number of elements in the state vector that are limited at that iteration. It is a rough measure of length. It cannot be larger than 200, the number of elements. Note the error corrected in the later iterations in figure 16.15, where Aspepgim becomes Aspergil as the state vector wanders through state space.

The data base can also generalize to a limited extent. When the network was "asked" what drug should be used to treat meningitis caused by a gram-positive bacillus, it responded penicillin. It had never learned about a meningitis caused by a gram-positive bacillus. It had learned about several other gram-positive bacilli and that the associated diseases could be treated with penicillin. The final state vector contained penicillin as the associated drug (figure 16.17). The other partial information cooperated to suggest that this was the appropriate output. This inference may or may not be correct, but it is reasonable given the past of the system. This might be considered a spurious attractor in that it was not learned, but was valuable anyway.

Extensions of the network architecture can reject answers. If, for example, the patient is known to be allergic to penicillin, that answer would have to be rejected. There are a number of ways to do this; the discussion of the BSB network in chapter 15 gave several. One straightforward way is to use Hebbian outer product learning with a negative sign, as we used for the Necker cube simulation in chapter 15.

It is possible to use a complex version of James's partial recall to look for a conjunction of information (figure 16.18). If the system was asked about either the side effect hypersensitivity or the disease urinary tract infection by itself, it gave one set of answers. When the data base was told urinary tract infection, the drug it retrieved was gentamicin, whose side effect is ototoxicity. Hypersensitivity, used as a probe in part II of figure 16.18, is a typical reaction to the penicillin class of drugs. Since penicillin is the most common drug in the data base, it is the one most strongly associated with hypersensitivity. It is not given (in this set of information) to treat urinary tract infections.

If the probe information included urinary tract infection *and* hypersensitivity, a drug that is associated with both is cephalosporin, and this is the

```
 1.    _____+bacMening_____    Check:   80
...
11.    _____+bacMening_____    Check:   90
...
21.    Co_____'+bacMening_en__i_     Check:  155
...
29.    Co_____'+bacMeningPenici_     Check:  174
30.    Co___e_'+bacMeningPenicil      Check:  177
```

Figure 16.17 A simple example of network generalization. The data base has no information stored explicitly about a gram-positive bacillus causing meningitis.

```
      Part I: Urinary Tract Infections

 1.    _____UrTrIn_____    Check:   48
...
21.    _____ -__cUrTrIn_____      Check:  108
...
31.    ___d___ -bacUrTrInC__lamm        Check:  147
...
41.    _r_____q -bacUrTrIn_e__am_       Check:  157
...
51.    _ro_e_q -bacUrTrIn_e__am_        Check:  162
...
61.    Prote__ -bacUrTrInGe_tamy        Check:  185
...
71.    Proteus -bacUrTrInGe_tamy        Check:  195


      Part II: Hypersensitivity

 1.    ____Hypersen_____         Check:   64
...
11.    _i__Hypersens_____e_____        Check:   81
...
21.    SiEfHypersensIj____e_____        Check:  161
...
31.    SiEfHypersensIn____e_____        Check:  171
...
41.    SiEfHypersensInj___e_____        Check:  174
...
51.    SiEfHypersensInje_Penicil         Check:  181


Part III. Hypersensitivity + Urinary Tract Infection

 1.    ____HypersenUrTrIn_____        Check:  112
...
11.    Q__dHypersenUrTrInC_____        Check:  126
...
21.    Q__dHypersenUrTrInCe__alo         Check:  174
...
31.    Q__dHypersenUrTrInCephalo         Check:  188
```

Figure 16.18 An example of more complex associative retrieval. Part I asks for information about urinary tract infections. Part II asks for information about the side effect hypersensitivity. Part III combines the two and asks for a drug to treat urinary tract infections that has the side effect of hypersensitivity. "Check" is the number of limited state vector elements.

choice of the system. (Ampicillin would also be a satisfactory answer.) Using two sets of information gave a different result than either by itself. Notice that the form of this state vector, in which a side effect and a disease occur simultaneously, never occurs in the learned vectors forming the data base, and in fact informally constructs a new field structure different from those given in figure 16.12. It is easy to perform such violations of logical structure.

Ohm's Law: Qualitative Physics

Considerable interest is displayed in the construction of artificial systems capable of qualitative reasoning about physical systems. There are several reasons for this. First, much human real world knowledge is of this kind, where information is not stored in propositional form but in a hazy intuitive form generated by extensive experience with real systems. Second, qualitative reasoning is particularly hard to handle with traditional AI systems because of its inductive and ill-defined nature. Third, it is an area in which distributed neural systems may be effective. Fourth, the ideal model for reasoning about complicated real systems may be a hybrid, partly rule driven and partly intuitive. We will discuss this further in chapter 17.

For an initial test of these ideas we constructed a set of state vectors representing the functional dependencies in Ohm's law; for example, what happens to E when I increases and R is held constant. These vectors were in the form of quasi-analog codings (figure 16.19). The system was taught according to our usual techniques and it learned to reproduce the patterns easily. The parameters of the system were unchanged from those in the drug data base simulation.

This data representation is often called a *bar code* and is used for continuous valued quantities, in this case, E, I, and R. As the bar moves from left to right, the position signals increasing magnitude of the parameter. We are interested in changes in magnitude, so a bar in the center codes constant value, a bar to the left means decrease in value, and a bar to the right means increase in value. We used a bar code for the radar classifier in chapter 15, and will use it in our representation of number in chapter 17.

sample state vector

$$E.. = = = ..I.... = = = R = = = ...$$

0	$\rightarrow$	$\leftarrow$
voltage	current	resistance
constant	up	down

```
E..===..I.....==R==......
E..===..I....===R===.....
E..===..I...===.R.===....
E..===..I..===..R..===...
E..===..I.===...R...===..
E..===..I===....R....===.
E.===..I==.....R.....==.
E==.....I==.....R..===...
E===....I===....R..===...
E.===...I.===...R..===...
E..===..I..===..R..===...
E...===.I...===.R..===...
E....===I....===R..===...
E.....==I.....==R..===...
E==.....I..===..R==......
E===....I..===..R===.....
E.===...I.===...R.===....
E..===..I..===..R..===...
E...===.I..===..R...===..
E....===I..===..R....===.
E.....==I..===..R.....==.
```

Figure 16.19 The factual information stored in the simulation of the qualitative behavior of Ohm's law.

Bar code representations are extremely effective at representing continuous quantities in neural networks. They are consistent with the topographical maps in the cerebral cortex, although simplified. Periods, '.', are place holders. Since periods are characters, they form patterns of 1s and −1s. These learned patterns have no zero elements. As before, the network was tested by putting in a partial state vector—that is, one field had all zeros in it—and letting the network fill in the zeros.

If we put in part of one of the learned vectors, the correct vector was recovered:

after 11 iterations

$$E===....I..===..R\_\_\_ \quad \rightarrow \quad E===....I..===..R===.....$$

↑
(initial 0s)

When a more complex relation was used, the system estimated answers. Suppose both current and voltage dropped, corresponding to the state vector,

after 6 iterations

$$E\_\_\_I===....R===..... \quad \rightarrow \quad E===....I===....R===.....$$

What did that imply about the voltage? From Ohm's law, I and R both are down, therefore

$$\downarrow I \downarrow R \rightarrow \downarrow\downarrow E.$$

Suppose the voltage went up and the current went down. Then,

after 5 iterations

$$E....===I===....R\_\_\_ \rightarrow E....===I===....R....===.$$

If the voltage is up and the current down, then from Ohm's law:

$$\frac{\uparrow E}{\downarrow I} \rightarrow \uparrow\uparrow R.$$

A major problem with this network and with intuitive systems of any kind is that Ohm's law is never uncertain. Given any two values, it will always predict the third. But intuition can get confused, and it practically always fails for some regions of parameters. For example, consider the case when both voltage and current decrease. The equation becomes

$$R\uparrow\downarrow \rightarrow \frac{E\downarrow}{I\downarrow}.$$

Resistance is not well determined here. The exact qualitative behavior will depend on the fine details of the system, and our network is not that precise. After 32 iterations, it cannot make a decision, and, in fact, never does.

after 32 iterations

$$E===....I===....R\_\_\_ \rightarrow 32.\ E===....I===....R.._.....$$

The bar code representation of input data allowed us to construct this qualitative system. Engineers need not throw out their calculators, because intuition is confused in tough cases.

Disambiguation and Common Association

An important part of reasoning by association is what might be called disambiguation by cooperation. We saw that a structure in which an input had several possible associations was a difficult construction for an associative network, as was known to Aristotle. Yet humans are effective when dealing with ambiguity, far more so than any artificial system we know about. Neural networks can handle it very well. Several examples in this chapter revolve around disambiguation, and the motion network in chapter 11 worked by integration of ambiguous inputs.

Let us consider a simple associative system, for example, the linear associator. Consider what happens when more than one output is associated with an input. Suppose in response to an input, $\mathbf{f}$, the system learns outer product associations with $\mathbf{g}_1$, $\mathbf{g}_2$, and $\mathbf{g}_3$, that is,

$$\mathbf{A} \propto \mathbf{g}_1 \mathbf{f}^T + \mathbf{g}_2 \mathbf{f}^T + \mathbf{g}_3 \mathbf{f}^T.$$

Then, if the input, $\mathbf{f}$, is input to $\mathbf{A}$, the output is

$$\mathbf{A}\mathbf{f} \propto \mathbf{g}_1 + \mathbf{g}_2 + \mathbf{g}_3.$$

Due to superposition, since this is a linear system, the actual response pattern is a sum of the three responses. The superposition of responses is only sometimes what we want. We want one response or the other, not a mixture. An input that caused a driver to turn left some of the time and right some of the time would lead to a straight ahead response, on the average, which is not always the right thing to do. (As a Texas politician commented, "The only things in the middle of the road are a yellow line and dead armadillos.") We need a nonlinear response selection mechanism; an attractor network like BSB or a Hopfield network would work.

But we can start to see how even the linear system can use partial information to reason cooperatively. Suppose we have a simple memory formed that has associated an input, $\mathbf{f}_1$, with two outputs, $\mathbf{g}_1$ and $\mathbf{g}_2$, and an input, $\mathbf{f}_2$, with two outputs, $\mathbf{g}_2$ and $\mathbf{g}_3$, so that

$$\mathbf{A}\mathbf{f}_1 \propto \mathbf{g}_1 + \mathbf{g}_2$$

and

$$\mathbf{A}\mathbf{f}_2 \propto \mathbf{g}_2 + \mathbf{g}_3.$$

Suppose we then present $\mathbf{f}_1$ and $\mathbf{f}_2$ together. We have

$$\mathbf{A}(\mathbf{f}_1 + \mathbf{f}_2) \propto \mathbf{g}_1 + 2\mathbf{g}_2 + \mathbf{g}_3,$$

with the largest weight for the common association. This obvious consequence of superposition lets us enhance the common association of $\mathbf{f}_1$ and $\mathbf{f}_2$. The cooperative effects described in several contexts above depend critically on the linearity of the memory, since things add up in memory.

James used a similar mechanism to explain how a forgotten item can be remembered by the concerted action of its associates. In explaining his diagram (figure 16.20), he said, "The whole process can be rudely symbolized in a diagram. Call the forgotten thing Z, the first facts with which we felt it was related *a*, *b*, and *c*, and the details finally operative in calling it up *l*, *m*, and *n*. Each circle will then stand for the brain process principally concerned in the thought of the fact lettered within it. The activity at Z will at first be a mere tension; but as the activities in *a*, *b*, and *c* little by little irradiate into

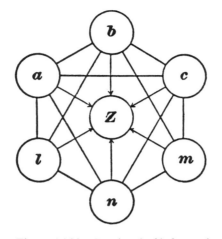

Figure 16.20 Retrieval of information by using its common associations according to James (1892).

l, *m*, and *n*, and as *all* these processes are somehow connected with *Z*, their combined irradiations upon *Z*, represented by the centripetal arrows, succeed in rousing *Z* also to full activity (pp. 242–243).

Disambiguation Example

Disambiguating ambiguous words is a major theoretical problem in psychology, a major practical problem in AI, and a problem particularly pronounced in language behavior. Ambiguity in language is everywhere. There is a good quantitative relationship between the length of an entry in the *Oxford English Dictionary* and word frequency, so that the most frequent words have the most different meanings. This is more curious than it at first appears. If we want to be clear and *unambiguous* we are told to use the simplest words. Yet, these are, in fact, the most ambiguous ones. Therefore, the simplest sentences are likely to have the most possible interpretations.

Worse, English is often ambiguous as to part of speech. It is difficult to do machine parsing when it is not clear which part of speech a word represents. As a well-known example, words of the proverb, "still waters run deep," could be nouns, verbs, or adjectives, not to mention that each has several meanings as a particular part of speech. Ambiguity as to part of speech is called *syntactic* ambiguity and ambiguity as to meaning is called *semantic* ambiguity. It is necessary to be able to handle both in an efficient manner for a machine to understand natural language. It was failure to appreciate this problem that lead to the dismal results of early attempts at machine translation.

```
BaseballGameBat BallDiamd
Vampire MythBat NiteDracu
Animal  LiveBat WingFlyng
Poker   GameBeerTablCards
Tennis  GameCortBallRackt
Dancing RichPrtyBallSocty
GeoShapeTwoDCrclSqreDiamd
GeoModelTreDSphrBallTetra
ExpJewelRichRubyOpalDiamd
```

Figure 16.21   Information stored in the disambiguation system.

Kawamoto (1985) suggested a neural network system that was capable of performing disambiguation of words using a number of techniques including adaptation. Successful preliminary use of a neural network for syntactic disambiguation with a large body of natural text is described in Benello, Mackie, and Anderson (1989). (Julian Benello was tragically killed in the bombing of Pan American flight 103 over Lockerbie, Scotland, in 1988.)

For this chapter we will demonstrate simple semantic disambiguation using some of the techniques suggested by Kawamoto. If one says to a resident of North America, *bat, ball,* and *diamond,* everyone knows quickly that we are talking about baseball. (This has been tested in many lectures with the expected results.) Baseball is only one possible association of each of these ambiguous words, yet we were able to pick out the predominant one quickly. If we give more information (it is a game), response time drops further and certainty of the answer increases. In general, hints help performance in both speed and accuracy.

We can solve the baseball example with a neural network. The input stimulus set is given in figure 16.21. A computer would probably handle this by some kind of combinatorial tree search (figure 16.22). Unfortunately, the more information we have, the more possible permutations of associations have to be explored. For the computer, hints give increased accuracy, but must hurt reaction time, since a growing number of potential associations must be checked. Worst of all, the rate at which the number of possible associations is growing is combinatorial. For example, if each word has 4 possible meanings, there are 16 possible combinations of associations for two words, 64 for three words, and 256 for four words. A sentence containing eight quadruply ambiguous words would have 65,536 possible combinations of meanings to check.

The stimuli are autoassociated using our usual BSB network, with 50% connectivity. In the nine stimuli, *Game, Bat, Ball,* and *Diamd* are triply redundant; that is, each is used by three different possible stimuli. For a test stimulus such as *Ball,* three possible final states contain *Ball,* each one equally

Meanings of Bat:	Animal Vampire Baseball	Meanings of Diamond:	Shape Jewel Baseball
Meanings of Ball:	Model Tennis Party Baseball	Meanings of Game:	Poker Tennis Baseball

	Bat	Ball	Diamond	Game
1.	Animal	Model	Shape	Poker
2.	Animal	Model	Shape	Tennis
3.	Animal	Model	Shape	Baseball
4.	Animal	Model	Jewel	Poker
5.	Animal	Model	Jewel	Tennis
6.	Animal	Model	Jewel	Baseball
7.	Animal	Model	Baseball	Poker
8.	Animal	Model	Baseball	Tennis
9.	Animal	Model	Baseball	Baseball
10.	Animal	Tennis	Shape	Poker
...				
18.	Animal	Tennis	Baseball	Baseball
19.	Animal	Party	Shape	Poker
...				
27.	Animal	Party	Baseball	Baseball
28.	Animal	Baseball	Shape	Poker
29.	Animal	Baseball	Shape	Tennis
...				
106.	Baseball	Baseball	Baseball	Poker
107.	Baseball	Baseball	Baseball	Tennis
108.	Baseball	Baseball	Baseball	Baseball

Figure 16.22 One way of looking for a common interpretation of many ambiguous bits of information is brute force combinatorics. Here are some of the possible interpretations of the ambiguous words in the example. If more information was provided, it would yield more combinations to search through.

correct since no more information was given. Which one is actually chosen by the simulation is somewhat arbitrary, and depends on the short-term history of the learning and on the structure of the particular connections.

In one simulation, when *Bat* was the test input, the output was *Vampire*, *Ball* gave rise to *Tennis*, and *Diamd* gave rise to *GeoShape*. The average number of iterations for the state vector to be completely limited was 98.5. The input and output vectors for a few conditions are shown in figure 16.23.

If a test stimulus was composed of two items, *Bat* and *Ball* or *Bat* and *Myth*, the appropriate answer was chosen, *Baseball* or *Vampire*, respectively, even though each word separately gave a different output. The average number of iterations for such an input was 38. If a test stimulus contained three ambiguous items, the average number of iterations was 20, and if all four ambiguous items were used as the input vector, only 14 iterations were required to produce an answer. Clearly, in this simulation additional information greatly decreased reaction time.

1. Single Ambiguous Words

	Input		Fully Limited Output	Number of Iterations
_____Bat _____	→	Vampire MythBat NiteDracu	81
_____Ball_____	→	Tennis GameCortBallRackt	105
_____Diamd	→	GeoShapeTwoDCrclSqreDiamd	134
_____Game_____	→	Poker GameBeerTablCards	68

2. Pairs of Ambiguous Words

_____Bat Ball_____	→	BaseballGameBat BallDiamd	30
_____Bat ____Diamd	→	BaseballGameBat BallDiamd	28
_____BallDiamd	→	BaseballGameBat BallDiamd	27
_____Game____Ball_____	→	Tennis GameCortBallRackt	91
_____GameBat_____	→	BaseballGameBat BallDiamd	28
Geo_____Diamd	→	GeoShapeTwoDCrclSqreDiamd	22

3. Other Associate Pairs

_____Bat Nite_____	→	Vampire MythBat NiteDracu	23
_____Bat Wing_____	→	Animal LiveBat WingFlyng	25
___Shape_____Diamd	→	GeoShapeTwoDCrclSqreDiamd	22

4. Triples of Ambiguous Words

_____Bat BallDiamd	→	BaseballGameBat BallDiamd	18
_____GameBat ____Diamd	→	BaseballGameBat BallDiamd	18
_____GameBat Ball_____	→	BaseballGameBat BallDiamd	25

5. Quadruple of Ambiguous Words

| _____GameBat BallDiamd | → | BaseballGameBat BallDiamd | 14 |

Figure 16.23 The neural network finds a common interpretation of numerous ambiguous bits of information by quite a different mechanism. Notice that the more information that is given, the faster the recall.

The point of this simulation is that disambiguation by context can take place in a rapid and natural way with a network architecture. A concept such as baseball can be formed from ambiguous pieces. Predictable reaction time patterns are related to degree of ambiguity when retrieving information.

Semantic Networks

It is possible to build more complex associative structures with slight extensions of these ideas. By making associations between state vectors, one can realize a simple semantic network. The example in figure 16.24 is based on the famous network of Collins and Quillian, shown redrawn in figure 16.7. Each node of the network in figure 16.23 corresponds to a state vector that contains related information that is present at the node.

A Simple Knowledge Network

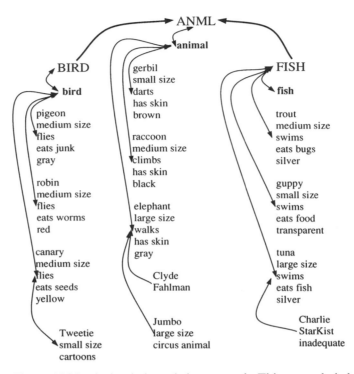

Figure 16.24 A simple knowledge network. This example is loosely based on the information in figure 16.7.

For example, the canary node contains the information that a canary is medium-sized, flies, is yellow, and eats seeds. This is connected by upward and downward links to the BIRD node, which says that canary is an example of the BIRD concept, which has the name bird. A strictly upward connection informs us that birds are ANMLs (with name animal). The network contains three examples of fish, birds, and animal species and several examples of specific creatures. For example, Charlie is a tuna, Tweetie is a canary, and Jumbo and Clyde are elephants.

The specific sets of associations that together are held to realize this simple network are given in figure 16.25. They were learned using the Widrow-Hoff error-correction rule. Two matrices were formed, one corresponding to the associations of the state vectors with themselves (autoassociation) and one corresponding to the association of a state vector with a different state vector (heteroassociation). The matrices were 50% connected.

When the matrix was formed and learning ceased, the system was interrogated to see if it could traverse the network and fill in missing information

```
BIRD_*_bird___fly_wormred       →       _____*_robin__fly_wormred
_____*_robin__fly_wormred       →       BIRD_*_bird___fly_wormred
BIRD_*_bird___fly_junk          →       _____*_pigeon_fly_junkgry
_____*_pigeon_fly_junkgry       →       BIRD_*_bird___fly_junkgry
BIRD_*_bird___fly_seedylw       →       _____*_canary_fly_seedylw
_____*_canary_fly_seedylw       →       BIRD_*_bird___fly_seedylw
ANML*__animal_dartskinbrn       →       _____*__gerbil_dartskinbrn
_____*__gerbil_dartskinbrn      →       ANML*__animal_dartskinbrn
ANML_*_animal_clmbskinblk       →       _____*_raccoonclmbskinblk
_____*_raccoonclmbskinblk       →       ANML_*_animal_clmbskinblk
ANML__*animal_walkskingry       →       _____*elephanwalkskingry
_____*elephanwalkskingry      →       ANML__*animal_walkskingry
BIRD_____       →       ANML_____
_____Clyde___Fahlman___       →       _____*elephanwalkskingry
_____*__Tweetie_cartoon___      →       _____*_canary_fly_seedylw
_____*Jumbo_____circus__      →       _____*elephanwalkskingry
FISH_____       →       ANML_____
FISH*__fish___swimfoodxpr       →       _____*__guppy__swimfoodxpr
_____*__guppy__swimfoodxpr      →       FISH*__fish___swimfoodxpr
FISH_*_fish___swimbugsslv       →       _____*_trout__swimbugsslv
_____*_trout__swimbugsslv       →       FISH_*_fish___swimbugsslv
FISH__*fish___swimfishslv       →       _____*tuna___swimfishslv
_____*tuna___swimfishslv      →       FISH__*fish___swimfishslv
StarKistCharlieinadequate       →       _____*tuna___swimfishslv
```

Figure 16.25 A realization of the knowledge of network in figure 16.24 by means of associated state vectors.

in an appropriate way. Figures 16.26 and 16.27 show simple disambiguation in which the context of a probe input *gry* leads to output of elephant or pigeon. A useful property of a semantic network is sometimes called property inheritance. This effect is shown in figure 16.28, where we ask for the color of a large creature who works in the circus who we find out is Jumbo. Jumbo is an elephant. Elephants are gray. Jumbo is gray.

In this early simulation, two matrices were used, an autoassociative one (Mx1) realizing a BSB attractor network, and a heteroassociative one (Mx2). When the autoassociative feedback system reached a stable state, the hetero-associative matrix was applied for five iterations. The network state vector can oscillate back and forth between nodes.

Complex Concept Models

In chapter 11 we discussed simple concept formation. But real concepts have a complex associative structure, including hierarchical arrangement. Concepts seem to be organized in a nested fashion: *robin-bird-animal-living thing*, as suggested by the hierarachy of the previous example. This structure allows information to be represented about increasingly general categories. The facts known about birds apply to canaries, as well as penguins, robins, larks, crows, and so on. The facts known about animals apply to birds,

```
Mx 2.       1.   ANML___animal_____gry   Check:    0
Mx 2.       2.   ANML___animal_____gry   Check:    5
...
Mx 2.      20.   ANML__*animal___l___i_gry    Check:  133
...
Mx 2.      28.   ANML__*animal_walkskingry    Check:  154

Mx 2.      37.   ANML__*animal_walkskingry    Check:  176
Mx 2.      38.   ANML__*animal_walkskingry    Check:  176
Mx 1.      39.   ANML__*_____nwalkskingry    Check:  128
Mx 1.      40.   _____*elephanwalkskingry    Check:  136
Mx 1.      41.   _____*elephanwalkskingry    Check:  150
Mx 1.      42.   _____*elephanwalkskingry    Check:  152
Mx 1.      43.   _____*elephanwalkskingry    Check:  152
Mx 2.      44.   _____*elephanwalkskingry    Check:  152
Mx 2.      45.   _____*elephanwalkskingry    Check:  152
Mx 2.      46.   _____*elephanwalkskingry    Check:  152
Mx 1.      47.   ANML__*_____nwalkskingry    Check:  128
Mx 1.      48.   ANML__*ani_a_nwalkskingry    Check:  160
Mx 1.      49.   ANML__*animal_walkskingry    Check:  170
Mx 1.      50.   ANML__*animal_walkskingry    Check:  173
```

Figure 16.26 The network retrieves information about gray animals.

```
Mx 2.       1.   BIRD___bird_____gry   Check:    0
...
Mx 2.      10.   BIRD___bird___f_____gry   Check:   76
...
Mx 2.      20.   BIRD_**bird___f___j__kgry    Check:  127
...
Mx 2.      32.   BIRD_**bird___fly_junkgry    Check:  149
Mx 1.      33.   BIRD_**_i__on_fly_junkgry    Check:  112
Mx 1.      34.   B____**pi_eon_fly_junkgry    Check:  120
Mx 1.      35.   _____**pigeon_fly_junkgry   Check:  122
Mx 1.      36.   _____**pigeon_fly_junkgry   Check:  125
Mx 1.      37.   _____*_pigeon_fly_junkgry   Check:  129
Mx 2.      38.   _____**pigeon_fly_junkgry   Check:  132
Mx 2.      39.   _____**pigeon_fly_junkgry   Check:  137
```

Figure 16.27 The network retrieves information about gray birds.

mammals, reptiles, and so on. These hierarchies allow us to draw inferences about objects. For example, if Tweetie is a bird, we deduce that it is an animal and has skin, as we can show with the network used in the last example. Figure 16.29 shows what happens when the network is probed with Tweetie. Such inferences are important, because we can now infer properties true of all animals to be true of Tweetie as well. Hierarchies are important means of organizing concepts and storing information.

Let us try to see how hard it is explicitly to construct a hierarchy. The set of stimuli shown in figure 16.30 has a hierarchical structure in that segments of the state vector to the right determine the segments to their left. Two separate hierarchies are stored in the matrix, one involving vehicles and one involving furniture (Anderson and Murphy, 1986; Anderson, Golden, and Murphy, 1986).

```
Mx 2.     1.    _____*_____circus___    Check:    0
...
Mx 2.    10.    _____*J_____circus___     Check:   65
...
Mx 2.    20.    _____*Jumbo____circus___      Check:   93
Mx 2.    21.    _____*Jumbo___circus___       Check:   94
Mx 2.    22.    _____*Jumbo____circus___      Check:   94
Mx 2.    23.    _____*Jumbo____circus___      Check:   97
Mx 2.    24.    _____*Jumbo____circus___      Check:   97
Mx 2.    25.    _____*Jumbo____circus___      Check:   97
Mx 1.    26.    _____*_____anw_____      Check:   67
Mx 1.    27.    _____*el_phanwa_ksk_ngr_      Check:  105
Mx 1.    28.    _____*elephanwalksk_ngr_      Check:  136
Mx 1.    29.    _____*elephanwalkskingr_      Check:  145
Mx 1.    30.    _____*elephanwalkskingry      Check:  148
Mx 2.    31.    _____*elephanwalkskingry      Check:  149
Mx 2.    32.    _____*elephanwalkskingry      Check:  149
Mx 2.    33.    _____*elephanwalkskingry      Check:  149
Mx 1.    34.    ANML___*_____nwalkskingry     Check:  133
Mx 1.    35.    ANML___*ani_a_nwalkskingry      Check:  160
Mx 1.    36.    ANML___*ani_al_walkskingry      Check:  165
Mx 1.    37.    ANML___*ani_al_walkskingry      Check:  171
Mx 1.    38.    ANML___*ani_al_walkskingry      Check:  173
```

Figure 16.28 Tell me about a large circus animal.

```
Mx 2.     1.    _____Tweetie_____      Check:    0
...
Mx 2.    10.    _____Tweetie_cart_____     Check:   37
...
Mx 2.    20.    _____*__Tweetie_cartoon___      Check:  112
...
Mx 2.    28.    _____*__Tweetie_cartoon___      Check:  120
Mx 1.    29.    _____**_____ef__r_____lw       Check:   68
Mx 1.    30.    _____*__anary_fly_seedylw        Check:  103
Mx 1.    31.    _____*_canary_fly_seedylw        Check:  127
Mx 1.    32.    _____*_canary_fly_seedylw        Check:  133
Mx 1.    33.    _____*_canary_fly_seedylw        Check:  134
Mx 2.    34.    _____*_canary_fly_seedylw        Check:  135
Mx 2.    35.    _____*_canary_fly_seedylw        Check:  135
Mx 2.    36.    _____*_canary_fly_seedylw        Check:  135
Mx 1.    37.    BIRD_*_____ry_fly_seedylw      Check:  112
Mx 1.    38.    BIRD_*_b_rd_y_fly_seedylw        Check:  141
Mx 1.    39.    BIRD_*_bird___fly_seedylw        Check:  143
Mx 1.    40.    BIRD_*_bird___fly_seedylw        Check:  151
Mx 1.    41.    BIRD_*_bird___fly_seedylw        Check:  152
```

Figure 16.29 Tell me about Tweetie.

```
VehiclesAirplaneJetPlane_      FurniturChairs  Ostuffed_
VehiclesAirplaneX15Rockt_      FurniturChairs  Diningrm_
VehiclesAirplanePropplne_      FurniturChairs  Swivelch_
VehiclesNewcars Porsche _      FurniturBgtablesWorkshop_
VehiclesNewcars Mercedes_      FurniturBgtablesKitchen _
VehiclesNewcars Chevrolt_      FurniturBgtablesClassrom_
```

Figure 16.30 Two simple hierarchies involving vehicles and furniture.

Table 16.1 Ratio of largest to tenth-largest eigenvalue

Number of Presentations	Largest Eigenvector	10th-Largest	Ratio
12	3.830	0.331	11.57
50	1.000	0.374	2.67
100	1.000	0.555	1.80
200	1.000	0.765	1.31
400	1.000	0.928	1.08

We can do a slightly more sophisticated analysis of this completely autoassociative system. We constructed the matrix using the Widrow-Hoff error-correcting procedure. In an autoassociator, the aim of error correction learning of an input vector, **f**, is to produce the same vector, **f**, as output; that is, to construct a matrix, **A**, so that

$$\mathbf{Af} = \mathbf{f}.$$

Therefore, error correction applied to a learned set of examples will, in the limit of perfect operation, produce a matrix in which all the learned vectors are eigenvectors of the system with value 1, *if this is possible*. If simple Hebbian learning without error correction is used, the eigenvalues can be large, and the eigenvectors will be related to the principal components of the system, as discussed in chapter 15. Therefore, we would expect the eigenvalue spectrum of the largest eigenvalues to contract toward 1 as learning by error correction proceeds. We can see this process in table 16.1, where we look at the ratio between the tenth-largest and the largest eigenvalue as learning progresses.

The eigenvectors also become more meaningful as learning progesses (figure 16.31), although they never become identical to the stimuli. They were interpreted as if they were state vectors, that is, as ASCII codes. The most reasonable sign of eigenvector was used, that is, the sign that gave the most wordlike interpretation.

Concepts are frequently arranged hierarchically, and we have structured the stimulus set to show this. Retrieval respects the hierarchy. If we put in one field, information to the left is usually reconstructed correctly, in whole or part, whereas ambiguous information to the right is slow to be reconstructed and sometimes is incorrect. General information can be retrieved, but it is often impossible to give a specific example.

Specific examples can be used to construct a common category as shown in figure 16.32. If *JetPlane* is input, the recall process reconstructs *Vehicles*

```
             Interpreted largest eigenvector

Number                  Interpretation                        Eigenvalue
of Presentations

      12                VuxmidesBetabpg!Cmtscdod_                 3.83
      50                VehiclesAirplaneProplne_                  1.00
     100                FurniturBfeir~!2GM'kossm_                 1.00
     200                FurniturChairs  Ostuffdd_                 1.00
     800                FurniturChairs  034Uffet_                 1.00
```

Figure 16.31 The dominant eigenvector, interpreted as a character string, after different numbers of learning presentations.

```
           Examples of Computations with a Hierarchy


Input                             Result

JetPlane                          Vehicles (16) Airplanes (15)
JetPlane + X15Rockt               Vehicles (17) Airplanes (16)
JetPlane + X15Rockt + Proplne     Vehicles (16) Airplanes (15)

JetPlane + Chevrolt               Vehicles (20)    --      (50)
```

Figure 16.32 The computation performed by this network can be manipulated by the way information is presented. All combinations—simple vector sums of state vectors—of planes are quickly found to be vehicles and airplanes. However if the state vector for *JetPlane* is added to the state vector for *Chevrolt*, the network produces only the most abstract common level in the hierarchy—*Vehicles*—as the final state of the network.

after 16 iterations and *Airplane* after 15. The sum of two specific examples of *Airplane*, *JetPlane* and *X15Rockt*, gives *Vehicles* after 17 iterations and *Airplane* after 16. Suppose we want to know what *JetPlane* and *Chevrolt* have in common. They are neither *Airplanes* or *Newcars*, but both are *Vehicles*. If we put in the average of *JetPlane* and *Chevrolt*, the system reconstructs the common category *Vehicle* after 20 iterations. Use of commonalities can automatically raise or lower the level of computation in this concept hierarchy. This hierarchy does not have to be traversed node by node, but nodes can be leapfrogged.

These examples demonstrate the complexity of behavior possible with even the simplest associative system. Association can give flexibility and power in operation; however, the proper way to use it, the proper data representations, and the proper network architectures are still not known. We think that this area is one of the frontiers of cognitive science.

We conclude with a final word from William James (1892) on the connection of associative systems and brain structure: "Truly the day is distant when physiologists shall actually trace from cell-group to cell-group the irradiations which we have hypothetically invoked.... The order of *presentation of the mind's materials* is due to cerebral physiology alone" (p. 243). We can hope that "distant" is perhaps less far away than it was in James's time.

References

J.A. Anderson (1986), Cognitive capabilities of a parallel system. In E. Bienenstock, F. Fogelmann, and G. Weisbuch (Eds.), *Disordered Systems and Biological Organization*. Berlin: Springer.

J.A. Anderson (1988), Categorization in a neural network. In W.V. Seelen (Ed.), *Theoretical Neurobiology*. Wenheim: VCH Verlagsgesellschaft MbH.

J.A. Anderson, R.M. Golden, and G.L. Murphy (1986), Concepts in distributed systems. In H. Szu (Ed.), *Optical and Hybrid Computing*. Bellingham, WA: S.P.I.E.

J.A. Anderson and G.L. Murphy (1986), Psychological concepts in a parallel system. *Physica, 22D*, 318–336.

J.R. Anderson (1976), *Language, Memory, and Thought*. Hillsdale, NJ: Erlbaum.

J.R. Anderson (1983), *The Architecture of Cognition*. Cambridge: Harvard University Press.

J. Benello, A.W. Mackie, and J.A. Anderson (1989), Syntactic category disambiguation with neural networks. *Computer Speech and Language, 3*, 203–217.

N. Chomsky (1959), A review of Skinner's verbal behavior. *Language, 35*, 26–58.

A. Collins and R. Michalski (1989), The logic of plausible reasoning: A core theory. *Cognitive Science, 13*, 1–50.

A. Collins and M. Qullian (1969), Retrieval times from semantic memory. *Journal of Verbal Learning and Verbal Behavior, 8*, 241–248.

J.A. Fodor and Z.W. Pylyshyn (1988), Connectionism and cognitive architecture: A critical analysis. In S. Pinker and J. Mehler (Eds.), *Connections and Symbols*. Cambridge: MIT Press.

A. Gilman, T.W. Rall, A.S. Nies, and P. Taylor (Eds.) (1990) *Goodman and Gilman's The Pharmacological Basis of Therapeutics*. New York: Pergamon Press.

D.O. Hebb (1949), *The Organization of Behavior*. New York: Wiley.

J. Hertz, A. Krogh, and R.G. Palmer (1991), *Introduction to the Theory of Neural Computation*. Reading, MA: Addison-Wesley.

G.E. Hinton (1981), Implementing semantic networks in parallel hardware. In G.E. Hinton and J.A. Anderson (Eds.), *Parallel Models of Associative Memory*. Hillsdale, NJ: Erlbaum.

G.E. Hinton (Ed.) (1991), *Connectionist Symbol Processing*, Cambridge, MA: MIT Press.

W. James (1892/1984), *Briefer Psychology*. Cambridge: Harvard University Press.

A.H. Kawamoto (1985), Dynamic processes in the (re)solution of lexical ambiguity. Doctoral thesis, Department of Psychology, Brown University, Providence, RI.

S. Kosslyn (1981), *Image and Mind*. Cambridge: Harvard University Press.

J. Lachter and T.G. Bever (1988), The relation between linguistic structure and associative theories of language learning: A constructive critique of some connectionist learning models. In S. Pinker and J. Mehler (Eds.), *Connections and Symbols*. Cambridge: MIT Press.

B.B. Mandelbrot (1982), *The Fractal Geometry of Nature*, San Francisco: W.H. Freeman.

J.L. McClelland and A.H. Kawamoto (1986), Mechanisms of sentence processing: Assigning roles to constituents. In J.L. McClelland and D.E. Rumelhart (Eds.), *Parallel, Distributed Processing*, Vol. 2. Cambridge: MIT Press.

J.L. McClelland and D.E. Rumelhart (Eds.) (1986), *Parallel, Distributed Processing*, Vol. 2. Cambridge: MIT Press.

D.A. Norman and D.E. Rumelhart (Eds.) (1975), *Explorations in Cognition*. San Francisco: Freeman.

S. Pinker and A. Prince (1988), On language and connectionism: Analysis of a parallel distributed processing model of language acquisition. In S. Pinker and J. Mehler (Eds.), *Connections and Symbols*. Cambridge: MIT Press.

D.E. Rumelhart and J.L. McClelland (1986), On learning the past tenses of English verbs. In J.L. McClelland and D.E. Rumelhart (Eds.), *Parallel, Distributed Processing*, Vol. 2. Cambridge: MIT Press.

D.E. Rumelhart, P. Smolensky, J.L. McClelland, and G.E. Hinton (1986), Schemata and sequential thought processes in PDP models. In J.L. McClelland and D.E. Rumelhart (Eds.), *Parallel, Distributed Processing*, Vol. 2. Cambridge: MIT Press.

B.F. Skinner (1957), *Verbal Behavior*. New York: Appleton-Century-Crofts.

J.P. Sutton and H.C. Breiter (1994), Neural scale invariance: An integrative model with implications for neuropathology. In *Proceedings of the 1994 World Congress on Neural Networks*. Hillsdale, NJ: Erlbaum.

J.B. Watson (1925), *Behaviorism*. New York: Norton.

Multiplication is mie vexation,
And Division is quite as bad,
The Golden Rule is mie stumbling stule,
And Practice drives me mad.
Student rhyme, before 1570
(Yeldham, 1936)

If you don't think too good,
don't think too often.
Attributed to Ted Williams

In this chapter we will analyze the problems involved in learning arithmetic, in particular, why it is considered difficult. In a neural network model for arithmetic learning, data representation for number contains both a symbolic and an analog component. The analog component uses a bar-coded topographical map of number magnitude. We discuss reaction time, error patterns, and false product experiments. The general problem of flexibility in neural network computation arises naturally in the context of arithmetic, and uses a small number of known facts to answer novel problems. An extension of neural network architecture to include programming patterns *allows a network to answer reliably questions involving number comparisons that were not specifically learned.*

Some of this material appears as a chapter by Anderson, Bennett, and Spoehr, in *Neural Networks for Knowledge Representation and Inference*, edited by Levine and Aparicio (1993). Early simulations were reported by Susan Viscuso as part of her doctoral thesis in Viscuso, Anderson, and Spoehr (1989), and also in Anderson, Rossen, Viscuso, and Sereno (1990). The co-authors played essential roles in this project, and many of the ideas and computer simulations are due to their efforts. Some other material is adapted from Anderson (1992). Financial support was provided by a grant from the McDonnell-Pew Program in Cognitive Neuroscience to J.A.; by the National Science Foundation under grants BNS-86-18675, DIR-89-07769, and BNS 90-23283; and by the Digital Equipment Corporation, Maynard, Massachusetts.

Let us analyze an important cognitive problem in a little detail using the techniques that we introduced in this book. We will discuss what is involved in teaching a neural network to learn arithmetic, a problem that is more difficult than one might think at first. Its implications suggest an area in which current neural networks are seriously and fundamentally flawed. We will finish our discussion by offering one way to extend network architecture and overcome a few of these difficulties. This chapter is highly speculative.

Sometime during the first years of elementary school, students are supposed to learn a few hundred arithmetic facts along with a few simple algorithms. Many youngsters find the facts and algorithms difficult to learn and easy to forget. But arithmetic learning is an important cognitive skill, is used routinely in daily life, and forms the core of advanced mathematics.

Why does the largest and most impressive massively parallel computer that we know of, the human brain, operate at a milliflop rate when it does arithmetic? Not only is it slow, it is also inaccurate. Simple calculations are prone to errors at the best of times. Even adults make over 7% errors on elementary arithmetic facts (Graham, 1987).

Neural networks are often said to display brain-style computation. We would like to know if this claim holds for arithmetic learning as well. If it does, we might expect it to be intrinsically difficult to teach arithmetic to a neural network.

Other aspects make arithmetic learning of general interest. Entities such as "number" are much more complex than they might at first appear. Numbers relate to each other in ways that are clearly rule governed. One of the most difficult problems in current research is reconciling the abilities of neural networks to recognize patterns with the symbolic and rule-based behavior that seems to form an important part (some hold the only part) of human cognition.

At the end of this chapter we will discuss one direction to proceed in future research: programming a neural network to answer a question that has not been asked before. We saw in earlier chapters that neural networks are good at generalization in that they respond to novel patterns based on similarity. However, rule-based and symbolic computation involves a more complex kind of generalization. To cite an example we will discuss later, can we build a system that can tell us that 9 is bigger than 8, even though that particular comparison had never been seen before? Can we use previously learned facts about individual numbers to draw a new conclusion? If we can, what kind of assumptions do we have to make to do it, and what limitations does the resulting system have?

Data Representation

The key assumption in making a neural network learn arithmetic, and one whose importance should be clear to readers of previous chapters, is the data representation for number. Our discussion of arithmetic will center around a question that was asked by McCulloch (1961/1965), in the title of a paper: "What is a number, that a man may know it, and a man that he may know a number?" Even a computer has lots of ways to represent a number internally: characters, words, floating point numbers, and integers. Perhaps the most primitive representation would be the one shown in figure 17.1, three generic neural network units signaling number by activity level.

Let us consider multiplication as performed by this network. We have two input units, one for each multiplicand, and one output unit, corresponding to the product. Number magnitude is proportional to activity level, for example, 8 at one input unit is represented by an activity of 8 spikes per second, 6 at the other unit is 6 spikes per second, and the output unit fires at 48 spikes per second, corresponding to the product. If this network were to learn addition instead of multiplication, the output unit would fire at 14 spikes per second to signal the answer.

This architecture would be good for addition because the generic network output unit is a linear integrator. Therefore the addition network could be realized by using linear units with connection strengths of 1, or a subtraction

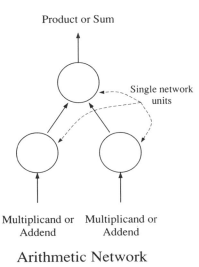

Arithmetic Network

Figure 17.1   A three-unit arithmetic network. Numerical quantities are represented by unit activities.

network with one connection strength of $+1$ and another of -1. The connection strengths could be learned from only two examples because there are two free weights in the network. If the network units were accurate linear integrators, the network would correctly generalize to all new addition problems. This simple network is an analog computer, since the output is computed by network interactions. The input-output relationships built into the properties of units themselves are doing the computation. This network generalizes far better than humans do.

Unfortunately, the network would not be so good at multiplication. It is possible to make good analog multipliers with precise electronic components; however, neurons and their interactions are not so precise. It would be necessary to adjust carefully the nonlinearities of model neurons to realize an accurate multiplier or divider. The units themselves are of limited accuracy, for reasons we discussed in earlier chapters. We need a different approach.

A great deal flows from the fact that neurons are not accurate devices. John von Neumann (1958) noted, "The nervous system is a complex machine which manages to do its exceedingly complex work on a rather low level of precision." The technical problem that arises immediately is that when a complex algorithm is realized by a series of operations performed in sequence, errors will amplify. A large part of the field of applied mathematics called numerical analysis is devoted to estimating the final accuracy of a computation. The reason computers often perform arithmetic operations at double or even quadruple precision is not because anyone is actually interested in this degree of precision, but to avoid unacceptable error amplification in the intermediate steps. Von Neumann then concluded, "Whatever language the central nervous system is using is characterized by less logical and arithmetic depth than we are normally used to." Even a simple computation with a digitial computer may involve thousands or millions of steps in series, each of which must be performed with great accuracy. Brains do not have that ability. Their answers must be computed in a small number of steps. Only Sherlock Holmes could put together a long chain of deductions and come out with the correct answer.

If problems are involved in computing arithmetic facts directly with the hardware, let us see if we can do it another way. We have already seen that a congenial computation for most neural networks is pattern association. This leads to our next model for arithmetic learning (figure 17.2). Here, a set of associations for the number facts is specifically learned on a case-by-case basis. This is close to the way that children actually are taught arithmetic, that is, individual facts that, together, form the various tables. Learning the facts simply involves getting the associations correct. We are not talking

Associative Arithmetic
Model

Output pattern representing
sum or product

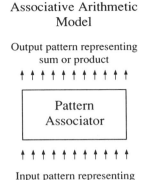

Input pattern representing
the addends or multiplicands

Figure 17.2 An associative arithmetic network. Numerical quantities are represented by large activity patterns.

about a large number of facts, at most a few hundred for all of elementary arithmetic.

Despite this simple structure and the limited number of facts, children still have problems learning them. This is remarkable, because arithmetic facts are orders of magnitude less complicated than the syntactic relations that are routinely used in the language of even young children. In addition, at the same age that children are having trouble learning arithmetic, they are learning vast amounts of other information. Around the second- or third-grade level they are learning several new words a day, and a word is a very complex cognitive object.

A good part of the problem seems to be with the structure of arithmetic, which is difficult for any associative system to deal with. Learning arithmetic by pure association, as in figure 17.2, leads to problems with *associative interference*. Figure 17.3 indicates why this is so, with an example from multiplication.

The real difficulty with associative computation is that a single input number is associated with many different possible output products. Most network associators have trouble when there are many possible outputs, and tend to produce an initial output that contains the sum of possible outputs. Most networks have units that are linear enough for small activities so they obey the superposition principal to some degree. Eventually, any one of a number of mechanisms (an attractor network, say) can be used to choose one or another of the possible outputs. If there are many potential associations, this process is slow and error prone. Graham (1987) and others argued from human data that associative effects are the primary mechanisms for both learning and errors.

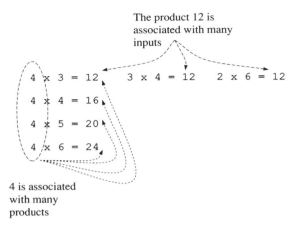

Figure 17.3 Associative arithmetic learning is difficult because of associative interference.

It is significant that both simple associators and children have no difficulty learning the multiplication tables for zero. The reason is obvious; there is no output ambiguity. Yet, the concept of zero and a notation for it were late entering Western mathematics. One and none were sometimes not even considered numbers in the Middle Ages; numbers were held to begin at 2. A medieval riddle can be answered only if it is realized that 1 is not a number (Yeldham, 1936).

I came to a tree where there were apples.
I ate no apples,
I gave away no apples,
Nor I left no apples behind me:
And yet I ate, gave away, and left behind me.

One way that we could aid a network in coping with the difficult associative structure of arithmetic is to provide information of another kind. A surprisingly large body of pertinent psychological data is available. For example, consider number comparisons. Suppose a subject is asked whether 85 is bigger than 86, or whether 85 is bigger than 14. The response time to perform number comparisons shows what is often referred to as a *symbolic distance* effect (Moyer and Landauer, 1967). The response time is long when the numbers compared are close together and short when they are far apart. Such two-digit number comparisons are graphed in figure 17.4.

The symbolic distance effect for this experiment is extremely large. For nearby numbers, comparison response time is nearly 600 msec, for distant numbers it falls below 400 msec. If a digital computer performed a number comparison it probably would not show a symbolic distance effect at

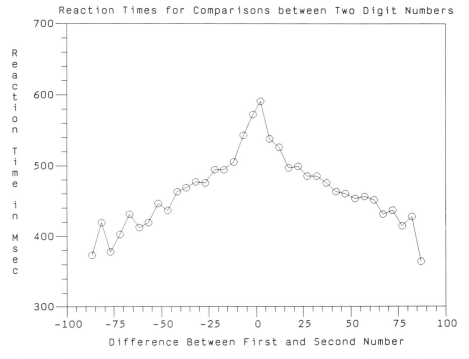

Figure 17.4   Reaction time for two-digit number comparisons. The x-axis gives the difference between the first and second numbers and the y-axis the reaction time in milliseconds. Data replotted from Link (1990).

all; inspecting the sign of the difference between the numbers would be adequate.

Such response time patterns are common in psychological studies of arithmetic. Another place where they apply is in comparisons of *sensory* magnitudes. That is, if a subject has to decide which of two weights is heavier, or which of two lights is brighter, or which of two sounds is louder, the response time pattern is similar. Response is longer if the two comparatives are nearly equal. The pattern observed for number comparisons suggests that the representation of number contains a part that acts like a weight or a light intensity. That is, a sensory magnitude is attached to what seems like an abstract quantity, number.

Instead of purely abstract symbols, humans seem to use a richer coding of number containing powerful sensory components. This aspect is deliberately developed by some methods of instruction about number. The Montessori schools, for example, try to teach number using rods of various lengths so a 3 rod is three times as long (and weighs three times as much) as a 1 rod.

Development of the Number Concept

Turning an abstraction such as a number into something concrete and sensory would seem at first to be primitive. However, children develop the analog aspect of number relatively late in cognitive development. The following discussion is based to some degree on work by Gelman and Gallistel (1978). Two-year-old children have number words that can be used in a fixed sequence that almost always involves a motor act, typically pointing and counting aloud. Our traditional sequence names for counting are one, two, three, and so on. Sometimes children have idiosyncratic names, but even these maintain a stable order; that is, never two, three, one. Slightly older children use the idea of order invariance, that is, it does not matter in what order one counts a group of objects, the answer will be the same.

In many respects two- and three-year-olds have an abstract and mathematically sophisticated view of integers (Gelman and Gallistel, 1978). Modern axiomatizations of integers depend heavily on successor relationships and concepts related to stable order and order invariance. Therefore it is surprising that many four-year-olds show an effect called *nonconservation* that seems at first to involve a regression in number ability.

Conservation is the fact, well known to adults, that equality of two sets means the items in the set can be matched one for one. Thus two sets of seven items, say, always have seven items no matter how they are arranged. Therefore number is *conserved* across arrangement. But, as Gelman and Gallistel (1979) commented, "Whatever relation exists between counting and conserving is not a strong one" (p. 229). A classic number nonconservation experiment is diagrammed in figure 17.5.

The experimenter takes seven objects of one kind and seven of another, squares and circles in figure 17.5, and lines them up. A four-year-old will say that the number of squares and circles is the same. Then with the child watching, and with the experimenter being as nondirective as possible, one row, say the one containing the squares, is spread out. Asked if there are more or fewer squares than circles, the four-year-old now says more squares. The spread out row is pushed together, and the child says fewer squares than circles.

This effect is not simply a matter of misunderstanding the meaning of equality. Our department has a videotape that we used to demonstrate nonconservation with a bright, articulate four-year, nine-month-old. After displaying the effect, and explaining it by pointing to the ends of the linear arrays in the different configurations (i.e., equating length with number), he was asked to count out loud the number of items in the two rows. The boy

A Conservation Experiment

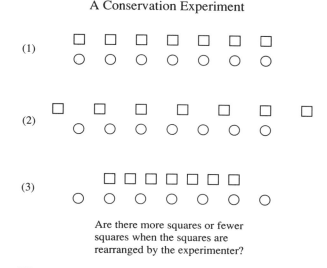

Are there more squares or fewer
squares when the squares are
rearranged by the experimenter?

Figure 17.5 A conservation experiment. The child is asked if the numbers of squares and circles in the first row are equal. With the child watching, the squares are rearranged, either spread apart (as in row 2) or pushed together (as in row 3). A five-year-old will often say that the number of squares is greater when the squares form a longer row, and smaller when the squares form a shorter row. Number is not conserved when the squares are rearranged.

was upset when the numbers turned out equal. He knew perfectly well he blew it.

Nonconservation is shown by many, perhaps most, four- and five-year-olds. It must be looked for, however, because casual conversation rarely reveals its presence. The effect ends, often abruptly, around age six. It is highly significant that some physical measures such as volumes also show nonconservation. For example, a child watching an experimenter pour water from one vessel into another of different shape will say the amount of water has changed.

Nonconservation has been studied for many years and discussed in hundreds of publications. That is not only because of its considerable intrinsic interest, but also because the effect is one of the cornerstones of Jean Piaget's child-development theories. The current consensus seems to be that the effect is indeed real, but it can be modified to some degree by the details of the experiment and the instructions to the child. Much of the discussion seems to agree that children's concept of number contains both a perceptual judgment and an identity rule, and children have trouble knowing which one to use in a particular instance (Acredolo, 1982). This conclusion is consistent with the number representation we will use for our simulations. We would conjecture that four-year-olds are starting to expand number representation to include the sensory aspects of number that adults use; they are building in sensory quantities, but at first they do not do it accurately. They have not properly incorporated the *perceptual invariances*; that is, they have not realized that number is not directly proportional to size, and that if you spread out an array, the linear size increases but the density of objects drops.

Additional sensory or analog information is incorporated after what seems at first to be a more abstract representation. The information is an elaboration of number rather than interference with an abstraction. The sensory components add something to our practical interactions with number.

Computer evolution has followed a somewhat similar course. From highly efficient, abstract, and hard-to-use number and logic crunchers, computers have evolved interfaces that are display panels full of bars, icons, and analog graphical aids. These complex interfaces take up a large amount of processing time and are difficult to program. Many users comment that programs often run slower with the new graphical interfaces than they did before, even as the computers themselves become faster. But they are willing to sacrifice raw performance because computers are so much easier to use. A large part of processing cycles is well spent providing a cognitive impedance matcher between user and computer.

Mathematics

Mathematics is supposed to be the most lawful and abstract of the sciences. Surely real mathematicians would not do anything as crude as linking an abstraction with a sensory quantity. It seems they do, however.

I first became interested in this issue when I wrote a paper with Professor Davis of Brown University's Applied Mathematics Department (Davis and Anderson, 1979). The point of the paper was that the theorem-proof method of teaching mathematics has had devastating effects. The reason is psychological. Real mathematicians do not think this way at all, but instead use a complex kind of memory and intuition when they try to understand mathematical systems. Proving theorems is the very last stage. The effects of the theorem-proof method of teaching are especially hard on users such as engineers and scientists. A frequent reaction is, "But I don't think like that ..." followed by the obvious conclusion, "Therefore I am inferior." Although this teaching method may have beneficial effects for the egos of mathematicians, confirming their distance from ordinary intellects, it produces loss of confidence and misunderstanding on the part of others. A highly recommended book on this topic is Davis and Hirsh (1985).

Jacques Hadamard (1945), a first-class mathematician himself, interviewed his peers about how they did mathematics. Most of those he talked to said they did not reason abstractly, but used visualization, kinesthetic imagery with imagined muscle motions, or even auditory imagery. Language-based, and formal, abstract reasoning were conspicuous by their rarity. Einstein was typical when he wrote to Hadamard, "[T]he words or the language as they are written or spoken do not seem to play any role in my mechanism of thought." Einstein (1951) stated, "For me it is not dubious that our thinking goes on for the most part without use of signs (words) and beyond that to a considerable degree unconsciously" (p. 9). Gleick (1992) gives a good example from Richard Feynmann: "Intuition was not just visual but also auditory and kinesthetic. Those who watched Feynmann in moments of intense concentration came away with a strong, even disturbing sense of the physicality of the process, as though his brain did not stop with the gray matter but extended through every muscle in his body" (p. 244).

Einstein made some other comments on observations of his own thought processes that bear noting, given other material presented in this book.

What, precisely, is "thinking?" When, at the reception of sense-impressions, memory-pictures emerge, this is not yet "thinking." And when such pictures form series, each member of which calls forth another, this too is not yet "thinking." When, however, a certain picture turns up in many such series, then—precisely through such return—it becomes an ordering element for such series, in that it connects series which in themselves are unconnected. Such an element becomes an instrument, a concept. I think the transition from free association or "dreaming" to thinking is characterized by the more or less dominating role which the "concept" plays in it. It is by no means necessary that a concept must be connected with a sensorily cognizable and reproducible sign (word); but when this is the case thinking becomes by means of that fact communicable. (1951, p. 7)

It is interesting to compare this quotation with the ideas about associative computation we brought out in the previous chapter. Both William James and Aristotle would have found much to agree with in this discussion of the use of an associative computer based on the manipulation of sense images.

Others have stated similar ideas about mathematical reasoning. This was a popular topic among mathematicians around the turn of the century. For example, Poincaré (1929), observed, "[L]ogic and intuition each have their necessary role. Each is indispensable. Logic, which alone can give certainty, is the instrument of demonstration; intuition is the instrument of invention" (p. 219). Perhaps intuition, which is surely based largely on the sensory systems, was more valued at that time than it is now, and was developed and honed as a tool of discovery. Hadamard was Poincaré's student, and that is perhaps why he developed his lifelong interest in mathematical cognition. For more examples of the use of nonverbal reasoning in mathematics and physics, see Miller (1984).

A Neural Network Data Representation

How can we make our number representation richer so as to incorporate some of this analog sensory-like information? One biologically inspired way would be to build a topographical map of number magnitude. A moving bar of activation on a topographical scale formed of individual elements would code the magnitude of the number, just as position on the cortical surface roughly codes visual position or frequency of a sound. (See chapter 10 for many examples of topographical coding of sensory parameters.) This, one might conjecture, is why it is hard to tell which is larger, 74 or 73. The two numbers weigh about the same, so we cannot distinguish them easily based on the analog code. It is unlikely that this is exactly the form of representation used by the nervous system, but perhaps it captures some of the essentials of whatever magnitude code is used. We used topographical codes earlier in this book, for example, in the concept forming model and object motion systems described in chapter 11, the radar simulation in chapter 15, and in the Ohm's law simulation in chapter 16.

For our simulations we constructed a two-part number representation that had both an arbitrary, symbolic part, which was sometimes a random vector and sometimes a coding of the number name, and a sensory part, a bar that moved from positions on the left of the state vector to positions on the right as number magnitude increased (figure 17.6). The part of the vector associated with the symbols consisted in some simulations of random 1s and −1s, and in others, character-based representations of the number name, as we used in chapters 7 and 16. The rest of the vector was set to zero. This means that the closeness or similarity of the different facts is determined by the overlap of the bar codes (figure 17.7). Correlation between nearby numbers can be controlled using bar width. The spacing of the bars was compressed, with the distance between the beginning of the bar and the symbolic part of the coding growing in a roughly logarithmic fashion with the magnitude of the individual operands and the answers.

One of the less obvious virtues of such a hybrid code is that it allows movement from an analog world to a symbolic world, since arbitrary patterns are associated with magnitudes and vice versa. Such a code allows

Hybrid Number Representation
Coding for number name
Moving bar of activity
Smaller Bigger
Segment of state vector representing a single number.

Figure 17.6 Schematic of the hybrid coding scheme used in the simulations. The moving bar of activity codes the sensory part of the number. As number magnitude increases, the area of activity moves to the right.

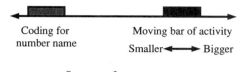

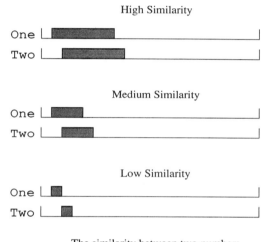

High Similarity
One
Two
Medium Similarity
One
Two
Low Similarity
One
Two
The similarity between two numbers can be controlled by the width of the bar.

Figure 17.7 Hybrid number coding allows a controllable degree of similarity between two numbers, depending on the degree of overlap between the bars.

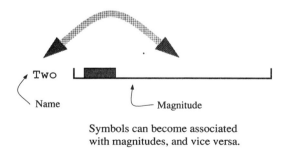

Two
Name
Magnitude
Symbols can become associated with magnitudes, and vice versa.

Figure 17.8 Hybrid number coding allows movement from a magnitude-based world to one with abstract (or at least arbitrary) names.

convenient hooks to symbol-based artificial intelligence, if necessary (figure 17.8). This associative connection between two fundamentally different representations is a powerful strategy. It allows for division of labor and complementarity; each part of the representation can do what it does best.

Number Systems

An excellent popular book on science emphasizing physics and mathematics for laymen was *One, Two, Three ... Infinity* by Gamow (1947). It took its title from the number system of a tribe who, it was claimed, had words only for the numbers one, two, three, and then something like "very big." In fact, all human languages have the ability to count to large numbers. The systems that probably suggested Gamow's title are common in Papua New Guinea where they make up 15% of all number systems. They are called type II number systems (Lancy, 1983). They are sometimes used alone, in conjunction with other systems, or for ceremonial purposes. The Kiwai type II number names have a simple structure (Lancy, 1983):

1 = na'u

2 = netowa

3 = netowa na'u

4 = netowa netowa

5 = netowa netowa na'u

. . .

Of course, this is a base 2 system, capable of representing any number. These binary number names are interesting from the point of view of the history of neural networks, since they solve the parity problem in one step.

Our number representation that assumes a moving area of activation on a topographical map may be realized directly for one human population. The Oksapmin of Papua New Guinea count by associating a number with a position on the body (figure 17.9). Lancy classified this as type I. The Oksapmin can count to 29. A finer-grained set of associations with body regions allows the Kewa to count to 68. They count higher numbers by grouping 68s as "men" and keeping track of the number of "men" involved. During commercial negotiations, the Kewa try to keep the number under discussion to a multiple of 68 so as to ease computation.

Oksapmin Counting System

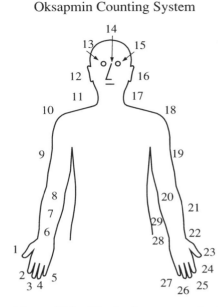

Figure 17.9   The body part numbering system used by the Oksapmin of Papua New Guinea. This system, in conjunction with the topographical cortical maps of the body surface known to exist, suggests that number representation using a moving area of activation has at least some plausibility. Redrawn from Lancy (1983).

The interesting thing about this number system is that it gives rise immediately to one version of a bar-coded topographical map of magnitude. Simple inspection of the maps of the body surface presented in chapter 10 (figures 10.14, 10.15, and 10.16) show that, since different body parts have different spatial locations on the map, this system presumably would as well.

Associations of number with body regions are not retricted to Papua New Guinea. The Venerable Bede, who lived in England around A.D. 700, devoted considerable space to describing a system that could count as high as a million by a combination of limb pointing and positioning. The flavor of this system can be seen in a translation of Bede's *Temporum Ratione* (Yeldham, 1926):

The reckoning is made first on the left hand, as follows: When you say One, bend the little finger of the left hand and place it on the middle of the palm. When you say Two, bend the fourth finger and place it likewise.... For Five, raise the fourth finger.... When you say Thirty, join the tips of the thumb and first finger in a loving embrace [30 signifies marriage according to Jerome].... To represent a Hundred, make the same sign on the right hand as you did for Ten on the left, Two Hundred on the right as Twenty on the left.... When you say Ten Thousand, put your left hand palm outward on the breast, with the fingers pointing upward to the neck.... When you say a Million, you clasp your hands with the fingers interlocked. (pp. 32–33)

This system uses body region to indicate place and any number up to a million can be represented unambiguously. Bede suggested it could be a secret way of communicating numbers in negotiations. In light of topographical cortical maps, this method would correspond to numerous separated number maps. This suggests a representation of number with, for example, 10s and 1s each having a separate bar code in a separate cortical region. McClosky and Lindemann (1992) assumed such a representation in their MATHNET simulation.

Other Work

Given the practical importance of arithmetic, the volume of experimental literature, and the amount of time spent teaching it in school, it is surprising that only a handful of attempts have been made to apply neural networks to arithmetic learning. Part of the reason is that modeling arithmetic itself is a problem of limited practical interest. No rational person would build an artificial neural network for the reliable performance of arithmetic. The motive for modeling it is primarily to understand human cognition. Any practical applications would come from understanding what is required for systems that can develop artificial intuition and mathematical creativity. Although such understanding is of great potential importance, it is also remote.

Another, perhaps more subtle, problem is that neural networks are really not good at arithmetic. They are slow to learn and make many errors. Anyone who thinks that the function of networks is to develop accurate input-output associations will be sorely disappointed by attempts at neural network arithmetic. Most scientists prefer to get correct answers. However, cognitive scientists like errors because they are diagnostic of the internal workings of a system. There are many very different ways of getting the correct answers, but error patterns are idiosyncratic. (One is reminded of the first sentence of Tolstoy's *Anna Karenina*: "Happy families are all alike; every unhappy family is unhappy in its own way.")

Three-digit multiplication can illustrate how to generate an iterative serial process using a neural network (Rumelhart, Smolensky, McClelland, and Hinton, 1986). Essentially, the physical appearance of the problem in standard form on the paper or blackboard triggers the first partial multiply, which is then physically written down. Then this new physical pattern triggers the next step, by association, involving a multiply and shift, and perhaps a carry, and so on, leading to something else being written down, leading to yet another associative step, and so on, until the problem is completed. This is an associative model for arithmetic that is perfectly compatible with classic stimulus → response association. The basic idea is to convert an abstract idea into a series of associatively linked, concrete physical steps. Neural networks are not necessary except to provide a possible linkage mechanism.

One neural network arithimetic simulation based on this approach simultaneously learned both the arithmetic facts themselves and the procedural operations on them (Cottrell and T'sung, 1991). Learning was aided, and probably made possible, by the data representation, which picked out important properties that were used to drive the iterative process. For example, the output representation contains an *action* field that determines the next step: carry, write, shift to the next column, or finish. The computation continues until the network signals that it is finished, so numbers of indefinite length can be added.

The network used a modified version of recurrent back-propagation that incorporated previous states into the current input state. The simulation was restricted to base 4 arithmetic for practical reasons. Otherwise the number of examples of both data and operations would have overwhelmed the computer. Even so, learning times were extremely long, requiring many tens of thousands of presentations to achieve acceptable accuracy.

The elementary number representation, a direct mapping of bit patterns to the state vector, suggests that the already long learning times would be much longer as the simulation moved toward base 10 arithmetic. By far the most interesting aspect of the simulation was the attempt to create a truly iterative system that could add number strings of indefinite length. Final network accuracy was not tested systematically, but the comment was made that 1 error was made on average for 10 novel test problems. The error rate was much lower on problems that were in the training set.

MATHNET, a network model for arithmetic fact learning, was specifically designed to model human performance (McCloskey and Lindemann, 1992). One version used a relatively standard back-propagation network. The data representation was similar to part of the one used for the model in this chapter. Number magnitudes were represented as bar codes. Two bar-coded numbers were used as input; the output also contained two bar-coded numbers, one for the 10s place, and one for the 1s place. This representation did not incorporate a symbolic part.

The bar code alone was sufficient to model a variety of experimental effects. One might question, as the authors did, whether the symbolic part of the representation is necessary at all. Our experience is that it does make the network work better and more reliably, although many experimental effects seem to depend largely on the bar-coded part of the state vector.

The examples of network programming at the end of this chapter become much more versatile when an arbitrary part of the number representation is present. The arbitrary part generates the answer to the problem and allows other forms of association to be used. Our speculation is that the real computational power of human cognition in these areas arises from the presence of a hybrid representation, part analog, derived from the sensory systems, and part symbolic (or at least arbitrary), connected to other forms of cognition, including ones that seem to be rule based. The utility of number is partially contained in its ability to move back and forth between these cognitive domains.

Campbell and Clark (1992) described a set of experiments they call "alpha-plication" in which children were taught arbitrary associations using letter combinations that reproduced in structure arithmetic facts. Children learned and retrieved these arbitrary associations in ways similar to those seen in experiments on arithmetic learning. Campbell and Clark agree that magnitude representations are important, but that other significant aspects are involved in learning arithmetic as well.

Based on the bar-coded representation, MATHNET reproduced a number of experimental effects. The authors provided a set of simulations of the effects of brain damage on arithmetic performance, and an impressive set of detailed discussions of error distributions and reaction times from humans and from the network. They did not analyze the representations developed in the hidden layer of the network, although this study is reported as being under way.

Multiplication Facts

Most of our simulations involved the multiplication tables. Multiplication is learned after addition and seems to be considered arbitrary by many children despite valiant pedagogical attempts to tie it to physical measures such as area. Direct counting strategies exist for answering addition problems that are used to some degree by many young children when they first learn addition. Very early addition learning is therefore sometimes algorithmic, and becomes more purely associative with experience. Association is apparently faster and more reliable than a counting algorithm, which says a good deal about human arithmetic performance. Counting strategies are difficult or impossible to carry over to multiplication, which, from the beginning, is much more of a rote association task.

Once we have the representation for an individual number, we can code a multiplication fact by associating the two muiltiplicands with a product.

Autoassociative
Representation of a Multiplication Problem

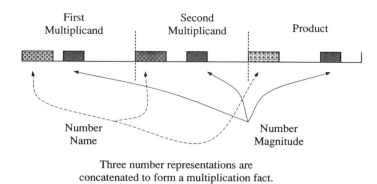

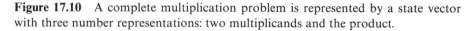

Three number representations are
concatenated to form a multiplication fact.

Figure 17.10   A complete multiplication problem is represented by a state vector with three number representations: two multiplicands and the product.

There are many ways to do this; since we used the BSB model, which is autoassociative, we simply concatenated three number representations to form a "fact" (figure 17.10). Sets of these patterns involving three numbers were what was learned by the network.

Simulations

One problem with this approach to arithmetic is that it requires very many units to code a number fact. The product location, ideally, would have to represent about 100 possible output products, the numbers from 1 to about 100. The size of state vector, several thousand units, required to code the complete times tables and all possible products, maintaining reasonable analog similarities, is enormous, and required use of a supercomputer in early simulations. (Modern work stations are able to cope with problems of this size.) Therefore, from the beginning, our simulations were of qualitative arithmetic, in which the system was asked to produce only ballpark estimates rather than precise answers.

Many variations of this number coding were tried, for example, lengthening and shortening the symbol and bar sections, changing the amount of compression, and changing the ratio of the relative amount of space allotted to the operand and answer fields. Whereas these changes do affect performance and accuracy to a limited degree, the qualitative behavior shown in the simulations is extremely robust. What seems to be essential to generate

Errors in Multiplication

two++++   *seven   ++++ =   ten ++++

                                  Correct is fifteen

five   ++++   *seven   ++++ =   fifty   ++++

                                  Correct is forty

Figure 17.11 Errors reconstructed in the qualitative arithmetic simulation. The correct answers are 15 and 40, and the network reconstructs answers 10 and 50.

the full range of effects is the combination of analog bar coding and an arbitrary pattern.

We will describe a typical simulation in which a BSB network was trained on a representative sample of multiplication facts. The data part of the vector, that is, the two operands, was always clamped, leaving only the answer field, the product, free to change. An activation threshold was used to keep the zero parts of the vector set to zero, since, especially in the false product simulations (discussed below), these elements would otherwise tend to wander away from zero. There was also a readout threshold for the answer field; that is, only activation over this threshold was shown and recorded.

Ever since our first simulations of arithmetic using hybrid coding, the overall pattern of results has remained remarkably consistent. First there are many errors, either wrong answers, or answers in which an attractor was not reached in a large number of iterations. Typically, a simulation reconstructs the correct answers about 70% of the time. That is, with considerable effort and a remarkable amount of computer time we created a student who earned a C in arithmetic. (This led to an interesting conversation with one of the administrators of a supercomputer center a few years ago, trying to explain why we believed several hours of supercomputer time were usefully spent getting the wrong answers to problems in elementary arithmetic.)

Second, errors are not random. Figure 17.11 shows typical error results from an early simulation (Viscuso, Anderson, and Spoehr, 1989). The errors are close to the correct answer. It is clearly the analog component that causes this error pattern. The pattern is also a consistent feature of human performance for elementary multiplication (Norem and Knight, 1930; Graham, 1987). Figure 17.12 presents a table of error data from children learning arithmetic in 1930. Figure 17.13 shows typical error data, similarly plotted, for an early series of simulations (Anderson, Rossen, Viscuso, and

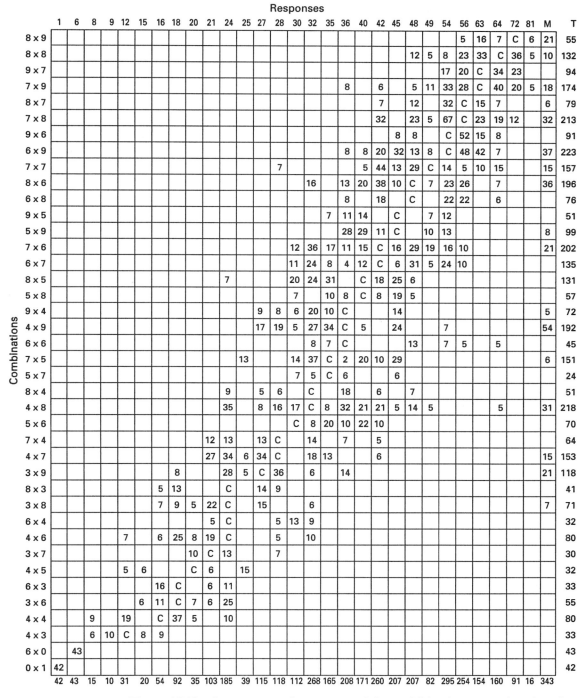

Figure 17.12   An error matrix constructed from children's errors when learning arithmetic. From Norem and Knight (1930).

Product

Problem:	0	1	5	10	15	20	30	40	50	60	70	80
4x0	C											
4x1			C	1								
4x2				C	1							
4x3				C								
4x4				1	C							
4x5				1		C						
4x6				1		C						
4x7							C		1			
4x8							C		1			
4x9								C		1		
5x0	C											
5x1			C	1								
5x2				C								
5x3				1	C							
etc.												

Figure 17.13 A similar form of error matrix showing errors produced in an early computer simulation. From Viscuso, Anderson, and Spoehr (1989). Reprinted by permission.

Sereno, 1990). These patterns are very robust as long as the data representation has an analog portion.

A slightly more detailed study of human error patterns is informative. Table 17.1 gives error patterns for adults for the problem 6 times 9 (Graham, 1987). Many errors are associative, that is, incorrect multiples of either 6 or 9. The most striking associative errors are 36 and 81, that is, 6^2 and 9^2, the strongest multiplicative associates of 6 and 9. A very strong effect is based simply on magnitude as well. For example, the most common error for both adults and children is 56, which is a multiple of neither 6 nor 9 but is the closest product number to 54, the correct answer. Errors that are not products, for example, 53 and 57, are rare. Graham reports that about 7.4% of errors made by adults are nonproduct errors; fifth-graders make the somewhat higher rate of 17.2%. The numbers in table 17.1 are errors per 1000 presentations of the problem. Therefore the total error rate for adults is greater than 25%, a remarkably high rate for a problem that was supposed to be learned in elementary school.

This type of error pattern strongly suggests that the human arithmetic algorithm is something like this: the answer to a multiplication is (1) familiar, that is, a product for some elementary problem; and (2) it is about the right size (figure 17.14). Arithmetic fact learning and retrieval is a process that combines memory with estimation. This is fundamentally different from the way that a digital computer would do arithmetic.

Table 17.1   Adult errors on the problem 6 times 9

Answer Number	Errors per 1000 Trials
30	8
32	4
36	58
40	0
42	17
45	17
48	8
54	Correct
56	79
63	50
72	8
81	5

From Graham (1987, table 7.2, p. 128).

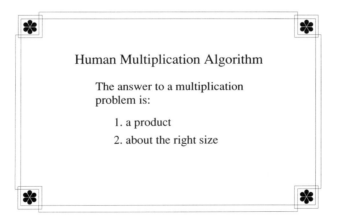

Figure 17.14   The rule suggested by human arithmetic performance.

```
        A False Product Near the Correct Answer:

Output:   two ++++                * two++++            =    ten ++++

                                   Accepts initial value of ten
                                   Correct answer is five




        A False Product far from the Correct Answer:

Output: three  ++++               * two++++            =    five++++

                                   Rejects initial value of eighty
                                   Fills in five (correct).
```

Figure 17.15 An example of a false product simulation.

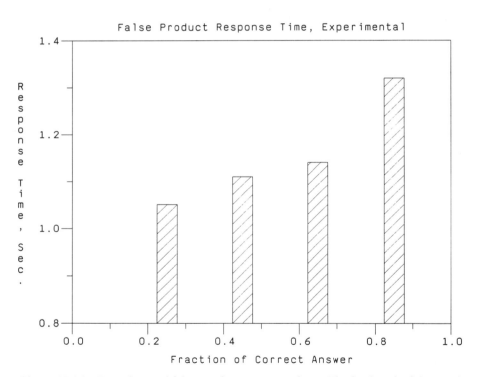

Figure 17.16 Experimental false product response times. The farther the false product is away from the correct answer, the more rapid the response time. The distances from the correct answer are measured as approximately 75%, 50%, and 25%. From data in Stazyk, Ashcraft, and Hamman (1982).

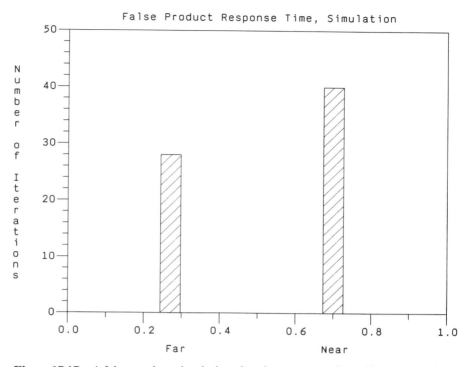

Figure 17.17   A false product simulation showing response times (times to reach the attractor) for near and far false products. Two near and two far false products were constructed for all the problems in the learning set of arithmetic facts.

False Products

We have many ways to see if we are on the right track in our model for arithmetic performance. For example, another way to compare our simulations to human data is in its ability to reject false products. Consider the two problems,

is 8 times 7 = 63 correct?

is 8 times 7 = 11 correct?

When humans do such problems they show a strong symbolic distance effect: the second answer can be rejected more quickly than the first because the incorrect answer is much farther from the correct one (Stazyk, Ashcraft, and Hamman, 1982). It is easy to simulate such problems. The false product is presented to a trained BSB network. Figure 17.15 shows some early false product simulations (Viscuso, Anderson, and Spoehr, 1988). The network is provided an input pattern that corresponds to either a correct answer to a problem or to a false product. We are using BSB, so if the network rejects an answer, it will overwrite it with the correct answer. That is, the false product is not a stable attractor in the BSB network, but the correct answer is. The network will accept close false products (2 times 2 = 10, when the correct answer is 5) but will reject and overwrite distant false products (9 times 8 = 5 with the correct answer, 70).

Our later simulations looked also at the network response time, which we defined in chapter 15 as the time required for BSB dynamics to reach an attractor. As we might expect, false products show a strong symbolic distance effect. Near false products are rejected more slowly than distant ones. Figure 17.16 shows an example of experimental response time data. A systematic simulation was done by constructing four false products for problems that the system had learned correctly. Two were close, for example, 5 times 2 = 20 and 5 times 2 = 30, and two were far, for example, 5 times 2 = 50 and 5 times 2 = 60. Figure 17.17 shows network response time data for the false products tested. This simulation displays a symbolic distance effect, as do the experimental data. The error data showed a similar pattern. The system accepted six near false products, but no far ones.

Other Effects

Several other experimental effects can be simulated. For example, humans respond more quickly when giving correct answers to arithmetic problems that they have just practiced. These priming effects are relatively short lived (Campbell, 1987; Stazyk et al., 1982).

To test the system for priming effects, it was trained on 32 facts, and then selectively primed on 24 problems that were initially answered correctly when probed with the problems before priming. To prime the system, a fact was presented and the weights were changed by an amount proportional to preconnection and postconnection activity; that is, an outer product matrix was formed and multiplied by a small priming constant, and the result was added to the overall weight matrix formed during learning.

Figure 17.18 illustrates the basic priming response time data. The primed facts exhibited the greatest decrease in response time. The priming generalizes in an interesting way: a substantial, although somewhat less dramatic, decrease in response time is seen answering problems that share an operand and an answer field with a primed problem. Similarly, a somewhat smaller decrease occurs for problems with the operands switched, and a small decrease for problems that share only an answer field.

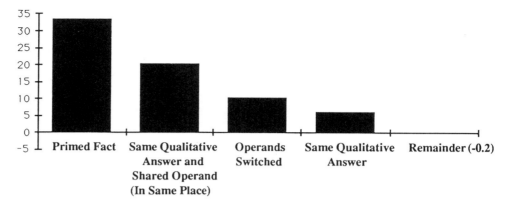

Figure 17.18   Results of simulated practice effects. Priming affects response time to many aspects of the problem, and can give decreases in response time to problems that share operands and results.

Product-Size Effect

An unexpected experimental result found in simple arithmetic is the general increase of response time with increase of the magnitude of the operands, and therefore also the answer. This is referred to here as the *product-size effect*, or *problem-size effect*. For example, the response time to a problem such as 2 times 3 is less on the average than to one such as 6 times 9. This can be explained in several ways. Campbell and Graham (1985) suggested that the effect results from different amounts of practice on different problems (all else being equal, the greater the practice, the faster the response times) and the order in which problems are learned (those learned first interfere with those learned later). Typically, smaller problems are learned first and practiced more. Practice and order effects can be modeled in an associative framework, but the two-part code we used in our simulations suggests an additional mechanism.

The effect may result from the pattern of bar codings. Codings for problems of increasing magnitudes may become successively more muddled together since the analog code is compressed to some degree for larger magnitudes, based on the observation that most sensory systems respond to the logarithm of the magnitude, as expressed in the Weber-Fechner law. Other psychophysical magnitude functions, for example, power law functions, would show an equivalent pattern at the level of approximation used in our simulations. Most of our simulations exhibited a product-size effect. Figure 17.19 shows one set of experimental data on the product-size effect, and figure 17.20 shows results from one simulation.

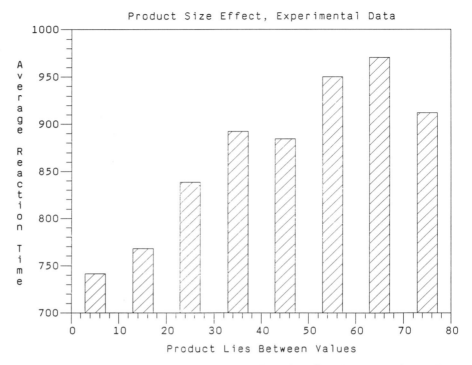

Figure 17.19 Experimental data for the product size effect. Response time to large products is longer than to small products. Data from Campbell and Graham (1985).

Other Generalization Effects

The model also generalizes when presented with completely novel problems. This has been explored in detail with operand interchanges: the system was trained on one operand order, and then presented with the same problem with the operands reversed. The response was correct 50% of the time, although slower (24.2 iterations to an attractor) than responses to the originally trained order (12.6 iterations). When the system provided specific incorrect answers, they were close to the correct answer. Presumably the system makes a reasonable generalization based on the similarity of the pattern of bars to a problem on which it has been trained. Such behavior is surprising because the system appears to extract, imperfectly, the rule of commutativity.

It is hard to tell how psychologically plausible this aspect of its performance is. One reason is that students usually practice both operand orders when learning a problem, although both orders may not be practiced equal-

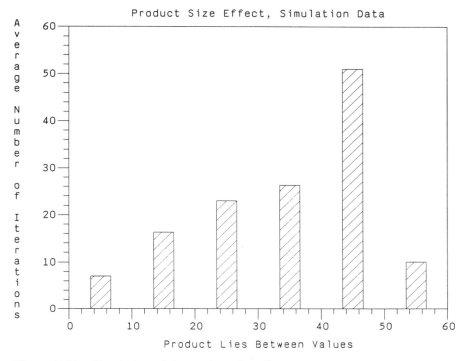

Figure 17.20   Simulation of the product size effect.

ly. According to anecdotal evidence, children do not always generalize well across operand order, and fail to understand commutativity until it is specifically pointed out to them. The simulations also have difficulty with commutativity, for example, generating a reasonable answer, but taking more time to generate correct commutative answers than to generate answers to problems on which they were trained. The rough knowledge of commutativity acquired by the model might give a useful bias when the system learns the other operand order.

Programming Brainlike Computation

As we mention throughout this book, the operation of neural networks is sometimes referred to as brainlike computing. Realistically, the arithmetic network does not do a very good job, and if it is viewed simply as a device to produce accurate arithmetic fact retrieval, it is a failure. However, the data representations apparently used by humans and modeled in these simulations suggest that in human, and network, performance, overall accuracy may be less important than estimation and flexibility. One of the most striking aspects of human cognition is its flexibility, that is, the same data can be used and reused in many ways. The ability to apply old facts to new problems may be more important than retrieval accuracy.

Let us consider the following quotation from Mark Twain: "We should be careful to get out of an experience only the wisdom that is in it—and stop there; lest we be like the cat that sits down on a hot stove-lid. She will never sit down on a hot stove-lid again—and that is well; but also she will never sit down on a cold one any more."

Given the way neural networks are usually trained, they may be only as wise as Twain's cat. Worse, the way that neural nets are usually formalized ensures that they may never be any smarter. Consider the way that they are most often presented in engineering texts, and that we have used in this book on many occasions. The problem usually takes the following form. We have a training set of classified examples of input and output patterns. We train the network using one of many learning rules so that the desired network output and the actual network output are close to each other based on some error function. If all is done properly, after a substantial period of learning, the resulting network will do well on the training set. Its ability to give correct answers to novel problems of the same type is a measure of its ability to generalize. Presumably, that cat could generalize well to different sizes and shapes of stove lids. This observation misses the point of the criticism, however.

Generalization has many aspects. It is not clear what the word actually means. Heated discussion about generalization on neural network email nets has failed to result in any accepted definition. In many of the exchanges it was clear from context that generalization really meant good interpolation, as discussed in chapter 13. That is, given some known data points, good generalization means that unseen data points, intermediate between seen points, are properly estimated or categorized.

Interpolation is an important problem in statistics, but there are other kinds of generalization. Intuitively, generalization means something like "you don't have to see all the examples to get the right answers." The real, and intrinsic, difficulty is that in almost any situation beyond the trivial, most generalizations will be wrong. Extrapolation is far more difficult than interpolation. A moment observing a child learning to speak will convince you of that.

After extended periods of training, humans can learn to become good at generalizing in limited domains. Much of the practical value of education consists in learning how, when, and how much to generalize. This is far more than simply seeing a number of examples. The entire notion of inference, where experience is applied correctly to novel problems, can be understood as an important form of generalization.

Even when what appears to be straightforward pattern classification is performed by humans, it is hard to generalize correctly. Deliberately chosen good and bad examples and explicit instruction are necessary. Structuring a network so it can be taught easily and accurately may be more important than constructing it so it learns well from a random set of examples.

Generalization: Fodor and Pylyshyn

It was the difficulty of training neural networks to make good generalizations on problems more complex than simple pattern recognition that led to the important criticisms of Fodor and Pylyshyn (1988). An interesting history in both cognitive science and neural networks involves what might be called arguments from cognitive adequacy. We have already run into two of them: the demonstration by Minsky and Papert in 1969 (chapter 8) that perceptrons could not learn certain properties of a figure and had trouble with some logical relations, and the argument by Chomsky that simple stimulus-response learning was inadequate to explain some language behavior (chapter 16). Fodor and Pylyshyn made equally potent criticisms about the computational adequacy of neural networks from the point of view of cognitive science and linguistics.

Their essential criticism is, surprisingly, one that an engineer would be happy to make: sometimes neural networks are such an inefficient way to compute that it would be foolish to build a brain (or an artificial cognitive system) like that. The authors are prepared to admit that the hardware of the brain is based on neurons and is in some sense physically connectionist. But, they claim, the kinds of computations actually done, however they are carried out in practice, reflect a very different, higher-level, and much more flexible computational strategy, one based on the manipulation of symbols.

Essentially all simple neural networks form part of an associationist tradition prominent in the history of psychology. As we discussed in chapter 16, associationists view mental life as being composed of often arbitrary associations between different events. Neural networks are part of this tradition; the point of weight-modification rules is usually to make sure that the right output patterns are produced by the appropriate input patterns. Fodor and Pylyshyn's work should be viewed not just as a criticism of neural networks but as part of an old battle that has been going on for centuries. As the authors stated, "[I]t's an instructive paradox that the current attempt to be thoroughly modern and 'take the brain seriously' should lead to a psychol-

ogy not readily distinguishable from the worst of Hume and Berkeley"
(p. 64).

Fodor and Pylyshyn contrasted neural network associators with what
they call the *classical* view of mental operation. In essence, it postulates "a
language of thought," that is, "mental representations have *a combinatorial
syntax and semantics*" (p. 12). This view is held by almost all linguists and is
dominant in virtually all branches of artificial intelligence. Unexpectedly,
the beauty of this highly abstract approach is in its practicality. Syntax
allows extreme flexibility and generality in mental operations. The great
power of the traditional digital computer arises in part from the fact that it
is deliberately designed to be an extreme example of this organization: a
programming language operating on data is the prototype.

Fodor and Pylyshyn used a linguistic example to demonstrate the computational flexibility of the classical architecture. Suppose we have a sentence
of the form *A and B* that we hold is true. An example they use is *John
went to the store and Mary went to the store.* The truth of this sentence
logically entails the truth of *Mary went to the store.* This arises from the rules
of logic and of grammar; however, it is not easy for a neural network to
handle this problem. A simple net could easily learn that *John went to the
store and Mary went to the store* is associated with *Mary went to the store.*
But the power of this approach arises from the fact that every sentence of
this form gives rise to the same result: the truth of the assertion *the moon is
high and the night is beautiful* implies *the moon is high.* Also, having learned
that *John went to the store and Mary went to the store* is true implies that
John went to the store is true. We have no trouble drawing the same conclusion from the related sentences, *Mary went to the store and John went to the
store,* and *John and Mary went to the store.* It is not necessary to learn many
examples. Given a sentence of the appropriate form, the correct conclusion
can be drawn, even by children.

Given the huge number of possible sentences that can be combined, *it
makes practical sense* to assume that a logical syntax exists to handle this
class of problems. In fact it would be hard to figure out how language could
function without some global rulelike operations, however implemented. At
the same time, language rules have many exceptions and come in varying
degrees of softness. For example, we used the combinatorial power of conjunction in the example, but Fodor and Pylyshyn gave some other sentences
with a different twist: *John and Mary went to the store* implies *John went to
the store,* but *the flag is red, white, and blue* does not imply *the flag is blue.*

However, peculiarities and special cases do not show that grammar is not
rule based. It may show instead that the language implementation we are
stuck with just may not be very good, perhaps because language is lately

evolved and full of bugs. It rarely provides absolutely clear-cut examples of combinatorial grammars in operation. It seems as if the real world is too imprecise and sloppy to be entirely pinned down by symbolic grammars. Some areas of language clearly lend themselves to neural networks. Fodor, in talks, has often mentioned that language up to the lexicon could be handled effectively by neural networks. In fact the lexicon, our detailed knowledge of what words mean, is an almost ideal candidate for the virtues of neural networks because an appropriate word can be produced in response to an immense number of novel situations that differ in detail.

The ability to understand immediately sentences or phrases that are new to the listener is difficult to explain on the basis of simple association. Fodor suggested a novel phrase, easily produced by a symbol-based syntax, *antimissile missile*, which has the clear meaning to most listeners of a missile that shoots down missiles. One more application of the operation leads to the next phrase, an *antiantimissile missile*, a missile that shoots down missiles that shoot down missiles, and so on. New and understandable combinations such as *antithief thief* might mean roughly what is meant by the proverb, set a thief to catch a thief. Examples like this are numerous, combinatorially numerous, and demonstrate exactly the point Fodor and Pylyshyn made. They are both novel and immediately understandable, suggesting ability to generalize beyond the power of networks that learn specific examples.

Attempts have been made to build neural networks that realize parts of the classic account (see Hinton, 1991). Even charitable observers, however, sometimes including creators of the models themselves, have noted that the networks are often a bit unreasonable and that the operations they perform seem unnatural. One might speculate that if a qualitative difference exists between human and animal cognition, it lies right here. Somehow, perhaps not very well or very efficiently, the human brain has developed a way of handling a few kinds of simple rulelike operations corresponding to the rudiments of a mental syntax based on symbol-like entities.

Fodor and Pylyshyn presented in clear form what may be the major technical problem confronted by neural networks in the future. Is it possible to build a simple and reliable network that can reproduce the kind of rule-governed behavior that does in fact seem to be part of human cognition? The advantages of a net with this ability are immensely practical, as it would allow for far greater computational flexibility and generalization than the current, simple networks can give. But this may not be an easy solution to find. Many animals have a sizable nervous system, but only our own, a single species out of millions, seems to be able to perform this class of operations.

Programming a Network: Bigger and Smaller

Many of the really interesting and significant generalizations, the ones most likely to be incorrect, and the ones that are the most important for human cognition, involve inferring relationships between objects where all examples of the relationship are not and cannot be seen. Let us extend our arithmetic model and speculate about how this might be done for the relationships "bigger" and "smaller." We use this problem as a very simple case of the more general problem raised by Fodor and Pylyshyn. If we consider the number of examples, there are around 100 single-digit comparisons, around 10,000 two-digit comparisons, around 1 million three-digit comparisons, and so on. It would not be possible to have a neural net see and learn all comparisons before it could answer questions about which of two numbers is bigger. Children do not learn comparisons this way.

The hybrid representation for number, besides producing systems that are qualitatively similar to observed human experimental data, allows flexible control structures to be built into the system. We will call use of these control structures programming, following the traditional computer model. In elementary arithmetic we have programming terms, for example, multiply, add, bigger, and smaller, that operate on number data. The programming term defines an operation or a relation between two numbers. These particular operations are part of the set of hardware machine instructions in most digital computers.

The first digital computers were programmed by making connections between logic and arithmetic units with plug boards and wires and with switches. Von Neumann (1945) and the group at the Moore School that developed the ENIAC, had the important insight that it was possible to represent programming instructions in the same way that data were represented: as patterns of numbers in the computer's memory. Although von Neumann sometimes gets the sole credit for the idea, the 1945 technical report apparently summarized and reflected the thinking of others in the Moore School group as well. It also made significant reference to the McCulloch-Pitts neuron and the brain as a computing device, as we mentioned in chapter 2.

Unfortunately, it is not so clear how to implement a program in a neural network. Is it an activity pattern (a vector), a set of connection weights (a matrix), or simply a gain parameter or some other global quantity, perhaps reflecting the level of a circulating biochemical? Or, more likely, is it all three at the same time?

We will investigate the potential utility of representing a network program by a pattern of activity, that is, a vector. A vector in a network obviously contains much less information about events than a matrix of connection strengths. As a pattern of activity, however, it is related in a direct way to network activity; and, used as a control structure, it does not need all the detailed information available to the system. Let us speculate about some of the potential properties of a basically vector-valued programming technique. Many models for attention also act like or use vector quantities, for example, the "searchlight of attention" suggested by Crick (1984). This is not a coincidence, since the functions of attention and network programming must be deeply related.

Let us program a network to answer two simple questions: which of two numbers is bigger, and which of two numbers is smaller? We will restrict ourselves to the numbers between 1 and 10. We must use information we have available to us in our number representation. The symbolic portion of the hybrid representation tells us nothing about relative size, but the analog part does, which suggests a way to do the comparison operation effectively. We want to feed the two state vectors into the network, with a programming state vector, bigger or smaller. The network should give an output state vector representing the bigger or smaller number as the answer.

The BSB network lets us choose one pattern relative to another, based on the relative amount of feedback through the connection matrix. In the analysis of BSB, amount of feedback is related to the different values for the eigenvalues of the connection matrix. To choose the larger of two values, therefore, all we have to do is arrange it so that smaller numbers have smaller feedback weights than larger numbers. This is easily accomplished by producing relative inhibition or excitation of different parts of the analog map, and not through detailed synaptic excitation patterns. All that is required is an excitatory weighting *vector* that is large on one side of the map and small on the other. Therefore, the programming term "bigger" merely has to be associated with this differential vector excitation pattern of the analog portion of the map. For our simulations, the weighting was multiplicative, not additive, so the activity of a unit was multiplied by the value of the weighting function. A possible schematic diagram is shown in figure 17.21, which presents a system that realizes a neural network program in which both the data and the control structures are activity patterns. If we put in the sum of the analog portions of the number vectors, the dynamics of the system should produce as the final state of the system the larger number.

It would be relatively easy to learn the differential activation pattern: it is related to differential success rates in the comparison process. With only

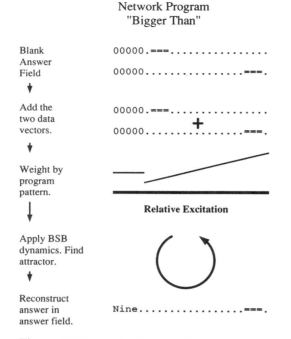

Figure 17.21   A simple neural network program to compare two numbers and generate as output the bigger one. The two data vectors are added together and weighted by the relative excitation shown. The BSB dynamics operate on the vector sum. The final attractor has the correct answer (9) in the number field. The correct answer was reconstructed by an estimation and weighting process, based on the bar code part of the representation.

random pairs of different digits between one and nine to compare, for example, nine is always bigger than other digits, one is never bigger, and five is bigger about half the time. Therefore if the magnitude of the differential activation term corresponded to frequency of success, it would have the right pattern for bigger. Presentation of only a small number of the many possible comparisons allows a reasonable estimation of the frequency of success. In addition, the system is very insensitive to the details of the weighting pattern so long as it is monotonic. The inverse pattern could be used for programming "smaller."

We can now ask two questions about the model: can we get the network to produce the right answers, and, do the results agree with the experimental data on response times for bigger and smaller? It was easy to get the network to work. The structure of the problem representation is sufficiently regular so that the larger number appeared reliably as the final state of the system.

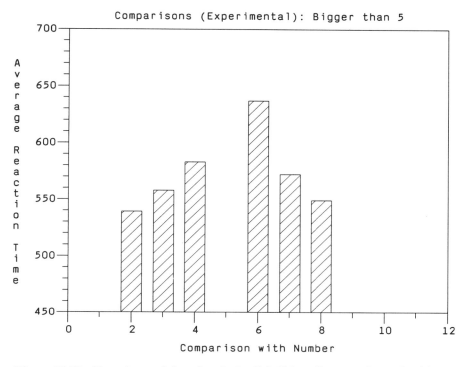

Figure 17.22 Experimental data for single-digit "bigger" comparisons. In this case the reaction time to judge whether a particular digit is bigger or smaller than five is given. From data in Banks, Fujii, and Kayra-Stuart (1976).

The network actually *constructed* the correct answer in the answer field using network dynamics. Figure 17.22 shows the experimental reaction time data for human subjects for single digit comparisons (Banks, Fujii, and Kayra-Stuart, 1976). Figure 17.23 shows the response time (iterations to the attractor) results of a representative simulation using this technique. In operation, the simulations are reliable—they nearly always produce the correct answer—and show the kinds of qualitative response time patterns, including the symbolic distance effect, that characterize human performance. In our simulations (figure 17.22), as in the experimental data (figure 17.23), errors appeared, and response times were longer when the two numbers were close together.

Because we assumed that the relative excitation of the map resulted from a frequency-of-success estimation process, we can fortuitously let someone else do part of the detailed fits of experimental data for us. Link (1990) proposed a frequency estimation model for two-digit number comparisons, as we just used. Link observed that 90s usually were larger compared with other numbers and teens usually were smaller. He then used parameters

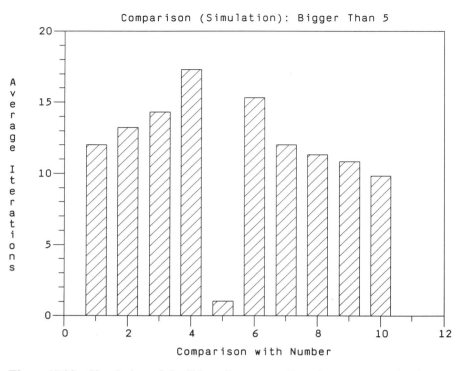

Figure 17.23 Simulation of the "bigger" program. Iterations measure the time required for the system to reach an attractor. These are all the single-digit comparisons with the digit five. Answers were correct in all cases. Note the symbolic distance effect, which is also shown by the experimental data available for this task (see figure 17.22).

based on the frequency estimates for what mathematical psychologists call a random walk response time model. Link's random walk model has formal similarities to the neural network response model we have discussed, since it integrates information and makes a final decision by entering an attractor state. The model gave good fits to his number comparison data.

Discussion and Speculation

We have shown that a simple network program can be implemented by differential weighting of the input data vector. In favorable cases the programming vector can be estimated by seeing relatively few examples of the output, *if* the task and the structure of the data are suitable. Easy programming is possible, but only in a limited domain controlled by the details of data representation.

Our number representation contained a spatial map of magnitude. Our claim, based on psychological data, is that strong evidence exists for such a map, although certainly not in such a simple form. Given a map, the computation of bigger and smaller becomes straightforward. It is possible to use the map to do other operations. For example, high–low poker players are familiar with the concept that might be called "most middle", that is, a useless hand that is neither high nor low. A programming pattern with a peak around five would be able to compare two numbers as to their "middleness." Other unfamiliar questions answered easily by both adults and children are of the form, what is a number larger than four and less than six? Such questions can be answered using combinations of the elementary programming patterns.

It would be much more difficult to do other reasonable computations using the number representations we propose. Suppose we asked questions based on the number names. If we ranked numbers by the length of their names, then one is less than five but three is greater than nine; or by position in the alphabet: eight is less than six but seven is greater than nine. These comparisons are low-level operations in some computer language implementations.

It is hard to define what flexible programming consists of. A relatively small number of stable attractors in a high-dimensional space is of value as a memory or pattern-recognition device, but it is rigid. What makes human cognition interesting is its combinatorial nature. Even the simple number comparisons show this, because we can generalize a computation to new pairs that were never learned. If other neural computations are done in a way at all similar to the frequency-estimation technique used here, it suggests that the possibilities for easy and flexible cognitive computation are limited to a significant extent by physical brain architecture. This is an intriguing and testable suggestion. The discussion of important aspects of language found in Lakoff (1987) is consistent with such an idea, as are the tight constraints on natural concepts that are obviously present and yet very hard to define precisely and logically.

If we assume that programming flexibility is desirable, and that differential vector weighting is one way of achieving it, we might speculate as to what data-representation architecture would best be able to achieve it. Clearly, such an architecture with a degree of spatial localization of important parameters would allow easy and flexible programs involving those parameters. A representation that was fully distributed, in which information of all types was distributed everywhere, would be harder to program. One that was very selective, for example, grandmother cells, would also be hard to deal with since there would be no natural spatial structure to work with to form generalizations.

Current models of the visual system are highly modular, with different parts of the vision computation (motion, form, color) being carried out by physically separate pathways (see chapter 10). Even a weak degree of spatial separation would allow a great deal of flexible programming based on crude vector-programming patterns. At the same time, the strong implication is that only certain programs would be easy, ones that the spatial structure of the representation encouraged and allowed. Animal psychology is full of examples of difficult and easy learning. Even humans find it hard to learn unnatural associations, for example, a linear array of buttons in which keys are randomly assigned numbers.

The number system we have discussed has other drawbacks. The accuracy of the network computation is poor because estimation and generalization seem to be at odds with high precision. Use of the analog representations requires huge numbers of computing elements. However, the brain seems to work in a regime where processing elements and connections are cheap, topographical maps are easy to construct, and powerful qualitative computations thereby become possible.

We discussed at length the importance of incorporating into the input data representation of number a sensory or analog part, which, together with an abstract (symbolic, arbitrary) part, produces a system that acts in simulation like experimental data from humans. For humans, mathematics and the abstract quantities associated with it are represented in rich and complex ways that contain a substantial sensory component. Such formally extraneous material strongly influences the kinds of errors that humans make, as well as their insights. This extrasymbolic material gives rise to some of the things that one might call intuition, understanding, insight, or common sense, if humans displayed it, and demonstrates how a large amount of previously learned information can cooperate to produce good answers to new problems. Extrasymbolic information and operations based on it may be responsible for more of the creative aspects of mathematics than formal symbol manipulation.

Symbol manipulation as done by humans rarely stands alone, and pure symbol manipulation seems to be rare even in domains such as mathematics. What seems to be done, if we are to take the leap of assuming the arithmetic simulation is typical of complex high-level cognition, is to form a fascinating hybrid computation; part symbolic and part sensory-based and intutitive, in which the parts mutually enhance and support each other. Since we live in a world that our sensory systems have evolved to interpret, we expect that our symbol systems are tightly coupled to our universe and are extremely effective in handling it. Perhaps true understanding of a complex system occurs when the abstract and the sensory-based components agree with each other, arriving at the same answer by very different means.

In artificial systems we have the potential to see how these ideas might be implemented, how they work best, and how they can be extended. In the past half-century of computer development we focused on the development and refinement of aids for formal symbolic methods of computation. Perhaps neural networks will let us develop a different class of artificial systems to help us with mechanical aids for our common sense. We may also be able to understand better our own cognitive abilities.

References

C. Acredolo (1982), Conservation-nonconservation: Alternative explanations. In C.J. Brainerd (Ed.), *Children's Logical and Mathematical Cognition*. Berlin: Springer.

J.A. Anderson (1992), Neural network learning and Mark Twain's cat. *IEEE Communications Magazine, 30*, 16–23.

J.A. Anderson, D.A. Bennett, and K.T. Spoehr (1993), A study in numerical perversity: Teaching arithmetic to a neural network. In D.S. Levine and M. Aparicio (Eds.), *Neural Networks for Knowledge Representation and Inference*. Hillsdale, NJ: Erlbaum.

J.A. Anderson, M.L. Rossen, S.R. Viscuso, and M.E. Sereno (1990), Experiments with representation in neural networks: Object motion, speech, and arithmetic. In H. Haken and M. Standler (Eds.), *Synergetics of Cognition*. Berlin: Springer-Verlag.

W.P. Banks, M. Fujii, and F. Kayra-Stuart, (1976), Semantic congruity effects in comparative judgements of magnitudes of digits. *Journal of Experimental Psychology: Human Perception and Performance, 2*, 435–447.

J.I.D. Campbell (1987), The roll of associative interference in learning and retrieving arithmetic facts. In J.A. Sloboda and D. Rogers (Eds.), *Cognitive Processes in Mathematics*. Oxford: Oxford University Press.

J.I.D. Campbell and J.M. Clark (1992), Cognitive number processing: An encoding complex perspective. In J.I.D. Campbell (Ed.), *The Nature and Origins of Mathematical Skills*. Amsterdam: North Holland.

J.I.D. Campbell and D.J. Graham (1985), Mental multiplication skill: Structure, process, and acquisition. *Canadian Journal of Psychology, 39*, 338–366

G.W. Cottrell and F.S. T'sung (1991), Learning simple arithmetic procedures. In J.A. Barnden and J.B. Pollack (Eds.), *Advances in Connectionist and Neural Computation Theory*, Vol. 1. *High Level Connectionist Models*. Norwood, NJ: Ablex.

F. Crick (1984), Function of the thalamic reticular formation: The searchlight hypothesis. *Proceedings of the National Academy of Sciences, 81*, 4586–4590.

P.J. Davis and J.A. Anderson (1979), Nonanalytic aspects of mathematics and their implication for research and education. *SIAM Review, 21*, 112–127.

P.J. Davis and R. Hirsh (1985), *The Mathematical Experience*. Birkhauser: Boston.

A. Einstein (1951), Autobiographical notes. In P.A. Schilpp (Ed.), *Albert Einstein: Philosopher-Scientist*. New York: Tudor.

J. Fodor and Z. Pylyshyn (1988), Connectionism and cognitive architecture: A critical analysis. In S. Pinker and J. Mehler, (Eds.), *Connections and Symbols*. Cambridge: MIT Press.

G. Gamow (1947), *One, Two, Three–Infinity*. New York: Viking.

R. Gelman and C.R. Gallistel (1978), *The Child's Understanding of Number*. Cambridge, MA: Harvard.

J. Gleick (1992), *Genius*. New York: Pantheon.

D.J. Graham (1987), An associative retrieval model of arithmetic memory: How children learn to multiply. In J.A. Sloboda and D. Rogers (Eds.), *Cognitive Processes in Mathematics*. Oxford: Oxford University Press.

J. Hadamard (1945), *The Psychology of Invention in the Mathematical Field*. Princeton, NJ: Princeton University Press.

G.E. Hinton (Ed.) (1991), *Connectionist Symbol Processing*. Cambridge: MIT Press.

G. Lakoff (1987), *Women, Fire, and Dangerous Things*. New York: Basic Books.

D.F. Lancy (1983), *Cross-Cultural Studies in Cognition and Mathematics*. New York: Academic Press.

S. Link (1990), Modeling imageless thought: The relative judgement theory of numerical comparisons. *Journal of Mathematical Psychology, 34*, 2–41.

M. McClosky and A.M. Lindemann (1992), MATHNET: Preliminary results from a distributed model of arithmetic fact retrieval. In J.I.D. Campbell (Ed.), *The Nature and Origin of Mathematical Skills*. Amsterdam: North Holland.

W.S. McCulloch (1961/1965), What is a number, that a man may know it, and a man that he may know a number? In W.S. McCulloch (Ed.), *Embodiments of Mind*. Cambridge: MIT Press.

A.I. Miller (1984), *Imagery in Scientific Thought: Creating 20th Century Physics*. Boston: Birkhauser.

M. Minsky and S. Papert (1969), *Perceptrons*. Cambridge, MA: MIT Press.

R.S. Moyer and T.K. Landauer (1967), Time required for judgement of numerical inequality. *Nature, 215*, 1519–1520.

G.M. Norem and F.B. Knight (1930), The learning of the one hundred multiplication combinations. In *National Society for the Study of Education: Report of the Society's Committee on Arithmetic*, Vol. 15. *NSSE Yearbook 219*.

H. Poincaré (1914), *Science and Method*. London: Thomas Nelson and Sons.

H. Poincaré (1929), *The Foundations of Science*. New York: Science Press.

D.E. Rumelhart, P. Smolensky, J.L. McClelland, and G.E. Hinton (1986), Schemata and sequential thought processes in PDP Models. In J.L. McClelland and D.E. Rumelhart (Eds.), *Parallel Distributed Processing, Volume 2*. Cambridge, MA: MIT Press.

E.H. Stazyk, M.H. Ashcraft, and M.S. Hamman (1982). A network approach to mental multiplication. *Journal of Experimental Psychology: Learning, Memory, and Cognition, 8*, 320–335.

S.R. Viscuso, J.A. Anderson, and K.T. Spoehr (1989), Representing simple arithmetic in neural networks. In G. Tiberghien (Ed.), *Advanced Cognitive Science: Theory and Applications*. Cambridge: Horwoods.

J. von Neumann (1945/1982), First draft of a report on the EDVAC. In B. Randall (Ed.), *The Origins of Digital Computers: Selected Papers* (3rd ed). Berlin: Springer.

J. von Neumann (1958), *The Computer and the Brain*. New Haven, CT: Yale University Press.

F.A. Yeldham (1926), *The Story of Reckoning in the Middle Ages*. London: George G. Harrap.

F.A. Yeldham (1936), *The Teaching of Arithmetic Through Four Hundred Years*. London: George G. Harrap.

Afterword

A series of appendixes, referred to in the chapters, are available over Internet or on a floppy disk that can be obtained from The MIT Press. In addition to the appendixes, we have also made available some of the complete programs and data sets used in the simulations in the book.

The programs be obtained directly from The MIT Press at the following address: Orders Department, The MIT Press, 55 Hayward Street, Cambridge, MA 02142. Please refer in a covering letter to the disk associated with James A. Anderson's *An Introduction to Neural Networks* and mention what format disk you need.

The fastest, cheapest and most convenient method of retrieving the appendixes and additional material is over Internet. The files are available by anonymous ftp from mitpress.mit.edu or 18.173.0.28. Give your full e-mail address as password, then cd pub/Intro-to-NeuralNets. The files are all in ASCII form.

A complete current table of contents and comments about the programs and the book are available in the file README.TXT.

The programs in the appendixes and used in this book are written in Pascal. The version of Pascal used is VMS Pascal, a product of the Digital Equipment Corporation. Compiled versions of the programs run correctly on DEC VAXstations. Some modification of the programs will usually be necessary to make them run on other machines. Some of the programs are large and computation intensive, and it may not be practical or even possible to run the compiled programs on smaller personal computers.

Some care was taken to make sure that the programs were sufficiently general that they could be converted easily to other workstations or to Turbo Pascal, available for many personal computers. They can also be (and have been) converted to the language C using one of the commonly available UNIX Pascal to C converters.

Index

Note: Figures and tables are indicated by italic *f* and *t*, respectively.

Abstractions, 460
Acetylcholine, 11
  in association memory, 161–162
  nicotinic, receptor for, 12*f*
ACT model, 556
Action potentials, 17, 18*f*
  function of, 32–33
  from *Limulus* optic nerve, 100–102
  oscillograms of, 102*f*
  presynaptic and postsynaptic, 37, 38–40
  in response to physical stimulus, 87–88
  slow, 46–47
  speed of, 33
Activity patterns, interactions between, 367–368
ADALINE, 209, 213–214, 240–242
  computation by, 242–249
Adaptability, brain and, 293–296
Adaptation
  corner shifting and, 517–518
  vector, 520–522
Adaptive Linear Neuron. *See* ADALINE
Adaptive mapping, 468–469
  algorithms, 459
    predictions by, 489
    sensitivity of, 485, 486*f*
  formation of, 469–470
Adaptive maps, 463–490
Adaptive multilayer network, 257
Adaptive pattern classification, 434–435

Addition
  matrices, 130–131
  neural network architecture for, 587–588
  vector, 67–69
ALCOVE model, 358
Algebra, linear, 64–65
All-or-none response, 17
Ambiguity, 570–575
Ambiguous motion, 381
Amplification, 289
Amplifiers, 150–152
Amplitudes, sine waves of, 180–182
Analogy, 344
Analysis tools, 1
Angle, between vectors, 75–77
Angle of resolution, minimal, 312–313
Annealing
  rate of, 420
  schedule, 420, 421–422
  simulated, 416–417, 420–421, 429–430
ANSI screen control, 111
Anti-Hebb adaptation, 500
Aperture problem, 378–391
  network model of, 392*f*
Aplysia
  abdominal ganglion, 8
  learning in, 147
Approximation, 21, 452–453
Area centralis, 310
Aristotelian concept, 353–354
Aristotle, 595
  on memory, 546–553

Arithmetic. *See also* Mathematics; Numbers
  error patterns in, 604–606, 607*f*
  experimental effects in, 610
  experimental literature on, 600–602
  Fodor and Pylyshyn on, 615–617
  generalization effects in, 612–613
  product-size effect in, 611–612
  programming brainlike computation for, 613–615
  programming network for, 618–622
  speculation on, 622–625
  teaching to neural networks, 585–625
Arithmetic network
  associative, 588–589
  three-unit, 587–588
Array, 66
ART model, 532
The Art of Computer Programming, 186
Artificial information retrieval system, 194–195
Artificial intelligence (AI) system, 4
  data representation in, 568–570
  potential of, 624–625
  symbol-based, 598
  symbol-processing, 228
Artificial neural networks, 345–346
ASCII characters, 63
  coding of, 289
ASCII list, 200
Association, 3, 159–161, 545
  capacity for storage of, 351–352
  character-based program, 202*f*
  computation of, 165–167
  correctness of, 250
  disambiguation and, 570–575
  input-output relationships in, 545–546
  late models of, 553–555
  multiple, 167–168
  pattern, 161–163
  storage of, 164–165
    prototype effect in, 361–362
  total, partial, and focalized, 562–564
Association-area cells, 284–285
Association units, 216, 217–218
Associative computation, 555–558
  complex concept models and, 577–582
  disambiguation and, 570–575

Ohm's law and, 568–570
  problems of, 589–590
  semantic networks and, 575–577
  simulations of, 558–568
Associative hierarchy, 577–582
Associative interference, 589
Associative learning, 147–152
  supervised, 176*f*
Associative memory, 143, 195, 404, 545
  Aristotle on, 546–553
  demonstration AMDEMOL in, 197–199
  program AMDEMO and, 195–197
  reasoning with, 551–553
Associative network, 359
Associative retrieval, 564–567
Associators
  applications of, 351–398
  neural network, 615–617
Asymmetrical inhibition, 379*f*
Asymmetry, 344
Asynchronous updating, 407–409
Attention, 343, 345
Attractor networks, 404
Attractors, 402, 404, 505–506
  basins of, 507–511
    size of, 511–512
    structure of, 513*f*
  connectivity and, 517
  network of, 403*f*
  response time to, 524–527
  spurious, 412
  stability of, 507–511
  structures in continuous network, 415*f*
Auditory cortex, tonotopic relations in, 318–319*f*
Auditory system, 318–321
Augmentation, 216
Autoassociation, 168–170, 204–206
Autoassociative connectivity, 405*f*
Autoassociative network, 410
  architecture of, 256*f*
  flexibility of, 559
  simple nonlinear neural, 493–541
Autoassociative reconstruction, 496–498
Automatic gain control (AGC), 179
Axon, 8

behavior, equation governing, 25
in brain mass, 463
branches of, 282–283
bundles of, 292–293
hillock of, 40
myelinated, 33, 34*f*
physics of, 32

Babbage, Charles, 2
Backpropagation, 239, 248–249, 255, 452
applications of, 265–275
in image compression, 270
limitations of, 276–277
necessity of, 255–265
slowness of, 271, 276
three-layer network, 453
Backward error propagation. *See* Backpropagation
Ballistic firing tables, 2
Bar code number representations, 568–569, 601, 602
Bar codes, 596
pattern of, 611
Bar distribution, 476–477*f*
Basin of attraction, 506
Bat cortex
concerned with constant-frequency, 334–338
concerned with frequency-modulation, 338–340
map of best distance in, 338–339*f*
Bat sonar system
effectiveness of, 339–340
prey countermeasures to, 339
Bayes classifier, 438
Bede, Venerable, number system of, 599–600
Behavioral complexity, 581
brain and, 293–296
Behaviorism, stimulus-response, 553–555
Bell Laboratories, digit-recognition system of, 271–275
Berra, Yogi, 281
Bessel function, 368
Best-fitting, 178
Best match, location of, 451

Bias, 150
Binary model neurons, 60
Binding problems, 289–291
Biological mapping, 487–490
Biological system, versatility of, 494
Blood flow, brain, 152
Boltzmann machine, 416–417, 420*f*
array of units in, 422–424
demonstration of, 426–431
process of, 425–431
program of, 422–431
BOOLEAN variable, 74
Boundary behavior, in BSB model, 507–511
Bouton, 14
Brain. *See also* Central nervous system; Cerebral cortex; Nervous system; Neuron
blood flow in, 152
capacity of, 351–352
computation in, 4–5
as computer, 179
connectivity problem of, 463–464
engineering considerations in, 3
local circuitry of, 299–304
memory storage in, 145–146
NMDA receptors in, 156–157
structures of, 292–299
Brain-state-in-a-box model. *See* BSB model
Brainlike computing, programming of, 613–615
Brown, Thomas, 153–154
BSB equation, decay rate of, 513–515
BSB model, 494–541
analysis of, 619
attractor basins and stability in, 507–511
autoassociated stimuli in, 573–574
clustering in, 531–541
connectivity in, 515–517
constants of, 566
corner shifting and adaptation in, 517–518
corner stability in, 506–507
learning in, 502–504
limit effects in, 511–512
Necker cube reversals in, 518–522

BSB model (cont.)
  response time in, 522–531
  simulation of, 515
  two-dimensional, 505*f*
  vector feedback in, 497*f*
BSB nonlinearity, 504–506
Burke, Edmund, 433

Cable equation, 25, 29
  solution of, 30*f*, 31*f*
Cabling, 463
Calcium ion
  cellular concentration of, 20–21
  flow of, 38
Cameras, 315
Capacitance, 27–28
Capacitor, 26–27
Capacity, 351–352
  in Hopfield network, 411–413
Carbon monoxide, 14
Cartesian coordinate system, 66*f*
Case-based reasoning, 460–461
Catastrophic unlearning, 277
Categories, real world, 435*f*
Categories and Concepts, 354
Categorization, 3
  activity pattern interactions of, 367–
    368
  classical approach to, 353–354
  dot pattern representation, 364–367
  learning networks in, 212–214
  models of, 353–354
    exemplar, 358–358
    prototype, 354–358
  multiple-member, 360–361
  prototype effect in, 361–362
  random dot pattern stimuli in, 364
  similarity experiment simulations,
    368–376
  simulations of, 362–363
Categorizers, 358–359
Cells
  assemblies of, 284–285
  input and output states for, 51*t*
Center embedding, 554–555
Central limit theorem, 188–189*f*
Central nervous system. *See also* Brain;
  specific structures

inhibitory neurons in, 149
neurons of, 5
spontaneous neural activity in, 150
Cerebral cortex
  area MT, 382–391
  area V1, 382–391
  auditory, 318–321
  cytoarchitectonic areas of, 295*f*
  depth of processing in, 296–299
  electrical activity in, 290–291
  functional regions of, 296
  homogeneous neural structure of, 327
  map of, 304–310
  modifiable connections in, 301–304
  motor output of, 321–323
  plasticity of, 306–309
  regions of, 293–296
    arrangement of units in, 299
    relative sizes of, 305*f*
  structure of, 292–293
  visual area of, 310–318
Chain rule for partial derivatives, 259
Character-based simulations, 199–201
  prototype effect in, 376–378
Character recognition, 271–275
Character strings, 200
  accuracy of, 201–204
  input and output of, 363–364
  matching, 528–529
  in simulations, 528*f*
Characteristic values, 136–138
Characters, retrieval accuracy and,
  201–202
Chemical sensors, 85–86
Chemical synapses, 11–12, 13*f*
Chesterton, G.K., 239
Chomsky, Noam, 554
Circuitry
  nervous system, 299–304
  very-large-scale integrated (VLSI),
    414–415
Classification, 3, 210
  pattern, 434–436
  in prototype simulation, 373–376
  response time in, 227–228
Classifiers. *See also* Generalization
  nearest neighbor, 433–461
Clustering, 531–532

emitter, 532–534
  neural network algorithms of, 534–537
  stimulus coding and representation of,
    537–541
  supervised and unsupervised, 531
Cognition, example-based reasoning in,
    459–461
Cognitive problems, in neural networks,
    586–625
Cognitive representations, 340–344
Cognitive science, 2
Collateral system, 303*f*
Colliculus
  peak activation of, 328–329
  topographic organization of, 325–327
Color
  names for, 355–356
  physiology of, 354–355
COLOSSUS, 2
Combinatorial problems, 288–289
Combinatorial syntax/semantics, 616–
    617
Communications overhead, 323–324
Comparison, asymmetrical, 344
Computation, 112–115
  associative, 545–582
  strategy for, 3–5
  theory of, 228–229, 323–325
Computer
  displays in, 111–112
  evolution of, 594
  first digital, 2
  sources of inefficiency in, 323–324
  Von Neumann, 51–52
Computer simulations, of similarity
    experiments, 368–376. *See also*
    Simulations
Computing "engine," 2
Computing system, organization of, 2–
    3
Computing unit, 163–165
Concept-based computation, 352–353
Concept formation theory, 194
Concept models, 460–461
  complex, 577–582
  exemplar-based, 459
Conditioned learning, 276–277
Conduction, speed of, 32–33

Confrontation naming, 211–212
Connectedness. *See* Connectivity
Connection matrix, 129, 166–167, 197
  strength of, 167–168
Connection strength, 55, 120
  in motion network, 396*f*
Connectionist computing unit, 283–285
Connectionist models, 556, 558
  features in, 340–341
Connectionist neurons, 54–60, 240
  in nearest neighbor calculation, 436–
    446
Connectivity, 231–238
  in adaptive maps, 463–490
  brain complexity and, 293–296
  in BSB model, 515–517
  between cortical regions, 296
  full, 465
  in Hopfield network, 405*f*
  limitations on, 232–233
  in motion detection, 387–391
  of nervous system, 304
  neuronal, 283–285
  partial, 516
  problem of, 463–464
  random, 516–517
  reciprocal, 495
  in sensory processing, 296–299
  visual demonstration of, 235*f*
Conservation, 592–493
Constant-frequency (CF) sonar, 329–
    332
  cortex region concerned with, 334–338
Constraint lines, 381
  in motion pattern solution, 382
Constraint satisfaction, 406
Continuous maps, 324
Continuous networks, 413–415
Continuous-wave sonar, 329
Cornell Aeronautical Laboratories, 227
Corner shifting, 517–518
Corner stability, 506–507
Correctness, 214
Correspondence problem, 379
Cortical column, 283
Cortical hypercolumn, 317–318, 392–
    393
Cortical magnification factor, 312

Cortical regions. *See* Cerebral cortex, regions of

Counting systems, 598–600. *See also* Numbers

Credit-assignment problem, 257–258

Crossover distortion, 151–152

Cross-talk, 171

Cubic spline, 454–455

Current flow, 26–27, 28

Data
  compression of, 267–271
  fusion of, 538
  overfitting of, 276
  TYPEs and RECORDs, 66–67
Data bases, 195
Data representation, 63, 398
  in AI systems, 568–570
  for numbers, 587–591, 596–598
  of random dot patterns, 364–367
  of response time, 527, 528–530
  techniques, 281–346
Decay rate, 513–515
Decision hyperplane, 222*f*
Decision surface, 219
Default meaning, 357–358
Deinterleaving network, 532–534
Delta notation, 261
Delta rule, 248
Dendrites, 8, 282–283
  passive membrane in, 40–42
  pyramidal cell, 158*f*
Dendritic spine, 42, 302*f*
Dendritic synapses, 299–300
Dendritic trees, 9*f*
Depolarization, 16
Depolarizing current, 16–17
Developmental constraints, 3
Diameter-limited connectivity, 232–233, 235*f*, 236
Differential activation pattern, 619–620
Digit-recognition system, 273*f*
Digital signal-processing (DSP) chips, 274
Digits, cortical maps of, 306, 308–309*f*
Dimensionality, 67, 68, 80
  histogram width and, 183–185
Directional derivative, 244–245

Directional selectivity, 384*f*, 385*f*

Disambiguation, 570–575

Distance computation, 70–71

Distinctive features, 340–341, 503

Distributed neural computation, 94–105

Distributed representations, 63–64, 289–299

Distributed storage, 449–452

Distribution, 286–289

DNA sequences, memory, 146

Doppler radar signal processing, 334–337

Doppler sonar, 329–332

Dot patterns
  prototype, 364, 365*f*
  representation of, 364–367

Dynamical system, energy minimum in, 412–413

Dynamics, Hopfield network, 406–409

Earlier layer, backpropagation to, 262–265

Echo frequency, 337

Eigenvalues, 136–138, 268, 270
  in bistable perception, 519*f*
  decay and recovery of, 520
  estimates of, 500–501
  large, 510–511
  largest, 498–499
  outer product and, 517
  shifting, 501–502, 510–511
  spectrum of, 501
    during learning, 503–504

Eigenvectors, 136–138, 268, 496
  dominant, 511–512, 581*f*
  dynamics of, 521–522
  in learning, 580
  outer product and, 517
  power computation of, 497–501

Einstein, Albert, 595

Electrical behavior, 14–15

Electrical events, synaptic, 37–42

Electrical response, 16–17

Electrical synapses, 11

Electronic amplifiers, 150–152

Element distributions, 185–187, 188–189*f*

Elementary features, 341
Emitter clustering, 532–534, 540
Encoder network, 269, 270f
End plate, 14
Energy
  average, 418
  decreased, 409
Energy functions, 414
 in Hopfield network, 417
Energy landscapes, 535f
  good and bad, 421–422
  in Hopfield network, 417–422, 425–
  426
Energy minimization, 405–406, 415–
  416, 417–422
Energy minimum
  finding, 417–422, 425–431
  fixed, 412–413
  local and global, 409, 417–418
  randomized, 411–412
Energy networks. *See* Hopfield
  networks
English language, associative
  computation of, 556–558
ENIAC, 2
Equilibrium, 22–23
Error
  correction of, 240, 250–252, 253–254,
  403
   Widrow-Hoff, 276
  criteria, 178–180
  definition of, 276
  in estimating speed, 394–397
  function, 177f, 178
  matrix, 605f
  measurement of, 248
  nonrandom, 604–606
  patterns of, 604–606, 607f
  reconstruction of, 604f
  reduction of, 242–249, 276
  signal, 178, 241, 250
   square of, 242
  square of
   gradient of, 259–260
   minimization of, 247–248
  summed, 263–265
  surface, 243
   gradient of, 244–249

local minima in, 276
  paraboloid, 248
  principal axes of, 504
vector, 178
  in backward pass through network,
  262–265
  length of, 180
  noise and, 536–537
ETANN chip, 415
Euclidean vector spaces, 65
Evolution, 3–4
Example-based reasoning, 459–461
Excitatory postsynaptic potential
  (EPSP), 39–40
  interaction with inhibitory
  postsynaptic potentials, 42–46
Excitatory synapses, 299–300
Exclusive OR, 255, 276
Exemplar-based concept models, 358–
  359, 459
  on continuum, 361f
Exemplar categorization, 354
Exoskeleton, 96–97
Experience, learning by, 433–434
Expert systems, 460
Exponential activity distribution, 368
Extracellular single unit potentials, 88–
  94
Extraocular muscles, 327
Eye movement, superior colliculus in,
  325, 327–329

Fading out, 562
False products, 609–610
  simulation of, 608f, 609–610
Faraday constant, 20
Feature
  analysis of, 340–341
  continuant, 341
  decomposition of, 342f
  distinctive, 503
  maps of, 274
  voicing, 341
Feedback, 497
  autoassociative matrix, 503
  connectivity in, 516–517
  during learning, 213f
  models of, 495, 497

Feedback (cont.)
  neural network, 170
  positive, 504–506
Feedforward connections, 218
Feedforward learning, 276
Feedforward network, 257–258, 453,
    454*f*, 459
  in backpropagation architecture, 259*f*
  three-layer, 265–266
Feynmann, Richard, 595
Field effect transistor (FET), 150
FILE
  bars.txt, 481–482, App L
  numbers.txt, 481–482, App L
  states.txt, 205, 254, App J, App M
  symbols.txt, 481–482, App L
  test.txt, 204, App I
Filter system, 192–193
Firing rates, 57–58
  stimulus intensity and, 59*f*
Flashbulb memory, 162
Flexibility, 558
Fodor and Pylyshyn critique, 615–617
Forgetful integrate-and-fire model, 52,
    54*f*
Fourier analysis, 141–142, 540
Fourier integral, 141
Fovea, 310, 489
Frequency
  coding, 24*f*, 25
  information, 538
  processing of, 180
Frequency band, center of, 334–338
Frequency-modulated (FM) sonar,
    332–334
  cortex region concerned with, 338–340
FUNCTION
  Inner_product, 73, App A
  Vector_cosine, 76–77, App A
  Vector_length, 71, App A
Function machine, 139

Galileo, 351
Gallistel, C.R., 592
Gaussian distribution, 187, 525–526
Gaussian function, 456, 458
Gelman, R., 592
Generalization, 3, 177, 363

effects of, 612–613
Fodor and Pylyshyn critique of, 615–
    617
forms of, 433–434
  grandmother cells and, 287–288
  in neural network programming, 614–
    615
  noise effect on, 377
Generator currents, 87
Generator potentials, 87
GENESIS program, 31
Gibbs distribution, 418–420
Glia, function of, 5
Global minimum, 243, 248, 417
  finding, 418*f*, 420
Global variables, 107–108
Glutamate, 155
Gradient, 244–246
  calculation of, 247–249
Gradient descent
  algorithm
    Widrow-Hoff, 240–242
  algorithms, 239–277
  in error computation, 264–265
  problems with, 275–277
Grandmother cells, 63–64
  generalizations of, 623–624
  linking hypothesis and, 560–561
  problems with, 287–289
  representations of, 286–288, 290*f*, 342
Gratings, moving, 383
Gray matter, 292
Gray scale, 121*f*
Gyri, 292

Hadamard, Jacques, 595
HAM model, 556
Hands, cortical areas controlling, 306
Hard locations, 448
Hardware, 5–7
Hebb, Donald, 148
Hebb learning rule, 148–149, 163–164,
    178–179, 248, 442, 536
  generalized, 160–161
  history of, 171–172
  in Hopfield network, 410–411
  outer product, 159–161
Hebb modification rule, 175

biological basis of, 152–159
Hebb rules, 143–145
Hebb synapses, 284
  outer product form of, 268
Hebb synaptic antilearning, 519–520,
  520
Henry, Patrick, 433
Heraclitus, 351
Heteroassociation, 168–170
Hidden layers, 255, 259, 452
  computation by, 456–458
  connection of, 256–257
  function of, 268–271
  linear, 270–271
  representation of, 267
Hidden unit, 231
High-dimensional maps, 473–480
  simulations using, 480–487
Hippocampus
  Hebbian synapses in, 153–154
  in learning, 152–155
  long-term potentiation of, 153
Hirnrinde, 292
History, computing system, 2–3
Hodgkin-Huxley equation, 52
Homunculus, 304–306, 309–310
Hopfield, John, 401
Hopfield networks, 402–404
  analysis of, 404–405
  association in, 545
  Boltzmann machines and, 416–417
  capacity and phase transition in, 411–
    413
  computing unit interactions in, 408*f*
  connectivity in, 515
  continuous, 413–415
  demonstrations of, 426–431
  dynamics example in, 422–426
  dynamics of, 406–409
  energy minimization in, 405–406
  finding minima in, 417–422
  Ising models, 409–410
  learning in, 410–411
  optimization using, 415–416
Hybrid number coding, 596–597, 604,
  618
Hypercolumn, 317–318, 392–393
Hyperpolarization, 16

Image analysis, character recognition
  in, 272–275
Image compression, 269–270
Imaging techniques, 315
Impulse discharge, 121*f*
Independent vectors, 197
Inefficiency, sources of, 323–324
Inferotemporal cortex, 315–316
Information
  in disambiguation system, 572–573
  representation of, 281–346
  retrieval of, 191–194, 194–195, 206–
    207, 564–568
  storage of, 191–194, 559–560, 561–564
  transfer of, 14
Information processing
  neuronal, 14–15
  synaptic integration in, 43–46
Inhibition
  asymmetrical, 379*f*
  lateral, 102–103, 105, 467*f*, 490
    examples of, 115–117
    program demonstrating, 106–115
    WTA network and, 117–124
  in *Limulus* ommatidium, 102–104
  strength of, 119
Inhibitory coefficient, 116–117, 129
Inhibitory neurons, in learning, 149
Inhibitory postsynaptic potential
  (IPSP), 39–40
  interaction with excitatory
    postsynaptic potentials, 42–46
  in learning, 149
Inhibitory synapses, 38, 299–300
Initialization, 108–111
Inner product, 72–73, 260
  computation of, 240–241
  distribution of values of, 184–185*f*
  element distribution in, 185–187, 188–
    189*f*
  properties of, 182–187
Input distribution patterns, 473, 474–
  475*f*, 476–477*f*
Input-output relationships, 57–60, 175–
  176, 402
  in association, 545–546
Input patterns, 175, 192
  distribution of, 470–472

Input patterns (cont.)
  in high-dimensional mapping, 482*f*
Input state vector, transformation of,
    452
Insulated region, 224–225
Insulation, 25, 33
Integrate-and-fire neuron, 52–53, 54*f*
Integrators, 52–54
INTEL neural network chip, 415
Intelligence
  versus behavioral complexity, 293–296
  neuron loss and, 5–6
Intensity-frequency space, 321
Interpolation, 454–459
Ion conductance, 11, 38–40
  in synaptic Hebbian modification,
    156–159
Ionic channels, 11
Ionic flow, during synaptic events, 38–
    39
Ions, 16–20
  cellular concentration of, 18–25
  equilibrium of, 22–23
  imbalance of, 16
IS-A link, 556
Ising models, 409–410
Iterative best-match location, 451
Iterative gradient descent procedure,
    239
Iterative learning algorithm, 239

James, William, 4, 148, 533–534, 558–
    559, 562–566, 571–572, 581–582,
    595
James rule. *See* Hebb rule

k-nearest neighbor classifier, 438–449
Karhunen-Loeve expansion, 267
Kinesthetic imagery, in mathematicians,
    595
Kiwai, type II numbers, 598
Klem, Bill, 351
Knight, Bruce, 52
Knowledge network, 576*f*
  realization of, 577*f*
Kohonen maps, 464–465
  adaptive program, 468–469
  architecture of, 466*f*

basic algorithms of, 465–468
distributions of, 471–472
formation of, 469–470
high-dimensional, 473–480
  simulations of, 480–487
pathologies of, 472–473

Language. *See also* Speech; Syntax
  associative rules for, 554–555
  neural network models for, 556–558
  rule-governed behavior in, 228–229
Large-scale maps, 464
Lateral geniculate nucleus (LGN), 310
Lateral inhibition, 102–103, 105
  computation and examples of, 115–
    117
  functions of, 490
  program, 467*f*
  program demonstrating, 106–115
  in sensory processing, 85–125
  WTA network and, 117–124
Le Cun, Yann, 255
Learned associations, 204*f*
Learning, 175–207. *See also* Memory
  associative, 147–152, 159–161, 195–
    199
  associative hierarchy in, 577–582
  by backpropagation, 276–277
  in BSB model, 502–504
  in categorization, 212–214
  of concepts, 352–353
  danger of, 161–162
  by experience, 433–434
  external control of, 162
  input-output relationships in, 175–176
  map forming in, 487–490
  neural networks in, 175–178, 276–277
  of object motion, 393–397
  outer product, 518
  perceptron, 220–228
  random noise, 536–537
  stimulus-response, 553–555
  supervised, 176*f*, 177–178, 209–210
  synaptic, 143–145
  synaptic modification with, 152–159
  synaptic strength and, 146–147
  testing after, 177*f*
  unsupervised, 209–210

Widrow-Hoff, 249–254
Learning rules
  in Boltzmann machine program, 430–431
  in Hopfield networks, 410–411
Learning threshold, 171
Learning vector quantizer, 443–446
Least mean squares algorithm, 503
Least mean squares criterion, 178
Length vector, 70–71
Lenses, graded-index, 98
Letter recognition. *See* Character recognition
Levy, William, 153
Libraries, 195
Light intensity, change in, 379
LIMIT function, 58
Limulus polyphemus
  biology of, 94–105
  circulatory system of, 97–98
  detection of movement by, 379
  distributed neural computation in, 94–105
  dorsal views of, 96*f*
  lateral eye of, 8, 25, 53, 57, 97*f*, 98–105, 379
  nervous system of, 95*f*
  neurophysiology of, 52
Linear algebra, 64–65
Linear analysis, principal components from, 268–269
Linear associator, 57
  computing unit in, 163–165
  degradation of, 411
  extension of, 240
  foundations of, 143–172
  object movement, 393–394
  output of, 201
  simulations of, 175–207
  two-layer, 453, 454*f*
Linear independence, 80–81
Linear networks, 139–142
  two-layer synapses of, 257*f*
Linear separability, 219
Linguistic computations, 554
Linking hypothesis, 560–561
LMS algorithm, 240–242
LNR book, 556

Load imbalance, 323
Local minimum, 243, 276, 417
  deep, 425–426, 430*f*
  finding, 418*f*
  stable, 431*f*
Localized coding, 286
Logarithm, 88
Logistic function, 57
Long-term depression, 159
Long-term memory, 144–145
Long-term potentiation, 153, 155
LVQ algorithms, 443–446
Lyaponov function, 406

Mammalian nervous system, physiologic inhibition in, 149
Map pathologies, 472–473, 476–477*f*
Matched filter, 192–194
Mathematical creativity, 600
Mathematics, 594–596. *See also* Arithmetic; Numbers
MATHNET, 600, 601–602
Matrix, 65, 129–130
  adding, subtracting, and multiplying by constant in, 130–131
  definition of, 129
  eigenvectors and eigenvalues in, 136–138
  multiplication of, 134–135
  multiplying vector and, 132–133
  notational convention of, 130
  simple operations of, 129–142
  transposing, 133–136
McCulloch, Warren, 49, 51–52
McCulloch-Pitts neuron, 49–52, 60, 229, 405*f*
Mean square error, 268
Mechanical stimulus, intensity of, 88–92
Membrane, 11
  capacitor, 27–28
  capacity, 16
  channel, 12*f*
  current, 27–28
  passive model of, 27*f*
  permeability to ions, 20*f*
  potentials of, 15–25, 54–55
    sodium conductance and, 23

Membrane (cont.)
  resistors and capacitors of, 26–27
Memorization, 176
Memory. *See also* Learning
  Aristotle on, 546–553
  associative, 143, 159–161, 195–199,
    404, 545
  elementary units of, 548
  hierarchy of, 551*f*
  long-term, 144–145
  in mental life, 460–461
  organization of, 403
  physical basis of, 145–146
  retrieval of, 450–451
  sensory qualities of, 546–547
  short-term, 144–145, 252–253
  sparse distributed, 446–452
  storage of, 449–450
  sum, 372*f*
  summed vector, 187–194, 196, 199
  synaptic learning and, 143–144
  synaptic modification in, 146–147
Memory-based reasoning, 460–461
Meno dialog, 551
Mental constructions, 559
Mental life, 402*f*
Mental operation, classical view of,
    616–617
Metabolism, 5–6
Metaphor, 344
Microelectrode, 11, 15*f*
Minsky and Papert analysis, 214, 228–
    231, 236–237, 277
Missing patterns, 471–472
MIT AI Laboratory, 228
Mitochondria, 14
Momentum, 264–265
Monotonic sigmoid function, 258
Monotonicity, violation of, 236
Moore School of Engineering,
    University of Pennsylvania, 2
Motion
  computations, 296
  detection of, 382–397
  determining speed and direction of,
    378–397
  direction of, 386*f*
  speed of, 387*f*

Motion network, 391–393
  connection strengths in, 396*f*
Motor cortex, connectivity of, 304
Motor movement, 291
Motor neuron patterns, 292
Motor output, 291–292
  distribution of, 321–323
Mountcastle, Vernon, 87
Mouth, cortical areas controlling, 306
Moving edge, 380
Moving objects, 378
  aperture problem of, 378–391
  motion network in, 391–393
  simulation of, 393–397
Multineuron units, 283–285
Multiple-member categories, 360–361
Multiple spike train data, 286
Multiplication
  facts of, 602–603
  learning of, 590
  in linear system, 139–140
  matrices, 131, 134
  network for, 587–588
  three-digit, 600–601
  vector and matrix, 72, 132–133
Muscle fibers, 291–292
Myelin
  as insulator, 33
  sheath, 34*f*

Natural data representations, 345–346
Nearest neighbor calculation, 436–
    446
Nearest neighbor classifiers, 433–461.
    See also Generalization
  assumptions of, 436
  efficiency of, 440*f*
Nearest neighbor techniques, 211, 358
Nearest neighbor variants, 438–440
Necker cube reversals, 518–522
Neighborhoods, formation of, 468
Neocortex, 293, 301
Nernst equation, 16, 22–23
Nerve cells. *See* Neuron
Nervous system. *See also* Brain; Central
    nervous system
  comparison to amplifiers, 150–152
  connectivity of, 304, 515–517

distributed representations of, 289–299

evolution of, 3

information representation in, 282–286

layered structures of, 292–299

local circuitry of, 299–304

mapping of, 304–310

organization of, 4

Nestor algorithm, 440–443

NestorWriter, 440

NETtalk, 265–267

architecture of, 266*f*

Network computation, 398

Network models, early, 209–238

Neural network. *See also* Boltzmann machine; Hopfield networks

architecture of, 175

capacity of, 175, 351–352

cells of, 5–6

chips, 414–415

computational limitations of, 228–229

computational power of, 558

data compression and principal components of, 267–271

data representation of numbers in, 596–598

engineering considerations in, 3–5

first computer simulations of, 285

function of, 2–3, 175–176

generalization in, 433–434

information representation techniques of, 281–346

in Kohonen mapping, 465–468

models of, 1

motion, 391–393

mysticism of, 277

nonlinear dynamical system theory and, 401–403

optimization using, 415–416

simple nonlinear autoassociative, 493–541

teaching arithmetic to, 585–625

Neural network associators, 615–617

Neural network clustering algorithms, 534–537

Neural network neuron, generic, 54–55

first stage of, 55–56

second stage of, 57–60

Neuromuscular junction, transmitter release at, 38

Neuron, 4

binary model, 60

classic, 7–14

components of, 8–14

electrical behavior of, 11, 14–15

feedback in, 495*f*

freeze fracture picture of, 10*f*

inability to divide, 5–6

inaccuracy of, 588

information representation and, 282–286

layered structures of, 292–299

maps of, 304–310

McCulloch-Pitts, 49–52, 60, 229, 405*f*

metabolic activity of, 5–6

neural networks in, 282–283

single, properties of, 1–34

single dendrite, 44*f*

slow potential, 46–47

slowness of, 25–31

spatial arrangement of responses, 304–310

spontaneous activity in, 150

theoretical models of, 48–60

two-dimensional structure of, 292

two-state, 49–52

two-value, 60

NEURON program, 31

Neuroscience, 1

Neurotransmitter receptors, 38

Neurotransmitters, 11–12, 14

in association memory, 161–162

release of, 38, 48

Nitric oxide

as neurotransmitter, 14

as retrograde transmitter, 157

NMDA

receptors in brain, 156–157

in synaptic Hebbian modification, 155–159

Noise, 47–48, 177

in associator output, 201

error vector length and, 536–537

learning and, 377

Noise-free neuron, 52

Nonconservation, 592–594
Nonlinear dynamic systems, 496–497
  theory of, 401–403
Nonlinear function, 58–59
Nonlinear network, 398
  autoassociative, 493–541
Nonlinear units, in nearest neighbor
  calculation, 437–438
Nonlinearity, 255–256, 258*f*, 588
  BSB model of, 504–506
Nonuniform input distribution, 474*f*
Nonverbal reasoning, 595–596
Norepinephrine, 161–162
Normalization, 73–75
Novel patterns, 177
Now-print order, 161–163
Nucleus, 8
Number representation, 587–591, 596–
  598, 601
  sensory part of, 624
  spatial map of magnitude in, 623
Number systems, 598–600
Numbers
  coding of, 603–604
  development of concept of, 592–594

Object movement, 378–397
Ohm's law, 16, 568–570
Oksapmin number system, 598–599
Ommatidium, 98
  inhibitory effects in, 102–104
  receptor cells in, 100–103
One, Two, Three ... Infinity, 598
Optimization, 415–416
Order-limited connectivity, 232–233
Orientation columns, 315
Orientation preferences, 316–317
Orthogonality, 77–80
Outer product
  Hebbian learning rule, 159–161
  linear associator, 171
  matrix, 135
Output function, 455
Output layer, 261–262
Output patterns, 175, 178–179
Output representation, 327
Output state vector, 452
  actual vs. desired, 197–199

length of, 179, 198*f*, 363*t*
Overcorrection, 251–252
Overfitting, 276

Paleocortex, 293
Parallel computer, sources of
  inefficiency in, 323–324
Parity, 436*f*
Parity bit, 403
Parker, David, 255
Pascal, 65
  compilers, 200
  computer programs in, 66–67
  program demonstrating lateral
    inhibition, 106–115
Patch clamping, 11
Patchy maps, 324
Pattern classifiers, 434–436. *See also*
  Generalization
Pattern recognition, 210–211, 434–436
  association in, 546
  confrontation naming and, 211–212
  with perceptron, 220–222
Pattern-recognition algorithm, 438
Patterns
  association of, 161–163
  interactions between, 367–368
  recall of, 410–411
Perception, multistable, 518–520
Perceptron, 118, 209–238
  architecture of, 216–218
  conceptual problems of, 227
  connectedness and, 231–238
  diameter-limited, 232–233, 235*f*, 236
  history of, 214–215
  learning, 220–222
  limitations of, 214–215, 231–238, 241–
    242
  Minsky and Papert attack on, 228–
    231
  multilayer, 219–220, 237
  order-limited, 232–233
  simplified, 218–220
  threshold logic units of, 215–216
  two-stage logical process of, 229–231
Perceptron convergence (learning)
  theorem, 220, 222–228, 239
Perceptrons, 214

Perceptual invariances, 594
Perikaryon, 40–41*f*
Peripheral nerve fiber response, 89–90
Peripheral vision, 464
Phase transitions, Hopfield, 411–413
Pheromones, 86
Phonemes, 340–341
  recognition of, 478–480
Phonetic typewriter, 478–480
Photoreceptors, 100
Physical similarity, 342–343
Physical stimulus, 85–86
  action potentials responding to, 87–88
  magnitude of, 88
  neural response to, 88–93
Physics, qualitative, 568–570
Piaget, Jean, 593
Pitts, Walter, 49, 51–52
Pixel array analysis, 272–274
Plato's dialog, 551
Po, Huang, 545
Point image concept, 327
  neural, 367
Poisson distribution, 48
Polynomial functions, 456
Polypeptides, 14
Population coding hypothesis, 322*f*
Posner-Keele concept-formation
    problem, 540
Postsynaptic frequency, 48
Postsynaptic potentials (PSP), 39–40
  interaction of, 42–46
Potassium equilibrium potential, 38
Potassium ion
  conductance of, 18
  equilibrium and conductance of, 23
  neuronal concentration of, 16
Power function, 88
Power law relation, 88–89
Power method, 497–498
  experiment with, 498–501
Practical pattern classification, 435
Practice, 227–228
Predefined functions, 200
Preprocessing, 345
Presynaptic action potentials, 38–39
  slow, 46–47
Priming effects, 610, 611*f*

Principal components, 503
  analysis of, 267–271
  meaning of, 269*f*
Probabilistic categorization, 354–358
PROCEDURE
  Add_matrices, 131, App C
  Add_Vectors, 69, 82, App A
  ClrScr, App A
  Compute_activities, App L
  Compute_inhibited_state_vector, 112,
    114–115, App B
  Constant_times_matrix, 131, App C
  Convergence_test, 114, App B
  Display_state_vector, 112, App B
  GoToXY, App A
  Initialize_parameters, 109, App B
  Initialize_state_vector, 110, 119,
    App B
  Limit_state_vector, 114, 120, App B
  Make_inhibitory_weights, App B
  Matrix_times_vector, 132, App C
  Normalize, 74, App A
  Outer_product, 136, 195, App C
  Plot_histogram, 183, App D
  Scalar_times_vector, 72, 82, App A
  Special_Purpose, 82
  Subtract_matrices, 131, App C
  Subtract_Vectors, 69, App A
  Widrow-Hoff, 252, App J
PROCEDURE vector, 66–83
Product-size effect, 611, 612*f*
  simulation of, 613*f*
PROGRAM
  AmCharDemo, 201–207, 362, 376–
    378, App I
  AMDEMO, 195–197, 199, App F
  AMDEMOL, 197–199, App G
  AMDEMONOGR, App F
  Boltz, 422–431 App K
  BSB25, 515–516, App M
  Inner_Product_Demo, 182–189,
    App D
  Lateral_inhibition, 107–108, App B
  Maps, 468–478, App L
  Mapnumber, 480–486, App L
  SVDEMO, 194–196, App E
  Test_vectors, 67, 83, App A
  WhChar, 253, App J

Programming
  flexible, 623–624
  network, 618–622
  of number simulation computation,
    613–615
Projective field, 389–391
Protein, neuron, 11
Prototype, 354–358, 460
  enhancement of, 374–376
Prototype-based nearest neighbor
    categorizer, 441–442
Prototype effect, 361–362
  in character-based simulations, 376–
    378
  in similarity simulation, 370–372
  in simulations, 362–363
Prototype model, 359*f*
  on continuum, 361*f*
  dot pattern simulation, 373–376
  random dot patterns, 364–367
Prototype techniques, 439–440
Prototype units, 441–442
Psychology, association in, 561–562
Pulse code-modulation system, 48
Pulsed radar, 329
Purkinje cells, cerebellar, 282–283
Pyramidal cells, 299, 301*f*
  in cat visual cortex, 300*f*
  dendrites of, 158*f*
  numbers of, 304
  recurrent collaterals in, 301, 303*f*
Pythagorean theorem, 70

Radar system, 329–332. *See also* Sonar
    system, bat
Radial basis functions, 456
Random dot patterns
  computer simulations of, 368–376
  stimuli, 364
Random variable distribution, 185–
    187
Random walk model, 621–622
Ranvier, nodes of, 33
Reaction time, 522–523
  computation of, 523–527
  data representation of, 528–530
  simulation of, 527–528
  simulation vs. theory of, 530–531

Reasoning
  with associative hierarchy, 551–553
  case-based, 460–461
  example-based, 459–461
  mathematical, 594–596
Recall. *See also* Memory
  accuracy of, 253–254, 410–411
    measuring, 177–178
  averaging operation in, 451*f*
  partial, 562–564, 566
Receptive field, 391
Receptor cells
  in ommatidium, 100
  specialized, 89–90
Receptors, 85
  specialized structure of, 86–87
  tactile, 87–91
Reciprocal connectivity, 495
Reciprocity, 344
Recognition
  character, 271–275
  decision, 199
  memory, 194
  pattern, 210–212, 434–436, 546
Recollection, 546. *See also* Memory;
    Recall
  as dynamic and flexible, 547–549
Reconstruction, 403–404
  autoassociative, 496–498
RECORD vector, 66–67
Rectangular matrix, 130
Recurrent collaterals, 300–301, 303*f*
Refractory period, absolute and
    relative, 23
Regenerative feedback process, 18
Relational similarity, 343
Repellors, 402
Representation, 63
  auditory system, 318–321
  of clustering, 537–541
  cognitive, 340–344
  distributed, 289–299
  motor, 321–323
  natural data, 345–346
  of random dot patterns, 364–367
    in computer simulation, 369–370
  structure built into, 557–558
  techniques, 63–64, 281–346

visual, 310–318
Rescorla-Wagner model, 276–277
Resistance, 26
Response time, 522–523
  in false product simulation, 609*f*
  patterns of, 591
Retina
  cell densities of, 310, 313
  circular, 106*f*
  complex processing of, 310
  connectivity in, 232–233
  peripheral, 312–313
  response of neurons of, 379–380
Retrieval, 191–194
  accuracy of, 201–202, 203–204
  network in, 206–207
Retrieved vector, 412
Retrograde flow, 8
Retrograde transmitter, 157
Rigid movement, 381
RNA molecules, 145–146
Root mean square value, 194
Rosenblatt, Frank, 214
Rule-forming, in language network
  model, 557
Rule-governed behavior, 228–229
Rules, 460

Sample covariance matrix, 268
Scalar quantity, 72
Scattered maps, 324
Screen control procedures, 111–112
Segmentation, 272
Self-organizing clustering system, 533–
  534
Self-similarity, 561
Semantic network, 549, 551
  associative computation in, 556–558,
  575–577
Semilinear function, 57, 59–60, 258
Senility mechanism, 537
Sense image, 546
Sensor fusion, 538
Sensory computation arrangements,
  324
Sensory information processing, 296–
  299
Sensory input, 150

Sensory magnitudes, 591
Sensory motor homunculus, 304–306
Sensory systems
  lateral inhibition and, 85–125
  plasticity in, 488–489
  specialization of, 85–86
Sensory transduction, 86*f*
Short-term memory, 144, 252–253
Sigmoid curve, 57
Sigmoidal nonlinear function, 413–414
Signal processing
  in bat sonar, 329–334
  Doppler, 334–337
Signal-to-noise ratio, 194–195
Similarity, 341–343
  computer simulations of, 368–376
  difficulty in judging, 343, 344*f*
  mapping based on, 490
  perceived, 365
Simplicity, 232
Simulated annealing, 416–417, 420–
  421, 429–430
Simulation
  arithmetic, 603–609
  of associative computation, 558–568
  of BSB model, 515
  character, 199–201, 376–378
  of object movement, 393–397
  prototype effect in, 362–363
  of response time, 527–528
  of similarity experiments, 368–376
  using high-dimensional maps, 480–
  487
Sine waves, 180–182
Single linear neuron, 190–191
Skin senses, cortical area of, 306
Skinner, B. F., 553–554
Skinner box, 553
Slow potential theory, 46–47
Socrates, 551
Sodium
  conductance of, 18, 23
  equilibrium of, 22–23
  neuronal concentration of, 16
Sodium channel, 155
Sodium pump, 16
Solution region, 222–223, 224–225, 227
Soma, 8

Soma membrane depolarization, transient, 45f
Somatosensory system map, 306, 307f
Sonar system, bat, 329–340
Space domain, 28
  time domain and, 29–31
Sparse distributed memories, 446–452
Spatiotemporal information processing, 29–31
Specificity, 286–289
Speech. *See also* Language
  NETtalk learning of, 265–267
  phonemes in, 340–341
  understanding of, 179
Speech-recognition system, 286–287, 478–480
Spinal motor neuron, 8
  synaptic action potential of, 37–38
  synaptic interactions in, 43–46
Spurious attractors, 412
Square matrix, 129–130
Squashing function, 57, 58
Squid axons, 17–18
Standard deviation, 185
Standing, Lionel, 144–145
State space representation, 229–231
State vectors, 65–66, 282
  associations between, 575–577
  changing of, 517–518
  decay rate of, 513–515
  storing factual information, 560f
Statistics, normal distribution of, 187
Stellate cells, 299–300
Step size, 112
Stimulus, 85–86
  coding of, 537–541
  intensity and firing rate, 59f
Stimulus-response learning, 553–555, 562
  in memory, 143
Stored vector, 412
Strengthening-by-use learning rule, 190
String reconstruction accuracy, 203–204
Subtraction
  matrices, 131
  vector, 69–70

Sulci, 292
Sum of squares, 178
Summed error computation, 263–265
Summed vector memory, 187–194, 199
  demonstration of, 196f
Superior colliculus
  in eye movement, 327–329
  model and map of, 326–327f
  sensory motor relationships in, 328f
  topographic organization of, 325–327
Superposition principle, 141
Supervised learning, 176, 177–178
  interpolation in, 454
  versus unsupervised learning, 209–210
Symbol-based artificial intelligence, 598
Symbol-based syntax, 617
Symbol-processing artificial intelligence, 228
Symbolic distance effect, 590–591
Symbols
  manipulation of, 624–625
  patterns of
    highly correlated, 485
    uncorrelated, 484
Symmetric matrix, 133–134
Synapsis, 11, 14
  activation of, 12
  chemical, 11–12, 13f, 39f
  electrical, 11
  electrical events of, 37–42
  Hebbian modification of, 171–172
  interaction of effects, 42–46
  interactions in *Limulus* ommatidium, 104–105
  modification in memory, 146–147
  modification threshold of, 157
  transmission mechanisms of, 37–40
Synaptic cleft, 14
Synaptic delay, 38–39
Synaptic inputs, linear addition of, 55–56
Synaptic learning, 148
  biological basis of, 152–159
  Hebb rule for, 143–145, 171–172
Synaptic strength, 55
Synaptic vesicles, 14
  neurotransmitters from, 38

Synergetics, 496–497
Syntactic ambiguity, 572
Syntax, combinatorial, 616–617

Tactile receptors, 87–91
Taylor series, 170–171
Template, 192
Terminal arborization, 8
Testing programs, for vector
    PROCEDUREs, 82–83
Thalamic connections, 293
Threshold, 23
Threshold device, 241
Threshold logic units (TLUs), 215–216,
    241
Time domain, 27–28
  space domain and, 29–31
Time quantum, 112
Tonotopic organization, 318–319*f*
Topographic organization structures,
    325–329
Topographical maps
  adaptive, 463–490
  of bat sonar system, 329–340
  of number magnitude, 596
Toroidal geometry, 425–426
Training set, 176, 177, 223
  error for, 178
Transfer function, 141–142
Transpose, 133–136
Twain, Mark, 614
Twain's cat, 614
Twisted map formation, 477*f*, 478*f*
Two-dimensional array, 478–479
Two-dimensional state space, 229–231
TYPE
  Matrix, 130, App C
  Neuron, 468, 480, App L
  Vector, 67, App A

Uniform distribution, 471–472
Unit hypersphere, 185
Unsupervised algorithms, 210

Vector, 56, 65
  adaptation of, 520–522
  addition and subtraction of, 67–70

angle between, 75–77
characteristic, 136–138
components of, 65–66
essential operations, 63–83
feedback, 497*f*
inner product of, 72–73
length of, 70–71
linear independence of, 80–81
memory, summed, 187–194, 196*f*,
    199
multiplication by constant, 72
multiplication with matrix, 132–135
normalization of, 73–75
notational and computer conventions
    of, 66–67
orthogonality of, 77–80
of partial derivatives, 245–246
state, 65–66
test programs of, 82–83
two-dimensional, 185
utility of in networks, 619
Widrow-Hoff, 249–252
Vector associator
  architecture of, 161*f*
  diagram of, 160*f*
Versatility, 494
Very-large-scale integrated (VLSI)
    chips, 463
Vestibular system, 93
Visual cortex
  areas of, 382–391
  hierarchy of, 298*f*
  mapping of, 316–318
  modeling of, 464
  neurons of, 382–383
  pyramidal cells in, 300*f*
  responses of single cells in, 314*f*, 315
Visual field
  in cats, 489
  location of, 327
  mapping of, 312*f*, 325–327
  receptive, 325–327
  representation, 311*f*
Visual grasp reflex, 325
Visual perception, connectedness in,
    231–238
Visual processing, 296–299, 313–318

Visual system
  character recognition in, 272–275
  computing object movement, 378–391
  cortical area of, 310–318
  Limulus, 98–105
  macaque, 297*f*
  modular models of, 624
  reciprocal connectivity and
    interconnectivity in, 495
Visualization, 546–547
  in linguistic coding, 549*f*
  in mathematicians, 595
Voicing, feature, 341
Voltage, 29–30
Voltage-to-frequency converter, 25
Von Neumann, John, 588
Von Neumann computer, 51–52

Walsh functions, 78–80, 499, 520
  32-dimensional, 520–521
Weakening-by-use system, 190
Weber-Fechner law, 611
Webster, Daniel, 493
Weight change computation, 262–265
Weight sharing, 274
Weight space, 222–223, 242–249
  error surface in, 243, 248
Weiss, Paul, 8
Weisstein, Naomi, 287
Werbos, Paul, 255
White matter, 293
Widrow-Hoff error-correction
    algorithm, 246–249, 534, 535*f*
  critical procedure for, 252–253
  to improve accuracy, 250
Widrow-Hoff gradient descent
    procedure, 240–242
Widrow-Hoff learning rule, 243, 249–
    252, 276, 398, 503–504, 515, 536
  demonstration of, 253–254
  single-layer, 262
  vector form of, 394
Widrow-Hoff vector, 249–252
Williams, Ted, 585
Winner-take-all (WTA) network, 117–
    124
  with constant light bias, 123*f*
  with two peaks, 124*f*, 125*f*

Words, storage and retrieval of, 449–
    452

X-OR problem, 229–231

Yeldham, 585
Yellow Volkswagen detector, 287–288

Zero crossing, 180–182